BIOCHEMISTRY AND FUNCTION OF PHAGOCYTES

ADVANCES IN EXPERIMENTAL MEDICINE AND BIOLOGY

Recent Volumes in this Series

BIOCHEMISTRY AND FUNCTION OF PHAGOCYTES

Edited by

F. Rossi

Istituto di Patologia Generale
dell' Università di Padova
Verona, Italy

and

P. Patriarca

Istituto di Patologia Generale
dell' Università degli Studi di Trieste
Trieste, Italy

PLENUM PRESS • NEW YORK AND LONDON

Library of Congress Cataloging in Publication Data

European Conference on Phagocytic Leucocytes
(2nd: 1980: Trieste, Italy)
Biochemistry and function of phagocytes.

(Advances in experimental medicine and biology ; v. 141)
"Proceedings of the second European Conference on Phagocytic Leucocytes, held in Trieste, Italy, September 15-18, 1980"—T.p. verso.
Bibliography: p.
Includes index.
1. Phagocytosis—Congresses. 2. Phagocytes—Congresses. 3. Biological chemistry—Congresses. I. Rossi, F. (Filippo) II. Patriarca, P. (Pierluigi) III. Title. IV. Series.
QR187.P4E87 1980 599.02′9 81-19251

DOI 10.1007/978-1-4684-8088-7

Proceedings of the Second European Conference on Phagocytic Leucocytes,
held in Trieste, Italy, September 15-18, 1980.

A Division of Plenum Publishing Corporation
233 Spring Street, New York, N.Y. 10013

PREFACE

This volume collects the papers presented at the "Second European Conference on Phagocytic Leucocytes" which was held in Trieste, September 15-18, 1980. The Conference was the continuation of the thread of a discourse initiated with the "First European Conference on Phagocytic Leucocytes," which was held also in Trieste in 1976. In fact, in both Conferences the biochemical aspects of the basic functions of phagocytes--that is movement, metabolism, secretion, and bactericidal activity--were mainly emphasized. In the Second Conference, two additional subjects were also dealt with, and those are the tumoricidal activity of phagocytes and the clinical aspects of phagocyte functions. In connection with the latter subject a "Round Table on the Clinical Application of Leucocyte Function Tests" was held during the Conference, and the proceedings are reported in this book in the hope that they may be useful to those who are interested in the relationship between phagocytes and diseases. The number of papers which are being published concerning the phagocytic process is such that one wonders whether publication of a new book in this field is necessary. However, the philosophy which inspired the editing of this book differs from that underlying publication of original articles in scientific journals. We started from the common notion that the real progress of our knowledge is not necessarily proportional to the amount of published material in that specific field. Additionally, a true progress in knowledge requires a careful verification of the validity of the information that one receives, a realistic analysis of the reasons of conflicting results and of controversies, and a severe criticism of erroneous or false problems which generate unuseful and confusing papers. The two European Conferences on Phagocytic Leucocytes were organized to meet those requirements. The worldwide participation in both Conferences (77 scientists from 15 countries in the First Conference, and 106 scientists from 14 countries in the Second Conference) guaranteed a broad exchange of opinions. Consequently, this book does not represent a mere addition of new papers to an already vast literature, but defines the present status of the art and the future trends of research in the field of phagocyte functions. We hope that, owing to the value of the participants and to the interest of their contributions, this volume can be useful to clarify problems and to stimulate new research projects.

Filippo Rossi
Pierluigi Patriarca

ACKNOWLEDGEMENTS

The editors are indebted to Drs. B.M. Babior, R.E.Basford, J. Breton-Gorius, R.A.Clark, L.R.DeChatelet, P.Elsbach, A.Fonnesu, R. Gallily, J.Hakim, R.A.Harkness, R.B.Johnston, H.U.Keller, A.Mantovani, C.McCall, G.Rotilio and A.W.Segal, who chaired the sessions of the Conference and edited the discussions, and to Drs. H.R.Hill and P.Patriarca for the preparation of the Round Table on the Clinical Application of Leucocyte Function Tests and for the editing of the relative proceedings.

Gratitude is also expressed to the members of the organizing committee Drs. P.Patriarca, G.D.Rottini, G.Zabucchi, P.Dri, R.Cramer, M.R.Soranzo, A.Dobrina of the Institute of General Pathology of the University of Trieste, and Drs. F.Rossi, G.Berton, P.Bellavite, G. deNicola of the Institute of General Pathology of the University of Padua, Sede di Verona. The Conference could not have succeeded without the financial assistance of the Medical Faculty of the University of Trieste, the National Council of Research (CNR) of Italy, the Azienda Autonoma di Soggiorno e Turismo of Trieste, the Banca Cattolica del Veneto of Trieste, the Dr.Ing.G.Terzano & C. S.p.A. of Milano and the Beckman Analytical S.r.l. of Milano.

Finally, a particular mention is due to the precious collaboration of Mrs. Milvia Gioria throughout the Conference and during the preparation of these proceedings.

CONTENTS

CYTOTOXIC ACTIVITY OF PHAGOCYTES

MICROBICIDAL ACTIVITY OF PHAGOCYTES

FACTORS AFFECTING PHAGOCYTE FUNCTIONS

CLINICAL ASPECTS OF PHAGOCYTE FUNCTIONS

THREE TYPES OF GRANULE FORMED IN GUINEA PIG AND RAT HETEROPHIL GRANULOCYTES

P. Brederoo and J. van der Meulen

Laboratory for Electron Microscopy, University of Leiden
2333 AA Leiden, The Netherlands

INTRODUCTION

It is generally accepted that in heterophil (in man neutrophil) granulocytes, two types of granule are formed successively during the development of these cells in the bone marrow. The process of formation and maturation of these two types of granule was well documented by Bainton and Farquhar (1966) in their studies on the development of heterophil granulocytes in the bone marrow of the rabbit. From these studies it became clear that during the promyelocyte stage azurophil granules are formed, whereas during the myelocyte stage and shortly thereafter, specific granules are produced. Both types of granule originate from vesicles pinching off from Golgi cisternae and changing into mature granules by condensation of their contents. Mature azurophil granules are characterized by a homogeneous matrix of higher electron density than that of specific granules. In the mature heterophil granulocytes furthermore, azurophil granules are considerably outnumbered by specific granules.

More information about the morphological heterogeneity of the two types of granule was later obtained in cytochemical studies (Bainton and Farquhar, 1968), which showed that azurophil granules possess both peroxidase and acid phosphatase and can therefore be considered primary lysosomes, whereas the specific granules have alkaline phosphatase and no lysosomal enzymes.

Besides these two types of granule, mention has been made in the literature of ellipsoid and spindle-shaped granules, characterized by the presence in their matrix of periodic structures lying parallel to the long axis of the granule. The fine structure varied from regular crystalloid to irregular fibrillar. Such observations have been made mainly in rat heterophils and human neutrophils (for references, see: Brederoo and Daems, 1978; see also: Brederoo and

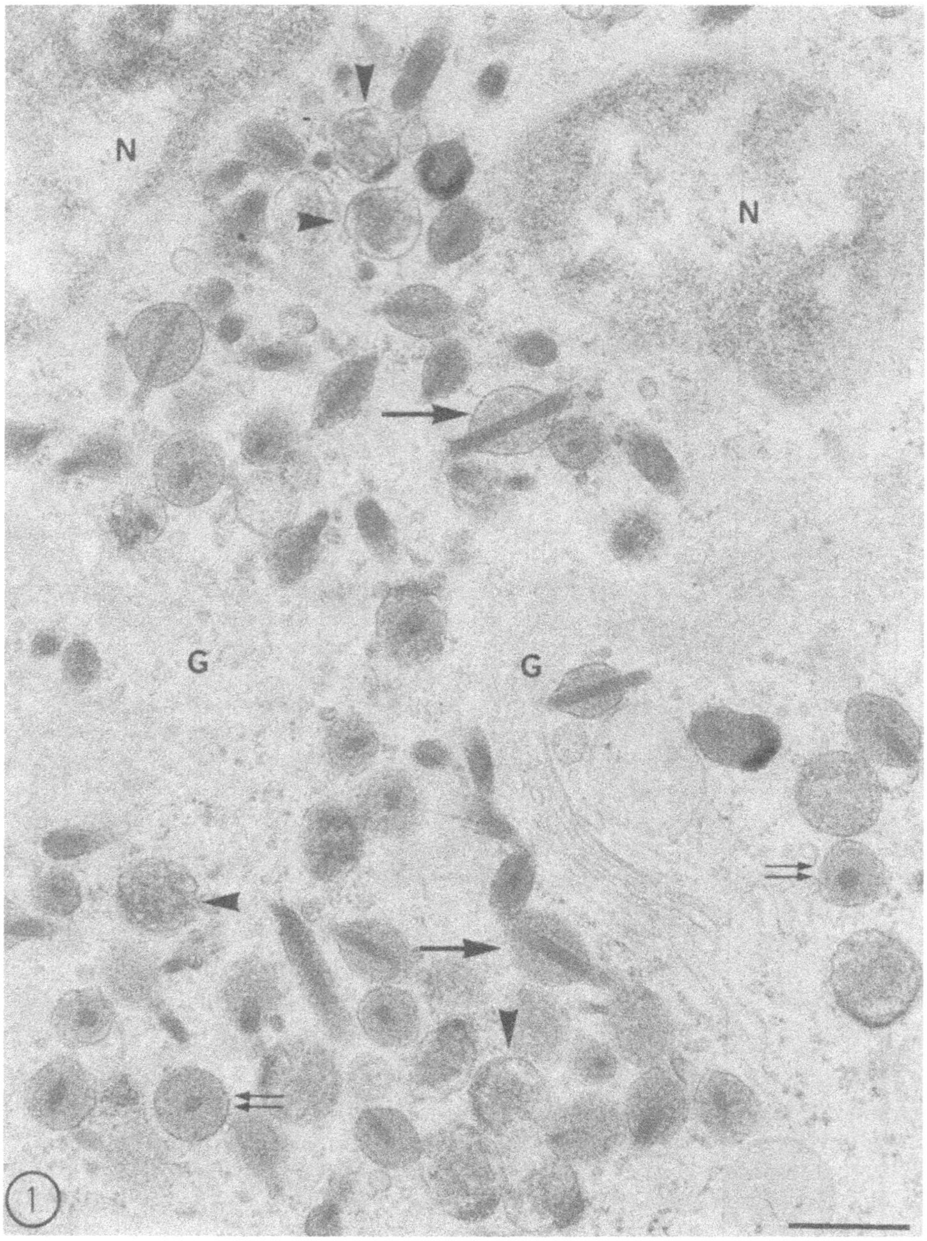

Fig. 1. Golgi area (G) of an early promyelocyte from a rat, showing numerous nucleated granules. Arrowheads point to vacuoles with a flocculent content. Almost-mature granules can be seen in longitudinal section (arrows) and cross-section (double arrows). N = nucleus. Bar: 0.5 μm.

van der Meulen, 1980). After incubation in a medium for the demonstration of the enzyme peroxidase, granules with an internal fine structure react in the same way as azurophil granules (Daems, 1968; Bainton et al., 1971; Breton-Gorius and Reyes, 1976; Brederoo and Daems, 1977).

On the basis of these observations it has been stated that the formation of two types of granule is not restricted to the rabbit, but is a general phenomenon in mammalian species. Furthermore, because both the azurophil and the fine structure-containing granules are peroxidase-positive, the latter have considered to be types or forms of the azurophil granule. In addition, study of the granules with fine structure has been hampered by the recommendation that the peroxidase reaction be used as a means to distinguish between azurophil and specific granules (Farquhar and Bainton, 1972).

This paper presents observations showing that these conclusions are not correct, at least as far as the guinea pig and the rat are concerned.

GUINEA PIG AND RAT HETEROPHILS

Bone marrow from femurs of male and female albino guinea pigs and rats was studied after fixation in osmium tetroxide alone as well as after fixation first in glutaraldehyde and then in osmium tetroxide. For details of the preparation for electron microscopy and the terminology used, the reader is referred to Brederoo and Daems (1978). Because the development of the heterophil of the guinea pig does not differ essentially from that of the rat heterophil, no distinction will be made here between the two species except where specially mentioned.

The steps in the process of development from promyelocyte via myelocyte, metamyelocyte, and band cell, to the mature heterophil of the guinea pig and rat are morphologically the same as those described for other mammals, including man. During maturation the cells become smaller and the nucleus changes in shape from slightly indented in the promyelocyte, via more and pronounced indentation in the myelocyte and metamyelocyte, to a multilobulated nucleus in the mature cell. After the myelocyte stage, the condensation of the nuclear chromatin increases and the amount of rough endoplasmic reticulum and Golgi cisternae decreases considerably. Throughout the development the total number of granules increases, and the mature cell shows a huge number distributed over the cytoplasm. Early stages of development, i.e., the promyelocyte and the myelocyte, are easily recognized due to the formation and maturation of typical granules. Formation of the specific granules starts in the myelocyte, as is the case in other mammals. However, in contrast to the other species, the guinea pig and rat were found to show two separate steps in the development of the promyelocyte, each characterized by the initiation and maturation of a typical granule. The production of granules containing fine structure, here called

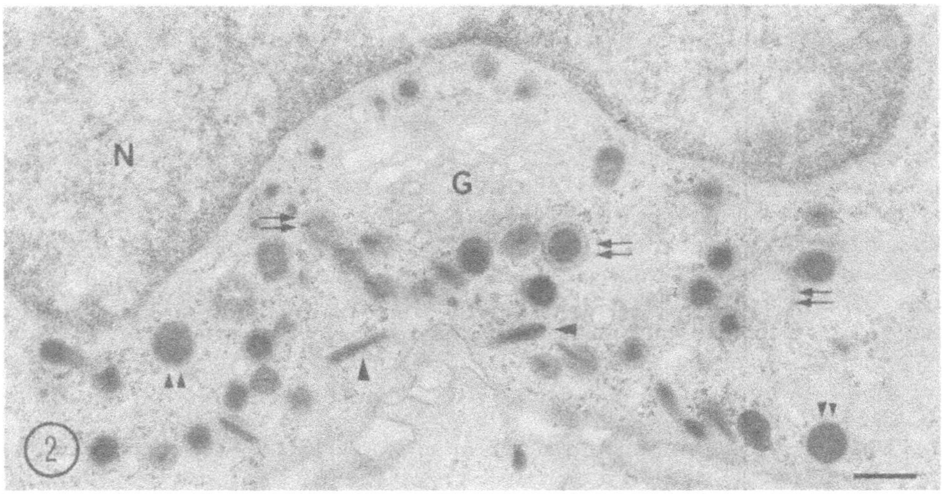

Fig. 2. Golgi area (G) of a late promyelocyte (rat). In addition to nucleated granules, azurophil granules can be seen in several stages of development (double arrows). Arrowheads point to mature nucleated granules, double arrowheads to mature azurophil granules. N = nucleus. Bar: 0.5 μm.

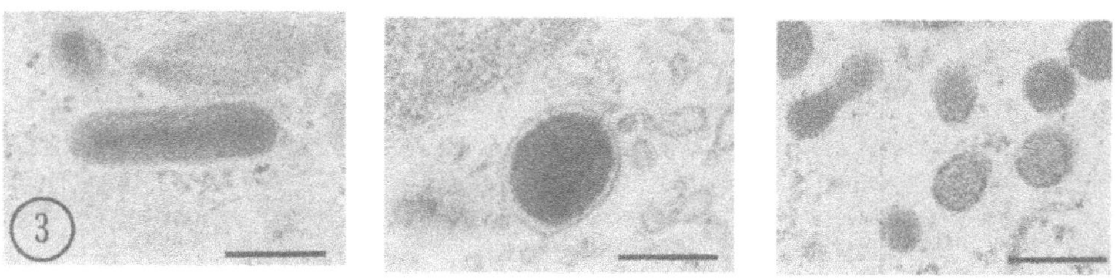

Fig. 3. The three types of granule formed successively in rat heterophils: the nucleated granules (left), azurophil granule (middle), and specific granules (right). Bar: 0.25 μm.

nucleated granules, starts in early promyelocytes, whereas the azurophil granules are formed in late promyelocytes.

Early promyelocytes are larger than late promyelocytes and contain only nucleated granules in different stages of maturation. In the Golgi area most of the precursors can be seen: vesicles and immature granules. Round vesicles situated near Golgi cisternae are electron lucent, whereas immature granules elsewhere in the Golgi area and the cytoplasm have an irregular flocculent content of intermediate electron density. Subsequently, the contents of the vesicles tend to concentrate, and a regular pattern starts to appear. Later the granules are ellipsoid and show a nucleoid in a flocculent matrix (Fig. 1). Mature granules are oval to spindle-shaped with the nucleoid lying parallel to the long axis of the granule. The nucleoid is crystalline in nature and shows a regular pattern with a periodicity of about 9 nm in the guinea pig and 5 nm in the rat.

Late promyelocytes are smaller than early promyelocytes and contain two types of granule, nucleated and azurophil. Almost all of the nucleated granules are in the mature state just described. Now the Golgi vesicles have a flocculent content which is variably reticulated. Immature azurophil granules have a dense core and a rim of intermediate density. They develop into the usually spherical azurophil granules with a homogeneously electron dense matrix (Fig. 2).

Myelocytes have mature nucleated granules in their cytoplasm and azurophil granules which are, in older cells, almost all in the mature state. The third type of granule, the specific granule, is now added to the granule population. Young specific granules, which arise from Golgi cisternae, have a flocculent matrix. Mature specific granules are spherical, elongated, or dumbbell-shaped, with a homogeneous or flocculent matrix of intermediate electron density. They have neither nucleoid or an electron-dense core.

Mature heterophils have a very large number of granules in the cytoplasm, the azurophil granules being very easy to recognize due to their highly electron dense matrix. However, the density of the matrix of nucleated granules closely resembles that of the specific granules. Especially in the guinea pig, the nucleoid is obscured by the matrix, which makes these granules very difficult to identify. In the rat, the nucleoid is more distinct and thus the nucleated granules are easier to recognize. In general, identification of granules is easier in osmium tetroxide-fixed cells than in cells fixed in glutaraldehyde followed by osmium tetroxide.

Table. Percentages of granules in mature heterophils from the bone marrow according to type and species.

granules	guinea pig	rat
nucleated	3.6	13.4
azurophil	9.3	10.2
specific	87.1	76.4
Total number	172.5	162.3

CONCLUSIONS

In heterophil granulocytes of the guinea pig and rat three morphologically identifiable granules are formed successively in different stages of cell maturation. Nucleated granules are formed in early promyelocytes, and specific granules in myelocytes. All three types of granule remain present during the maturation of the cells and can be found in the mature heterophil (Fig. 3).

REFERENCES

Bainton, D.F., and Farquhar, M.G., 1966, Origin of granules in polymorphonuclear leukocytes. Two types derived from opposite faces of the Golgi complex in developing granulocytes, J. Cell Biol. 28:277.

Bainton, D.F., and Farquhar, M.G., 1968, Differences in enzyme content of azurophil and specific granules of polymorphonuclear leukocytes. II. Cytochemistry and electron microscopy of bone marrow cells, J. Cell Biol. 39:299.

Bainton, D.F., Ullyot, J.L., and Farquhar, M.G., 1971, The development of neutrophilic polymorphonuclear leukocytes in human bone marrow, J. Exp. Med. 134:907.

Brederoo, P., and Daems, W.Th., 1977, A new type of primary granule in guinea pig heterophil granulocytes, Cell Biol. Int. Rep. 1:363.

Brederoo, P., and Daems, W.Th., 1978, The ultrastructure of guinea pig heterophil granulocytes and the heterogeneity of the granules. I. The development in the bone marrow, Cell Tiss. Res. 194:183.

Brederoo, P., and van der Meulen, J., 1980, Granule formation in rat heterophil promyelocytes, in: "Electron Microscopy 1980". Procs. 7th European Congress on Electron Microscopy, volume 2, P. Brederoo and W. de Priester, eds. The Seventh European Congress on Electron Microscopy Foundation, Leiden.

Breton-Gorius, J., and Reyes, F., 1976, Ultrastructure of human bone marrow cell maturation, Int. Rev. Cytol. 46:251.

Daems, W.Th., 1968, On the fine structure of human neutrophilic leukocyte granules, J. Ultrastruct. Res. 24:343.

Farquhar, M.G. and Bainton, D.F., 1972, Cytochemical studies on leukocyte granules, in: Histochemistry and cytochemistry 1972. Proc. 4th Internat. Congress, T. Takeuchi, K. Ogawa, S. Fujita, eds, Japan Society of Histochemistry and Cytochemistry, Kyoto.

DISCUSSION

Edited by Charles E. McCall, M.D.

"Morphologic Aspects of Neutrophil Maturation in Bone Marrow"

Presented by Dr. P. Brederoo

Question 1 - Dr. Rossi

Do you have some information about the enzymatic contents of the granules such as phosphatases, proteases, etc.?

Answer - Dr. Brederoo

Not yet. We only know at this moment that the nucleated granules react in the same way as azurophil granules after incubation in the assay for myeloperoxidase activity.

Question 2 - Dr. Breton-Gorius

I have only a comment about the heterogeneity of granules in human neutrophil promyelocytes. I agree that several subgroups of azurophils can be identified not only by their distinct morphology but also by their different contents. We have previously reported that patients with congenital myeloperoxidase deficiency, only elongated granules which contain a crystal, exhibit a peroxidatic activity while round dense azurophils are unreactive. In the classification of granules, azurophil granules are also called primary granules. In fact, it was shown that catalase containing granules represent the first type of granules produced during maturation.

LEUCOCYTE ACTIVATION AND THE ASSESSMENT OF LEUCOCYTE LOCOMOTION AND CHEMOTAXIS

H.U. Keller, M.W. Hess and H. Cottier

Department of Pathology
University of Berne
3010 Berne, Switzerland

INTRODUCTION

Many investigators showed that chemotactic factors induce a large variety of neutrophil responses such as cell aggregation, changes in adhesion to the substratum, chemokinesis, chemotaxis, exocytosis and the respiratory burst leading to the generation of oxygen derivatives[1,2,3,4,5]. Several of the studies showed a very close correlation between these responses to synthetic oligopeptides[1,2]. No exception was found in which the biological responses to a given chemotactic factor have been dissociated[2]. This was considered as evidence for the existence of a single population of multifunctional receptors[1]. Other investigators found either no relationship or at least no close parallelism[6,7,8].

The question as to whether or not exposure of leucocytes to cytotaxins leads to a uniform and parallel activation of many functions by means of a common pathway is of general interest in leucocyte physiology. It may also have some bearing on the assessment of leucocyte activation in general and of leucocyte locomotion and chemotaxis in particular. If the responses can not be dissociated from one another, all respective assays would basically be equivalent and mutually exchangeable.If this is not the case, it would nevertheless be of interest to know whether they are totally dissociated or at least partially related. This question is of particular interest with respect to functions like chemokinesis and chemotaxis which are directly involved in controlling leucocyte accumulation. Parallel

Table 1. Induction of crawling movements by f-Met-Leu-Phe

Medium	% neutrophils showing crawling movements[a]
2 % HSA - Gey's	1.4
2 % HSA - Gey's + f-Met-Leu-Phe (10^{-8}m)	80.6

[a]floating cells (11)

synergistic effects can render directional locomotion most efficient, whereas antagonistic effects of individual functions will make the analysis difficult and the cell accumulation is presumably not optimal. In the present paper we are going to analyse the question as to whether there is a strict parallelism between chemotaxis and any of the other leucocyte responses.

RESULTS AND DISCUSSION

Crawling movements and chemotaxis

Neutrophils in the peripheral blood are presumably more or less spherical. Leucocytes in vitro may either be spherical and immobile or alternatively show active crawling movements which give the cell a polarized appearance. Chemotactic factors such as f-Met-Leu-Phe can transform spherical neutrophils into cells which perform crawling movements (Table 1). Therefore, it is conceivable that chemotactic factors have synergistic effects on crawling movements and orientation to produce optimal locomotion,provided the control cells are spherical.

However, antagonistic effects on crawling movements, chemokinesis and chemotaxis, respectively, have been observed. Human gamma globulin prepared by the method of Kistler and Nitschmann[9] efficiently stimulates the induction of crawling movements,provided the neutrophils are in suspension. As soon as the cells make contact with the substratum they spread, become strongly adherent and loose their polarized shape which is typical for crawling cells (Fig. 1). At the same time they become immobilized. It appears that the gamma globulin pre-

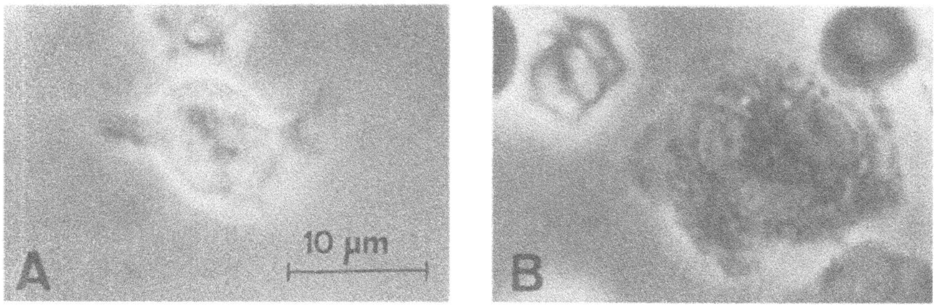

Fig. 1. Disappearance of gamma globulin-induced crawling movements and polarisation following cell contact with glas

A. Polarized human leucocyte floating in a medium containing 2 % gamma globulin[9] in Gey's solution.

B. The same cell after having settled on the substratum.

paration exerts an antagonistic effect on the leucocyte functions in question. It stimulates the locomotor machinery of the neutrophil; it also stimulates adhesion to glas and this is in turn associated with inhibition of crawling movements and locomotion of neutrophils. At present there is no evidence that this gamma globulin preparation is capable to induce chemotactic migration or orientation[10]. It may be worthwile to further investigate whether the expression of orientation is obscured by excessive spreading and adhesion to the substratum or whether there is a true dissociation between induction of crawling movements and chemotaxis.

Chemokinesis and chemotaxis

No strict general correlation between chemokinesis and chemotaxis has been established. Many experiments suggest that the two phenomena can,at least to some extent,be dissociated. It has been reported that many chemokinetic factors which either stimulate or inhibit locomotion lack detectable chemotactic activity. Some chemotactic factors were found to lack chemokinetic activity at chemotactic concentrations whereas others inhibited or stimulated the rate of locomotion. It is even conceivable that the effect of one and the same mediator may vary with the test conditions (for review see[12,13]). Furthermore, different cytotaxins inducing a similar degree of orientation in a visual assay were found to differ with respect to their chemokinetic activity. The synthetic peptide f-Met-Leu-Phe had a marked negative chemokinetic effect at high concentrations. This effect was small if not absent with $C5a_{des\ Arg}$[14]. Partial deactivation can be observed without concomitant chemokinetic effect[6].

The lack of correlation is not surprising. Chemokinesis can be produced by a variety of mechanisms including changes in adhesion and stimulation or inhibition of the locomotor machinery. F-Met-Leu-Phe can stimulate the induction of crawling movements (Table 1) and it interferes at high concentrations with locomotion,probably as a result of increased cell adhesion to the substratum[14]. In this particular instance the cytotaxin induces chemokinesis by means of two different mechanisms and with antagonistic effects. In summary, the findings suggest that chemotaxis can not be used as a reliable correlate for chemokinesis and vice versa.

Adhesion to solid substrata and chemotaxis

The hypothesis of Berthold which has been recently revived by Carter[15] suggests that the direction of locomotion is exclusively determined by adhesive forces between cells and the substratum. Though it has been demonstrated that cytotaxins bind to various substrata[16,17], there is no direct proof that the gradient of adhesion causes chemotactic orientation. Cytotaxins were found to induce orientation under conditions where they produced no detectable change in adhesion to the substratum. Furthermore, gradients of adhesion produced no detectable orientation[10]. Thus, there is no proof that chemotaxis of neutrophils is indeed a special case of haptotaxis. The large majority of the investigations lead to the conclusion that cytotaxins stimulate adhesion to solid substrata[4,14,18,19]. Others suggest, that they may also decrease adhesion or be indifferent at chemotactic concentrations[5,9,20]. There is no reliable correlation between the strenght of adhesion or a gradient of adhesion and chemotaxis. Measuring adhesion to solid substrata can not replace orientation assays.

Directional locomotion and chemotaxis

Most assays of "chemotaxis" measure in fact stimulated locomotion or directional locomotion provided morphological orientation in response the chemical in question has been demonstrated (for terminology see[21]). This pertains to various forms of the filter technique or the agarose method. They measure complex responses consisting of chemotactic orientation on the one hand and of locomotion including chemokinesis on the other hand. Adhesion is indirectly involved because chemokinetic effects can in part be produced by changes in adhesion[20,22]. If one takes into account that there is no strict parallelism between chemotaxis on the one hand and adhesion, chemokinesis or stimulation of motility (crawling movements) on the other hand, it seems unlikely that a strict correlation between chemotactic orientation and directional or stimulated locomotion can be established. The response in terms of cell accumulation is strongly influenced by chemokinesis and changes in adhesiveness. The presence of substance with negative chemokinetic activity can totally inhibit the response to potent cytotaxins[23].

Different cytotaxins were found to produce diverging chemokinetic effects. This affects the overall response measured in terms of directional locomotion[14]. The assessment of stimulated or directional locomotion is no reliable correlate for chemotaxis. Orientation assays involving visual observation should be used if one really wants to quantitate chemotaxis. The assessment of stimulated or directional locomotion is nevertheless of great interest because this complex reaction is probably a relevant model in vitro for leucocyte emigration in vivo.

Exocytosis and chemotaxis

A complete dissociation between directional locomotion and exocytosis has been observed in equine neutrophils. The peptide f-Met-Leu-Phe can induce exocytosis but not chemotaxis[8]. Human neutrophils respond to this peptide by chemotaxis as well as exocytosis, but there are marked differences in the dose-response curves. This lead to the conclusion that although the same chemotactic stimulation can cause both secretion and chemotaxis ,the number of receptors stimulated and the sequence of events after stimulation may be quite different[7]. One should also keep in mind that substantial stimulation of exocytosis but not of chemotaxis requires the presence of cytochalasin B or of surface stimuli[24].

In summary, many results suggest that leucocyte functions such as crawling movements, adhesion to the substratum, chemokinesis, chemotaxis, exocytosis and metabolic burst are not strictly correlated and the effects are not always synergistic. Antagonistic effects are produced in many situations. Therefore, the measurement of one single individual function or of one of the complex activities such as directional locomotion which is regulated by motility, adhesion, chemokinesis and chemotaxis, are not necessarily representative for an individual function like chemotaxis. Any complete and analytic assessment of the functional state of a leucocyte population requires the measurement of all individual functions involved.

The finding that different functions can at least in some situations be dissociated does not exclude that a single type of multifunctional receptor is involved in triggering multiple activities. The events after stimulation may be quite different but the mechanisms of transduction are not sufficiently known. Further experiments are needed to investigate this very important aspect.

SUMMARY

Leucocyte activation by different agents does not result in a uniform pattern of parallel responses. Some activities have synergistic and others antagonistic effects on directional locomotion in vitro. No reliable correlation between chemotaxis on the one hand and other leucocyte activities such as crawling movements, adhesion to solid substrata and chemokinesis has been firmly established. Therefore, the assessment of chemotaxis by measuring orientation is presumably the only reliable quantitative measure for chemotaxis.

ACKNOWLEDGEMENT

The study was supported by the Swiss National Science Foundation.

REFERENCES

1. Becker, E. L., A multifunctional receptor on the neutrophil for synthetic chemotactic oligopeptides, J. Reticuloendothelial Soc., 26 (Suppl.):701 (1979).
2. Kreutzer, D. L., O'Flaherty, J. T., Orr, W., Showell, H. J., Ward, P. A., and Becker, E. L., Quantitative comparisons of various biological responses of neutrophils to different active and inactive chemotactic factors, Immunopharmacol., 1:39 (1978).
3. Weissmann, G., Smolen, J. E., and Korchak, H. M., Release of inflammatory mediators from stimulated neutrophils, New Engl. J. Med. 303:27 (1980).
4. Fehr, J., and Dahinden, C., Modulating influence of chemotactic factor-induced cell adhesiveness on granulocyte function, J. Clin. Invest., 64:8 (1979).

5. Smith, R. P. C., Lackie, J. M., and Wilkinson, P. C., The effects of chemotactic factors on the adhesiveness of rabbit neutrophil granulocytes, Exp. Cell Res., 122:169 (1979).
6. Keller, H. U., Wissler, J. H., Hess, M. W., and Cottier, H., Distinct chemokinetic and chemotactic responses in neutrophil granulocytes, Europ. J. Immunol., 8:1 (1978).
7. Cramer, E. B., and Gallin, J. I., Localization of submembranous cations to the leading end of human neutrophils during chemotaxis, J. Cell Biol., 82:369 (1979).
8. Snyderman, R., and Pike, M. C., N-formylmethionyl peptide receptors on equine leukocytes initiate secretion but not chemotaxis, Science, 209:493 (1980).
9. Kistler, P., and Nitschmann, Hs., Large scale production of human plasma fractions, Vox Sang., 7:414 (1962).
10. Keller, H. U., Wissler, J. H., and Ploem, J., Chemotaxis is not a special case of haptotaxis, Experientia, 35:1669 (1979).
11. Keller, H. U., and Cottier, H., Crawling-like movements and polarisation in non-adherent leucocytes, Cell Biol. Int. Reports 5:3 (1981).
12. Keller, H. U., Hess, M. W., and Cottier, H., Granulocyte chemokinesis and chemotaxis, Symposium on Cellular Interactions, Elsevier (in press)
13. Keller, H. U., Relationship between leucocyte adhesion to solid substrata, locomotion, chemokinesis and chemotaxis, in: The Biology of the Chemotactic Response, J. M. Lackie and P. C. Wilkinson, eds., Cambridge University Press (in press).
14. Keller, H. U., Wissler, J. H., and Damerau, B., Diverging effects of chemotactic serum peptides and synthetic f-Met-Leu-Phe on neutrophil locomotion and adhesion, Immunology 1981, in press.
15. Carter S. B., Haptotaxis and the mechanism of cell motility, Nature (Lond), 213:256 (1967).
16. Dierich, M. P., Wilhelmi, D., and Till, G., Essential role of surface-bound chemoattractant in leukocyte migration, Nature, 270:351 (1977).
17. Wilkinson, P. C., and Allan, R. B., Chemotaxis of neutrophil leucocytes towards substratum-bound protein attractants, Exp. Cell Res., 117:403 (1978).
18. O'Flaherty, J. T., Kreutzer, D. L., and Ward, P. A., Chemotactic factor influences on the aggregation, swelling, and foreign surface adhesiveness of human leucocytes, Amer. J. Path., 90:537 (1978).
19. Smith, C. W., Hollers, J. C., Patrick, R. A., and Hassett, C., Motility and adhesiveness in human neutrophils, J. Clin. Invest., 63:221 (1979).

20. Lackie, J. M., and Smith, R. P. C., Interactions of leukocytes and endothelium, in: Cell Adhesion and Motility (Third Symposium of the British Society for Cell Biology), A. S. G. Curtis and J. D. Pitts, eds., Cambridge University Press, p. 235 (1980).
21. Keller, H. U., Wilkinson, P. C., Abercrombie, M., Becker, E. L., Hirsch, J. G., Miller, M. E., Ramsey, W. S., Zigmond, S. H., A proposal for the definition of terms related to locomotion of leucocytes and other cells, Clin. exp. Immunol., 27:377 (1977).
22. Keller, H. U., Barandun, S., Kistler, P., and Ploem, J. S., Locomotion and adhesion of neutrophil granulocytes: Effects of albumin, fibrinogen and gamma globulins studied by reflection contrast microscopy, Exp. Cell Res. 122:351 (1979).
23. Keller, H. U., Wissler, J. H., Damerau, B., Hess, M. W., and Cottier, H., The filter technique for measuring leucocyte locomotion in vitro. Comparison of three modifications, J. Immun. Meth., 36:41 (1980).
24. Henson, P. M., Zanolari, B., Schwartzman, N. A., and Hong, S. R., Intracellular control of human neutrophil secretion, I. C5a-induced stimulus-specific desensitization and the effects of cytochalasin B., J. Immunol., 121:851 (1978).

DISCUSSION
Edited by H.U. Keller

Rossi, F.: What is the mechanism of leucocyte adhesion to a surface? Is it similar to that of fibroblasts? Do you have some information on the mechanism of leucocyte adhesion to endothelial cells during the inflammatory response?

Keller, H.U.: Various factors such as electrochemical changes, divalent cations, the metabolic activity of the cell and binding of factors from the medium to the cell surface or the substratum are involved in adhesion. But to my knowledge the precise physiochemical forces involved in different forms of adhesion have not been determined even though hypothetical models exist. The focal adhesions which are a characteristic feature in fibroblasts have not been observed in efficiently moving leucocytes. The mechanism of adhesion to endothelial cells has been studied by Smith, Lackie and Wilkinson who found that the cytotaxin-induced adhesive changes occurred in the leucocytes rather than endothelial cells.

Gallily, R.: Can you explain the different effects of some agents on cell locomotion by a different degree of blockage and/or saturation of the chemotactic receptor sites on the cell surface?

Keller, H.U.: It may well be that blockage and saturation of receptors is involved. I know of no evidence to show that f-Met-Leu-Phe and $C5a_{des\ Arg}$ block or saturate the respective receptor sites to a different degree and that this is related to their effects on locomotion. This point needs to be further investigated.

DISTRIBUTION OF ACTIN-BINDING PROTEIN AND MYOSIN IN NEUTROPHILS DURING CHEMOTAXIS AND PHAGOCYTOSIS

Niels H. Valerius, Olle I. Stendahl, John H. Hartwig and Thomas P. Stossel

Hematology-Oncology Unit, Massachusetts General Hospital Harvard Medical School, Department of Medicine, Boston Massachusetts 02114, USA, Statens Seruminstitut Department of Clinical Microbiology, Rigshospitalet (State University Hospital), Copenhagen, Denmark, and Department of Medical Microbiology, University of Linkoping, Linkoping, Sweden

INTRODUCTION

Polymorphonuclear leukocytes (PMN) are highly motile cells which are vital for the defense of higher organisms against invasion by microorganisms. Important elements of their function are intimately related to their ability to move. They crawl towards microorganisms, attracted by chemotactic factors, during which process their shape becomes fusiform or triangular. When they reach their prey they can ingest it by extending pseudopods around it, which eventually fuse to form a closed phagocytic vacuole. During this process the granules are mobilized to get into contact with the plasma membrane to release lysosomal enzymes into the phagosome or to the surrounding medium. All these functions require motility.

There is increasing evidence that contractile proteins localized to the organelle-excluding cortical cytoplasm provide the motor force for locomotion and phagocytosis. Table 1 summarizes the characters of the contractile proteins identified in PMN or macrophages.

Actin is a major protein in PMN, constituting about 10% of the total protein content. Transmission electron microscopy has shown that actin-containing filaments constitute the principal structure within the pseudopods which extend from the cell as part of phagocytosis and chemotaxis [1,2]. Actin has also been identified in these areas of PMN by immunofluorescence [3]. Biochemical analysis

Table 1. Contractile Proteins Identified in Macrophages or PMN

Protein	% of total protein	Molecular weight	Properties
Rigidity (Gelation)			
Actin	10	42,000	5-7 μ in double helical filaments.
Actin-binding Protein (ABP)	1	500,000	Crosslinks F-actin in vitro to form a gel.
Gelsolin [4]	<1	180,000	Severs actin filaments reversibly in the presence of Ca^{++}.
Contraction			
Myosin [5]	1	500,000	Contracts actin in the presence of Mg^{++} and ATP.

has indicated that other contractile proteins interact with actin during phagocytosis [6]. Actin-binding protein (ABP) acts in vitro by cross-linking actin filaments to form a gel lattice , while myosin can contract this gel. In contrast to actin, ABP and myosin comprise only a small fraction of the total protein content in PMN and cannot readily be identified by transmission electron microscopy. However, in order to postulate that ABP and myosin are important for cell motility their presence should be demonstrated in the areas of the cell where motility takes place. We have therefore examined the distribution of ABP and myosin in rabbit PMN during chemotaxis and phagocytosis by utilizing indirect immunofluorescence.

MATERIALS AND METHODS

Anti-rabbit alveolar macrophage myosin and anti-rabbit alveolar macrophage ABP goat antiserum was prepared, and IgG was purified by standard techniques. The antigens were electrophoretically pure on sodiumdodecylsulfate polyacrylamide gels, and the specificity of the antibodies was shown by the criteria of immunoelectrophoresis against pure antigens and whole cell extract as well as by absorption. Controls with pre-immune serum did not stain the cells.

Fluorescein-labelled yeast particles were heat-fixed to glass slides. The PMN were suspended in 33% fresh rabbit serum in Krebs-Ringers phosphate buffer. After the slides had cooled a drop of the cell suspension was put on the slide and covered with a cover-slip. After incubation for 5-10 min at room temperature the slides were rinsed in phosphate-buffered saline (PBS) to float the cover-slips and then immersed into cold (-20°C) acetone for 7 min. The slides were then washed in PBS and incubated with anti-rabbit macrophage ABP goat IgG or anti-rabbit macrophage myosin goat IgG for 30 min at 37°C in a moist chamber. The slides were then washed in PBS for 45 min and incubated with rhodamine-conjugated anti-goat IgG rabbit antibody for 30 min at 37°C. The slides were finally washed in PBS for 2 h, mounted with 90% glycerol in PBS and sealed with Permount. The slides were examined in a fluorescence microscope using appropriate filters for viewing rhodamine and fluorescein fluorescence.

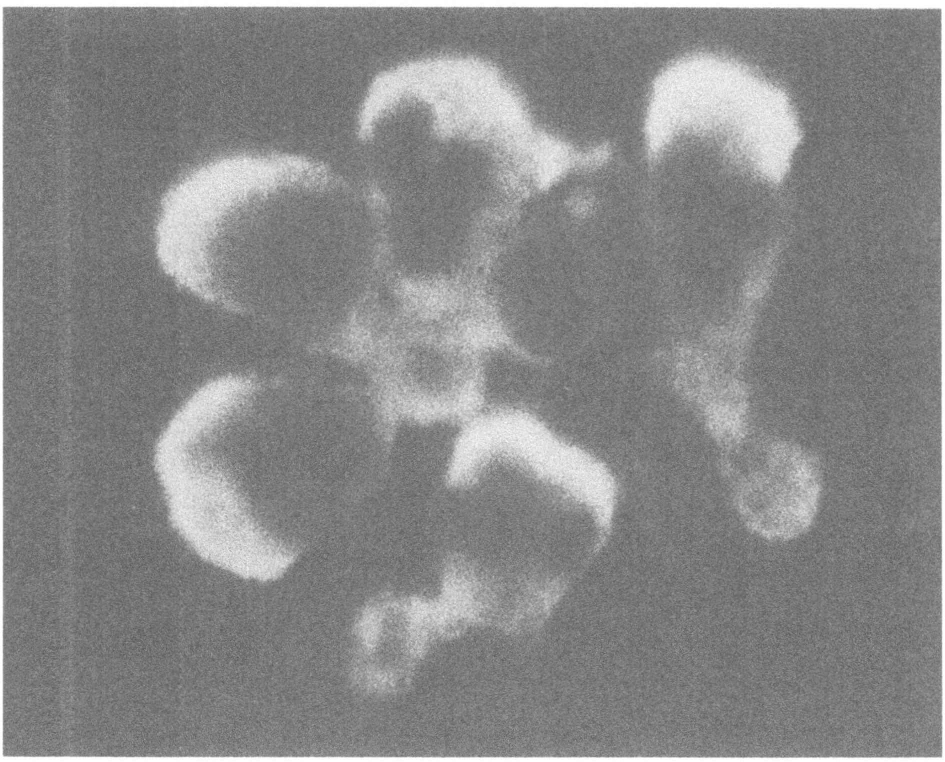

Fig. 1. Fluorescence photomicrograph of acetone-fixed polarized PMN, labelled with anti-ABP IgG and stained with rhodamine-labelled anti-IgG.

RESULTS AND DISCUSSION

Fig. 1 shows the distribution of anti-ABP fluorescence in polarizing PMN. A strong concentration of fluorescence is seen in the anterior lamellipod. Also the tail portion of the cells shows a slight concentration of fluorescence. Fig. 2 illustrates the distribution of anti-ABP fluorescence in a PMN fixed while crawling towards a clump of fluorescein-labelled yeast cells. The staining is localized to the lamellipod and to the tail portion of the cell, while the central part of the cell containing the nucleus and granules is almost devoid of staining, as is the cortical cytoplasm in this area of the cell. Fig. 3 shows a PMN ingesting a yeast particle. The fluorescence is now concentrated to the tips of the pseudopods and to a lesser extent to the tail portion of the cell.

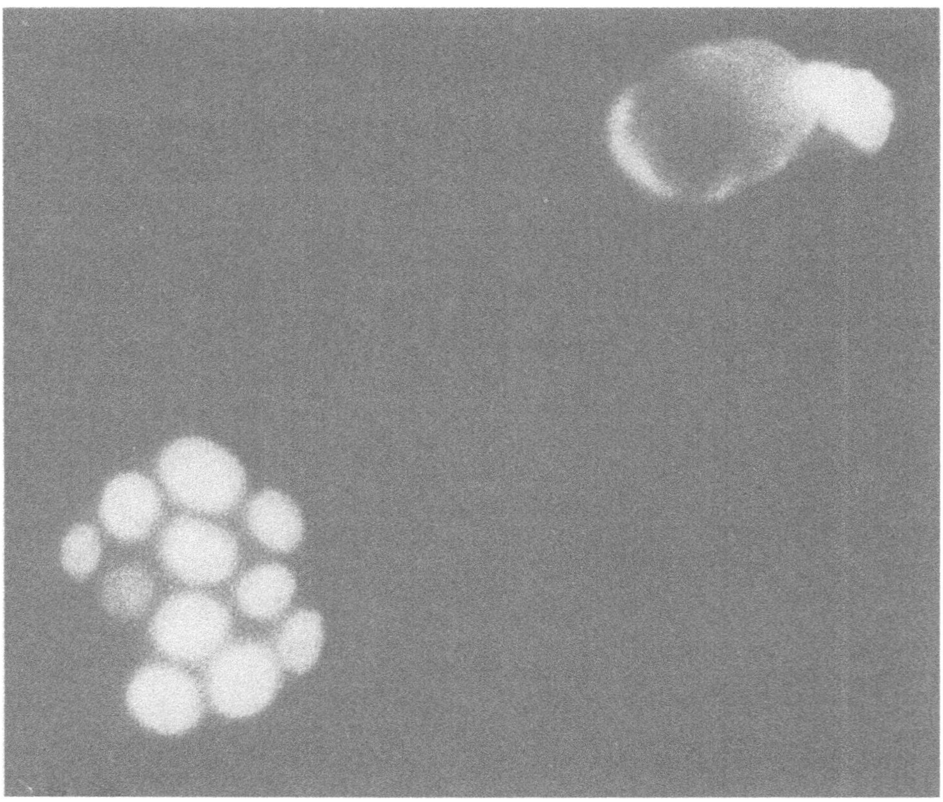

Fig. 2. Fluorescence photomicrograph of an acetone-fixed PMN migrating towards a clump of fluorescein-labelled yeast cells, reacted with anti-ABP IgG and stained as described in Fig. 1. Photographic exposure was made for both rhodamine and fluorescein fluorescence.

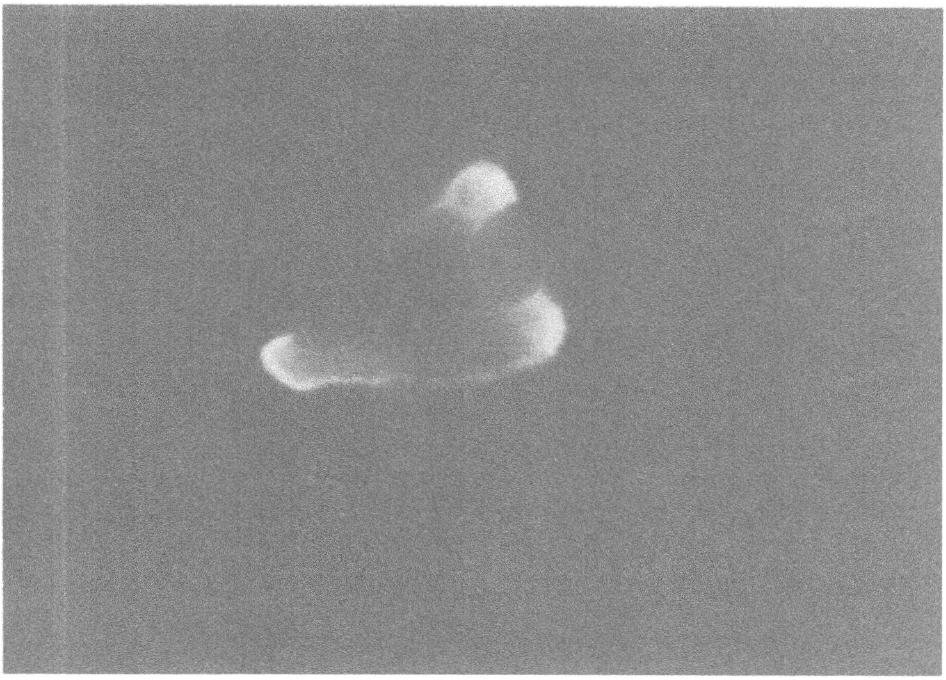

Fig. 3. Fluorescence photomicrograph of a PMN acetone-fixed during ingestion of a yeast cell and stained for ABP. Photographic exposure was made for rhodamine fluorescence only.

Fig. 4 shows the result of antimyosin staining of PMN fixed during migration toward and during ingestion of a yeast cell. Like the staining pattern seen after ABP-staining, the fluorescence is concentrated in the anterior lamellipod and in the tips of the pseudopods embracing the yeast particle. It is noteworthy that the central part of the hollow of the forming phagosome shows no staining. In contrast to staining for ABP little or no fluorescence is visible in the tail portion of the cells stained for myosin.

Both anti-ABP and antimyosin staining showed a uniform distribution throughout the cytoplasm, only excluding the nucleus, in resting PMN (not shown here). After completion of the ingestion of the yeast particles most PMN rounded up. In such cells the fluorescence for both ABP and myosin showed a homogenous distribution in the cytoplasm similar to that observed in resting cells.

We have here demonstrated that staining for ABP and myosin in phagocytosing and locomoting PMN concentrates in the lamellipods and pseudopods. It is important to bear in mind that these

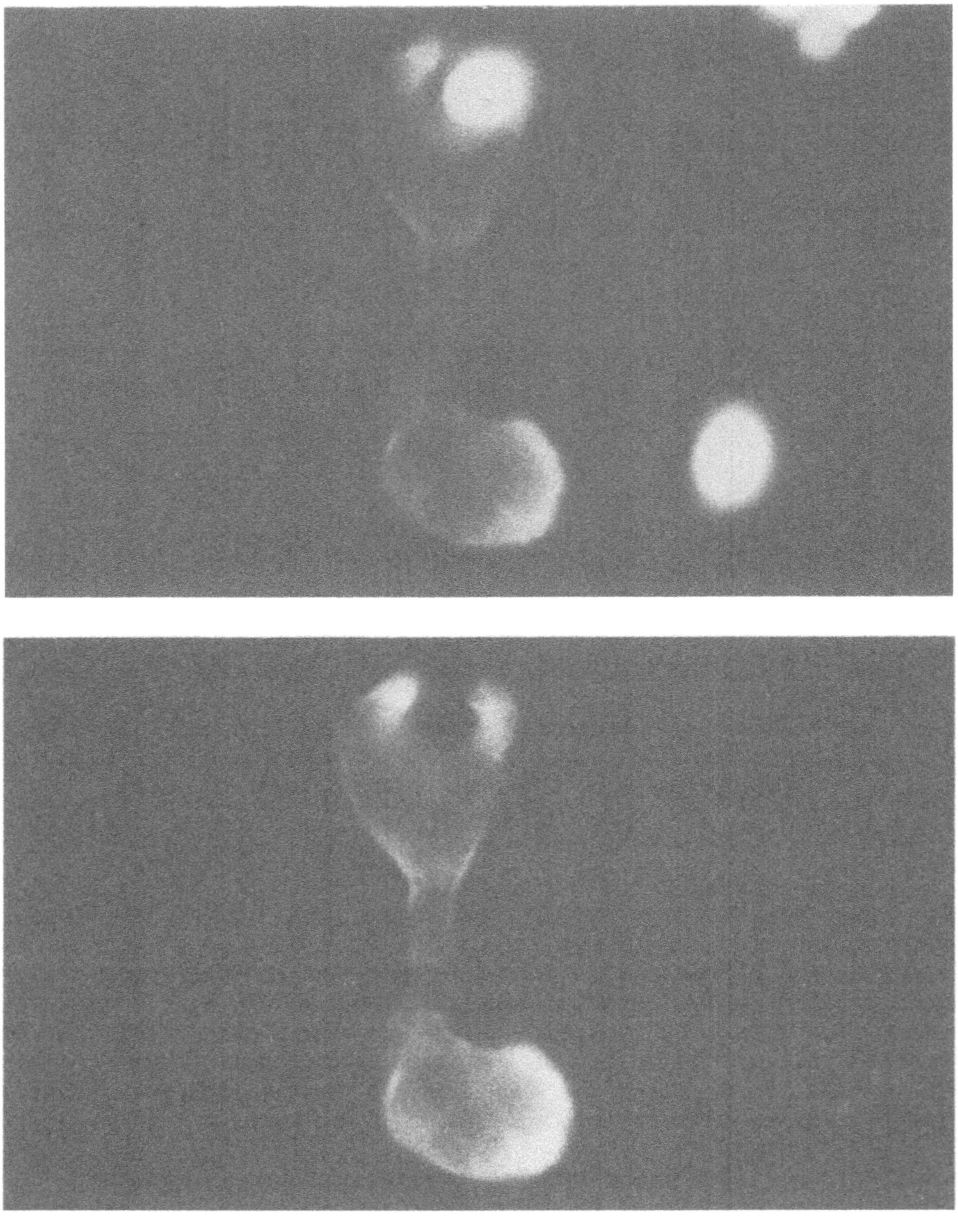

Fig. 4. Fluorescence photomicrographs of PMN acetone-fixed while crawling towards and ingesting fluorescein-labelled yeast particles, reacted with antimyosin IgG and stained with rhodamine-labelled anti-IgG antibody. The photograph at top was exposed to both fluorescein and rhodamine fluorescence, and the photograph at bottom to rhodamine fluorescence only.

structures represent the thinnest areas of the PMN. If the stained material was homogenously distributed in the cell during chemotaxis and phagocytosis one would expect to find a weaker staining in these regions. We therefore interpret the opposite finding to represent a very strong concentration of myosin and ABP in the pseudopods and lamellipods. The concentration of myosin and ABP to these areas enables the molecules to interact with actin-containing filaments in the region of the motile process of the cell. The observation thus further supports the notion that assigns a role to these proteins in PMN motility.

The most likely source of recruitment of ABP and myosin is adjacent areas of the cortical cytoplasm. This theory fits well with the observation in phagocytosing PMN that the staining is concentrated to the tips of the pseudopods while the bottom of the cup-shaped hollow where the initial particle-cell contact took place, at this later stage is completely devoid of staining.

Fig. 5 illustrates schematically how this redistribution of ABP and myosin is compatible with our current model for leukocyte movement. In the resting cell (top drawing) the cortical cytoplasm is gelled by actin filaments cross-linked by ABP. The average length of the filaments is uniform throughout the cortical cytoplasm as is the rigidity of the gel. Myosin molecules are dispersed in the actin lattice to generate the force for movement. Stimulation of the membrane by an opsonized particle (second drawing from top) causes an increase of the Ca^{++} concentration in adjacent regions of the cortical cytoplasm. This may result from transport of Ca^{++} across the membrane or release of Ca^{++} from intracellular storages. The increase in the concentration of Ca^{++} activates a Ca^{++}-regulated protein, gelsolin, which severs actin filaments to short segments, causing a localized decrease in the rigidity of the gel. The contraction of myosin now draws the short actin filaments and ABP from the area with lower rigidity towards areas with higher rigidity. This cytoplasmic streaming causes formation of pseudopods containing actin, ABP, and myosin. Other regions of the membrane get into contact with the particle to be ingested, and the series of events can now take place at these points (third drawing from top). The dissolution of the cortical cytoplasmic gel (bottom drawing) permits granules to get in contact with the plasma membrane to release their contents of lysosomal enzymes into the forming phagosome.

In conclusion while myosin provides the force for movement, directionality of movement occurs as a result of changes in the rigidity of the cortical gel due to concentration gradients of Ca^{++} and mediated via the Ca^{++}-regulated protein gelsolin. This theory fits well with the data presented in this study, showing a concentration of ABP and myosin in the pseudopods of phagocytosing neutrophils.

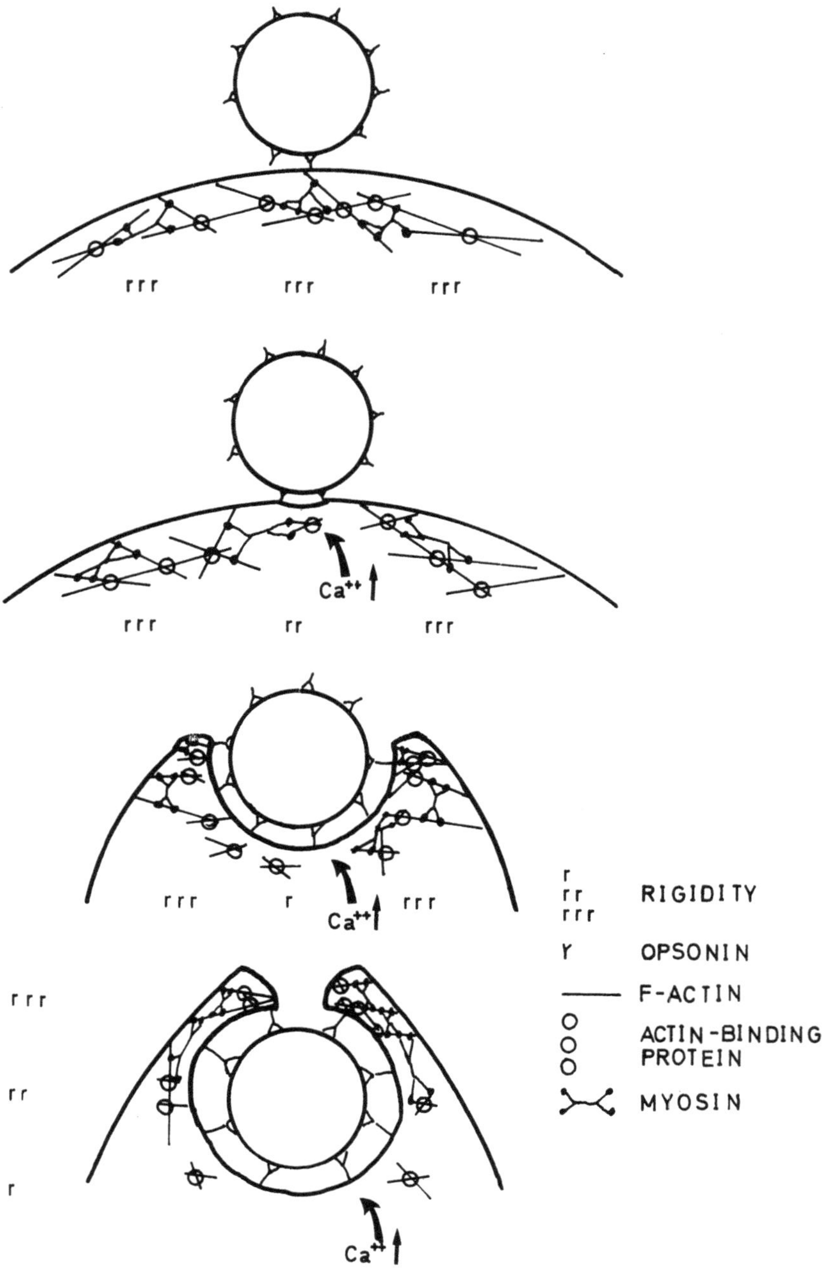

Fig. 5. Schematic drawing of the interaction of contractile proteins in PMN during phagocytosis. For further explanation see text.

REFERENCES

1. D. G. Keyserlingk, Elektronmikroskopische Untersuchung über die Differenzierungsvorgänge im Cytoplasma von segmentierten neutrophilen Leukozyten während der Zellbewegung, Exp. Cell Res. 51:79 (1968).
2. P. L. Moore, H. L. Bank, N. T. Brissie, and S. S. Spicer, Phagocytosis of bacteria by polymorphonuclear leukocytes. A freeze-fracture, scanning electron microscope, and thin section investigation of membrane structure, J. Cell Biol 76:158 (1978).
3. J. M. Oliver, J. A. Krawiec, and E. L. Becker, The distribution of actin during chemotaxis in rabbit neutrophils, J. Reticuloend. Soc. 24:697 (1978).
4. H. L. Yin and T. P. Stossel, Control of cytoplasmic actin gel-sol transformation by gelsolin, a calcium-dependent regulatory protein, Nature 281:583 (1979).
5. T. P. Stossel and T. D. Pollard, Myosin in polymorphonuclear leukocytes, J. Biol. Chem. 248:8288 (1973).
6. T. P. Stossel and J. H. Hartwig, Interactions of actin, myosin, and a new actin-binding protein of rabbit pulmonary macrophages. II Role in cytoplasmic movement and phagocytosis, J. Cell Biol. 68:602 (1976).
7. L. A. Boxer and T. P. Stossel, Interactions of actin, myosin and an actin-binding protein of chronic myelogenous leukemia leukocytes, J. Clin. Invest. 57:964 (1976).

Supported by a grant from The Council for Tobacco Research, USA. Dr. Valerius was supported by the Danish Medical Research Council, the Ebba Celinder's Foundation, and a Fogarty Foundation Fellowship. Dr. Stendahl was supported by a Travelling Fellowship from the Swedish Medical Research Council.

DISCUSSION
Edited by H.U. Keller

Romeo, D.: What is known about the subcellular distribution of actin-binding protein (ABP)? What fraction of the total cellular ABP is confined to the cytoplasm, and what mechanism can be postulated to explain the concentration of cytoplasmatic ABP in the pseudopods of the neutrophil?

Valerius, N.H.: Actin-binding protein (ABP) is primarily localized in the cortical cytoplasm. It is not clear as yet if part of it is bound to the plasma membrane. It has been reported that cytoplasmic extracts from phagocytozing alveolar macrophages contain more ABP than extracts from not-phagocytozing macrophages. This suggests that ABP may be released from the plasma membrane into the cytoplasm during phagocytosis. The present study showed that ABP staining in the tail portion of moving PMN was more intense than myosin staining. This observation may also suggest that some ABP is linked to the membrane. We think that the formation of pseudopods could result from translocation of actin filaments and ABP by the action of myosin. Myosin exerts a contraction on actin filaments cross-linked by ABP. A localized increase of the Ca^{2+} concentration could activate gelsoline to sever actin-filaments. The resistance against contraction diminishes at this point, because gel strength is dependent on the length of the filaments. The diminished resistance results in movement of ABP-actin filaments towards the areas with higher resistance against contraction. This cytoplasmic movement produces the shape changes of the PMN seen during phagocytosis, and accounts for the concentration of ABP myosin and actin in the pseudopods.

INTERACTION BETWEEN NEUTROPHILS AND MEDIATORS OF INFLAMMATION

S. Spisani, E. Vicenzi & S. Traniello

Istituto di Chimica Biologica, Università degli Studi di Ferrara - 44100 Ferrara - Italy

INTRODUCTION

In an inflammatory reaction three basic components are recognized: Injury to tissues, inflammatory mediators and tissue responses. Trauma causes tissue injury, this leads to the appearance of local inflammatory mediators which induce the tissue reactions, responsible in turn for the inflammatory symptoms.

As suggested by Ferreira[1], inflammatory mediators are any endogenous substances whose levels increase at the site of the trauma in association with the appearance of at least one tissue response.

Mediators of inflammation can be classified into three groups, according to their mode of action:

1) Direct-acting via a specific receptor (histamine, serotonine, prostaglandins and slow-reacting substance of anaphylaxis);
2) Lytic enzymes of plasma or cell origin, which cause direct damage to the integrity of tissue components (proteolytic enzyme may produce chemotactic peptides like kinins);
3) Chemotactic factors which stimulate cell migration to the site of inflammation (complement system and coagulation system).

We considered only the first group of compounds, these, although structurally unrelated, exert a central role in the development of the inflammatory response. The mediators modulate the granulocytic functions in several ways, since neutrophils are a prominent component of many inflammatory reactions.

MODULATION OF NEUTROPHIL MOTILITY

In studies from our laboratory on human polymorphs, by using modified Boyden assay system[2], we found that the two mediators histamine and serotonin were not chemotactic agents for neutrophils. But, both these amines exhibit a positive chemokinetic effect at very low concentration[3]. Serotonin increased random locomotion without affecting chemotaxis. The stimulation induced by serotonin could be removed by treatment with cyproeptadine, a specific serotonin antagonist (Fig. 1).

Histamine increased random locomotion and the sti-

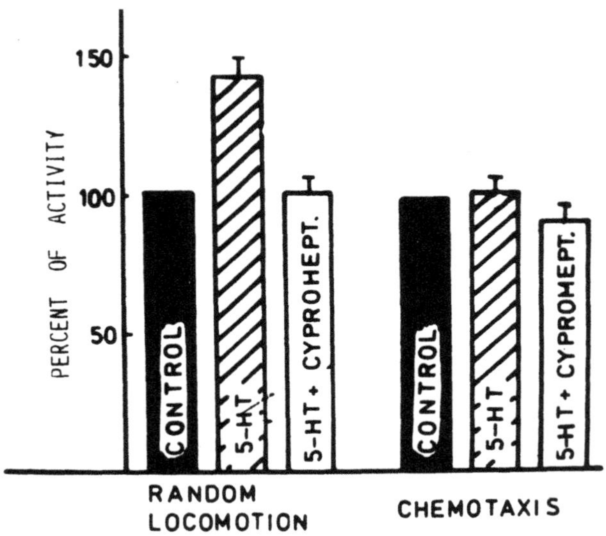

Fig. 1. Effect of serotonin on random locomotion and chemotaxis of human neutrophils.
Serotonin was added in chemotaxis chamber at a final concentration of 10^{-8} M. The receptor blocking agent, cyproeptadine, was added together with the amine at a concentration of 10^{-4} M. Results are expressed as percentage of control and values are mean $\pm$S.E. of 6 experiments.

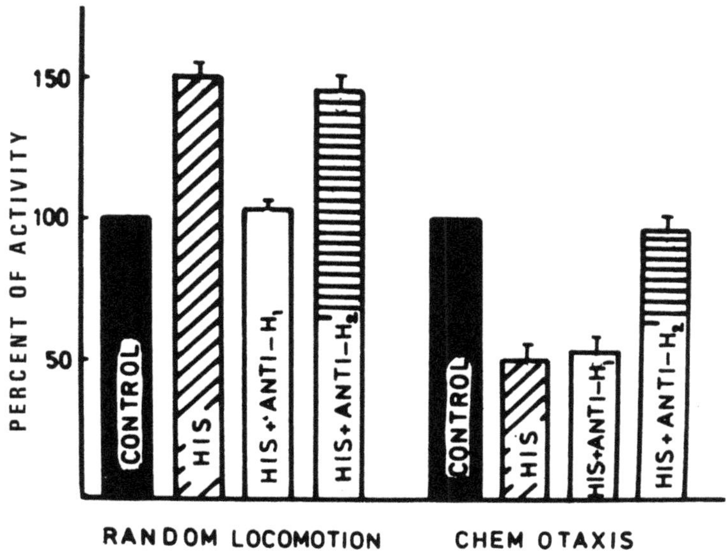

Fig. 2. Effect of histamine on random locomotion and chemotaxis of human neutrophils.
Histamine was added in the chemotactic chamber at a final concentration of 10^{-8} M. The receptor blocking agents, promethazine, diphenylhydramine and cimetidine, were added together with the amine at a concentration of 10^{-4} M.
Results are expressed as percentage of control and values are mean $\pm$S.E. of 6 experiments.

mulation could be removed by treatment with H1 specific antagonists (such as promethazine, carbinoxamine, diphenylhydramine). The H2 antagonist was ineffective. The same histamine concentration inhibited chemotaxis, and the inhibition could be removed by treatment with the H2 specific antagonist (such as cimetidine). The H1 antagonists were ineffective (Fig. 2).

These observations with granulocytes may serve to clarify in part the role of amines in inflammation. In fact, these drugs may have either an amplifying effect on the inflammatory response or a modulating effect on the inflammatory process, inhibiting neutrophil chemotaxis and the subsequent release of lysosomal content[4].

Also the secretion of endogenous hormones, like prostaglandins or other derivatives of arachidonic acid, such as HHT and HETE, may serve to regulate the inflammatory reaction of neutrophils[5,6]. Prostaglandins, mainly the E and F types, are in fact found in the inflammatory exudates[7]. Using the same cell model, we attempted to clarify the role of PGs in respect to neutrophils, since phagocytosis of polymorphs and macrofages is accompanied by release of prostaglandins E and F[8]; moreover, these PGs seem implicated in leukocyte chemotaxis. In fact, in some inflammatory experimental systems the appearance of PGs coincided with PMN leukocyte accumulation[9]. On the contrary, in other systems the levels of PGs peaked at the time when only few PMN cells were present[10].

We examined the effect of prostaglandins on human PMN cells. We found that both PGs were not chemotactic factors for human polymorphs; $PGF_{2\alpha}$ did not affect ran-

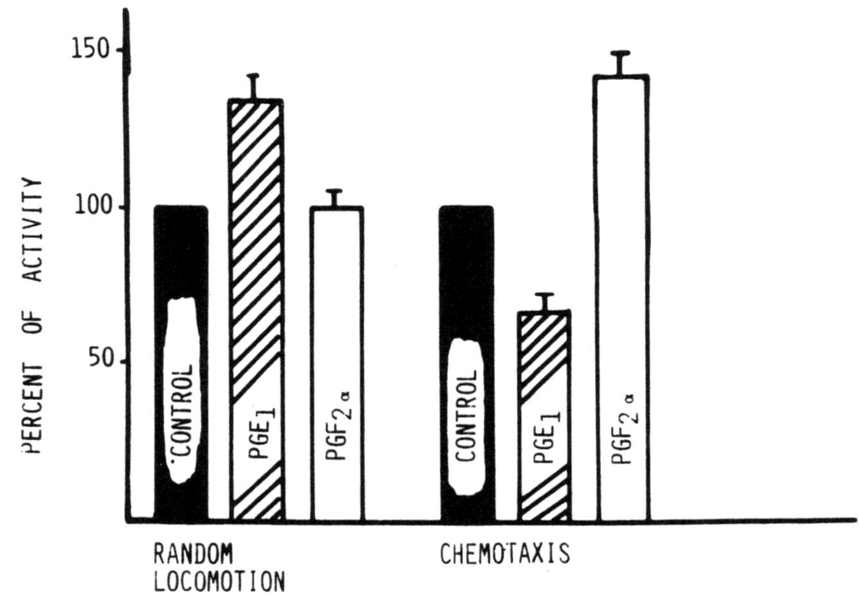

Fig. 3. Effect of PGs on random locomotion and chemotaxis of human neutrophils.
PGE_1 and $PGF_{2\alpha}$ were added in chemotaxis chamber at a final concentration of 10^{-5} mg/ml.
Results are expressed as percentage of control and values are mean $\pm$S.E. of 7 experiments.

dom locomotion but enhanced chemotactic migration, while PGE_1, weakly but significantly, stimulated random movement and reduced the chemotactic response of neutrophils (Fig. 3).

The capacity of PGs to be only in some cases a chemotactic agent was explained by Turner[11], who pointed out that "aged" PGE_1 could become chemotactic for polymorphs because of a presumed modification of prostaglandin structure. We explored the molecular basis in relationship between activity and structure using a PGE_1 analogue. This analogue (Fig. 4), in which a carbon atom of the cyclopentane ring was replaced by a nitrogen atom and the hydroxyl in position eleven was eliminated, dispayed a strong chemotactic activity for human PMN cells in the range of 10^{-4} - 10^{-2} mg/ml.

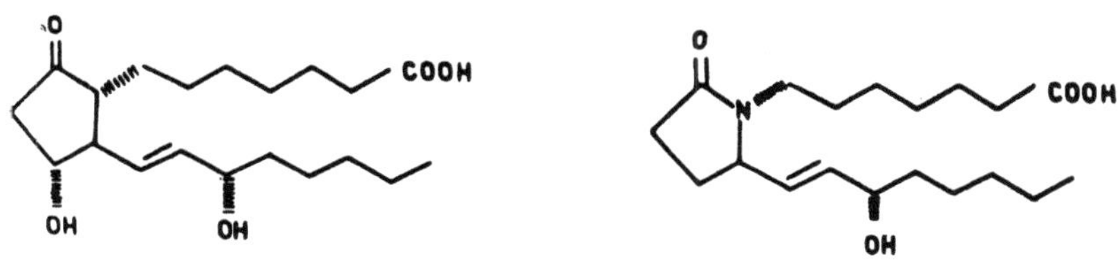

Fig. 4. PGE_1 and its analogue prepared by total organic synthesis[12].

The failure to detect chemotactic activity of PGE_1 towards neutrophils, may support the findings of Gordon[13] concerning the role of prostaglandins in the exudates. PGE_1 does not exert a direct effect on directional movement of cells, but its major action as inflammatory mediators is accomplished either by blocking the release of lymphokines from lymphocytes or by depressing chemotactic responses to known chemotactic factors [14], as suggested also by our results summarized in Fig. 5. This provides a negative feed-back mechanism for regulating the extent and the duration of the process.

The observation by Váne and co-workers[15,16] that non steroidal anti-inflammatory drugs inhibit prostaglandin biosynthesis, prompted us to demonstrate indirectly the mediation of motile functions by intracellular PGs. In fact, the in vitro treatment of polymorphs with anti-inflammatory agents completely abolished chemotactic activity, without variation of the random locomotion[17].

CONCLUSION

Taking all the data together, it is possible to speculate about an intracellular mechanism, common to different mediators of inflammation. Agents which cause a cellular increase of cyclic AMP, like histamine through the H2 receptor, the E series of prostaglandins and epinephrine, inhibit neutrophil chemotaxis and stimulate random motility. Substances increasing the level of cyclicGMP, such as histamine through the H1 receptor, serotonin, PGF, imidazole and ascorbic acid, enhance the chemotactic response (Fig. 6). So that, the final effect on the motile function of neutrophils will depend on the balance of the two cyclic nucleotides[18-20]. In particular cyclic AMP stabilizes the aggregation of microfilaments involved in spontaneous locomotion, whereas cyclic GMP, with Ca ions, control the assembly of the microtu-

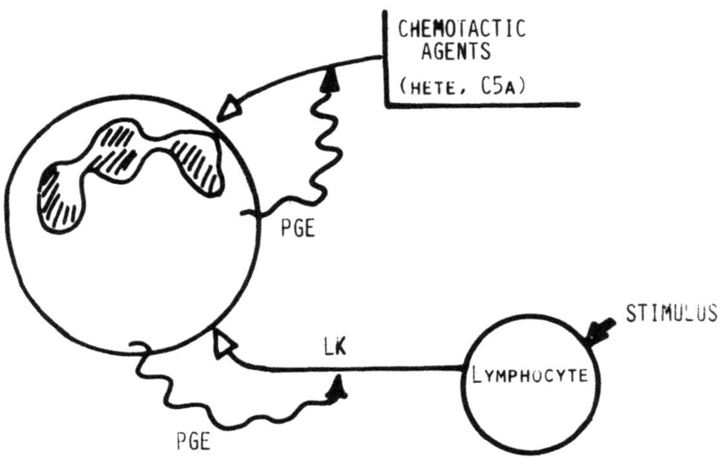

Fig. 5. Modulation of chemotaxis and release of granule enzymes in neutrophils by PGs.

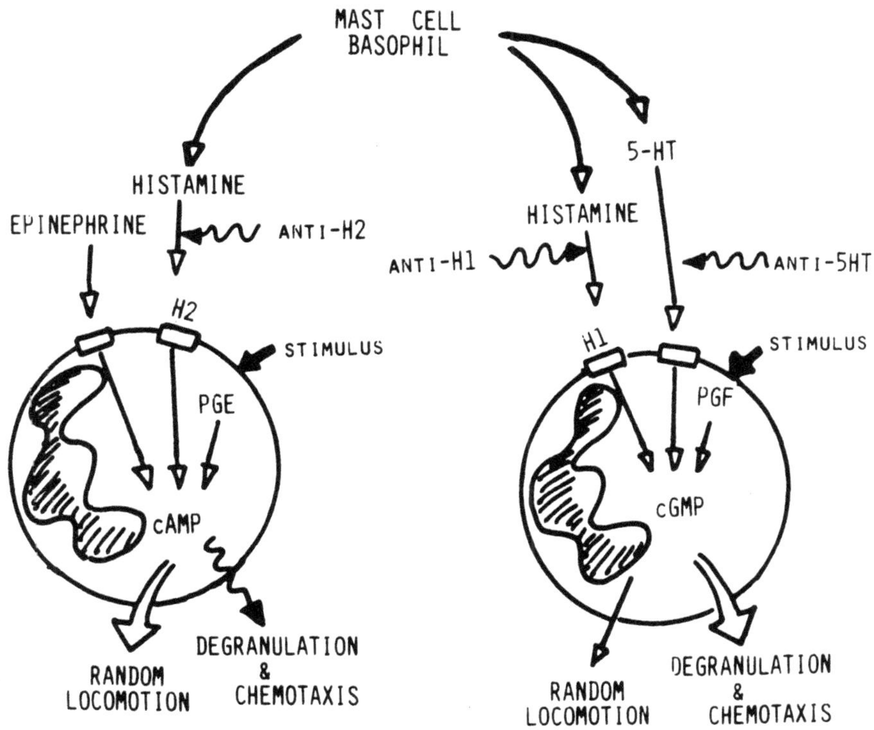

Fig. 6. Modulation of movement and release of granule enzymes in neutrophils by inflammatory mediators.

bules system, correlated with the cell's ability to activate directional movement and to release lysosomal contents. The degranulation after cell recruitment into the lesion site, amplifies the inflammatory process, while the inhibition of chemotaxis and lysosomal enzyme release reduce the inflammatory process.

REFERENCES

1. S.H. Ferreira, Prostaglandins, in: "Chemical messengers of the inflammatory process", J.C. Houck, ed., Elsevier/North-Holland Biomedical Press, Amsterdam (1979).
2. P.C. Wilkinson, "Chemotaxis and inflammation", Churchill-Livingstone, Edinburgh (1974).
3. S. Spisani and S. Traniello, Effect of mediators of inflammation on human neutrophil motility, in: "Advan-

ces in inflammation research", Vol. 1, G. Weissmann, B. Samuelsson, R. Paoletti, eds., Raven Press, New York (1979).

4. D. Rigal, J.C. Monier and S. Souweine, The effect of histamine on leukocyte migration test in man, Cell. Immunol. 46: 360 (1979).
5. F.H. Valone and E.J. Goetzl, Immunologic release in the rat peritoneal cavity of lipid chemotactic and chemokinetic factors for polymorphonuclear leukocytes, J. Immunol. 120: 102,(1978).
6. E.J. Goetzl and R.R. Gorman, Chemotactic and chemokinetic stimulation of human eosinophil and neutrophil polymorphonuclear leukocytes by 12-L-hydroxy-5,8,10-heptadecatrienoic acid (HHT), J. Immunol 120: 526 (1978).
7. I.L. Bonta and M.J. Parnham, prostaglandins and chronic inflammation, Biochem. Pharmacol. 27: 1611 (1978).
8. G.A. Higgs, E. McCall and L.J.F. Youlten, A chemotactic role for PGs release from polymorphonuclear leukocytes during phagocytosis, B. J. Pharmacol. 53: 539 (1975).
9. A.J. Anderson, W.E. Broklehurst and A.L. Willis, Evidence for the role of lysosomes in the formation of prostaglandins during carrageenin induced inflammation in the rat, Pharmacol. Res. Commun. 3: 13 (1971).
10. M. Glatt, B. Peskar and K. Brune, Leukocytes and prostaglandins in acute inflammation, Experientia 30: 1257 (1974).
11. S.R. Turner and W.S. Lynn, Lipid molecules as chemotactic factors, in: "Leukocyte chemotaxis: Methods, physiology and clinical implications", J.I. Gallin and P.G. Quie, eds., Raven Press, New York (1978).
12. A. Barco, S. Benetti, G.P. Pollini, P.G. Baraldi, D. Simoni and C.B. Vicentini, Synthesis of an 11-deoxy-8-aza prostaglandin E_1 intermediate, J. Org. Chem. 44: 1734 (1979).
13. D. Gordon, M.A. Bray and J. Morley, Control of lymphokine secretion by prostaglandins, Nature 262: 401 (1976).
14. P.A. Ward, T.E. Hugli and D.E. Chenoweth, Complement and chemotaxis, in: "Chemical messengers of the inflammatory process", J.C. Houck, ed., Elsevier/North Holland Biomedical Press, Amsterdam (1979).

15. J.R. Vane, Inhibition of prostaglandin synthesis as a mechanism of action for aspirin-like drugs, Nature, New Biol 231: 232 (1971).
16. J.R. Vane, Prostaglandins in the inflammatory response, in: "Inflammation mechanism and control", I.H. Lepow and P.A. Ward, eds., Academic Press, New York (1972).
17. S. Spisani, G. Vanzini and S. Traniello, Inhibition of human leukocyte locomotion by anti-inflammatory drugs, Experientia 35: 803 (1979).
18. G.E. Hatch, W.K. Nichols and H.R. Hill, Cyclic nucleotide changes in human neutrophils induced by chemoattractants and chemotactic modulators, J. Immunol. 119: 450 (1977).
19. I. Rivkin, J. Rosenblatt and E.L. Becker, The role of cyclic AMP in the chemotactic responsiveness and spontaneous motility of rabbit peritoneal neutrophils, J. Immunol. 115: 1126 (1975).
20. R. Anderson, A. Glover and A.R. Rabson, The in vitro effects of histamine and metiamide on neutrophil motility and their relationship to intracellular cyclic nucleotide levels. J. Immunol 118: 1690 (1977).

A SHORT TRANSIENT INCREASE IN CYCLIC ADENOSINE 3', 5'-MONOPHOSPHATE LEVELS OF NEUTROPHIL GRANULOCYTES FOLLOWING EXPOSURE TO CHEMOTACTIC FACTORS

A. Naef and H.U. Keller

Department of Pathology
University of Berne
3010 Bern, Switzerland

INTRODUCTION

Several studies which have been performed some years ago showed that increased cAMP levels are associated with inhibition of directional locomotion[1,2,3,4,5]. The increased cAMP levels were detected 2 minutes after chemotactic stimulation or later. Extracellular dibutyryl cAMP inhibited the cytotaxin-induced directional locomotion[2,6,7,8]. The results led to the conclusion that, though cAMP is not in the main sequence of events triggering the chemotactic response, increases in cAMP may modulate directional locomotion.

Recent studies demonstrated a short transient increase in cAMP levels following stimulation with cytotaxins[9,10,11,12,13]. Therefore, the question as to whether cAMP might nevertheless be involved in the transduction of the chemotactic signal has been reconsidered. The following experiments represent a preliminary attempt to clarify this question.

METHODS

Cell preparation: Leucocytes from healthy human donors or healthy horses were prepared by a two step procedure. In the first step heparinized blood (10 units heparin/ml) was layered on Ronpacon-Methocel. In a second step the leucocyte-rich supernatant was separated into granulocytes (pellet) and mononuclear cells and platelets (interface) by a Ficoll-Ronpacon gradient[14]. The pellet containing 96 % or more granulocytes was then washed twice and resuspended in Gey's solution containing 2 % HSA (Behringwerke, Marburg, FRG).

Stimulation of granulocytes and cAMP-assay: The cell suspension ($5 \cdot 10^7$ granulocytes/ml) was placed in a siliconised glass tube and gently stirred at 37°C in a water bath. Cells were exposed to the reagent in question by addition of the appropriate volume of a 100-fold concentrated stock solution. Samples of 40 µl were mixed with 160 µl 1.4N perchloric acid at the time indicated and immediately vortexed. After centrifugation the supernatant was neutralized by addition of 6^N K_2CO_3, centrifuged once more and diluted at least ten times with sodium acetate buffer (pH 6.2). The cAMP content was assayed by radioimmunoassay (cAMP $[^{125}I]$ RIA kit, New England Nuclear) according to the protocol for acetylated samples.

Assessment of PMN directional locomotion: The chemotactic activity was measured in triplicate by the two filter count method[15] using modified Boyden chambers, which were incubated at 37°C for 1 hour or 3 hours.

RESULTS AND DISCUSSION

Granulocyte cAMP levels following primary exposure to chemotactic and non-chemotactic factors

Studies with Dictyostelium discoideum suggest that cyclic nucleotides are involved in the chemotactic response on two levels. They can act as extracellular chemotactic signals and they also seem to play a role in the intracellular transduction of the chemotactic response[16,17]. Our studies on human granulocytes suggest that the increase in cAMP levels within the first minute following chemotactic stimulation is associated with the cells. Only about 5 to 10 % of the cAMP measured is associated with the medium (to be published). This raises the question as to whether the cAMP peak is relevant for the transduction of the chemotactic signal in leucocytes. It was therefore of interest to investigate to which extent the chemotactic response correlates with the cAMP peak appearing 1 minute following stimulation.

It has been found that several cytotaxins including CAT 1.5.1.[9], f-Met-Leu-Phe[10,11,12,13] and C5a[11] produce such a transient cAMP peak. There was, however, one exception. It has been reported that the peptide Gly-His-Gly is chemotactic[18,19] without producing a transient cAMP peak[20]. Our experiments confirmed the absence of the cAMP peak following stimulation with

Table 1. Effect of human serum albumin (HSA), CBZ-Phe-Met and f-Met-Leu-Phe (FMLP) on neutrophil granulocyte locomotion

Test solution	Medium for cell suspension	Incubation time	% neutrophils migrated
Gey's solution	Gey's solution	3 h	0.01 ± 0.02[a]
2 % HSA in Gey's solution	2 % HSA in Gey's solution	3 h	1.24 ± 0.23[a]
2 % HSA in Gey's solution	2 % HSA in Gey's solution	1 h	0.06 ± 0.1[a]
2 % HSA + CBZ-Phe-Met 10^{-4}M	2 % HSA in Gey's solution	1 h	0.06 ± 0.05[a]
2 % HSA in Gey's solution	2 % HSA in Gey's solution	1 h	0.06 ± 0.1[a]
2 % HSA + FMLP $5 \cdot 10^{-8}$M	2 % HSA in Gey's solution	1 h	8.9 ± 1.1[a]
2 % HSA in Gey's solution	2 % HSA in Gey's solution	1 h	0.1 ± 0.07[b]
2 % HSA + FMLP $5 \cdot 10^{-8}$M	2 % HSA in Gey's solution	1 h	0.1 ± 0.04[b]

[a] human neutrophils

[b] equine neutrophils

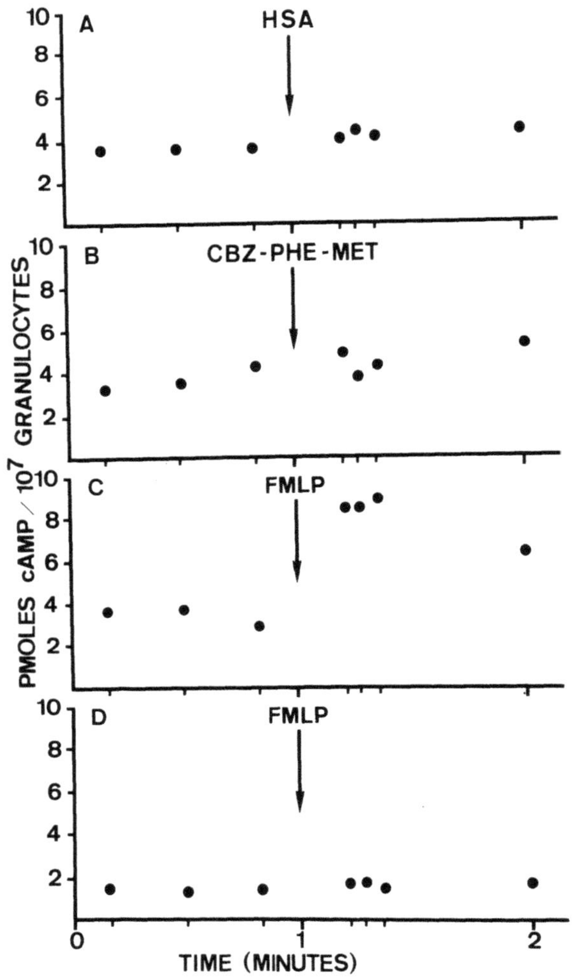

Fig. 1. Effect of human serum albumin (HSA), CBZ-Phe-Met and f-Met-Leu-Phe on cAMP levels of neutrophil granulocytes

Neutrophils were suspended in Gey's solution (A) or Gey's solution containing 2 % HSA (B,C,D). A: Human neutrophils exposed to 1 % HSA (final concentration). B: Human neutrophils exposed to CBZ-Phe-Met (final concentration 10^{-4}M). C: Human neutrophils exposed to f-Met-Leu-Phe (final concentration $5 \cdot 10^{-8}$M). D: Equine neutrophils exposed to f-Met-Leu-Phe (final concentration $5 \cdot 10^{-8}$M). Data represent the mean of 7 (A), 3 (B), 16 (C) and 3 (D) individual experiments.

Gly-His-Gly, but the material also lacked chemotactic activity. Various other agents which lack chemotactic activity have also been tested. Chemokinetic HSA (Table 1) which is not chemotactic produced no cAMP peak (Fig. 1A). The non-chemotactic antagonist of f-Met peptides, CBZ-Phe-Met (Fluka AG, Buchs, Switzerland), also failed to elicit a cAMP peak, whereas the chemotactic f-Met-Leu-Phe produced such an effect, at least in human granulocytes (Table 1; Fig. 1B and C). In contrast, equine granulocytes failed to respond chemotactically to f-Met-Leu-Phe[21]. They exhibited no increase in cAMP levels within the first minute of exposure to f-Met-Leu-Phe in concentrations of $5 \cdot 10^{-8}$M (Table 1; Fig. 1D) or 10^{-10}M, 10^{-9}M, 10^{-7}M, 10^{-6}M and 10^{-5}M (not shown). In contrast, $C5a_{des\ Arg}$ (Dr. B. Damerau, Max-Planck-Institut für experimentelle Medizin, Göttingen, FRG) which is chemotactic for equine neutrophils produced a transient cAMP peak in equine granulocytes (to be published). The finding that f-Met-Leu-Phe induces a release of lysosomal enzymes from horse granulocytes[21] but no increase of cAMP levels within the first minute (Fig. 1D) indicates that those two phenomena are probably unrelated. In summary, the findings suggest a close correlation between the chemotactic response and the appearance of a cAMP peak within the first minute. Whether all cytotaxins without exception elicit a cAMP peak requires further investigations. Furthermore, there is no proof for a causal relationship.

Secondary stimulation of human granulocytes

A primary exposure of granulocytes to cytotaxins can reduce the chemotactic response to a secondary stimulation with the same cytotaxin. We have investigated whether this phenomenon which has been called "deactivation"[22] is correlated with changes in the cAMP peak. In a first series of experiments (manuscript in preparation) neutrophils were exposed to $C5a_{des\ Arg}$ ($6{,}7 \cdot 10^{-8}$M) or f-Met-Leu-Phe ($5 \cdot 10^{-8}$M) for one minute, washed and restiumulated 30 minutes later with the same cytotaxin that had already been used for primary stimulation. These cells showed unimpaired directional locomotion whereas the cAMP peak was decreased by about 62 % and 32 %, respectively. Thus, the chemotactic responsivness is intact, even under conditions where the cAMP peak is significantly reduced. Neutrophils were partially deactivated after a primary exposure to a higher concentration of $C5a_{des\ Arg}$ ($3{,}3 \cdot 10^{-7}$M) for 30 minutes and the transient cAMP peak was reduced by about 56 %. So far it has not been possible to achieve complete deactivation with respect to the chemotactic response and to test cAMP levels in such cells.

SUMMARY

Stimuli which are chemotactic for the respective neutrophils always produce a transient increase in cAMP within the first minute whereas the chemokinetic or other non-chemotactic stimuli tested have no such effect. The question as to whether the transient increase of cAMP levels plays any direct role in the transduction of a chemotactic stimulus cannot be answered. Restimulation experiments indicate that the extent of the cAMP response is not a reliable indicator for the chemotactic responsiveness of neutrophils; there is no quantitative correlation between these two parameters. Further detailed kinetic and quantitative studies of the cAMP system would be needed to clarify this point.

ACKNOWLEDGEMENTS

The study was supported by the Swiss National Science Foundation.

REFERENCES

1. Hatch, G. E., Nichols, W. K., and Hill, H. R., Cyclic nucleotide changes in human neutrophils induced by chemoattractants and chemotactic modulators, J. Immunol., 119:450 (1977).
2. Rivkin, I., Rosenblatt, J., and Becker, E. L., The role of cyclic AMP in the chemotactic responsiveness and spontaneous motility of rabbit peritoneal neutrophils. The inhibition of neutrophil movement and the elevation of cyclic AMP levels by catecholamines, prostaglandins, theophylline and cholera toxin, J. Immunol. 115:1126 (1975).
3. Till, G., Kownatzki, E., Seitz, M., and Gemsa, D., Chemokinetic and chemotactic activity of various prostaglandins for neutrophil granulocytes, Clin. Immunol. Immunopathol., 12:111 (1979).
4. Gallin, J. I., Sandler, J. A., Clyman, R. I., Manganiello, V. C., and Vaughan, M., Agents that increase cyclic AMP inhibit accumulation of cGMP and depress human monocyte locomotion, J. Immunol. 120:492 (1978).
5. Hill, H. R., Cyclic nucleotides as modulators of leukocyte chemotaxis, in: Leukocyte Chemotaxis, J. I. Gallin and P. G. Quie, eds., Raven Press, New York, p. 179 (1978).
6. Tse, R. L., Phelps, P., and Urban, D., Polymorphonuclear leukocyte motility in vitro. VI. Effect of purine and pyrimidine analogues: possible role of cyclic AMP, J. Lab. Clin. Med., 80:264 (1972).

7. Giroud, J. P., and Roch Arveiller, M., In vitro effect of cyclic nucleotides on velocity and chemotaxis of neutrophil granulocytes, Biomedicine, 27:246 (1977).
8. Estensen, R. D., Hill, H. R., Quie, P. G., Hogan, N., and Goldberg, N. D., Cyclic GMP and cell movement, Nature, 245:458 (1973).
9. Keller, H. U., Gerisch, G., and Wissler, J. H., A transient rise in cyclic AMP levels following chemotactic stimulation of neutrophil granulocytes, Cell. Biol. Int. Reports, 3:759 (1979).
10. Jackowski, S., and Sha'afi, R. I., Response of adenosine cyclic 3'5'-monophosphate level in rabbit neutrophils to the chemotactic peptide formyl-methionyl-leucyl-phenylalanine Mol. Pharmacol., 16:473 (1979).
11. Simchowitz, L., Fischbein, L. C., Spilberg, I., and Atkinson, J. P., Induction of a transient elevation in intracellular levels of adenosine-3'5'-cyclic monophosphate by chemotactic factors; an early event in human neutrophil activation, J. Immunol., 124:1482 (1980).
12. Smolen, J. E., Korchak, H. M., and Weissmann, G., Increased levels of cyclic adenosine-3'5'-monophosphate in human polymorphonuclear leukocytes after surface stimulation, J. Clin. Invest., 65:1077 (1980).
13. Marx, R., Bass, D., and McCall, C., Correlation of stimulation or suppression of chemokinesis and chemotaxis of human neutrophils (PMNL) with early changes in cyclic AMP, Fed. Proc., 38:4970 (1979).
14. Bøyum, A., Separation of leucocytes from blood and bone marrow, Scand. J. clin. Lab. Invest., 21:1 (1968).
15. Keller, H. U., Gerber, H., Hess, M. W., and Cottier, H., Studies on the regulation of the neutrophil chemotactic response using a rapid and reliable method for measuring random migration and chemotaxis of neutrophil granulocytes, Agents and Actions, 6:326 (1976).
16. Gerisch, G., Maeda, Y., Malchow, D., Roos, W., Wick, U., and Wurster, B., Cyclic AMP signals and the control of cell aggregation in Dictyostelium discoideum, in: Developments and Differentiation in the Cellular Slime Moulds, Cappuccinelli and Ashworth, eds., Elsevier/North-Holland Biomedical Press, p. 105 (1977).
17. Gerisch, G., Fromm, H., Huesgen, A., and Wick, U., Control of cell-contact sites by cyclic AMP pulses in differentiating Dictyostelium cells, Nature, 255:547 (1975).
18. Spilberg, I., Mandell, B., Mehta, J., Sullivan, T., and Simchowitz, L., Dissociation of the neutrophil functions of exocytosis and chemotaxis, J. Lab. Clin. Med., 92:297 (1978).

19. Spilberg, I., Mandell, B., and Hoffstein, S., A proposed model for chemotactic deactivation. Evidence for microtubule modulation of polymorphonuclear leukocyte chemotaxis, J. Lab. Clin. Med., 94:361 (1979).
20. Weissmann, G., Smolen, J. E., and Korchak, H. M., Release of inflammatory mediators from stimulated neutrophils, New Engl. J. Med., 303:27 (1980).
21. Snyderman, R., and Pike, M. C., N-formylmethionyl peptide receptors on equine leukocytes initiate secretion but not chemotaxis, Science, 209:493 (1980).
22. Ward, P. A., and Becker, E. L., The deactivation of rabbit neutrophils by chemotactic factor and the nature of the activatable esterase, J. Exp. Med., 127:693 (1968).

DISCUSSION
Edited by H.U. Keller

Rossi, F.: Membrane activating agents including chemotactic factors induce many molecular events such as changes in membrane potential, ion fluxes etc. What is your opinion on the sequence of events leading to activation of leucocyte movement?

Naef, A.: Smolen, Korchak and Weissmann presented evidence to suggest that changes in membrane potential precede the rise in cAMP levels. This is in turn followed by release of lysosomal enzymes and the generation of the superoxyde anion. It remains to be assessed whether this time sequence is an invariable feature of any kind of leucocyte activation and whether these events play a role for the transduction of chemokinetic or chemotactic signals. There exist numerous hypothetical models to link these events but not enough data to give an opinion on what is really happening. Membrane depolarisation and the transient cAMP peak occur before or during chemotactic orientation. They are more likely to be instrumental in the transduction of chemotaxis than other events like superoxide generation or enzyme release which seem to take longer to develop.

Esmann, V.: The significance of the finding that the cAMP peak is decreased by 2/3 of the normal might be questioned. The cAMP level was quite high in both cases and within the range where it causes in our hands a complete dissociation of cAMP dependent protein kinase. The increase in cAMP may be a dead-end event.

Naef, A.: The results obviously do not answer the question as to whether cAMP is involved in the transduction of chemotaxis. The fact that we could not establish a quantitative correlation between the cAMP peak and the chemotactic responsiveness of the cells would be compatible with the view that the transient cAMP peak is a dead-end event. Further work is needed to clarify this point.

Elsbach, P.: Leukocytes respond to multiple signals with a wide range of reactions, including membrane permeability changes, depolarisation, release of divalent cations, alterations in phospholipids, changes in cyclic nucleotide levels, production of mediators derived from arachidonic acid, release of granule contents and locomotion. I wonder if it is justifiable to select just two events and to try to link them.

Naef, A.: We are by no means trying to link the cAMP peak to chemotaxis or motility. However, there are good reasons for trying to find out whether such a link has to be considered. The transient rise in cAMP is one of the early events following chemotactic stimulation. It takes place during the time required for chemotactic orientation. It is conceivable but not proven that the cAMP peak is related to other events. An interesting hypothesis is that it might play a role in myosin phosphorylation.

Patriarca, P.: Is there any evidence that proteases are involved in leucocyte motility?

Keller, H.U.: It is tempting to speculate that proteases might be involved in the detachment from the substratum during cell locomotion. Several investigators, including Dr. Gallin, showed that degranulation and release of lysosomal enzymes modulate cell adhesion and the chemotactic response. Dr. Ward and other investigators demonstrated that proteases may generate chemotactic activity and that neutral proteases released from activated leucocytes may inactivate cytotaxins.

NEUTROPHIL MIGRATION AND ORIENTATION UNDER AGAROSE: FINDINGS IN PATIENTS WITH THE IMMOTILE CILIA SYNDROME AND EFFECTS OF CYTOCHALASIN B AND VINBLASTINE

Jan Palmblad, Ann-Mari Udén, Nikos Venizelos and Björn Afzelius

Depts. of Medicine IV & Clin. Bact., the Karolinska Institute at Södersjukhuset, S-10064 Stockholm and Wennergren Institute, Univ. of Stockholm, Sweden

INTRODUCTION

The technique measuring the migration of PMN neutrophil granulocytes under agarose is simple and inexpensive. Usually results are expressed as the distance migrated by the leading front cells[1], but it is also possible to quantitate number of migrating cells[2,3] and observe their morphology[4]. Finally, modifications have been presented for discriminating chemokinesis (enhanced random migration) from chemotaxis (directed migration) by a checker-board analysis[2], or by incorporation of migration stimulating agents, cytotaxins, into the agarose[5].

The aim of this presentation was to show that PMN migration under agarose can be assessed by several different measures of locomotion, including orientation of migrating cells, a new and simple method, which may allow distinction between chemotaxis and chemokinesis. To obtain information relevant for this question, we have utilized the dissambling effects of vinblastine on microtubules and the microfilament disrupting effects of cytochalasin B. Finally, we studied the migration and orientation of neutrophils from patients with the immotile cilia syndrome[6]. In this disorder, lack of dynein arms or radial spokes on and between microtubules makes cilia and sperms immotile, leading to infertility and increased incidence of infections in the respiratory tract[7]. Since microtubular function is essential for the direction finding of neutrophils[8], it was regarded important to include an assay for this function.

MATERIAL AND METHODS

Patients and donors. The 8 patients with a documented absence of dynein arms or radial spokes of microtubuls in sperms and/or cilia, have been detailed previously[6,7]. None were acutely ill or on a drug regimen. The controls were healthy blood bank donors and members of the laboratory staff.

The preparation of leukocytes has been presented previously[6]. In some experiments healthy donor leukocytes were pretreated for 15 min. in 37°C with varying concentrations of vinblastine or cytochalasin B dissolved in saline or dimethylsulphoxide, 1 %.

PMN migration assays. The stimulated and spontaneous locomotion of PMNs were assayed with a modification of methods measuring PMN migration under agarose[1,2,5,6]. Briefly, a 1 % agarose solution (Sigma), with a final albumin concentration of 0.08 % was allowed to gel in plastic Petri dishes (Falcon Plastics, Oxnard, Calif., USA). Series of 3 wells were punched in the agarose. The diameter of one well was 3 mm, and the distance between the centers of two adjacent wells was 5 mm. 10 µl of a leukocyte suspension, containing approximately 10 x 10^9 PMNs/1 was filled in the central well. Chemotactic factors, either a sterile E. coli bacterial filtrate (BF) or pooled human AB serum were simultaneously filled in the outer wells. Appropriate control media were added to the inner wells. To stimulate chemokinesis, BF was added to the liquid agarose solution, at a final concentration of 10 %[5]. The dishes were incubated at 37°C in an atmosphere of 5 % CO_2. After fixation with methanol and staining with hematoxylin the agarose was removed and migration quantified:

(1) The distance migrated by the leading front PMNs was measured by microscopy; the results are given in mm.

(2) The number of cells migrating toward the cytotaxin and control wells was also counted[2,6]. From these measurements, we determined the location of the grid space harbouring the highest number of cells (the location of the maximal cell density) and the grid space to which 50 % of all counted cells had migrated (location of the limit for 50 % of the cells).

(3) Finally, the orientation of migrating PMNs was assessed after fixation and staining by microscopy[6,8,9]. The PMNs were considered oriented when the nuclei were in the rear of the cell and the anterior lamellopodia located within a 90° sector open towards the cytotaxin or control media containing wells.

RESULTS

Effect of incubation time. It was found that PMN migration was measurable already after 30 min. when BF was used as a stimulant (Fig. 1), and maximal distance for the leading front was reached within 2-3 hours, whereafter it remained unchanged.

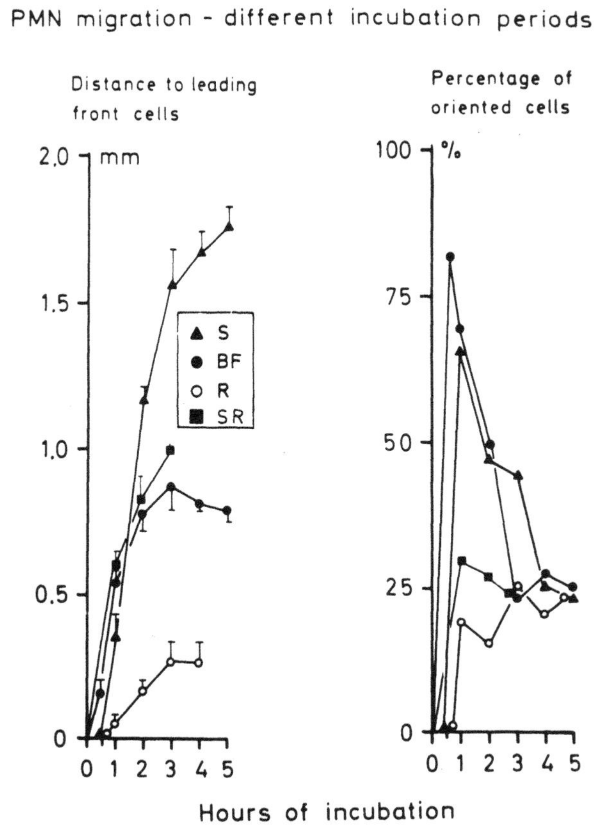

Fig. 1. The effect of varying the incubation time on migration measures.

The degree of orientation was highest in the 30 min. sample, but decreased subsequently and assumed values representative for spontaneously moving cells in the 3-hour specimens (Fig. 1 and 2A). When serum was used as cytotaxin rather similar results were obtained. One difference was that a striking and continous progression was observed for the leading front distance, persisting for the 5 hours of incubation (Fig. 1). Also, the degree of orientation was high initially and persisted so longer, being statistically significantly higher in the 3-hour specimen, compared both with BF-stimulated and spontaneously moving cells.

When BF was incorporated into the agarose, allowing PMNs to move in a chemokinetic way in an equal concentration of BF, neutrophils were found to be elongated which contrasts to the rounded forms of spontaneously moving cells, but being similar to cells engaged in a chemotactic locomotion (Fig. 2 B). The leading front

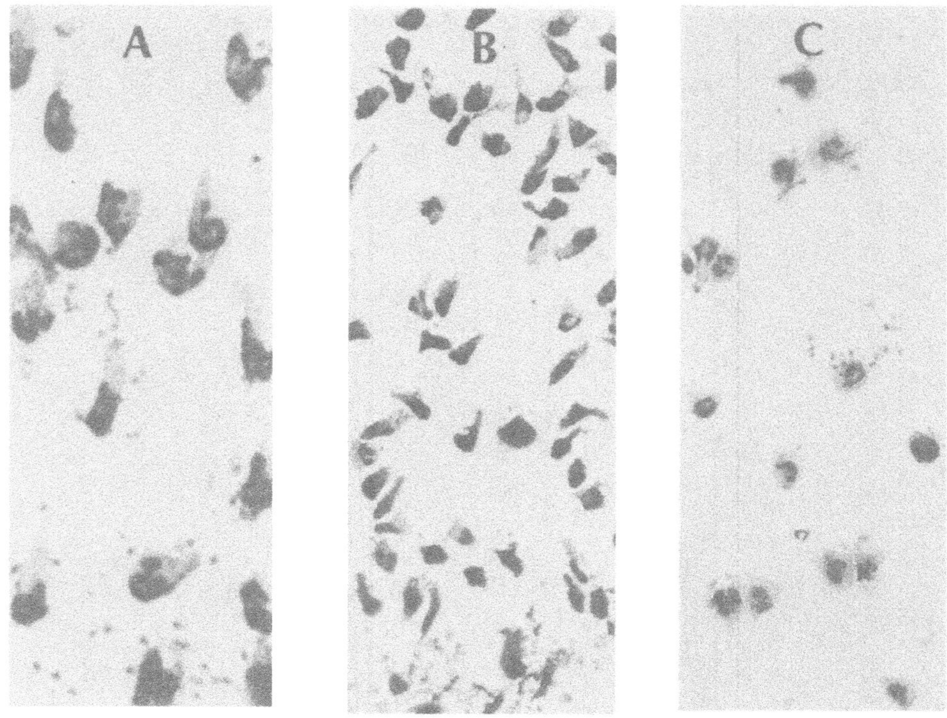

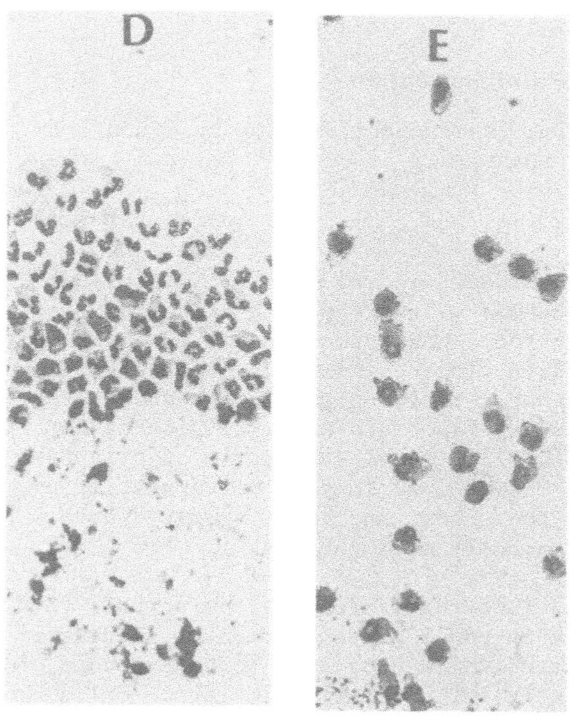

Fig. 2. Micrographs of PMNs migrating towards the cytotaxin or control wells situated at the top of the micrographs. A. PMNs moving for 1 h. towards a BF containing well. B. PMNs moving for 1 h. at random, under agarose into which BF was incorporated (chemokinesis). C. PMNs moving spontaneously for 1 h. D. Cytochalasin B treated cells (5 ug/ml) incubated for 1 h. with serum as cytotaxin. E. Vinblastine treated cells (10 ug/ml) incubated for 1 h. with BF as cytotaxin.

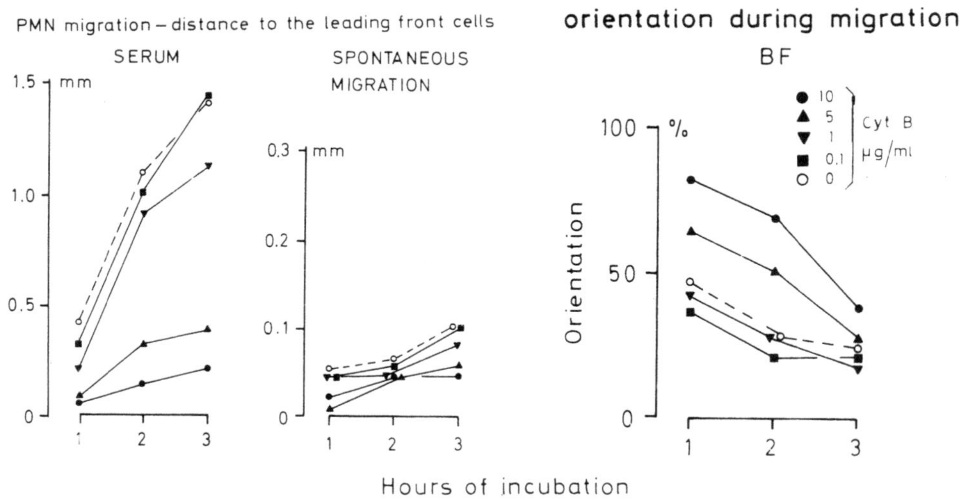

Fig. 3. PMN migration after incubation with cytochalasin B.

distance increased over time (Fig. 1), being rather parallel to that found when BF had been added into the cytotaxin well, i.e. the chemotactic response. The degree of orientation was close to that for spontaneously moving cells and did not change considerably over time (Fig. 1).

Spontaneously moving cells exhibited a round contour, with few lamellopodia and rear-oriented nuclei (Fig. 2C). Also, migration distance remained low. Mean orientation percentages were

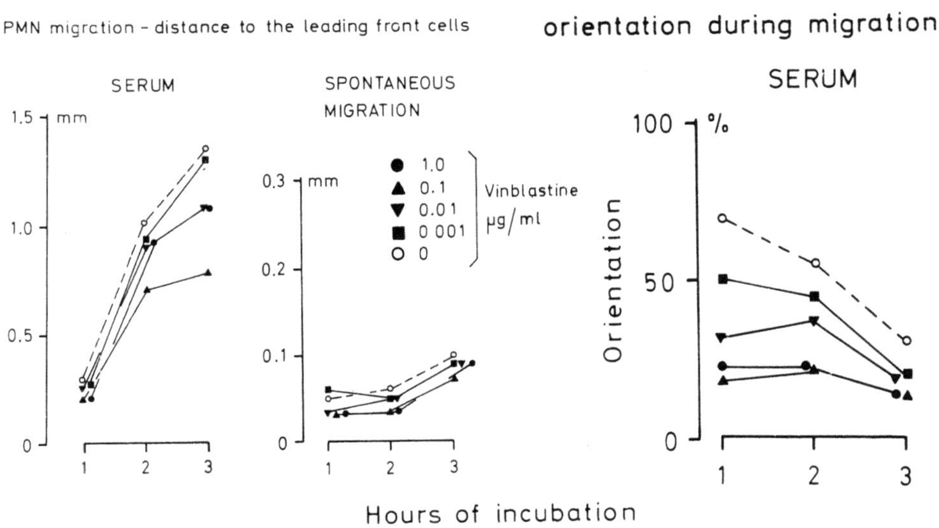

Fig. 4. PMN migration after incubation with vinblastine.

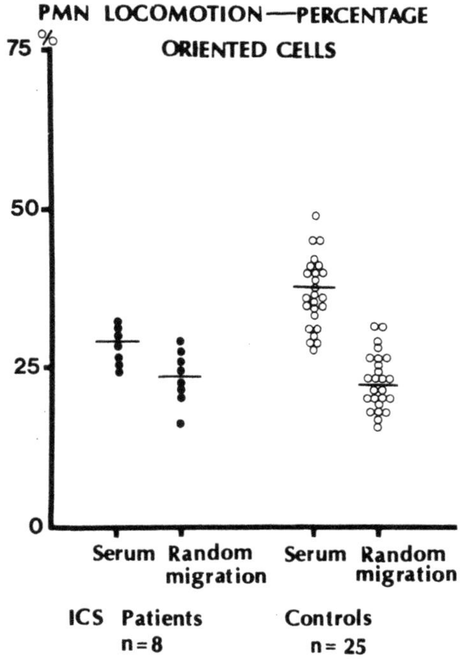

Fig. 5. Percentage migrating PMNs from immotile cilia syndrome patients and healthy controls exhibing distinct orientation in the cytotaxin gradient.

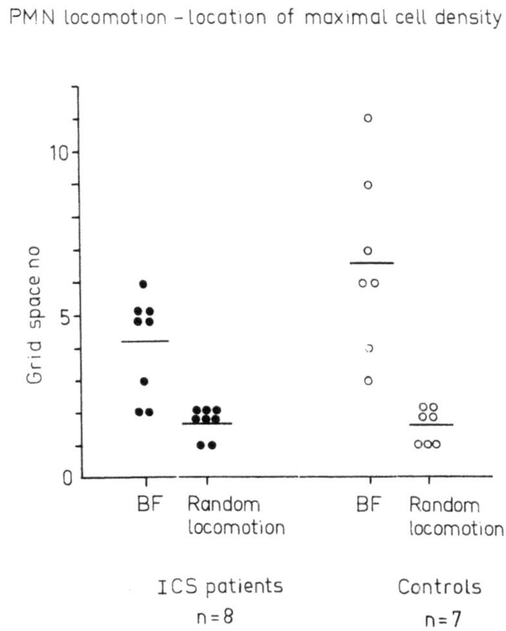

Fig. 6. The distance to the location of maximal PMN density for immotile cilia syndrome patients and healthy controls.

slightly lower than 25 % during the whole incubation period (Fig.1).

The effect of cytochalasin B. A dose dependent inhibition of the progression of the leading front distance was found both for stimulated and spontaneously migrating cells (Fig. 3). The morphology of PMNs was also changed (Fig. 2 D). Neutrophils had the nuclei in the rear and were oriented distinctly. As shown in Fig.3 this degree of orientation was also related to the cytochalasin B concentration in a dose-dependent way.

The effect of vinblastine. Inhibitions of the leading front distances were noted in a dose-dependent manner for stimulated migration, but not for spontaneous migration (Fig. 4). Also there was a dose-related inhibition of cell orientation, where incubation with the highest concentration of vinblastine was accompanied by appearance and orientation percentages comparable to spontaneously moving cells, i.e. 15-20 % (Fig. 4 and 2E).

Immotile cilia syndrome patients. Neutrophils from the patients showed a statistically significant ($p<0.01$) decrease in their ability to orient towards the well containing serum compared with controls, when measured after three hours of incubation[6] (Fig. 5). Also, the location of maximal cell density was closer to the PMN containing well for neutrophils from the patients than for controls, when BF was used as cytotaxin (Fig. 6). Half of the patients showed values outside the reference values for the distance to the leading front[6]. However, PMN adherence, bactericidal ability and chemiluminescence was normal. No dynein arms were identified on microtubules, neither in patients nor in controls[6]. Thus, neutrophils showed defects in the ability to orient in a cytotaxin gradient, which also could be followed by impairments of length of migration.

DISCUSSION

The present study has shown that information is easily achieved concerning several variables from a single determination of PMN migration under agarose. In addition to the distance to the leading front cells, we have demonstrated here that the degree of PMN orientation during migration in the cytotaxin gradient is readily determined. By assessments of these variables, it was found that both used cytotaxins elicited a directed movement during the first part of the incubation time, i.e. a chemotactic response which was followed by a continous decrease of cell orientation suggesting that PMNs thereafter moved in a chemokinetic manner. When PMNs were incubated with vinblastine, direction finding was equal to spontaneously moving cells, i.e. chemokinesis predominated, and the foremost PMNs were unable to migrate as far as the untreated cells. Finally, incubation with cytochalasin B was associated with an impairment of the cells to move towards the cytotaxin containing well, whereas direction finding was preserved.

It has not been appreciated previously that chemotaxis and chemokinesis may occur in a time-associated manner for a certain cytotaxin. For distinguishing chemotaxis from chemokinesis the time and resource consuming checker-board technique has utilized end point determinations and with time lapse micrography expensive equipment is required. The assay presented herein is both cheap and easily performed and has shown to be valuable in the analysis of, e.g. PMNs from patients with the immotile cilia syndrome[6].

Based on the present findings, it is suggested that, when mainly chemotaxis is to be studied by the agarose technique, short incubation times should be preferred e.g. 1/2 - 2 hours, i.e. when the orientation of cells is most pronounced. In many clinical investigations, however, it might instead be desirable to determine whether the locomotive ability of PMNs from a patient is adequate. Longer incubation periods may then allow better discrimination. Also, if impairments are noted a failure of direction finding or locomotive ability could be documented, but these are easier discernable with shorter incubation periods.

One example of disease where locomotive functions are impaired is the immotile cilia syndrome (ICS) characterized by immotile cilia in the respiratory tract, immotile sperms and often situs inversus (i.e. Kartagener s syndrome), leading to a decreased resistance to infectious agents entering the respiratory tract, and also to infertility. Ultrastructural abnormalities have been suggested to be the cause of the immobility, such as absence of dynein arms or radial spokes between the microtubules[7]. The present study has shown that additional defects may also be present. The abnormally short distance migrated by PMNs from 4 out of 8 tested ICS patients[6] and the significantly lower mean degree of orientation in the cytotaxin gradient point at a more widespread defect of the cytoskeleton than previously anticipated. The defective PMN migration and orientation might also help to explain the increased susceptibility to infectious agents in the respiratory tract of ICS patients, eventually leading to i.a. bronchiectasies.

SUMMARY

This study has shown that the agarose technique is well suited for determination of whether impaired PMN migration is due to defective direction finding of deficient locomotive ability. By assessments of the degree of orientation of neutrophils together with measures of migration distances, we demonstrated that vinblastine treated normal neutrophils and PMNs from patients with the immotile cilia syndrome mainly move in a chemokinetic manner, whereas cytochalasin B hampers locomotive ability, but leaves direction finding intact. Hence, information whether chemotaxis or chemokinesis predominates is easily obtained in one single assay system.

REFERENCES

1. R.D. Nelson, P.D. Quie and R.L. Simmons, Chemotaxis under agarose: a new and simple method, J. Immunol. 115: 1650 (975).

2. W. Orr and P.A. Ward, Quantitation of leukotaxis in agarose, J. Immunol. Methods. 20: 95 (1978)

3. H. Repo, Leukocyte migration agarose test for the assessment of human neutrophil chemotaxis, Scand. J. Immunol. 6: 203 (1977).

4. R.D. Nelson, V.D. Fiegel and R.L. Simmons, Chemotaxis of human PMN neutrophils under agarose: morphologic changes, J. Immunol. 117: 1676 (1976).

5. H. Repo, A.A.I. Kostiala and T.U. Kosunen, Use of leucocyte migration under agarose to study spontaneous and directed locomotion of leucocytes, Immunol. 35: 539 (1978)

6. B.A. Afzelius, L. Ewetz, J. Palmblad, A.-M. Udén and N. Venizelos, Structure and function of neutrophil leukocytes from patients with the immotile cilia syndrome, Acta Med. Scand. 208: 145 (1980).

7. R. Eliasson, B. Mossberg, P. Camner and B.A. Afzelius, The immotile cilia syndrome, New Engl. J. Med. 297:1 (1977).

8. H.L. Malech, R.K. Root and J.I. Gallin, Structural analysis of human neutrophil migration, J. Cell. Biol. 75: 666 (1977).

9. S.H. Zigmond and J.G. Hirsch, Leukocyte locomotion and chemotaxis, J. Exp. Med. 137: 387 (1973).

DISCUSSION
Edited by H.U. Keller

Johnston, R.B.: I am not aware that neutrophils have dynein arms or a counterpart. Perhaps you have some indication that they do. In any event, did you consider that the decreased neutrophil migration could be a secondary phenomenon. For example, increased serum IgA or ingestion of immune complexes have been associated with decreased neutrophil migration.

Palmblad, J.: We have not been able to demonstrate dynein arms on microtubules of granulocytes from patients or controls. This is understandable since it would require crosscutting a very large number of the rather rare and widely dispersed microtubules. Infections are among the factors that could cause a reduction of the distance migrated by the majority of neutrophils, but I do not think that this can explain decreased direction finding by these leucocytes. We carefully avoided to take blood from these patients during or shortly after acute infections. Furthermore, the erythrocyte sedimentation rate and serum electrophoresis disclosed no elevated levels of acute phase proteins.

Siccardi, A.: We studied four cases of Kartagener's syndrome. Two of them were brothers. The microtubular defect was demonstrated by electron microscopy in all four cases. Defects in neutrophil chemotaxis were found in three cases. One of the two brothers had the defect and recurrent infections, while the other had normal chemotaxis and no infections. This suggest that the condition causing defective chemotaxis in leucocytes from patients with Kartagener's syndrome is inherited separately from the microtubular defect.

Palmblad, J.: First, the immobile cilia syndrome which includes the Kartagener triad, is most certainly a heterogenous disorder. Therefore, we cannot expect that all patients exhibit identical findings. Second, although the neutrophils from half of our patients did not move as far as control cells, the predominant phenomenon was their inability to orient in the cytotaxin gradient. This could result in a reduction of the distance migrated by the cells forming the leading front. Our studies suggest that it might be worthwhile to measure also the numbers of cells along the migration path and to determine the location of maximal cell density.

Gallily, R.: The source of agarose often plays a role in cell locomotion. The type of plastic dish used might also affect cell movement. Can you comment on these two problems?

Palmblad, J.: We use the Sigma agarose, EEO Medium agarose and Falcon plastic petri dishes. However, the latter vary considerably in quality. Some batches have to be discarded because migration can not be assessed after fixation.

Hakim, J.: You separated neutrophil locomotion under agarose in two consecutive phases that you related to time. Do you think that the occurrence of the second phase is time-dependent or concentration-dependent?

Palmblad, J.: We related the transition of chemotaxis into chemokinesis to time because it occured when incubation time was prolonged. This does not imply that time is the cause. The phenomenon could for instance be due to instability of the cytotaxin gradient, inactivation of cytotaxins or other processes.

A DIFFERENTIAL RESPONSIVENESS OF *IN VITRO* DIFFERENTIATING MONONUCLEAR PHAGOCYTES FROM BONE MARROWS OF NORMAL AND INFLAMED MICE TO LYMPHOKINES AND POLY I·POLY C

Isia Bursuker and Rachel Goldman

Department of Membrane Research
The Weizmann Institute of Science
Rehovot 76100, Israel

INTRODUCTION

Immunological adjuvants, as well as a variety of chronic infections, stimulate the reticuloendothelial system and induce a population of activated macrophages exhibiting potent bacteriostatic and tumoricidal capacities.[1,2] Cytotoxic macrophages can also be induced *in vitro* by the activation of macrophages with non-specific stimuli, such as endotoxin, Poly I·Poly C, or with soluble mediators released by stimulated lymphocytes.[3-5] Ruco and Meltzer found that peritoneal macrophages, induced *in vivo* by acute inflammatory agents, are at least 10 times more responsive to activation by lymphokines than normal-resident peritoneal macrophages.[6] The macrophages, that are mobilized to inflammatory and tumor growth sites, are mature monocytes in transit in peripheral blood from their site of generation - the bone marrow. Lohmann-Matthes *et al.*[7] showed that macrophages, differentiating *in vitro* from bone marrow cell precursors, can be activated by lymphokines for tumor cell cytotoxicity.

Macrophages populating acute and chronic inflammatory sites differ in biochemical, morphological and functional parameters.[1,8-10] In a recent study,[11] we have shown that mononuclear phagocytes differentiating *in vitro* from bone marrows of acute and chronic inflamed mice differ from those derived from bone marrows of non-treated mice in several parameters. The differences are compatible with those observed in the corresponding peritoneal macrophage populations. This finding led to the suggestion that at least some of the functional differences observed in inflammatory exudate macrophages, as compared to resident macrophages, are determined already at the level of bone marrow precursor cells.

In the present study, we compared the ability of *in vitro* differentiating bone marrow derived mononuclear phagocytes (BMDMP) from normal, thioglycollate- and CP-inflamed mice, to undergo non-specific activation for inhibition of tumor cell proliferation by Poly I·Poly C and lymphokines derived from *in vivo Corynebacterium parvum* (CP)-sensitized lymphocytes incubated *in vitro* with the antigen.

MATERIALS AND METHODS

Media, Sera and Conditioned Medium

Dulbecco's phosphate-buffered saline (PBS), fetal calf serum (FCS), heat-inactivated fetal calf serum (HIFCS), Dulbecco's modified Eagle's medium with 100 u/ml penicillin, 100 μg/ml streptomycin (DMEM), were obtained from GIBCO (Grand Island, N.Y.). Horse serum (HS) was obtained from Bio-Lab (Jerusalem, Israel). Conditioned medium (CM) was collected from secondary cultures of BALB/c mouse embryo fibroblasts cultured for 5 days in DMEM supplemented with 10% HS. The medium was centrifuged to remove detached cells and stored at -24°C until use.

Collection and Cultivation of Bone Marrow and Peritoneal Exudate Cells

Bone marrow and peritoneal exudate cells were collected from BALB/c strain male mice (6 to 8 weeks old). Inflammation was induced by i.p. injection of : (a) *Corynebacterium parvum*, 0.2 ml/mouse of a heat-killed strain, CN 6134 (7 mg/ml), obtained from Burroughs Wellcome Ltd., Beckenham, England; (b) Thioglycollate, 3 ml/mouse of a solution (2.98 g/100 ml) obtained from Difco Labs, Detroit, MI.

Peritoneal exudate cells and bone marrow cells were collected, as described.[11]

Cell suspensions consisting of 10^5 and $2x10^5$ peritoneal exudate macrophages (resuspended in DMEM containing 10% HIFCS) and bone marrow cells (resuspended in DMEM containing 10% horse serum and 15% conditioned medium) were plated into flat-blottomed 96-well Linbro plates (IS-FB-96, Flow Labs, Irvine, Scotland) in a final volume of 0.2 ml. Peritoneal macrophages and bone marrow cells were cultivated for 24 h and 6 days, respectively, before initiation of the specified treatments.

Preparation of Macrophage-Activating Factor (MAF)

Mice were immunized by i.p. injection of 1.4 mg of CP (of the above specifications). Animals were sacrificed 10 days post-CP injection. Spleen lymphocytes (10^7/ml) were resuspended in DMEM containing 14 μg CP/ml and incubated for 24 h at 37°C. Subsequent to removal of cells by centrifugation, 450xg for 10 min at 4°C, the supernatant was filtered through a millipore filter (0.45 μm).

In Vitro Activation of Mononuclear Phagocytes (Peritoneal Macrophages and BMDMP

Mononuclear phagocyte monolayers were washed in PBS and incubated with either 10% or 20% MAF in DMEM or 50 μg/ml of Poly I·Poly C (P-L Biochemicals, Inc., Milwaukee, Wisconsin) for the specified time. The monolayers were washed in PBS and the cytotoxic potential of the cells assessed as described below.

Assessment of Growth Inhibition

Golden Hamster embryo cells transformed *in vitro* by dimethylnitrosamine (a cell line, HDMNA),[12] maintained under tissue culture conditions (DMEM containing 10% FCS), were used as target cells. Growth inhibition was assessed using the methylene blue binding assay as described.[9] Target cells in a final volume of 0.2 ml (5000 cells/well) were plated on the adherent mononuclear phagocyte monolayers. Controls of mononuclear phagocytes, cultured in the absence of target cells and treated with the specified agents, were included. The plates were incubated for 48 h at 37°C in a humid CO_2 incubator (10% CO_2 in air). Dye absorbance by the fixed cell cultures was measured at 660 nm (A_{660}). The percent of growth inhibition (GI) was calculated as follows :

$$\%\ GI = \left(1 - \frac{A_{t+m} - A_{to} - A_m}{A_t - A_{to}}\right) \times 100$$

where A = absorbance at 660 nm of t - targets at the end of the assay, to - targets at zero time (targets were plated on a separate plate and fixed 2 h after plating), m - mononuclear phagocytes cultured without targets, t+m - cocultures of targets and mononuclear phagocytes.

Percent activator (MAF or Poly I·Poly C) induced GI was defined as :

% activator induced GI =

$$\frac{\%\ GI\ after\ activator\ treatment - \%\ GI\ of\ nontreated\ BMDMP}{100\% - \%\ GI\ of\ nontreated\ BMDMP} \times 100$$

Inflammation induced growth inhibition (IGI) index was defined as follows :

$$IGI = \frac{MAF\ induced\ growth\ inhibition\ of\ BMDMP\ from\ inflamed\ mice}{MAF\ induced\ growth\ inhibition\ of\ BMDMP\ from\ normal\ mice}$$

RESULTS

Activation of BMDMP and Peritoneal Macrophages by CP-Induced MAF

Peritoneal macrophages, as well as BMDMP, inhibit the proliferation of HDMNA target cells to a certain extent (Table 1). Incubation of both cell types with MAF led to enhancement of the growth inhibition potential of these effector cells. The effect was dependent on the concentration of both the effector cells and MAF (Table 1).

Differential Activation by MAF of BMDMP from Normal and Inflamed Mice

The growth inhibition of HDMNA targets by nonactivated and MAF activated BMDMP from normal and thioglycollate- or CP-inflamed mice harvested on days 2, 4, 7 and 9 post-stimulus injection was assayed. When bone marrows were harvested 2 days after stimulus injection BMDMP from normal and inflamed mice exhibited a comparable response to MAF treatment (Fig. 1). When the bone marrows were harvested 4 days post-stimulus injection the BMDMP of both thioglycollate- and CP-inflamed mice showed an increased response to MAF, BMDMP from CP-inflamed mice were activated to a higher extent than those of thioglycollate-inflamed mice. This phenomenon was observed at both 10% (Fig. 1) and 20% (not shown) MAF concentrations. A residual increased response to MAF, as revealed by the capacity to inhibit tumor cell proliferation, of BMDMP from CP-inflamed mice and thioglycollate-inflamed mice, as compared to BMDMP from normal mice, persisted for at least 9 days post-stimulus injection.

Table 1. MAF-Mediated Activation[a] of BMDMP and Peritoneal Macrophages.

Effector Cells	Effector Cells/Well[b] ($A_{660\ nm}$)	% Growth Inhibition of Tumor Cells by Effector Cells Incubated in : DMEM	10% MAF	20% MAF
BMDMP	0.071	38	67	76
BMDMP	0.201	53	82	89
Peritoneal Macrophages	0.096	34	49	65
Peritoneal Macrophages	0.202	51	85	95

[a]Effector cells were incubated with MAF for 24 h prior to interaction with target cells.
[b]Content assessed at the end of the assay.

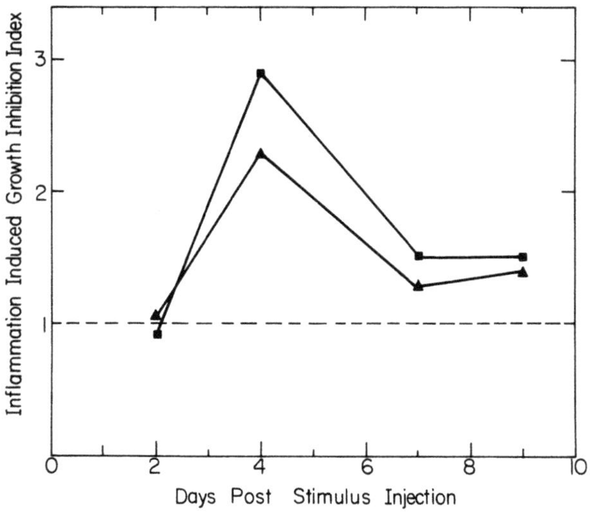

Fig. 1. Temporal changes in the differential activation of BMDMP from normal (---), thioglycollate-inflamed (▲-▲) and CP-inflamed (□—□) mice. BMDMP populations were incubated with 10% MAF for 24 h, prior to interaction with target cells. Bone marrow cells were harvested on the specified days. The GI exhibited by the nonactivated BMDMP populations was the same within each assay (a specified day post-stimulus injection) and varied from 30-40% on different days post-stimulus injection. A representative experiment out of three experiments. Each experiment was carried out in quadruplicate.

Kinetics of BMDMP Activation by MAF

The dependence of the degree of BMDMP activation by MAF on the time of interaction with the lymphokine is given in Fig. 2. A gradual increase in the growth inhibitory activity of BMDMP from normal and inflamed mice was observed up to at least 24 h of incubation with MAF. The cellular content of the various BMDMP populations activated for the specified periods with MAF was not changed during the treatment and during the 48 h of culture that followed the treatment (the time of growth inhibition assay) (Fig. 2). Since in all the growth inhibition assays we have deducted the value of dye binding by BMDMP cultured in the absence of target cells from the dye binding obtained by cocultures of BMDMP and target cells, we have also assessed the retention of viable BMDMP in those cultures at

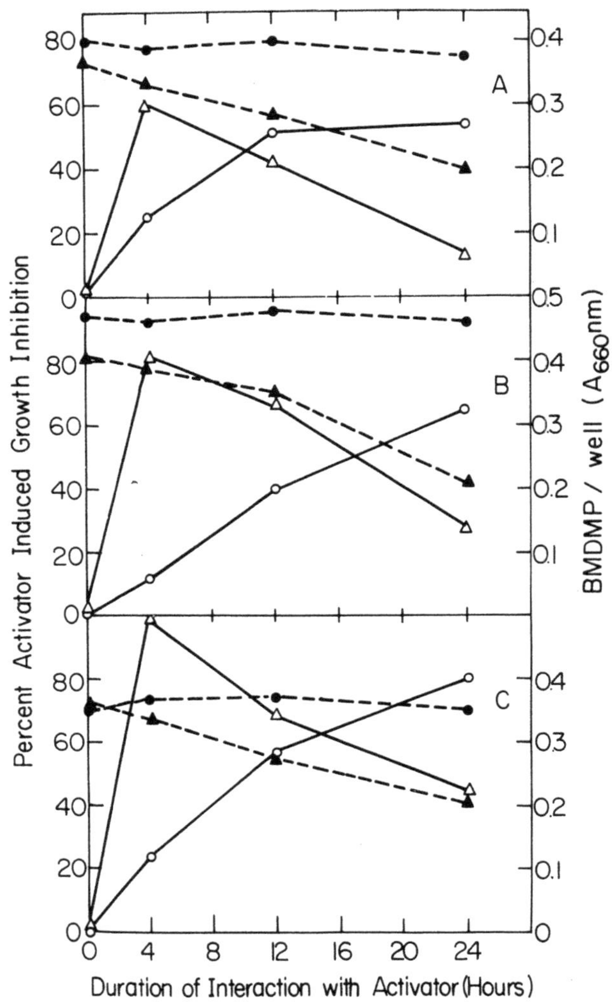

Fig. 2. Kinetics of activation of BMDMP by MAF and Poly I·Poly C. BMDMP from normal (A), thioglycollate-inflamed (B) and CP-inflamed (C) mice were incubated with 20% MAF (circles) or with Poly I·Poly C (50 μg/ml) (triangles) for the specified time, prior to interaction with target cells. (O-O) and (Δ-Δ), percent activator induced growth inhibition; (●--●) and (▲--▲), $A_{660\ nm}$ of dye-binding by BMDMP monolayers cultured in the absence of target cells.

the end of the 48 h assay period. For this purpose we labelled the BMDMP with ^{51}Cr (2 h, 2 μCi of ^{51}Cr/ml of DMEM containing 10% HIFCS at 37°C) and compared the retention of ^{51}Cr in the adherent cell monolayer of BMDMP cultured with and without target cells. About 90% of the ^{51}Cr-label of BMDMP cultured in the absence of target cells was retained in the cultures containing target cells.

Activation of BMDMP by Poly I·Poly C

The kinetics of activation by Poly I·Poly C of BMDMP from normal and inflamed mice is shown in Fig. 2. A 4-hour incubation period with 50 μg/ml of Poly I·Poly C resulted in augmentation of the growth inhibitory capacity of BMDMP from normal, as well as from inflamed mice. The activating effect diminished when the BMDMP cultures were incubated with Poly I·Poly C for longer periods (12 h and 24 h). The latter effect most probably stems from the marked reduction in the number of adherent effector cells during the assay period. The methylene blue binding capacity of BMDMP monolayers treated with Poly I·Poly C for 24 h prior to assay decreased by 50% at the end of the assay period (see Fig. 2). From Fig. 2 it appears that a 4-hour incubation period with Poly I·Poly C is optimal for BMDMP activation. Fig. 3 presents the results of three independent experi-

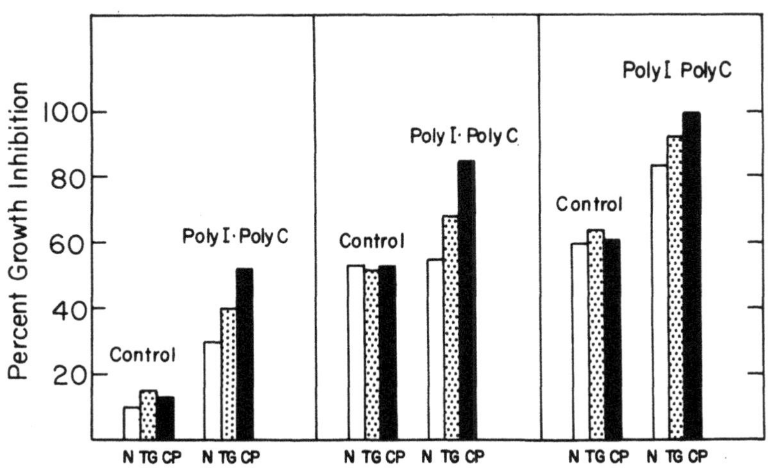

Fig. 3. Poly I·Poly C-induced activation of BMDMP from normal and inflamed mice. BMDMP were incubated with Poly I·Poly C (50 μg/ml), A - for 24 h, and B and C - for 4 h prior to target cell addition. Bone marrows from thioglycollate (TG) and CP-inflamed mice were harvested 5 days post-stimulus injection. N - normal bone marrow.

ments, in each of which the growth inhibition activity of the non-treated BMDMP derived from bone marrows of normal and inflamed mice was comparable. In each of the three experiments, we obtained a differential response to Poly I·Poly C of the BMDMP from normal and inflamed mice, i.e. that obtained for BMDMP from CP-inflamed mice surpassing the response obtained with BMDMP from thioglycollate-inflamed and normal mice. BMDMP from thioglycollate-inflamed mice gave an intermediate response to that obtained by the other populations.

DISCUSSION

Extensive studies as to the mechanism by which *in vivo* administered CP activates macrophages for cytotoxicity towards tumor cells have suggested both a direct stimulating effect, as well as an immunological mediated process.[13] Indeed, CP immunized spleen cells produce MAF that activates peritoneal macrophages[14] upon *in vitro* incubation with CP.

In the present report, we show that CP-induced MAF is also a potent inducer of BMDMP capacity to inhibit tumor cell proliferation, that the degree of growth inhibition depends on the time of preincubation of BMDMP with MAF and on the concentration of the latter. This finding is corroborating recent work describing the *in vitro* activation of BMDMP by lymphokines generated by Concanavalin A-stimulated lymphoblasts.[7]

Several lines of evidence exclude the possibility that the observed differential augmentation in growth inhibition of the three BMDMP populations was due to natural killer activation by interferon either endogenous or contaminating the MAF preparation.[15] (a) MAF-activated BMDMP from normal and inflamed mice exhibited the same differential cytotoxic activity against P 815 (not shown) target cells shown to be resistant to the effector function of natural killer cells.[16] (b) Treatment with mouse interferon of the three BMDMP populations did not result in an augmentation of their cytostatic effect on HDMNA (unpublished). Recently we found that BMDMP from inflamed mice exhibit a higher antibody-dependent cellular cytotoxicity (ADCC), a higher chemiluminescence activity and a lower sensitivity to inhibitory effect of prostaglandins than those exhibited by BMDMP from normal mice (submitted). The correlation between the differential activity of BMDMP from normal and inflamed mice in two pathways of cytotoxicity to tumor cells (MAF-induced and ADCC) and the alteration in biochemical and functional parameters of BMDMP from inflamed mice, as compared to BMDMP from normal mice,[11] strongly support the notion that some of the properties by which inflammatory macrophages differ from normal-resident macrophages may originate at the level of bone marrow precursors.

In the present study, we used CP-induced MAF. It was possible, therefore, that the fact that BMDMP from CP-inflamed mice expressed a higher responsiveness than BMDMP from normal and thioglycollate-inflamed mice stems from a specific affinity of the MAF to these cells. In order to differentiate between this possibility and the suggestion that the BMDMP have an intrinsic differential capacity to respond to *in vitro* activators, we studied the activation of BMDMP by a known peritoneal macrophage activator, Poly I·Poly C.[3] Poly I·Poly C was shown to be a potent activator of BMDMP and to have the same differential activating pattern on BMDMP of normal and inflamed mice as that exhibited by MAF.

Ruco and Meltzer[6] have shown that macrophages recovered from the peritoneal cavaties of mice inflamed with sterile irritants (acute inflammation) are more responsive to lymphokines than the resident peritoneal macrophage population. The authors suggest that the lymphokine-responsive cells in both the resident and inflammatory populations were the young peroxidase-positive blood derived mononuclear phagocytes, and that their percent increased in the inflammatory exudate. Our results suggest that the increased responsiveness to MAF of peritoneal inflammatory macrophages is not only a correlate of the percent of newly-arriving blood monocytes in the exudate population but it can also be an inherent property of these blood borne monocytes.

Borashchi and Meltzer[17] suggest that the *in vivo* activation of macrophages during a chronic inflammation (BCG) is the outcome of several sequential steps : (a) the recruitment of antigen-reactive lymphocytes and lymphokine-responsive mononuclear phagocytes by signals generated at the site of inflammation; (b) the *in situ* production of lymphokines that subsequently prime the localized macrophages; and (c) the interaction of the primed macrophages with certain signals in the immediate environment leading to the expression of cytotoxicity towards tumor cells. To this sequence of events we suggest to add a step in which chronic inflammation induces changes at the level of the bone marrow macrophage precursors, resulting most probably in the appearance of these highly-responsive mononuclear phagocytes at the inflammatory site.

REFERENCES

1. J. B. Hibbs, H. A. Chapman and J. B. Weinberg, The macrophage as an antineoplastic surveillance cell : Biological perspectives, J.Ret.Soc. 24:549 (1978).
2. G. B. Mackaness and R. V. Blanden, Cellular immunity, Progr. Allergy 11:89 (1967).
3. P. Alexander and R. Evans, Endotoxin and double-stranded RNA render macrophages cytotoxic, Nature New Biol. 232:76 (1971).

4. I. J. Fidler, Activation *in vitro* of mouse macrophages by syngeneic, allogeneic or xenogeneic lymphocyte supernatants, J.Natl.Cancer Inst. 55:1159 (1975).
5. L. P. Ruco and M. S. Meltzer, Macrophage activation for tumor cytotoxicity : Induction of tumoricidal macrophages by PPD and BCG-immune mice, Cell.Immunol. 32:203 (1977).
6. L. P. Ruco and M. S. Meltzer, Macrophage activation for tumor cytotoxicity : Increased lymphokine responsiveness of peritoneal macrophages during acute inflammation, J.Immunol. 120:1054 (1978).
7. M.-L. Lohmann-Matthes, B. Kolb and H.-G. Meerpohl, Susceptibility of malignant and normal target cells to the cytotoxic action of bone-marrow macrophages activated *in vitro* with the macrophage cytotoxicity factor (MCF), Cell.Immunol. 41:231 (1978).
8. Z. Bar-Shavit, A. Raz and R. Goldman, Complement (C3b) and Fc receptor mediated phagocytosis of normal and stimulated mouse peritoneal macrophages, Eur.J.Immunol. 9:385 (1979).
9. R. Goldman and Z. Bar-Shavit, A dual effect of normal and stimulated macrophages and their conditioned media on target cell proliferation, J.Natl.Cancer Inst. 63:1009 (1979).
10. M. L. Karnovsky, J. Lazdins, D. Drath and A. Harper, Biochemical characteristics of activated macrophages, Ann.N.Y.Acad.Sci. 256:266 (1975).
11. I. Bursuker and R. Goldman, Quantitative and functional differences induced in bone marrow derived mononuclear phagocytes by inflammatory stimuli, J.Ret.Soc. 25:533 (1979).
12. E. Huberman, S. Salzberg and L. Sachs, The *in vitro* induction of an increase in cell multiplication and cellular life span by the water-soluble carcinogen dimethyl-nitrosamine, Proc.Natl.Acad.Sci.USA 59:77 (1968).
13. R. Bomford and G. H. Christie, Mechanisms of macrophage activation by CP : II. *In vivo* experiments, Cell.Immunol. 17:150 (1975).
14. R. T. Cullen and A. Ghaffar, Possible mechanisms underlying the induction of cytotoxic macrophages by CP : *In vitro* induction of cytotoxicity in normal macrophages by immune lymphocytes, J.Ret.Soc. 24:339 (1978).
15. M.-L. Lohmann-Matthes, W. Domzig and J. Roder, Promonocytes have the functional characteristics of natural killer cells, J.Immunol. 123:1883 (1979).
16. J. C. Roder, M.-L. Lohmann-Matthes, W. Domzig, R. Keissling and O. Haller, A functional comparison of tumor cell killing by activated macrophages and natural killer cells, Eur.J. Immunol. 9:283 (1979).
17. D. Boraschi and M. S. Meltzer, Defective tumoricidal capacity of macrophages from A/J mice : II. Comparison of the macrophage cytotoxic defect of A/J mice with that of lipid A-unresponsive C3H/HeJ mice, J.Immunol. 122:1592 (1979).

THE OXYGEN-DEPENDENT ANTIBODY-DEPENDENT CELL-MEDIATED CYTOTOXICITY OF HUMAN MONOCYTES AND NEUTROPHILS

Niels Borregaard and Knud Kragballe

The Department of Medicine and Infect. Dis.
and the Department of Dermatology
Marselisborg Hospital, DK-8000 Århus C. Denmark

Introduction

Antibody-dependent cell-mediated cytotoxicity (ADCC) refers to the killing af target cells by affector cells in a process dependent on the presence of an antibody, which is specific for antigens on the target cell membrane, and for which the effector cell carries receptors[1]. Using antibody sensitized erythrocytes as target cells both monocytes and neutrophils are able to function as effector cells[1-3]. The process of ADCC by these effector cells involve contact between target cell and effector cell with triggering of a phagocytosing response of the effector cell and lysis of the target[4,5].

The mechanism responsible for the lysis of target cells by monocytes and neutrophils has been the subject of some debate. A role of oxygen was indirectly implied by Clark and Klebanoff[6] in a study of ADCC by neutrophils, since ADCC was found to be significantly depressed, when neutrophils from patients with chronic granulomatous disease (CGD) were investigated. These neutrophils are characterized by the inability to generate reduced oxygen radicals[6] at the plasma membrane when stimulated to phagocytose.

In a study on monocytes, Fleer et al.[7] argued against reduced oxygen radicals participating in ADCC since monocytes from CGD patients were as effective as target cells as were monocytes from normals, despite full expression of the same defect as found in neutrophils from these patients[8]. Evidence was given, that ADCC was dependent on discharge of lysosomal enzymes. Recently Katz et al.[9] and Hafeman and Lucas[10] confirmed the findings of Clark and Klebanoff[6] in a study of both monocytes and neutrophils from CGD patients, and normal neutrophils anaerobically.

Methods: The antibody used was heat inactivated human hyperimmune serum with IgG-antibodies against erythrocyte membrane antigens. Both anti-A, anti-B (ABO-system) and anti-D (Rhesus-system) (monocytes only) were used with identical results. Details are given in ref. 11-13.

Results

If a role of oxygen radicals is to be implicated in ADCC, there must be a correlation between ADCC, phagocytosis and the respiratory burst of phagocytosis. Fig. 1 shows, that this was established by following the parameters: lysis, phagocytosis, degranulation and acti-

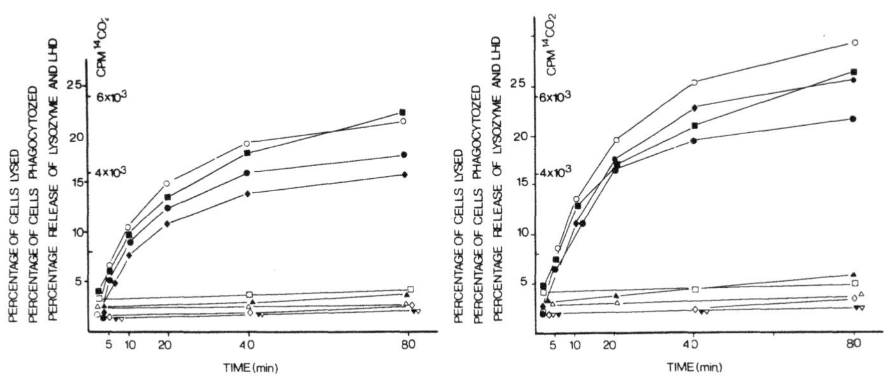

Fig. 1 A and B. Time course of cytolysis, phagocytosis, and postphagocytic events. In A the effector cells were monocytes and in B neutrophils. Effector 1 target cell ratio was 1:1, and the cell concentration was 10^6 cells/ml. Cytolysis (O), phagocytosis (●), Cpm of $^{14}CO_2$ from [1-^{14}C] glucose, stimulated (◆), unstimulated (◇) cells (anti-B serum has been omitted). Cpm of $^{14}CO_2$ [6-^{14}C] glucose, stimulated cells (▼), unstimulated cells (▽). Lysozyme liberation, stimulated cells (■), unstimulated cells (□). Lactate dehydrogenase liberation, stimulated cells (▲), unstimulated cells (△). Each figure given is the mean of six individual experiments. SEM: for cytolysis ⩽ 0.8%, for phagocytosis ⩽1.2%, for lysozyme ⩽2.2%, for LDH ⩽0.3%, for $^{14}CO_2$ ⩽250 Cpm. Reproduced from ref. 11 with permission of the journal.

vation of the hexose monophosphate shunt. It is seen, that neutrophils are somewhat more effective in the ADCC reaction than monocytes.

The evidence, presented here for the involvement of reduced oxygen radicals, or more specific, hydroxyl radicals, in ADCC, is based on five independent approaches:

1) ADCC in CGD patients
2) ADCC under anaerobic condition
3) ADCC in atopic dermatitis
4) Effect of mannitol on ADCC
5) Lysis by Xanthine-Xanthine oxidase

ADCC and CGD

ADCC by monocytes and neutrophils from CGD patients and carriers of the X-linked form of CGD was found to be depressed, as also observed by others[6,9,14], however, phagocytosis and degranulation during ADCC was normal, which indicates, that the depression of ADCC is due to the inability of these cells to generate oxygen radicals. The depression in ADCC closely followed the percentage of abnormal cells, as observed in Table I, where ADCC is related to an activity score used to

Table 1.

NBT-slide[a]	O_2-consumption[ax]	O_2^- liberation[a]		$1\text{-}^{14}C$ glucose ox.[ax]		Score[b]	ADCC, % Cr release	
test	neutro	mono	neutro	mono	neutro	mono + neutro	mono	neutro
27	30	25	30	41	23	29	11.8	13.9
36	40	37	33	42	51	39	10.2	13.5
14	12	12	12	15	10	13	7.8	10.7
9	6	2	4	10	5	6	8.3	10.4
11	14	5	6	10	8	9	7.4	10.5
63	54	52	62	63	56	58	13.3	16.6
53	56	42	54	67	63	56	12.6	15.1
48	37	35	52	36	40	40	12.7	16.5
82	80	78	85	83	80	81	15.4	19.4
52	44	57	41	45	37	46	12.5	15.2
81	78	80	84	77	73	79	17.9	22.1
65	69	57	65	73	68	66	16.2	19.0
78	79	77	71	81	91	80	16.3	20.7
87	85	77	75	85	81	82	17.5	21.6
79	80	72	79	85	81	79	16.6	20.2
0	0	0	0	0	0	0	6.4	9.3
0	0	0	0	0	0	0	6.7	9.2

a) Results were expressed in percent of the mean activity as determined in 17 normal controls. NBT slide-test: 99.5±0.4% positive cells; O_2 consumption: 4.1±0.2 f moles/min/neutrophil; O_2^--liberation: 2.6±0.1 f moles/min/monocytes. 13.0±0.1 f moles/min/neutrophil. $1\text{-}^{14}C$ glucose oxidation: 3.3 x 10^3±0.1 cpm/30 min/10^5 monocytes. 5.0x10^3±0.2 cpm/30 min/10^5 neutrophils.

x) Value for unstimulated cells has been subtracted.

b) The mean of the 6 determined activities.

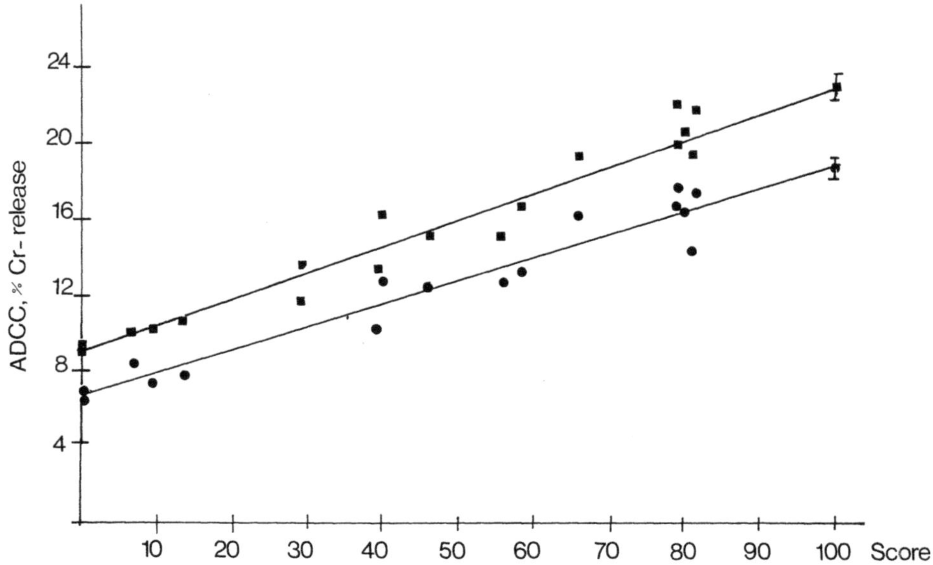

Fig. 2. Relation between oxygen dependent activity score as determined in Table I and ADCC. Monocytes (●). Neutrophils (■).

assess the percentage of abnormal cells, based on NBT slide test[15], O_2 consumption, O_2^- liberation and [1-^{14}C] glucose oxidation of stimulated cells[16]. It is calculated, that ADCC by both monocytes and neutrophils from CGD patients is reduced to 40% of the normal values(Fig.2).

ADCC during anaerobiosis

Under anaerobic conditions ADCC by normal monocytes and neutrophils is reduced to values close to those obtained by CGD cells under aerobic conditions (Table II). The depression of ADCC during anaerobiosis was calcelled, when oxygen was reintroduced, and did not effect phagocytosis or degranulation, thus showing, that the depression of ADCC was not due to a depression of the energy metabolism of the effector cells. This was confirmed by the observation, that inhibitors of mitochondrial function, such as azide, cyanide and 2,4-dinitrophenol, did not result in depression of ADCC by any of the cell types investigated, table III. This therefore indicates, that the depression observed during anaerobic conditions is due to diminished generation of oxygen radicals by the effector cells under these conditions.

ADCC and atopic dermatitis

Both monocytes and neutrophils from patients with severe atopic dermatitis show a diminished capacity for ADCC corresponding to 75% of the capacity of normal cells (Fig. 3). This depression is not due to reduced capacity for reduction of oxygen, as the capacity for superoxide liberation was normal, when induced by PMA, in contrast to the findings with CGD cells (cf. above). Also, in contrast to what is observed on CGD cells, cells from patients with atopic dermatitis show a depression of phagocytosis, degranulation and activation of the hexose monophosphate shunt during the ADCC reaction (Fig. 3). The reduction in these parameters correspond as a mean reasonably well to the reduction in the capacity for ADCC by both monocytes and neutrophils. However, only phagocytosis and the activity of the hexose monophosphate shunt could be correlated to the depression of ADCC in the individual cases. This observation speaks against lysosomal enzyme discharge as being the effector of ADCC, and is compatible with the idea of reduced oxygen species being of significance for ADCC, since the electrons used in the reduction of oxygen at the plasma membrane is thought to originate from the oxidation of glucose in the hexose monophosphate shunt[17].

Table II.
Effect of anaerobiosis on cytotoxicity of monocytes and neutrophils

	Monocytes	Neutrophils
Aerobic conditions		
Cytolysis,% ^{51}Cr liberated	17.0±1.4	22.5±1.8
Phagocytosis,% cells phagocytozed	14.1±2.2	18.0±2.4
Lysozyme liberation,% of total enzyme	11.2±2.4	15.1±2.1
Cpm of $^{14}CO_2$ from [1-^{14}C] glucose	3.16×10^3± 0.27	5.1×10^3± 0.43
Anaerobic conditions		
Cytolysis,% ^{51}Cr liberated	9.2±1.7	13.5±1.6
Phagocytosis,% cells phagocytozed	14.2±1.7	17.3±2.5
Lysozyme liberation,% of total enzyme	11.8±2.1	15.7±2.3
Cpm of $^{14}CO_2$ from [1-^{14}C] glucose	0.7×10^3± 0.16	0.9×10^3±0.21

Incubations were for 30 min. Effector and target cells were made anaerobic and used for both the aerobic and anaerobic experiments. For the aerobic experiment cells were made aerobic by a buffer shift after standing under anaerobic conditions when the anaerobic experiment was carried out. Figures are mean±SEM for six individual experiments.

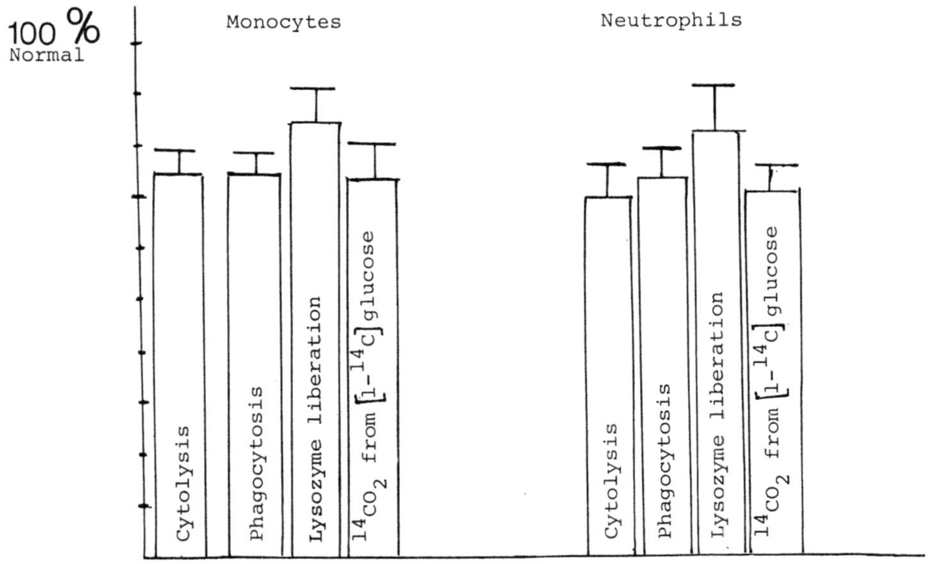

Fig. 3. Depressed function of neutrophils and monocytes from patients with severe atopic dermatitis. Incubations were made for 30 min as described in ref. 13.

Since binding of target cells to effector cells was found normal as judged from rosette studies, we ascribe the defective activation of the effector cells from patients with atopic dermatitis as being due to a depressed transducer function in the plasmamembrane, leading to decreased activation of the membrane oxidase and consequently to depressed ADCC.

Table III. Effect of blockers of mitochondrial function on ADCC

Additions	Monocytes Cytolysis	Phagocytosis	Neutrophils Cytolysis	Phagocytosis
None, control	100	100	100	100
Sodium azide, 1mM	144 ± 5.8	102 ± 6.2	157 ± 6.2	93 ± 5.4
Potassium cyanide, 1mM	115 ± 4.8	98 ± 4.9	112 ± 4.2	98 ± 3.9
2,4-Dinitrophenol, 10μM	103 ± 2.1	105 ± 2.9	100 ± 2.4	99 ± 5.2

Incubations were for 30 min. Figures are mean ± SEM for six experiments

Table IV

Effect of Scavengers of H_2O_2 and O_2^- on ADCC

Additions	Monocytes		Neutrophils	
	Cytolysis	Phagocytosis	Cytolysis	Phagocytosis
None	100	100	100	100
SOD 50 µg/ml	94±6.7	100±3.8	90±5.9	98±6.4
Catalase 4000 U/ ml	95±2.4	97±5.1	105±2.6	101±5.6
Cytochrome c, 0.2 mM	92±1.9	94±3.7	102±2.0	97±3.5

Incubations were for 30 min. Figures are mean ± SEM for six experiments.

Inhibitor studies

In order to identify which of the reduced oxygen species, O_2^-, H_2O_2, OH· known to be formed by the effector cells[18-23], was of significance for ADCC, the effect of inhibitors and scavengers of these radicals were assessed during the ADCC reaction. Addition of neither 200 µM cytochrome c, 50 µg/ml superoxide dismutase nor 4000 U/ml of catalase affected the ADCC by neither monocytes nor neutrophils (Table IV). Similar observations were made by Clark and Klebanoff and Hafeman and Lucas in their study of neutrophil ADCC [6]. However, Capsoni et al. did find some depression of neutrophil ADCC by superoxide dismutase[14] although far from the depression observed in CGD cells.

We interpret these findings as an expression of defective contact between the cytochrome c, superoxide dismutase and catalase and the activated membrane of the effector cell, which is in intimate contact with the target cell membrane[4], and therefore excludes these proteins from the site of generation and action of the reduced oxygen species.

More direct evidence for the participation of oxygen radicals was given by the studies with the hydroxyl radical scavenger mannitol (table V), which was found

Table V. Anaerobic conditions were similar to those reported in table II.

	Monocytes				Neutrophils			
	Aerobic		Anaerobic		Aerobic		Anaerobic	
Addition	Cytolysis	Phagocytosis	Cytolysis	Phagocytosis	Cytolysis	Phagocytosis	Cytolysis	Phagocytosis
None	100	100	100	100	100	100	100	100
Mannitol, 25 mM	65 ± 12.7	95 ± 4.0	96 ± 3.3	99 ± 5.2	71 ± 3.2	92 ± 4.1	92 ± 3.4	93 ± 3.9
Sodium azide, 1 mM	144 ± 5.8	102 ± 6.2	95 ± 5.0	102 ± 6.6	157 ± 6.2	93 ± 5.4	93 ± 5.1	101 ± 6.4

to significantly reduce the ADCC of normal monocytes and neutrophils under aerobic conditions without affecting phagocytosis, degranulation or hexose monophosphate shunt activity. Also, stimulation of the oxygen consumption by IgG-latex of both monocytes and neutrophils was normal in the presence of 25 mM mannitol, which was the concentration used during ADCC (data not shown). This therefore indicates, that mannitol does not inhibit the generation of reduced oxygen radicals during the ADCC reaction, and thus suggests, that it is through its action as a hydroxyl radical scavenger, that inhibition of ADCC by mannitol is achieved. This is supported by the finding, that mannitol is without effect on ADCC, when CGD cells were effector cells, and when ADCC was performed by normal cells during anaerobic conditions. In order to roule out, that ADCC was effectuated by H_2O_2 in a peroxidase reaction, the activity of myeloperoxidase and catalase was blocked by 1 mM azide. This resulted in increased ADCC during aerobic conditions, without affecting ADCC under anaerobic conditions. This is in accordance with the interpretation, that the generation of hydroxyl radicals are of importance for ADCC, since the formation of these radicals is enhanced by azide as shown by Green et al.[24] and Rosen and Klebanoff[25].

Lysis by xanthine-xanthine oxidase

Finally, in order to further establish the role of oxygen in lysis of erythrocytes, we investigated the lysing capacity of a hydroxyl radical generating system, xanthine-xanthine oxidase. It was observed (table VI), that this system effectuated lysis of erythrocytes. The observation, that this lysis is strongly inhibited by both superoxide dismutase, catalase and mannitol, indicates, that in this system, lysis is dependent on the action of hydroxyl radicals and that interaction with lysosomal enzyme is not a necessary prerequisite for lysis.

Table VI

Lysis of erythrocytes by xanthine-xanthine oxidase.

Supplements	^{51}Cr release %
Xanthineoxidase +	
Xanthine	26.7±0.8
+ catalase 1500 U/ml	1.3±0.1
+ S O D 400 U/ml	9.8±0.2
+ Mannitol 50 mM	4.4±0.2
+ Boiled Catalase	24.3±1.2
+ Boiled S O D	25.3±0.9

Incubations of 10^6 erythrocytes per ml with 0.025 U/ml of xanthine oxidase and 1 mM xanthine. Incubations were for 1 hour.

Boiled catalase and superoxide dismutase are enzymes, that have been heated at 95°C for 5 min. Figures are mean± SEM for three experiments. Reproduced from ref. 11 with permission of the journal.

REFERENCES

1) Holm, G. (1972) Lysis of antibody-treated human erythrocytes by human leukocytes and macrophages in tissue culture. Int. Arch. Allergy Appl. Immunol. 43, 671-682
2) MacDonald, H.R., Bonnard, G.D., Sordat, B. and Zawodnik, S.A. (1975) Antibody-dependent cell-mediated cytotoxicity: Heterogeneity of effector cells in human peripheral blood. Scand. J. Immunol. 4, 487-497
3) Levy, P.C., Shaw, G.M. and L Buglio, A.F. (1979) Human monocyte, lymphocyte and granulocyte antibody-dependent cell-mediated cytotoxicity toward tumor cells. I. General characteristics of cytolysis. J. Immunol. 123, 594-599
4) Fleer, A., van Schaik, M.L.J., van dem Borne, A.E.G., Jr., and Engelfriet, C.P. (1978) Destruction of sensitized erythrocytes by human monocytes in vitro. Effects of cytochalasin B, hydrocortisone and colchicine. Scand. J. Immunol. 8, 515-524
5) Penford, P.L., Greenberg, A.H. and Roitt, I.M. (1976), Characteristics of the effector cells mediating cytotoxicity against antibody-coated target cells. III. Ultrastructural studies. Clin. Exp. Immunol. 23, 91-97
6) Clark, R.A. and Klebanoff, S.J. (1977) Studies on the mechanism of antibody-dependent polymorphonuclear leukocyte-mediated cytotoxicity. J. Immunol. 119, 1413-1418
7) Fleer, A., Roos, P., van dem Borne, A.E.G. Kr. and Engelfriet, C.P. (1979). Cytosome activity of human monocytes towards sensitized red cells is not dependent on the generation of reactive oxygen species. Blood. 54, 407-411
8) Rodey, G.E., Park, B.H., Windhorst, D.B. and Good, R.A. (1969) Defective bactericidal activity of monocytes in fatal granulomatous disease. Blood, 33, 813-820
9) Katz, P., Simone, C.B., Henkart, P.A. and Fauci, A.S. (1980) Mechanism of antibody-dependent cellular cytotoxicity. J. Clin. Invest. 65, 55-63
10) Hafeman, D.G. and Lucas, Z.J. (1979) Polymorphonuclear leukocyte-mediated, antibody-dependent, cellular cytotoxicity against tumor cells: Dependence on oxygen and the respiratory burst.
11) Borregaard, N. and Kragballe, K. (1980) Role of oxygen in antibody-dependent cytotoxicity mediated by monocytes and neutrophils, J. Clin. Invest. 66, 676-684.

12) Kragballe, K., Borregaard, N., Brandrup, F., Koch, C. and Johansen, K.S. (1980). Relation of monocyte and neutrophil oxydative metabolism to skin and oral lesions in carriers of chronic granulomatous disease. Clin. Exp. Immunol. In press.
13) Kragballe, K. and Borregaard, N. (1980). Mechanisms of decreased antibody-dependent cytotoxicity mediated by monocytes and neutrophils in atopic dermatitis. Acta Dermatovener. (Stockholm), in press.
14) Capsoni, F., Meroni, P.L., Ciboddo, G.F. and Colombo, G. (1979) Cytotoxic mechanism in white blood cells. N. Engl. J. Med. 300, 344
15) Newburger, P.E., Cohen, H.J., Rotchild, S.B., Hobbins, J.C., Malawista, S.E. and Mahoney, M.J. (1979) Prenatal diagnosis of chronic granulomatous disease. N. Engl. J. Med. 300, 178-181
16) Borregaard, N., Johansen, K.S. and Esmann, V. (1979) Quantitation of superoxide production from normals and three types of chronic granulomatous disease. Biochem. Biophys. Res. Commun. 90, 214-219
17) Holmes, B., Page, A.R. and Good, R.A. (1967) Studies of the metabolic activity of leukocytes from patients with a genetic abnormality of phagocytic function. J. Clin. Invest. 46, 1422-1432
18) Johnston, R.B. Jr., Keele, B.B., Misra, M.P., Lehmeyer, J.E., Webb, L.S., Baehner, R.L. and Rajagopalan, K.K. (1975) The role of superoxide-anion generation in phagocytic bactericidal activity. J. Clin. Invest. 55, 1357-1372
19) Johnston, R.B. Jr., Lehmeyer, J.E. and Guthrie, L.A. (1976) Generation of superoxide and chemiluminiscense by human monocytes during phagocytosis and in contact with surface-bound immunoglobulin G. J. Exp. Med. 143, 1551-1556
20) Weiss, S.J., King, G.W. and LoBuglio, A.F. (1979) Evidence for hydroxyl radical generation by human monocytes. J. Clin. Invest. 60, 370-373
21) Tauber, A.I. and Babior, B.M. (1977) Evidence for hydroxyl radical generation by human monocytes. J. Clin. Invest. 60, 374-379
22) Root, R.K. and Metcalf, J.A. (1977) H_2O_2 release from human granulocytes during phagocytosis. Relation to superoxide anion formation and cellular catabolism of H_2O_2: Studies with normal and cytochalasin B-treated cells. J. Clin. Invest. 60, 1266-1279
23) Reiss, M. and Roos, D. (1978) Differences in oxygen metabolism of phagocytosing monocytes and neutrophils. J. Clin. Invest. 61, 480-

24) Green, M.J., Hill, H.A.O., Okolow-Zubkowska, M.J. and Segal, A.W. (1979). The production of hydroxyl and superoxide radicals by stimulated neutrophils - measurements by EPR spectroscopy. FEBS Lett. 100, 23-26
25) Rosen, H. and Klebanoff, S.J. (1979) Hydroxyl radical generation by polymorphonuclear leukocytes measured by electron spin resonance spectroscopy. J. Clin. Invest. 64, 1725-1729

DISCUSSION
Edited by A. Mantovani

De Chatelet

Your data on the involvement of hydroxyl radical with the intact cell system rely entirely on inhibition by mannitol. The specificity of this inhibitor is open to question and I do not think any firm conclusions can be drawn regarding mechanisms, particularly in the absence of inhibition by catalase or SOD in the intact cell system.

Borregaard

I agree that the indication for hydroxyl radicals participating in ADCC is indirect.

Roos

In a recent article in Blood, Lo Buglio showed that the cytotoxicity of neutrophils, monocytes and lymphocytes is highly dependent on the kind of antibody used to sensitize the erythrocytes, the amount of antibody per target cell and the use of monolayers or cell suspensions. Therefore, you have to define the system you are using with respect to these parameters. Also, if you compare your results with those reported before, you have to refrain from comparing non-comparable systems.

The system used by Fleer in our laboratory (Blood, 1979) consisted of monocytes incubated in suspensions with anti-D-sensitized erythrocytes (non-complement-binding IgG alloantibodies). In that system, the target cells were lysed extracellularly in a non-oxygen radical-dependent system. If, on the other hand, he used anti-A-coated erythrocytes, much more antibody was bound, and the targets were lysed intracellularly in a process that was oxygen-dependent.

In the anti-D-system, lymphocytes and neutrophils are completely inactive, monocytes and especially macrophages are very active.

Borregaard

The antibody used was an anti-D antibody and this was effective in activating both monocytes and neutrophils.

Johnston

This should be the proper audience to ask what is now known about non-oxidative mechanisms of cytolysis. Your data indicated a role for oxygen metabolites, but only to a level of 40-60%. This implies that 60-40% of lysis was non-oxidative, if I understand your system correctly. Do you or others have data on these non-oxidative mechanisms?

Borregaard

I do not have any data to answer your question. However I can immagine that the process of phagocytosis per se might cause some lysis (^{51}Cr release) of erythrocytes by the stress it causes on the erythrocytes. However this is very difficult to assess since it would require a cell type which phagocytoses normally, but which does not degranulate.

Roos

In relation to Dr. Johnston's question on oxygen-independent mechanisms, I can only quote again from the work of Fleer et al. (Scand.J.Haematol., 1979). He found a correlation between the extent of extra-cellular lysis and the release of lysosomal enzymes. That is, he found a decrease in lysis as well as in the release of lysosomal enzymes by hydrocortisone and colchicine, and in reverse, an increase in both parameters by cytochalasin B. Thus, in a very indirect manner, he found evidence that lysosomal enzymes might be released and lyse the extracellular target cell.

Venge

Just one more short comment on non-oxidative cytotoxic mechanisms. Some years ago we published in Transfusion that the chymotrypsin- like cationic protein isolated from polymorphonuclear leukocytes is very effective in lysing erythrocytes.

Segal

Can you explain how the indicator of cytotoxicity, radioactive chromium, escapes from the pahgocytic vacuole of the ingesting cell? Is it possible that the release of ^{51}Cr is greater from normal than from CGD neutrophils?

Borregaard

We do not know to what extent ^{51}Cr is released from the phagocytic vacuole, but clearly a significant portion is retained in the pahgocyte as this forms the basis of the measurement of phagocytosis. In order to answer the second part we have to differentiate between the escape ^{51}Cr from the phagocytic vacuole and the primarily extracellular release.

MACROPHAGE ACTIVATION FOR TUMOR CYTOTOXICITY : REACTIVITY OF PERITONEAL AND BONE MARROW MACROPHAGES

Luigi P. Ruco°, Monte S. Meltzer*, Edward J. Leonard*, Shogo Tomisawa*

°Second Institute of Pathological Anatomy, Experimental Pathology Unit, University of Rome, Viale Regina Elena 324, 00161 Rome, Italy

*Immunopathology Section, Laboratory of Immunobiology, National Cancer Institute, National Institute of Health, Bethesda, Maryland, 20205, USA

Peritoneal macrophages recovered from mice infected with intracellular parasites such as Mycobacterium bovis, strain BCG, Toxoplasma gondii or Listeria monocitogenes exert non specific cytotoxic activity against tumor cells in vitro (1,2). Macrophage cytotoxicity is selectively expressed against neoplastic cells. Tumor cells, derived from viral or chemical carcinogen-induced or spontaneous syngeneic and non syngeneic tumors, were found to be quantitatively more susceptible to macrophage cytophatic effects than were non neoplastic cells. Tumor cell killing requires cell to cell contact; soluble cytotoxic factors released by macrophages or by macrophage-tumor cell interactions have not been described. A cinemicrographic analysis revealed that macrophage-tumor cell interactions are characterized by repeated and relatively short (2 hr.) contacts inducing permanent tumor cytostasis and degenerative morphological changes (3). The development of a cytotoxicity assay based on radioisotope release facilitated quantitative analysis of macrophage tumoricidal activity (4). Tumor cell death was evaluated as release of tritiated thymidine from prelabeled target cells into supernatant culture fluids, BCG-activated macrophages induced significant release of ^{3}HTdR from tumor cells by 24 hours; however, optimal measurement of tumor cytotoxicity (50% total counts) was observed after 48-72 hours. Macrophages from uninfected mice or from oil-, starch- or thioglycollate induced peritoneal exudates had little effect on labeled tumor monolayers.

Studies performed to elucidate the mechanisms leading to formation of activated macrophages in vivo suggested that the immune system was probably involved in the afferent phase of macrophage activation (5). In fact it was shown that: i) Activated macrophages could be induced by soluble antigen (PPD) in previously immunized mice (BCG), but not in normal mice. ii) PPD induction of tumoricidal macrophages was evident by at least 10 days after BCG immunization, that is, in concomitance with positivization of delayed hypersensitivity reactions. iii) PPD treatment was effective in non immune mice receiving adoptive transfer of BCG-immune spleen cells.

Similar results were obtained also in an in vitro system (6). In fact, macrophages from untreated mice could be rendered tumoricidal in vitro by exposure to cell-free supernatant fluids from PPD stimulated BCG-immune spleen cell cultures (lymphokines). The development of an in vitro system allowed further analysis of intermediary reactions leading to formation of tumoricidal macrophages.

Characterization of macrophage activation factor (MAF) in supernatant of antigen stimulated leucocyte cultures by Sephadex G-100 chromatography indicated that MAF elutes as a single peak with a molecular weight of about 55,000 daltons; the elution pattern, however, is too broad for a single entity and suggests a limited heterogeneity within the MAF species. In addition, MAF activity is heat labile (80% loss of activity at 56°C for 30 min.), destroyed at pH 4 or 10 (90% loss after 16 hr.) and susceptible to proteolytic enzymes (7).

The time course for lymphokine activation in vitro is shown in Fig.1. Cytotoxic activity was evident after 4 hr. of incubation in lymphokines, increased to maximal levels by 8-12 hr., then progressively decreased to untreated levels by 24-36 hr. Loss of tumoricidal activity was not due to macrophage death in culture: no change in vital dye uptake or in phagocytic capacity was observed over

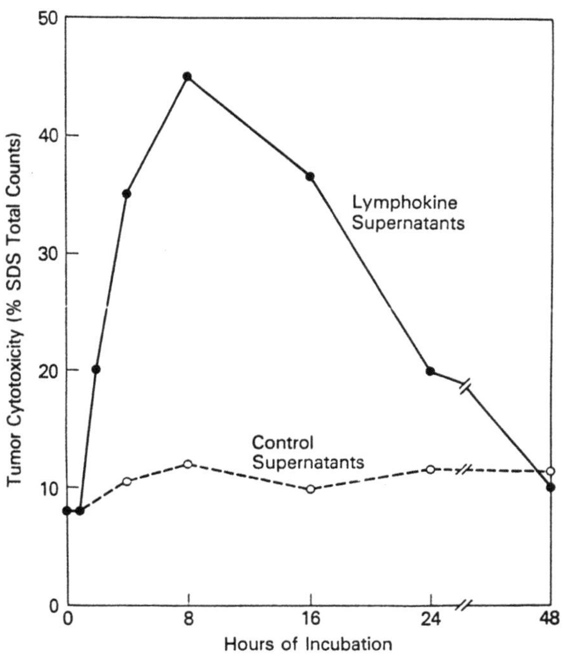

Fig. 1. Tumor cytotoxicity by lymphokine-activated macrophages: time course lymphokine activation. Adherent PC were incubated in lymphokines for various times, washed and cultured with ^{3}HTdR-labeled target cells for 48 hours. Cytotoxicity was estimated at 48 hours by measurement of ^{3}HTdR release.

this time period.Loss of tumoricidal activity was also not due to depletion or destruction of active lymphokines: replacement of lymphokine supernatants at 16 hr.did not alter this decay.Thus,tumoricidal activity of lymphokine-activated macrophages was short-lived <u>in vitro</u>.This loss of activity,although not due to cell death or lymphokine depletion,was irreversible.Identical decay of tumoricidal activity <u>in vitro</u> was observed with macrophages activated <u>in vivo</u> by BCG infection (1,2).Capacity of macrophages to be activated by lymphokines <u>in vitro</u> for tumor cytotoxicity was short-lived as well.Macrophages cultured <u>in vitro</u> for various times prior to the addition of lymphokines gradually lost capacity to be activated for tumor cytotoxicity (Fig.2).This loss of lymphokine responsiveness with time in culture was also irreversible and not due to cell death.

Macrophage population is functionally heterogeneous. Resident peritoneal macrophages are peroxidase negative, but most of macrophages present at inflammatory sites are young peroxidase positive cells which derive from recently migrated blood monocytes (8).<u>In vivo</u> experiments demonstrated that tumoricidal macrophages recovered 24 hr. after PPD injection in BCG-immune mice were mostly peroxidase positive.Changes in peroxidase cytochemistry,however,were entirely non specific sequela of inflammation and not of immune response per se;in fact,following PPD injection, identical changes occurred in control mice or in BCG-immune mice (5).Since these immigrant mononuclear phagocytes are the most likely partecipants in macrophage-lymphokine interactions we examined the response of inflammatory macrophages to lymphokines (9).Resident peritoneal macrophages or inflammatory macrophages recovered from mice injected intraperitoneally with 1 ml FCS 24 hr. earlier were pulsed with lymphokines <u>in vitro</u> and tested for tumor cytotoxicity (Fig.3).Inflammatory macrophages were found more responsive to lymphokines than equal numbers of resident macrophages.Comparison of dose and time responses by peritoneal exudate and resident cell populations to dilutions of lymphokines demonstrated that differences in lymphokine responses were quantitative.In fact,cytotoxicity curves for exudate and resident macrophages were superimposable after multiplication by a constant factor.This factor,about 10,may represent the relative difference in lymphokine-

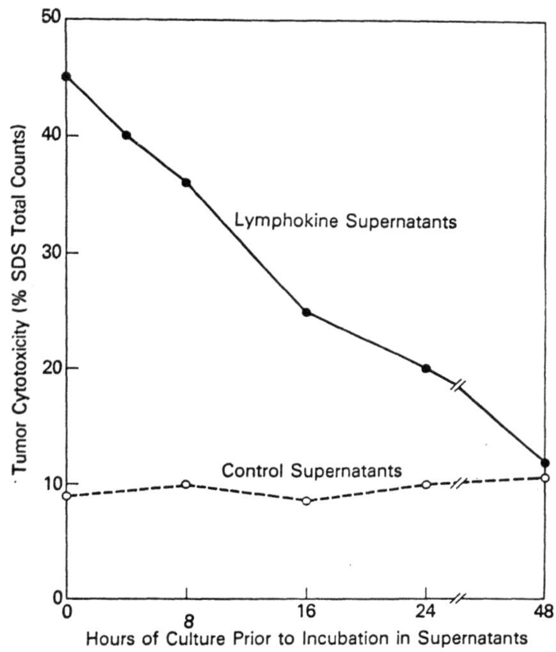

Fig. 2. Tumor cytotoxicity by lymphokine-activated macrophages: effect of time in culture prior to lymphokine activation. Adherent PC were cultured in medium for 0-48 hours, then incubated in lymphokines for 4 hours. Cultures were washed and incubated with ^{3}HTdR-labeled target cells for 48 hours. Cytotoxicity was estimated at 48 hours by measurement of ^{3}HTdR release.

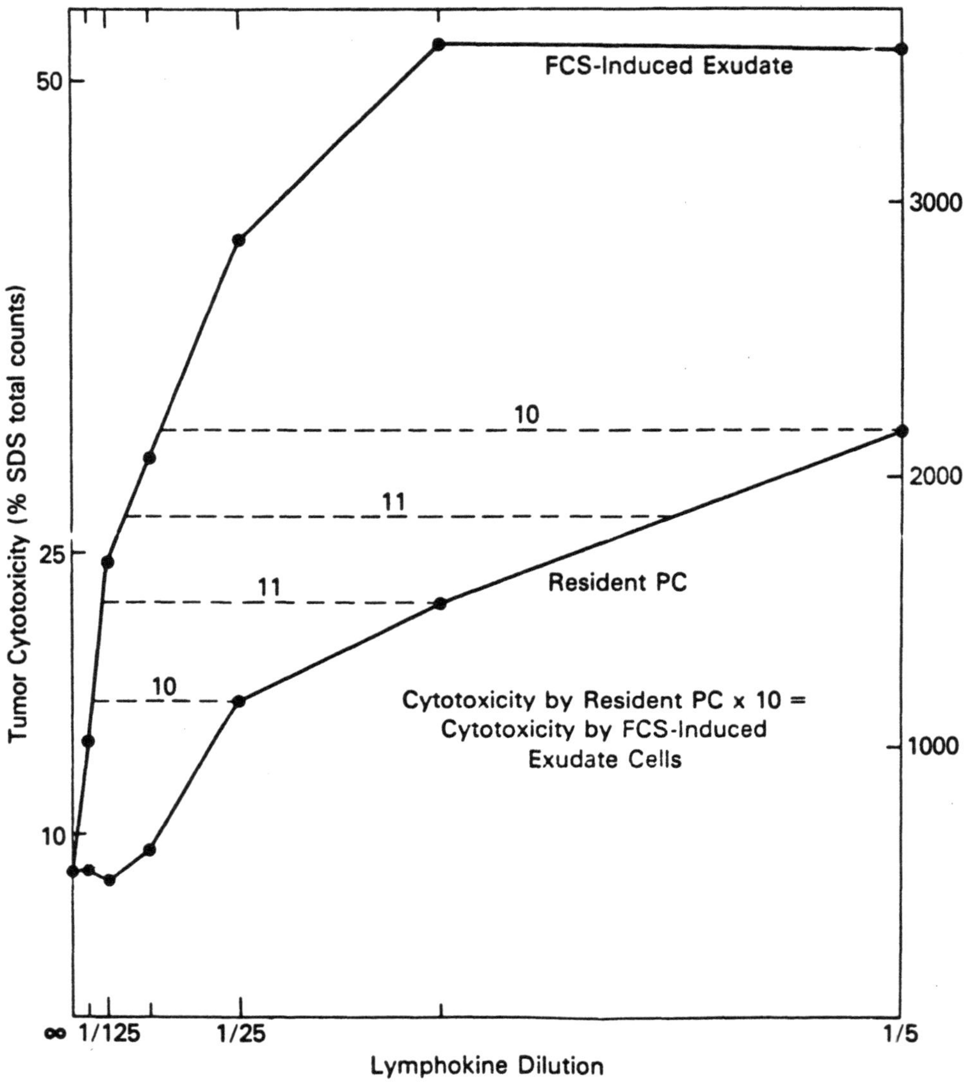

Fig. 3. Tumor cytotoxicity by lymphokine-activated macrophages: lymphokine responsiveness of resident and inflammatory macrophages PC were collected from untreated mice and from mice treated ip with fetal calf serum 1 day previously. Cells were adjusted to an equal macrophage concentration and assayed for lymphokine-induced tumoricidal capacity. Dashed line shows factor that arithmetically superimposes the resident PC point upon the FCS-induced exudate curve.

responsive cells between the two populations.Peritoneal exudate cells elicited by FCS injection were used as the source of macrophages for a series of experiments so that changes in lymphokine responsiveness could be compared with the reported changes in mononuclear phagocyte populations.Peritoneal cells were collected from groups of mice 0 to 6 days after intraperitoneal injection of FCS. Peritoneal macrophages were examined for peroxidase cytochemistry,phagocytic capacity and lymphokine activation for tumor cytotoxicity (Fig.4).Results from these experiments indicated that changes in peroxidase cytochemistry and lymphokine responsiveness were closely correlated throughout this time period.Experiments performed in irradiated mice provided additional evidences that the increased lymphokine responsiveness of peritoneal exudate cells was dependent upon influx of bone marrow derived peroxidase positive mononuclear phagocytes (Fig.5).Mice were irradiated 1 day prior to injection of a sterile irritant. Peritoneal cells were collected 1 day later.In mice receiving whole body x-irradiation,percent peroxidase positive cells during inflammation decreased from 40% to 3%.The ability of macrophages to respond to lymphokines showed a corresponding decrease.

Bone marrow precursors of macrophages are radiosensitive.Peripheral blood monocytes have a turnover time of about 30 hr. and resident macrophages are radioresistant. Results from experiments with irradiated mice suggest that monocytes and/or macrophage precursor cells represent the lymphokine responsive cell population.

Tissue culture techniques for bone marrow macrophages have been recently described(10).Bone marrow macrophage precursor cells cultured *in vitro* in the presence of growth factors (L-cell conditioned medium) proliferate and differentiate to the mature macrophage stage.Percentage of peroxidase positive macrophages and lymphokine reactivity were tested on bone marrow cells which were cultured *in vitro* for 3-7 days. A progressive decrease in percentage of peroxidase positive cells was observed between day 3 and 7 as previously described.However,also in this instance,decrease in peroxidase positive cells was associated with decreased lymphokine responsiveness.

Thus,our results indicate that macrophage activation for tumor cytotoxicity is the final result of a cascade

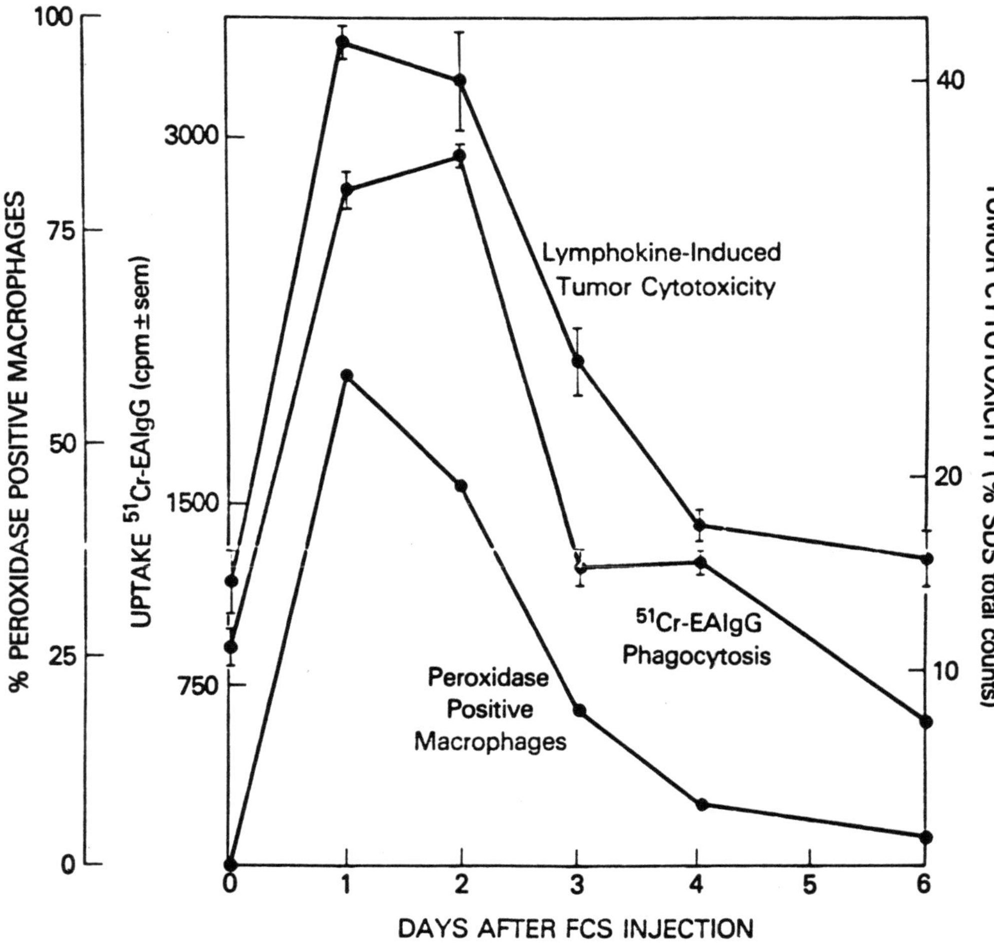

Fig. 4. Peritoneal macrophage function during acute inflammation. PC were collected from mice treated with 1 ml fetal calf serum ip 0-6 days previously and assayed for phagocytosis of ^{51}Cr-labeled antibody-coated sheep erythrocytes, peroxidase cytochemistry and response to lymphokine induced activation.

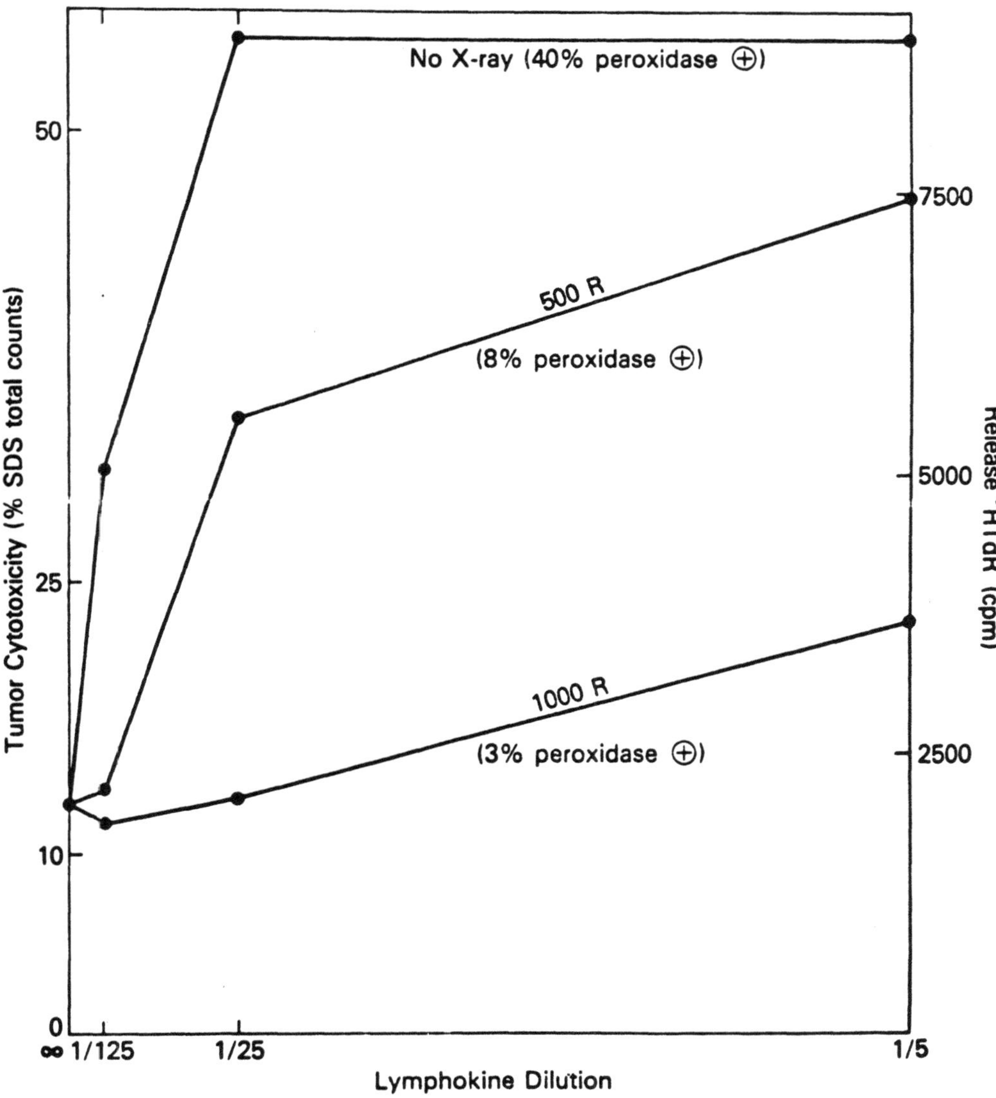

Fig. 5. Tumor cytotoxicity by lymphokine-activated macrophages: effect of X-irradiation on lymphokine-induced macrophage activation.

of short-lived intermediary reactions (11). These events include recruitment of reactive cells at the inflammatory site and interaction of this cell population with factors released by antigen stimulated immune lymphocytes. These events are short-lived. Reactive cells differentiate to a non reactive stage within 24 hr. and either *in vivo* BCG activated or *in vitro* lymphokine-activated macrophages mantain their status only for few hours. The short lived nature of both macrophage tumoricidal capacity and capacity of precursor cells to be activated by lymphokines may function as negative feedback mechanisms in immune reactions. Finally, the close association between frequency of macrophages with intracytoplasmic peroxidase positive granules and lymphokine reactivity of the cell population may indicate that lysosomal peroxidase is somehow involved in lymphokine activation for tumor cytotoxicity. If this is the case, cell degranulation and incapacity of mature macrophages to form new granules might, at least partially, account for rapid decay of both macrophage responsiveness to lymphokines and tumoricidal activity.

REFERENCES

1. J.B.Jr.Hibbs, The macrophage as a tumoricidal effector cell: A review of *in vivo* and *in vitro* studies on the mechanism of the activated macrophage non specific cytotoxic reaction, *in*: "The Macrophage in Neoplasia", M.A. Fink ed., Academic Press, New York (1976)
2. R.P.Cleveland, M.S.Meltzer, and B.Zbar, Tumor cytotoxicity *in vitro* by macrophages from mice infected with Mycobacterium bovis, strain BCG, *J.Natl.Cancer Inst.* 52:1887 (1974)
3. M.S.Meltzer, R.W.Tucker, and A.C.Breuer, Interaction ofBCG activated macrophages with neoplastic and non-neoplastic cell lines *in vitro*: Cinemicrographic analysis, *Cell. Immunol.* 17:30 (1975)
4. M.S.Meltzer, R.W.Tucker, K.K.Sanford, and E.J.Leonard, Interaction of BCG-activated macrophages with neoplastic and non-neoplastic cell lines *in vitro*: Quantitation of the cytotoxic reaction by release of tritiated thymidine from prelabeled target cells, *J.Natl.Cancer Inst.* 54:1177 (1975)

5 . L.P.Ruco and M.S.Meltzer,Macrophage activation for tumor cytotoxicity:Tumoricidal macrophages by PPD in BCG immune mice,Cell. Immunol.32:203 (1977)

6 . L.P.Ruco and M.S.Meltzer,Macrophage activation for tumor cytotoxicity:Induction of tumoricidal macrophages by supernatants of PPD stimulated Bacillus Calmette Guerin-immune spleen cell cultures, J. Immunol.119:889 (1977)

7 . E.J.Leonard,L.P.Ruco,and M.S.Meltzer,Characterization of macrophage activation factor,a lymphokine that causes macrophage to become cytotoxic for tumor cells, Cell. Immunol.41:347 (1978)

8 . R.Van Furth,J.G.Hirsch,and M.E.Fedorko,Morphology and peroxidase cytochemistry of mouse promonocytes,monocytes and macrophages,J.Exp.Med.132:794 (1970)

9 . L.P.Ruco and M.S.Meltzer,Macrophage activation for tumor cytotoxicity:Increased lymphokine responsiveness of peritoneal macrophages during acute inflammation, J. Immunol.120:1054 (1978)

10. T.J.L.M.Goud,C.Schotte,and R.Van Furth,Identification and characterization of the monoblast in mononuclear phagocyte colonies grown in vitro, J.Exp.Med.142:1180 (1975)

11. L.P.Ruco and M.S.Meltzer,Macrophage activation for tumor cytotoxicity:Development of macrophage cytotoxic activity requires completion of a sequence of short lived intermediary reactions, J. Immunol.121:2035 (1978)

DISCUSSION

Edited by A. Mantovani

Gallily

How would you explain the prolonged expression of macrophage tumoricidal activity *in vivo*, on the basis of your *in vitro* studies which demonstrate a short duration of this activity.

Ruco

We have data suggesting that the persistence of activated tumoricidal macrophages in the pritoneal cavity of mice intraperitoneally injected with BCG is mediated through recruitment of new cells from the blood stream.

Gallily

Did you find a correlation between the subset of Ia-positive macrophages and tumoricidal activity?

Ruco

I have not studied this problem; however, Daniela Mannel has recently obtained some prelimianry data suggesting a positive correlation between Ia positivity and tumoricidal activity.

De Chatelet

Having recently questioned the utility of inhibitors, I would now like to reverse myself and ask whether you have investigated the mechanism of cidal activity by the use of inhibitors such as catalase, SOD, mannitol, etc.

Ruco

No, we have not.

Patriarca

Do resident macrophages treated with lymphokines develop peroxidase?

Ruco

No, they do not. Once the resident macrophages have lost their peroxidase positive granules they are no longer able to synthetize new ones.

Patriarca

Did you study the subcellular distribution of peroxidase in cytotoxic macrophages?

Ruco

No we did not. We studied peroxidase cytochemistry only at the light microscopy level and we could only detect intracytoplasmic peroxidase positive granules.

Clark

Is the enhanced killing by cells enriched in peroxidase-positive macrophages mediated by a peroxidase-dependent mechanism, for example, the peroxidase-hydrogen peroxide-halide system?

Ruco

Carl Nathan has reported that the oxidative system is involved in the macrophage-mediated tumor cell killing. However, we have not tested this hypothesis in our system.

Rossi

Is there any difference between the killing mechanisms of resident activated macrophages and that of inflammatory (i.e. peroxidase positive), macrophages?

Ruco

Our data do not permit answering this question; however, it cannot be excluded that in the resident macrophage population the cytotoxic reaction is mediated by the 1-2% peroxidase positive cells which are normally present to allow the population turnover. In fact, we do not have any information about the required effector: target ratio to obtain tumor cell lysis.

Rossi

In conclusion you suggest that only peroxidase positive macrophages are able to kill tumor cells.

Ruco

Yes, this is our suggestion.

CYTOTOXICITY ON TUMOR CELLS OF HUMAN MACROPHAGES:

FUNCTIONAL STATUS OF TUMOR-ASSOCIATED EFFECTOR CELLS

Alberto Mantovani, Claudio Bordignon, Andrea Biondi,
Martino Introna and Paola Allavena

Istituto di Ricerche Farmacologiche "Mario Negri"
Via Eritrea, 62 - 20157 MILAN, Italy

INTRODUCTION

Cells of the monocyte-macrophage series are a major component of the lymphoreticular infiltrate of rodent and human tumors (1-8). Several studies have focused on the in vitro cytotoxicity of rodent mononuclear phagocytes on tumor cells and on the functional status of tumor-associated macrophages (TAM) from murine neoplasms (3,9-21). In contrast, little attention has been given to the interaction of human monocytes and macrophages with neoplastic cells, and human TAM have not been characterized. Here we will summarize results on the tumoricidal activity of human mononuclear phagocytes and we will discuss issues of current interest in our laboratory.

NATURAL AND ACTIVATED CYTOTOXICITY OF HUMAN MONOCYTES AND MACROPHAGES

Monocytes and macrophages were isolated by adherence as previously described (22-32). Cytolytic activity was measured as release of $[^3H]$ thymidine from prelabelled TU5 tumor cells (23-32). Cells of the monocyte-macrophage series isolated from various anatomical sites (blood, nonneoplastic ascites, peritoneal exudates, early lactation milk, benign ovarian cysts) had appreciable levels of spontaneous cytotoxicity on tumor cells, in the absence of any deliberate stimulation (26-29). Natural cytotoxicity of monocytes or macrophages was not dependent on measurable contamination of the culture system with endotoxin (26-29, 32).

Several lines of evidence suggest that effector cells involved in natural cytotoxicity of human adherent cell preparations were not contaminating non-adherent NK cells and that they belonged to the monocyte-macrophage series.

1) After through depletion of monocytes, non-adherent peripheral blood mononuclear cells rich in NK activity had essentially no cytotoxic activity against TU5, a target cell extremely susceptible to the natural cytolytic activity of adherent cell preparations (26,32). Conversely, adherent cells had little cytolytic and cytostatic capacity on K 562 target cells, a specific target for NK activity (23,26,32).

2) Silica particles markedly inhibited monocyte or macrophage-mediated cytotoxicity. Conversely treatment with a potent and specific anti-T cell serum and complement, which inhibited NK activity, did not affect monocyte tumoricidal activity (26).

3) Upon in vitro culture in medium with human serum, NK activity is greatly diminished (33). In contrast, human monocytes maintained in vitro under similar experimental conditions had cytotoxicity levels similar to or higher than fresh adherent effector cells (28,31).

4) The organ distribution of NK and macrophage-mediated cytotoxicity is quite different. Peritoneal non-adherent mononuclear cells had NK activity extremely low when compared to peripheral blood lymphoid cells. Conversely, peritoneal adherent macrophages had natural cytotoxicity against TU5 similar or higher than peripheral blood monocytes (32).

A variety of chemical or biological stimuli can enhance the cytotoxicity on tumor cells of rodent mononuclear phagocytes (15-17) and we have been investigating the regulation of cytotoxicity of human mononuclear phagocytes. Exposure to lymphokine supernatants, interferon (IF) preparations and endotoxin (Table 1) enhanced the cytolytic activity of human adherent effector cells (28-32, 34, 35). The evidence that the effector cells were in fact monocytes a macrophages was as discussed above for natural cytotoxicity (30-32).

The IF preparations used in the above mentioned studies were not, or only partially, purified. Activation of human monocyte cytotoxicity could therefore be due to a contaminant of these preparations. However, human IF preparations activated human but not murine, adherent cells and viceversa. Moreover an anti-human leukocyte IF immunoglobulin blocked the enhancing activity of human leukocyte IF (Mantovani and Herberman, unpublished data). These data strongly suggest that activation of human mononuclear phagocytes is in fact related to IF, but conclusive elucidation of this point will require testing IF preparations purified to homogeneity, a work which is currently in progress.

Table 1. Stimuli Capable of Augmenting Human Macrophage Tumoricidal Activity

1) Supernatants of human lymphocytes cultured with:		
	- antigen (PPD)	
	- mitogen (PHA)	
	- *C.parvum*	CN 6134
		CN 5888
2) Interferon from:		
	- fibroblasts	
	- leukocytes	
	- lymphoblastoid cells	
3) Endotoxin		

The expression of natural or activated cytotoxicity is not a property of all cells of the monocyte-macrophage series. Cells of different anatomical origin and/or at different maturation stages vary widely in cytotoxic potential. Macrophages from lung alveoli did not lyse TU5 targets and were only spontaneously growth inhibitory (29). Alveolar macrophages showed a weak, but significant, enhancement of cytotoxicity when exposed in vitro to lymphokines, but were totally unresponsive to IF (29).

In vitro culture of human monocytes for periods exceeding 10 days results in mononuclear phagocyte cultures consisting mainly of epithelioid and multinucleated giant cells. These had little spontaneous cytotoxicity and little responsiveness to stimuli (28, 31), but were capable of mediating antibody-dependent cytotoxicity (23,36). It is tempting to speculate that similar cells present at sites of chronic inflammation are similarly defective in direct tumoricidal activity.

TUMORICIDAL ACTIVITY OF TUMOR-ASSOCIATED MACROPHAGES (TAM) FROM OVARIAN TUMORS

Macrophages were isolated from ascitic and solid ovarian tumors (22-25) and their non specific cytotoxicity was measured using TU5 tumor cells as targets. Blood monocytes from ovarian cancer patients had cytolytic activity somewhat lower ($p<0.01$) than control effectors but, a wide overlap between the two groups was observed (25) when the capacity of monocytes to mature into macrophages (MP) was assessed as previously described (37), lower MP values tended to be observed in blood of ovarian cancer patients, but values were widely overlapping with controls

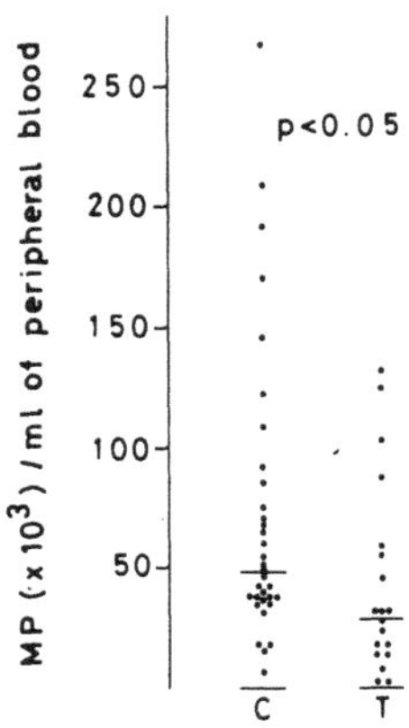

FIG. 1. Peripheral blood macrophage precursors in ovarian cancer patients

(Fig. 1). Similar results were obtained when the baseline cytotoxicity of TAM was investigated (25).

Various stimuli (IF, lymphokines and endotoxin) were tested for their capacity to activate macrophages from ascitic and solid (less frequently) ovarian neoplasms. Stimuli augmented the cytotoxicity of TAM, the levels of stimulation being similar to those of control peritoneal macrophages (data not shown). In a few patients macrophages from the ascitic and solid neoplasm were tested in parallel (Table 2). In two subjects, activation was similar for ascitic or solid TAM (No. 2 and 21), but TAM from 2 additional solid tumors (No. 27 and 29) were less responsive than effectors from ascites.

In an effort to further investigate the tumor-host relationship at the tumor site, the susceptibility of primary ovarian carcinoma cultures to macrophage cytotoxicity was investigated (25,32). Ovarian tumor cells from different donors were heterogeneous in their interaction with mononuclear phagocytes, a fraction of the patients having tumor cells resistant to lysis by unstimulated effectors (25,32). Growth of these macrophage-resistant tumor cells was usually stimulated by macrophages (25,32). A similar heterogeneity of susceptibility of ovarian carcinoma cells to macrophage cytotoxicity was observed when IF or lymphokine stimulated effectors were used (Table 3).

CONCLUSIONS AND SPECULATIONS

In humans cells of the monocyte-macrophage series have appreciable levels of spontaneous cytotoxicity against selected tumor target cells. Several factors regulate the levels of tumoricidal activity and exposure to lymphokine supernatants, type 1 IF or

Table 2. Responsiveness of Macrophages from Solid Ovarian Carcinoma to IF and Lymphokines

Patient No.	Macrophages from	Percentage of specific lysis ± S.D.		
		-	IF	Lymphokines
2	Ascites	23.2±2.1	44.0±2.4[a]	NT[b]
	Solid tumor	24.6±3.8	46.3±2.2[a]	NT[b]
21	Ascites	12.7±5.3	16.7±0.6[a]	17.1±2.6
	Solid tumor	10.3±0.5	14.9±0.4[a]	14.3±1.0
27	Ascites	17.2±1.3	68.7±2.1[a]	43.1±2.1[a]
	Solid tumor	3.6±0.8	4.1±0.7	8.6±1.5[a]
29	Ascites	11.1±1.0	59.5±8.1[a]	NT[b]
	Solid tumor	9.8±0.3	13.9±0.1[a]	NT[b]

Cytolytic activity of macrophages from ascites or solid tumor was tested against TU5. IF concentration was 1000 units/ml and lymphokines were diluted 1/3.

[a] $p < 0.05$ compared to unstimulated effectors
[b] Not tested

endotoxin enhances cytotoxicity. In addition, the maturation stage and/or the anatomical site from which mononuclear phagocytes are obtained, modulate the expression of spontaneous cytotoxicity and responsiveness to stimuli.

The *in vivo* relevance of human monocyte or macrophage-mediated cytotoxicity remains a matter of speculation. Cells of the monocyte-macrophage series are virtually ubiquitous in organs and tissues. Their natural tumoricidal capacity, preferentially expressed on transformed versus untrasformed target cells, could provide a mechanism of elimination of nascent tumor cells, not involving specific recognition by lymphocytes and subsequent

Table 3. Cytolytic Activity of Activated Mononuclear Phagocytes on Primary Ovarian Carcinoma Cultures

Target cells	Specific lysis (%)		
	-	IF (1000 u/ml)	Lymphokines (1/3)
TU5	28.2	74.6[a]	52.2[a]
Ovarian carcinoma No.2	27.0	25.9	24.2
" " 5	23.0	32.6[a]	48.4[a]

[a] $p < 0.01$

activation of mononuclear phagocytes through specific release of mediators. Accordingly it is tempting to speculate that the protection against breast cancer conferred by early pregnancy may in part be related to an influx in the lactating mammary gland of macrophages with appreciable levels of spontaneous cytotoxicity against tumor cells (29).

In established human and rodent neoplasms, macrophages can constitute a major component of the lymphoreticular infiltrate (1-8,23). Macrophages isolated from ascitic ovarian tumors can modulate the in vitro growth of tumor cells and express cytolytic activity, cytotoxicity levels being similar or lower than those of peritoneal macrophages from control subjects. In considering the possible in vivo role of tumor-associated macrophages in the control of advanced human neoplasia, the susceptibility of tumor cells to attack by mononuclear phagocytes has to be taken into account. In fact we have observed that primary ovarian carcinoma cultures are considerably heterogeneous in their susceptibility to macrophage cytotoxicity. It is therefore conceivable that, provided tumor cells are susceptible to attack by mononuclear phagocytes, macrophages may act as a mechanism of restraint of ovarian tumor growth at least at this anatomical site (peritoneal cavity) where macrophage concentration relative to tumor cells is relatively high (23). Conversely, in the presence of resistant ovarian tumor cells, it is unlikely that infiltrating mononuclear phagocytes, with cytotoxic capacity similar or lower than control, have any inhibitory effect on tumor cell proliferation, and neoplastic growth may actually be stimulated as observed in vitro.

ACKNOWLEDGEMENT

This work was supported by Grant 1R01 CA 26824 from the National Cancer Institute, USA and by Progetto Finalizzato "Controllo della Crescita Neoplastica" from CNR, Rome, Italy (No. 79.00643.96).

REFERENCES

1. R. Evans, Replication of Riley's plasma enzymes elevating virus in tissue culture: The importance of the cellular composition, J. Gen. Virol. 1:363 (1967).
2. R. Evans, Macrophages in syngeneic animal tumours, Transplantation 14:468 (1972).
3. R. Evans, Tumor macrophages in host immunity to malignancies, in: "The Macrophage in Neoplasia", M.A. Fink, ed., Academic Press Press, New York (1976).
4. H.F. Pross, and R.S. Kerbel, An assessment of intratumor phagocytic and surface marker-bearing cells in a series of auto-chthonous and early passaged chemically induced murine sarcomas. J. Natl. Cancer Inst. 57:1157 (1976).

5. I. Carr, Macrophages in human cancer: A review, in: "The Macrophage and Cancer", K. James, B. McBride and A. Stuart, eds., University of Edinburgh, Edinburgh (1977).
6. G.W. Wood, and K.A. Gollahon, Detection and quantitation of macrophage infiltration into primary human tumors with the use of cell-surface markers, J. Natl. Cancer Inst. 59: 1081 (1977).
7. J-L. Svennevig, M. Lövik, and H. Svaar, Isolation and characterization of lymphocytes and macrophages from solid malignant human tumours, Int. J. Cancer 23:626 (1979).
8. I. Lauder, W. Aherne, J. Steward, and R. Sainsbury, Macrophage infiltration of breast tumours: A prospective study. J. Clin. Pathol, 30: 563 (1977).
9. H.T. Holden, J.S. Haskill, H. Kirchner, and R.B. Herberman, Two functionally distinct anti-tumour cells isolated from primary murine sarcoma virus-induced tumours, J. Immunol. 117: 440 (1976).
10. P. Puccetti, and H.T. Holden, Cytolytic and cytostatic antitumour activities of macrophages from mice injected with murine sarcoma virus, Int. J. Cancer 23:123 (1979).
11. J.S. Haskill, ADCC effector cells in a murine adenocarcinoma. I. Evidence for blood-borne bone marrow-derived monocytes, Int. J. Cancer 20: 432 (1977).
12. S.W. Russell, and A.T. McIntosh, Macrophages isolated from regressing Moloney sarcomas are more cytotoxic than those recovered from progressing sarcomas, Nature 268:69 (1977).
13. S.W. Russell, W.F. Doe, and A.T. McIntosh, Functional characterization of a stable, noncytolytic stage of macrophage activation in tumors, J. Exp. Med. 146: 1511 (1977).
14. A. Mantovani, Effects on in vitro tumor growth of murine macrophages isolated from sarcoma lines differing in immunogenicity and metastasizing capacity, Int. J. Cancer 22:741 (1978).
15. R. Evans, and P. Alexander, Mechanisms of extracellular killing of nucleated mammalian cells by macrophages, in: "Immunobiology of the Macrophage", D.S. Nelson, ed., Academic Press, New York (1976).
16. J.B. Hibbs, The macrophage as a tumoricidal effector cell: A review of in vitro and in vivo studies on the mechanism of the activated macrophage nonspecific cytotoxic reaction, in: "The Macrophage of Neoplasia", M.A. Fink, ed., Academic Press, New York (1976).
17. R. Keller, Cytostatic and cytocidal effects of activated macrophages, in: "Immunobiology of the Macrophage", D.S. Nelson, ed., Academic Press, New York (1976).
18. M.S. Birbeck, and R.L. Carter, Observations on the ultrastructure of two hamster lymphomas with particular reference to infiltrating macrophages, Int. J. Cancer 9: 249 (1972).
19. S.A. Eccles, and P. Alexander, Macrophage content of tumours in relation to metastatic spread and host immune reaction.

Nature 250:667 (1974).
20. J.M. Zarling, and S.S. Tevethia, Transplantation immunity to simian virus 40-transformed cells in tumor-bearing mice. II. Evidence for macrophage participation at the effector level of tumor rejection, J. Natl. Cancer Inst. 50:149 (1973).
21. A. Mantovani, R. Giavazzi, N. Polentarutti, F. Spreafico, and S. Garattini, Divergent effects of macrophage toxins on growth of primary tumors and lung metastases in mice, Int. J. Cancer 25:617 (1980).
22. A. Mantovani, Separation of mononuclear phagocytes by adherence on microexudate-coated plastic, in: "Manual of Macrophage Methodology", H.B. Herscowitz, H.T. Holden, J.A. Bellanti, and A. Ghaffar, eds., Marcel Dekker, New York, in press.
23. A. Mantovani, G. Peri, N. Polentarutti, G. Bolis, C. Mangioni, and F. Spreafico, Effects on in vitro tumor growth of macrophages isolated from human ascitic ovarian tumors, Int. J. Cancer 23:157 (1979).
24. A. Mantovani, P. Allavena, C. Sessa, G. Bolis, and C. Mangioni, Natural killer activity of lymphoid cells isolated from human ascitic ovarian tumors, Int. J. Cancer 25:573 (1980).
25. A. Mantovani, N. Polentarutti, G. Peri, Z. Bar Shavit, A. Vecchi, G. Bolis, and C. Mangioni, Cytotoxicity on tumor cells of peripheral blood monocytes and tumor-associated macrophages in patients with ascites ovarian tumors, J. Natl. Cancer Inst. 64:1307 (1980).
26. A. Mantovani, T.R. Jerrells, J.H. Dean , and R.B. Herberman, Cytolytic and cytostatic activity on tumor cells of circulating human monocytes, Int. J. Cancer 23:18 (1979).
27. A. Mantovani, A. Tagliabue, J.H. Dean, T.R. Jerrells, and R.B. Herberman, Cytolytic activity of circulating human monocytes on transformed and untransformed human fibroblasts, Int. J. Cancer 23:28 (1979).
28. A. Mantovani, Z. Bar Shavitz, G. Peri, N. Polentarutti, C. Bordignon, C. Sessa, and C. Mangioni, Natural cytotoxicity on tumor cells of human macrophages obtained from diverse anatomical sites, Clin. Exp. Immunol. 39: 776 (1980).
29. C. Bordignon, R. Avallone, G. Peri, N. Polentarutti, C. Mangioni, and A. Mantovani, Cytotoxicity on tumor cells of human mononuclear phagocytes: defective tumoricidal capacity of alveolar macrophages, Clin. Exp. Immunol. 41:336 (1980).
30. J. Jett, A. Mantovani, and R.B. Herberman, Augmentation of monocyte-mediated cytolysis by interferon , Cell. Immunol. 54:425 (1980).
31. A. Mantovani, J.H. Dean, T.R. Jerrells, and R.B. Herberman, Augmentation of tumoricidal activity of human monocytes and macrophages by lymphokines, Int. J. Cancer 25:691 (1980).

32. A. Mantovani, G. Peri, N. Polentarutti, P. Allavena, C. Bordignon, C. Sessa, and C. Mangioni, Natural cytotoxicity on tumor cells of human monocytes and macrophages, in "Natural Cell-Mediated Immunity Against Tumors", R.B. Herberman, ed., Academic Press, New York (1980).
33. J.R. Ortaldo, G.D. Bonnard, and R.B. Herberman, Cytotoxic reactivity of human lymphocytes cultured in vitro, J.Immunol. 119: 1351 (1977).
34. D.J. Cameron, and W.H. Churchill, Cytotoxicity of human macrophages for tumor cells. Enhancement by human lymphocyte mediators, J. Clin. Invest. 63:977 (1979).
35. D.J. Cameron, and W.H. Churchill, Cytotoxicity of human macrophages for tumor cells: Enhancement by bacterial lipopolysaccharides (LPS), J. Immunol. 124:708 (1980).
36. A. Mantovani, V. Caprioli, P. Gritti, and F. Spreafico, Human mature macrophages mediate antibody-dependent cellular cytotoxicity on tumour cells, Transplantation 24:291 (1977).
37. G.A. Currie, and D.W. Hedley, Monocytes and macrophages in malignant melanoma. I. Peripheral blood macrophage precursors, Br. J. Cancer 36:1 (1977).

DISCUSSION

Edited by A. Mantovani

Tedesco

Do you see any reduction in the background tumoricidal activity of human macrophages when they are grown for sometime in culture?

Mantovani

Monocytes maintained *in vitro* for up to 10 days under suitable culture conditions maintain spontaneous cytotoxicity and responsiveness to lymphokines and interferon. Later, when cultures consist primarily of epithelioid and giant cells, little baseline cytotoxicity and little responsiveness to stimuli was observed.

Ruco

Did you test functions other than tumoricidal activity in peritoneal and alveolar macrophages?

Mantovani

In previous studies, we observed that peritoneal macrophages could mediate ADCC on tumor cells, and they were better effectors than blood monocytes. We have no data as far as lung macrophages are concerned.

Ruco

Were the lymphokine activation experiments performed by pulsing the cells or by adding the lymphokines during the assay?

Mantovani

We observed activation of human mononuclear phagocytes both by preincubating the effectors with stimuli and by adding stimuli to the assay system. The levels of activation were comparable.

STUDIES ON THE RECOGNITION OF XENOGENEIC CELLS BY NONIMMUNE MACROPHAGES. II. SEPARATE SIGNALS TRIGGERED CYTOSTASIS AND CYTOLYSIS

Ruth Gallily and Shmuel Cabilly

The Lautenberg Center for General & Tumor Immunology
Hebrew University-Hadassah Medical School
Jerusalem, Israel

INTRODUCTION

Macrophages are capable of a wide spectrum of foreign recognition which enables them to carry out their role in protecting the organism against foreign invasion. Their discrimination between foreign material and self components is expressed by their ability to recognize as foreign and engulf viruses, bacteria, protozoa and inert particles. Macrophages have also been found to exert a cytostatic and cytotoxic effect on a variety of target cell types. Following activation of macrophages by a variety of conditions and agents (chronic infections of the host, lymphokines, endotoxins and serum factors), these active cells become tumoricidal. They acquire the capacity to discriminate between normal and malignant cells and are capable of destroying the tumor cells (1,2). It was also reported that immune or armed macrophages, derived from specifically immune hosts were cytostatic and/or cytotoxic *in vitro* against specific tumor or normal cells (3-5).

Our aim in this study was to ascertain whether macrophages are capable of recognizing and killing foreign cells without either prior immunization or activation. We argued that macrophages are descendents of primitive phagocytic cells which recognize and destroy foreign cells in the invertebrate (molluska, annelides, arthropods and echinoderms) (6). It seemed reasonable to assume that some primitive properties of the invertebrate phagocyte still reside in the mammalian macrophage. We further argued that if such a property is present it might be manifested in an experimental system in which a wide phylogenetic distance exists between effector and target cells.

To study this hypothesis we employed as target cells, fibroblasts derived from chicken embryos whereas the effector cells were peritoneal macrophages of C57Bl/6 mice. Macrophages were usually collected after thioglycollate elicitation with neither previous activation nor immunization.

RESULTS AND DISCUSSION

In the first series of experiments the cytostatic effect of murine macrophages on chicken target cells was studied. Macrophages and fibroblasts were cocultivated for 24 or 48 h and ^{3}H-thymidine (^{3}HTdR) was added to the mixed cultures during the final 3 h of the experiment (7). As can be seen in Fig. 1A the incorporation of ^{3}HTdR by chicken fibroblasts was markedly inhibited when cocultivated with murine macrophages (solid line, ratio E:T - 5:1) -- compared to fibroblasts cultivated alone (broken line). Proliferation of allogeneic BALB/c fibroblasts and syngeneic C57Bl/6 fibroblasts (ratio 5:1) was not affected at all following interaction with C57Bl/6 macrophages (Fig. 1B,C). Cytostasis was also expressed by unelicited macrophages and by macrophages depleted of T lymphocytes following treatment with anti-Θ serum.

Thus, murine macrophages recognize chicken fibroblasts as foreign and following interaction with these cells inhibit their proliferation - a phenomenon defined as cytostasis.

The next set of experiments demonstrated that nonimmune nonactivated macrophages are capable of killing chicken target cells (xenolysis). Chicken fibroblasts were cultured in flasks and labeled with 1 μgCi/ml ^{3}HTdR for 24 h during the log phase of growth. The labeled cells were removed and subcultured in 30 mm Petri dishes for 24 h. Macrophages were then added and cocultivated proceeded for 24 or 48 h. Lysis was assessed by measuring the radioactivity of fibroblasts cultured alone or with macrophages (residual labeling). In the xenogeneic mixed cultures of C57Bl/6 macrophages and chicken fibroblasts (ratio 8:1), destruction of target cells was apparent after 24 h of cocultivation (Fig. 2A). However, lysis was not seen when allogeneic or syngeneic target fibroblasts were cocultivated under the same conditions with macrophages (Fig. 2B,C). Hence, macrophages recognize and destroy xenogeneic target fibroblasts without previous activation or immunization.

We further investigated the phylogenic range of macrophage foreign recognition and killing capacity by determining the potential of the macrophages to kill fibroblasts derived from various species. As can be seen in Table 1, C57Bl/6 macrophages kill avian fibroblasts of three species studied with profound efficiency, whereas rat and human fibroblasts were destroyed to a smaller degree (35-36%). In these experiments human fibroblasts were lysed when E:T ratio was 20:1, whereas rat fibroblasts were destroyed when E:T ratio was 10:1.

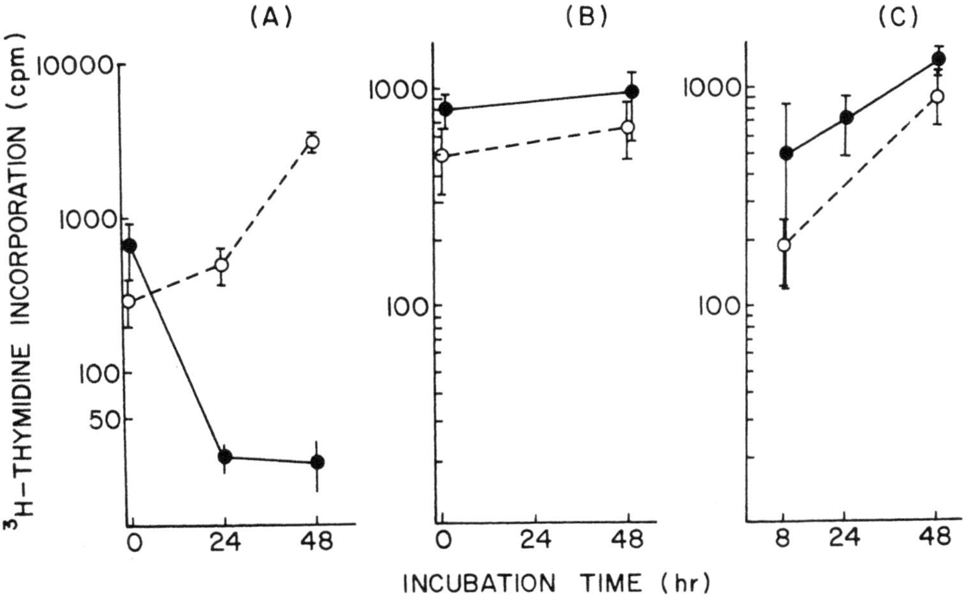

Fig. 1. Macrophage-mediated cytostasis incorporation of ^{3}HTdR by: (A) chicken fibroblasts; (B) BALB/c fibroblasts; and, (C) C57Bl/6 fibroblasts. Fibroblasts cultivated alone (o----o), and with C57Bl/6 macrophages (●——●). Ratio of E:T - 5:1.

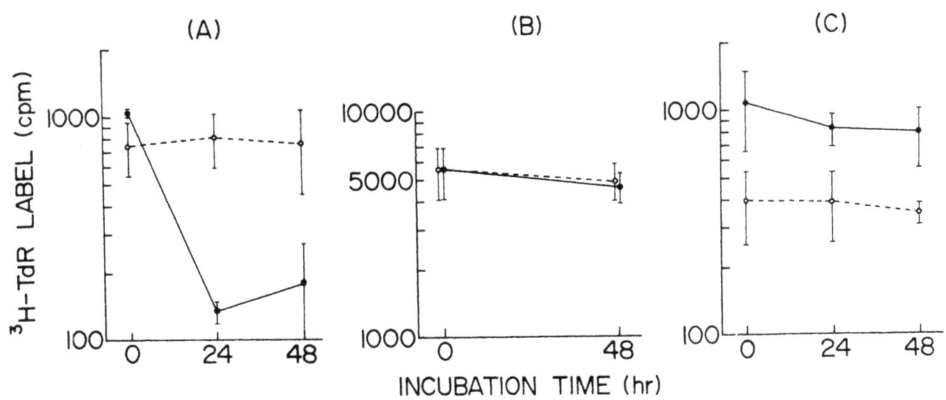

Fig. 2. Macrophage-mediated cytolysis. Fate of ^{3}HTdR prelabeled target cells: (A) chicken fibroblasts; (B) BALB/c fibroblasts; (C) C57Bl/6 fibroblasts. Fibroblasts cultivated alone (o----o), fibroblasts cultivated with C57Bl/6 macrophages (●——●). Ratio E:T (A) 8:1; (B) 6:1; (C) 15:1.

Table 1. Cytolysis of Foreign Fibroblasts by C57Bl/6 Macrophages

Origin of Target Fibroblasts	Number of Experiments	Effector:Target (E:T) Ratio	Fibroblast[a] Lysis %
C57Bl/6	10	10:1	0 ± 11
BALB/c	3	10:1	6 ± 0.5
Rat	6	10:1	35 ± 12
Human	4	20:1	36 ± 16
Chicken	20	10:1	75 ± 20
Chicken (SPEFAS)	3	10:1	69 ± 6
Duck	3	10:1	89 ± 5
Goose	3	10:1	82 ± 7

[a]After 24 h; mean values ± S.D.

It should be noted, however, that the human fibroblasts in our experiments, were much bigger than the embryonic rat cells; accordingly they might inherently serve as "tougher" targets.

To find our whether macrophages from other murine strains possess similar xenolytic activity, we incubated macrophages from various inbred strains of mice with chicken fibroblasts. As can be seen in Table 2, macrophages from each of the 7 strains studied had the capacity to kill the xenogeneic cells, although this killing varied in its intensity. As can also be seen in this Table, the binding efficiency of macrophages to the target cells correlate well with this killing intensity. These findings indicate that genetic background controls the xenolytic activity of macrophages. It is suggested that the binding of macrophages to targets represents a foreign recognition phase which might, under specific conditions, regulate the efficiency of target cell lysis. Indeed, morphological studies of the interaction between mouse macrophages and chicken fibroblasts demonstrated a close contact and aggregate formation between effectors and targets (see Fig. 3b).

We believe that contact between the effector macrophages to

target cells is a prerequisite for macrophages to exert their cytostatic and cytolytic activities against the targets. It was of interest therefore, to determine whether close contact between macrophages and syngeneic fibroblasts can, by itself, trigger cytostatic and/or cytolytic activity of macrophages, or whether an additional signal of foreign recognition is required for macrophages to express these potentials. To explore this problem, we artificially bound C57Bl/6 macrophages to C57Bl/6 fibroblasts by either Con A or rabbit anti-macrophage serum (AMS). Con A agglutinates cells by their D-mannose and D-glucose residues, while AMS agglutinates mouse cells by their common species antigens.

As can be seen in Table 3, macrophages artificially bound to syngeneic fibroblasts (by either Con A or AMS) markedly inhibited incorporation of ^{3}HTdR by syngeneic fibroblasts. However, this cytostasis was not followed by cytolysis (as seen in Table 4). These findings indicate that intimate binding by itself causes only cytostasis. Cytolysis on the other hand, necessitates a previous foreign recognition step for its expression. Hence, the two events, cytostasis and cytolysis, are triggered by two separate signals.

Table 2. Binding to and Lysis of Chicken Fibroblasts by Macrophages of Various Strains of Mice.

Strain	H-2 Haplotype	Adherance of MØ[a] to Fibroblasts %	Relative[b] Fibroblast Lysis %
C57Bl/6	b	87	100
BALB/c	d	68	73
CBA	k	72	77
AKR	k	72	72
A/J	a	66	66
C3H/CRGL	k	58	66
C3H/HeJ	k	51	63

[a]Expressed as percentage of ^{51}Cr labeled macrophages adhered to plastic.

[b]Expressed as percentage of lysis by C57Bl/6 macrophages.

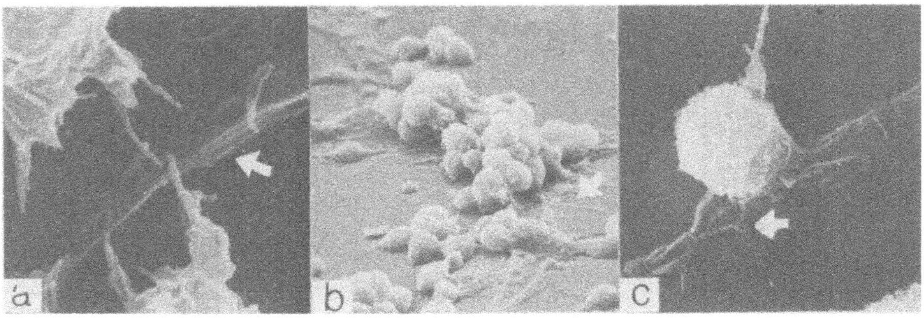

Fig. 3. Interaction of murine macrophages with chicken fibroblasts in mixed cultures.
(a) unidirectional macrophage filopodia towards the targets (2-3 h of cocultivation x 5000); (b) clusters of macrophages around and above the target fibroblasts (x 650); (c) lysis of target fibroblasts incubated with macrophages (24 h of cocultivation x 2000).

To shed some light on the mechanisms of foreign recognition and killing by macrophages, we adopted several approaches. To find out whether lymphocytes or their products assist in xenolysis, we treated peritoneal exudate cells with ALS. Killing was unaffected by this treatment. Similarly, nude macrophages deprived of T cells killed the targets as did pure population of macrophages (obtained following treatment of PEC with versane-trypsin and cultivation for 24 h in vitro in serum-free medium). These data indicate that macrophages possess the capacity to recognize and kill foreign normal cells without the help of lymphocytes or their products.

Is this foreign recognition mediated via viruses? The answer is no, since SPEFAS chicken fibroblasts free of viruses are destroyed by the macrophages (Table 1). It was also demonstrated that macrophages had to be viable in order to kill, since lysates of macrophages lack xenolytic activity. Furthermore, no cytotoxic substance is secreted by macrophages to the media following interaction with avian fibroblasts as supernatants of the mixed cultures lack any cytotoxic property.

We tried to block hypothetical foreign recognition sites on macrophage surfaces by adding anti-H-2 serum and AMS to the macrophages prior to and during their interaction with avian fibroblasts. This treatment did not affect xenolysis. Similarly, the addition of various carbohydrates to the mixed cultures did not decrease the killing. (The added compounds were: glucose, fructose, galactose, mannose, rhamnose arabinose, xylose, N-acetyl-D-glucosamine in concentration of 10 mM).

Table 3. Cytostatic Effect of Macrophages Artificially Bound to Syngeneic Fibroblasts

Cells in Culture[a]	Treatment	Cultures[b] ^{3}HTdR Incorporation[c] (cpm ± S.D.)	Inhibition %
fibroblasts	---	49275 ± 1046	
fibroblasts macrophages	---	55877 ± 3170	0
fibroblasts	Con A	53434 ± 3415	
fibroblasts macrophages	Con A	11754 ± 2295	78
fibroblasts	NRS	33732 ± 4178	
fibroblasts macrophages	NRS	33132 ± 3698	2
fibroblasts	AMS	45003 ± 5716	
fibroblasts macrophages	AMS	8110 ± 5716	82

[a] 3×10^5 mouse fibroblasts/plate. Macrophage-fibroblasts (E:T) 10:1.

[b] Mean of triplicates.

[c] After 20 h of cocultivation, 2 μCi/ml ^{3}HTdR was added for 3 h to the cultures.

To explore further the xenolytic activity we noted that macrophages derived from C3H/HeJ mice (a strain which does not respond to LPS), efficiently killed chicken fibroblasts (Table 2); whereas, they lack tumoricidal activity (8). Furthermore, unlike macrophage tumoricidal activity which requires specific serum factors for its expression (9), macrophage xenolytic activity could be demonstrated in each of the sera studied (rat, guinea pig, rabbit, goat, bovine, horse and human). It was also observed that macrophages retained their xenolytic potential after growing *in vitro* for 5 days in medium without serum, whereas activated macrophages lost their tumoricidal expression following *in vitro* cultivation for 24 or 48 h.

Table 4. Lack of Cytotoxic Effect of Macrophages Artificially Bound to Syngeneic Fibroblasts

Cells in[a] Culture	Treatment	Cultures[b] Residual Label[c] (cpm ± S.D.)	Cultures[b] Percent Cytotoxicity
fibroblasts	---	18531 ± 865	
fibroblasts macrophages	---	19580 ± 1635	0
fibroblasts	Con A	18382 ± 2366	
fibroblasts macrophages	Con A	18936 ± 1876	0
fibroblasts	NRS	108081 ± 29556	
fibroblasts macrophages	NRS	90746 ± 3082	16
fibroblasts	AMS	121614 ± 2877	
fibroblasts macrophages	AMS	111289 ± 14553	8

[a] 3×10^5 mouse fibroblasts/plate. Macrophages-fibroblasts (E:T) 10:1.

[b] Mean of triplicates.

[c] Cells were harvested 20 h after cocultivation of macrophages with labeled fibroblasts.

All these data support the concept that macrophage-mediated xenolysis differs in many respects from the tumoricidal potential of activated macrophages.

Additionally, our data do not support the claim that macrophage lysosomal enzymes participate actively in macrophage-mediated killing. Treatment of macrophages with agents known to affect the activity and secretion of these enzymes did not interfere with the killing. The agents tested were gold salts (1000 μg/ml), hydrocortisone (100 μg/ml) and colchicine (100 μg/ml). The only tested agent which inhibited xenolysis was cytochalasin B in concentration of 10 μg/ml. Cytochalasin B probably exerts its effect by inhibiting both macrophage locomotion and extension of cytoplasmic filopodia towards the target fibroblasts.

Indeed, morphological studies by SEM demonstrated that 1-2 h after initial interaction of macrophages with the targets, filopodia were unidirectionally extended towards the targets establishing a firm contact with the target cell membranes (Fig. 3a). After being formed, this contact could not be detached, either by trypsin or cytochalasin B. Following filopodia contact, macrophages are "pulled" toward the fibroblasts forming aggregates (Fig. 3b) which precedes target lysis (Fig. 3c). Such aggregates were not observed when macrophages were cocultivated with allogeneic or syngeneic fibroblasts.

In conclusion, our findings demonstrate that contact between effectors and targets is prerequisite for cytolysis to occur, although contact of effector to target alone is not sufficient to trigger lysis. For lysis to occur, a foreign recognition signal is required to activate the xenolytic potential of macrophages. Furthermore, our study shows that macrophages have an innate capacity to recognize phylogenetically distant target cells without any assistance from lymphocytes or their products.

We suggest that this activity is a clue to the primitive origin of macrophages. This activity, which is manifested by the invertebrate phagocytes prevails in the mammalian macrophages during evolution, probably because it assists the body in its defense against foreign invaders which might include multicellular parasites.

REFERENCES

1. J.B. Hibbs, Jr., Discrimination between neoplastic and non-neoplastic cells _in vitro_ by activated macrophages, J. Natl. Cancer Inst. 53:1487 (1974).
2. W.F. Piessens, W.H. Churchill, Jr. and I.R. David, Macrophages activated _in vitro_ with lymphocyte mediators kill neoplastic but not normal cells, J. Immunol. 114:293 (1975).

3. R. Evans and P. Alexander, Mechanism of immunologically specific killing of tumor cells by macrophages, Nature (Lond.) 236:168 (1972).
4. M.L. Lohmann-Matthes, H. Schipper and H. Fischer, Macrophage-mediated cytotoxicity against allogeneic target cells in vitro, Eur. J. Immunol. 2:45 (1972).
5. R. Gallily and H. Eliahu, Mechanism and specificity of macrophage-mediated cytotoxicity, Cell Immunol. 25:245 (1976).
6. E.L. Cooper, ed.; Invertebrate Immunology. Contemporary Topics in Immunology, Vol. 4, Plenum Press, New York and London (1974).
7. S. Cabilly and R. Gallily, Studies on the recognition of xenogeneic cells by nonimmune macrophages. I. Destruction of chicken fibroblasts in vitro by murine macrophages, Cell Immunol. 29:54 (1977).
8. L.P. Ruco, M.S. Meltzer and D.L. Rosenstreich, Macrophage activation for tumor cytotoxicity: control of macrophage tumoricidal capacity by the LPS gene, J. Immunol. 121:543 (1978).
9. J.B. Hibbs, Jr., R.R. Taintor, H.A. Chapman, Jr. and J.B. Weinberg, Macrophage tumor killing: Influence of the local environment, Science 197:279 (1977).

GENETIC ASPECTS OF MACROPHAGE ACTIVATION FOR TUMOR CYTOTOXICITY

Diana Boraschi

Istituto di Ricerche Farmacologiche "Mario Negri"
MILANO, Italy

It is well established that macrophages play a major role in the resistance to tumors and infections[1]. In view of this fact, special importance is assumed by the understanding of the mechanism leading macrophages to the aquisition of cytotoxic activity for tumor cells or parasites. Recently, several studies on macrophage activation for tumor cytotoxicity provided evidence that development of macrophages with nonspecific tumoricidal activity is not the result of a single interaction between cell and activation signal, yet it requires the completion of a sequence of reactions[2-4].

An important tool for a deeper analysis of the intermediary reactions of macrophage activation is provided by genetic models of macrophage defective mouse strains. The first genetic model studied in this sense was the C3H/HeJ mouse. This mouse is refractory to many biological effects of the lipid-A moiety of bacterial lipopolysaccharides (LPS)[5]. Furthermore, C3H/HeJ macrophages do not become cytotoxic for tumor cells following *in vivo* or *in vitro* stimulations which are active for macrophages of the genetically related but lipid-A responsive C3H/HeN strain[6]. However, C3H/HeJ macrophages can develop tumoricidal activity following the interaction with two activation stimuli, given in a precise sequence[7]. The importance of C3H/HeJ mice as tool for the analysis of the activation mechanisms prompted us to the identification of other strains with defective macrophage activation. The study of all such strains would have provided useful information not only on the mechanism,but also on the genetic control of macrophage activation.

A survey of 38 mouse strains demonstrated that there are at least three groups of mice which are not able to develop cytotoxic macrophages after *in vivo* stimulation with BCG (Table 1).

The first group includes the C3H/HeJ strain and other two lipid-A unresponsive strains (C57BL/10ScCR and C57BL/10ScN). The other groups are the A-derived strains (A/J, A/HeJ, etc.) and the P (P/J, P/N) mice. The macrophage defects of A/J and P/J have been analyzed and compared to that of C3H/HeJ mice in the attempt to characterize both phenotypically and genetically the two new defective strains. Macrophages of A/J and P/J mice, as well as cells from C3H/HeJ mice, failed to develop tumoricidal activity following a variety of *in vivo* and *in vitro* stimuli that were active for macrophages of responder C3H/HeN mice (Table 2).

It has been described that inflammatory reactions and the accumulation of lymphokine-responsive mononuclear phagocytes at the inflammation site are necessary events for macrophage activation *in vivo*[11]. Inflammatory events were therefore examined in the defective strains. As shown in Table 3, inflammatory macrophages of A/J and P/J strains were compared to cells of C3H/HeN and C3H/HeJ mice for their ability to respond to chemotactic stimuli, to phagocytize IgG-opsonized erythrocytes and to express the cytochemical marker peroxidase. The presence of such characteristics is consist-

Table 1. Macrophage tumoricidal activity among mouse strains

High Activity		Variable	Low Activity
AKR/N	C58/LwN	A/WySn	A.BY/Sn
B10.A/ScSn	DBA/1J	BALB/cAnN	A.θAKR/N
B10.D2/nSn	DBA/2N	BDP/J	A/HeJ
B10.D2/oSn	NBL/N	RIII/AnN	A/HeN
CBA/CaHN	NIH Swiss		A/J
CBA/N	NGP/N		AL/N
C3HeB/FeJ	NH/LwN		
C3H/HeN	NZB/N		C3H/HeJ
C57BL/6CR	NZW/N		C57BL/10ScCR
C57BL/6J			C57BL/10ScN
C57BL/6N			
C57BL/10KaLwN			P/J
C57BL/10J			P/N
C57BL/10Sn			

Peritoneal macrophages from mice inoculated i.p. 7 days previously with viable BCG were cultured for 2 h at 37°C. Non-adherent cells were then removed by repeated washing and adherent cells were cultured for 48 h in presence of ^{3}H-TdR prelabeled tumor cells at a ratio of 20:1. Cytotoxicity was estimated by measurement of ^{3}H-TdR release at 48h[8].

Table 2. Defective tumoricidal activity of macrophages from C3H/HeJ, A/J and P/J mice

Activation stimulus	Macrophage tumoricidal activity			
	C3H/HeN	C3H/HeJ	A/J	P/J
In vivo				
BCG	100	7	6	3
C.parvum	100	10	4	3
Pyran	100	3	10	4
In vitro				
Lymphokines	100	7	9	7
PHA	100	9	9	10
ConA	100	8	4	6
LPS	100	4	4	11

Results are from a representative experiment. For *in vitro* activation,adherent peritoneal cells were exposed for 4h to lymphokines (1/3 dilution),PHA (20 μg/ml),ConA(2.5 μg/ml) or LPS (100 ng/ml;for C3H/HeJ 10 μg/ml)[6,9,10]. The activating stimulus was washed off prior to the addition of target cells. For *in vivo* activation,BCG($5x10^6$ viable CFU),*C.parvum* (1.4 mg) or Pyran Copolymer(50 mg/kg)were inoculated i.p. 7 days previously[9,10]. Macrophage to target ratio was 20:1.Cytotoxicity values,measured as ^{3}H-TdR release from prelabeled tumor cells at 48h, are expressed as percentage of C3H/HeN activity.

ent with the development of normal inflammatory responses *in vivo* [10,11,14]. Inflammatory events seem to take place normally in A/J mice,since the inflammation characteristics studied are comparable to those of control strains. In contrast,P/J macrophages showed a profound abnormality in phagocytic capacity. This phagocytic defect suggests that inflammatory events are at least partially abnormal in P/J mice: thus,their macrophage tumoricidal defect may be different from those of C3H/HeJ and A/J mice.

The second step of the macrophage activation process was then analyzed,i.e. the production by lymphocytes of activating factors and the interaction of such factors with inflammatory macrophages[4]. Table 4 shows that lymphocytes of C3H/HeJ and of A/J mice produced a fully active macrophage activating lymphokine(MAF)[7,13] .

Table 3. Inflammatory macrophages in defective strains

Mouse strain	% Peroxidase positive cells[a]	^{51}CrEAIgG phagocytosis	Chemotactic response[c]
C3H/HeN	42	21.2	5.0
C3H/HeJ	51	19.7	4.1
A/J	65	16.1	8.0
P/J	55	0.6	4.8

Results are from a representative experiment, where inflammation was induced by i.p. injection of BCG 7 days previously.

[a]Peroxidase positive macrophages were identified by the method of Kaplow[12].

[b]Macrophages were assayed for phagocytic capacity as previously described[13]. Phagocytosis was expressed by the following formula :

$$\frac{\text{cpm in inflammatory macrophages}}{\text{cpm in unstimulated macrophages}}$$

[c]Ability of macrophages to migrate in response to the chemotactic lymphokine LDCF (diluted 1/4) was measured as previously described[10].

Chemotactic response was expressed by the following formula:

$$\frac{\text{migrated cells in presence of LDCF}}{\text{migrated cells in absence of LDCF}}$$

In contrast, P/J lymphocytes showed a profound defect in MAF production[15]. Thus, P/J mice have been tested for the ability of producing other two lymphokines: the chemotactic factor (LDCF) and immune interferon (IF). Whereas LDCF production was normal, P/J mice appeared defective also for the production of interferon. The selective deficiency of P/J mice in the production of MAF and IF, both important in the induction of macrophage cytolytic activity[16,17], suggests that their macrophage tumoricidal defect might be due in part also to inability to produce activating factors.

The development of activated tumoricidal macrophages requires, beside lymphokine production, also the presence of reactive mononuclear phagocytes[11]. Such cells, by interacting with lymphokines, would become able to react with a second activation signal and thus express cytolytic activity[11]. Hence, A/J and P/J macrophages were tested for their ability to adequately react to the sequence of activation stimuli. Results presented in Table 5 demonstrate that macrophages of A/J mice, prepared by the interaction with a first signal (MAF *in vitro* or BCG *in vivo*), could develop cytotoxic activity when exposed to a second signal (e.g. LPS). A/J macrophages were not different in this from C3H/HeJ cells[7].

Table 4. Lymphokine production in defective strains

Mouse strain	Lymphokine activity[a]		
	MAF[b]	LDCF[c]	IF[d]
C3H/HeN	100	100	100
C3H/HeJ	108	n.t.	n.t.
A/J	105	n.t.	n.t.
P/J	7	101	5

Results are from a representative experiment.
[a]Lymphokine activities were measured in supernatants of spleen lymphocytes incubated with ConA for 48h[10].
[b]Macrophage activating factor (MAF) production was measured and expressed as described in Table 2.
[c]Chemotactic activity was expressed as percentage of C3H/HeN activity .
[d]Interferon was measured by viral plaque reduction assay. IF activity was expressed as percentage of C3H/HeN values.
n.t. not tested

Table 5. Macrophage tumoricidal activity after multiple activation stimuli

In vivo treatment	*In vitro* treatment	Tumoricidal Activity of Macrophages from:			
		C3H/HeN	C3H/HeJ	A/J	P/J
PBS	none	16	6	0	5
	MAF	100	11	10	10
	LPS	106	14	0	8
	MAF + LPS	106	90	110	105
BCG	none	100	7	16	2
	MAF	100	105	115	7
	LPS	100	94	117	7
	MAF + LPS	104	109	n.t.	9

Results are from a representative experiment.
Tumor cytotoxicity values were expressed as percentage of C3H/HeN MAF- or BCG-induced activity,as described in Table 2.
n.t. not tested

Whereas cells from P/J mice were also activated by a sequence of signals provided *in vitro*, macrophages from BCG-infected P/J mice were refractory to any further stimulation. Here again, while C3H/HeJ and A/J defects are phenotypically similar, P/J macrophages show distinct features. Through functional analysis it was therefore impossible to distinguish between C3H/HeJ and A/J macrophage defects.

It has been described that the tumoricidal defect of C3H/HeJ macrophages is controlled by a gene closely linked or identical to the *Lps* gene (responsible for unresponsiveness to LPS)[18]. Therefore three functions controlled by the *Lps* gene were analyzed in A/J mice. As shown in Table 6, macrophage tumoricidal activity and

Table 6. Functions controlled by the *Lps* gene in defective strains

Mouse strain	Macrophage tumoricidal activity[a]	LPS-induced lymphocyte proliferation[b]	LPS toxicity to macrophages[c]
C3H/HeN	100	100	100
C3H/HeJ	3	3	3
A/J	8	5	96
P/J	2	97	87

[a]Tumor cytotoxicity by macrophages of BCG-infected mice was measured and expressed as described in Table 2.
[b]Spleen lymphocyte proliferation was measured by ^{3}H-TdR incorporation as previously described[10,13] and expressed as percentage of C3H/HeN response.
[c]Macrophage death *in vitro* in presence of LPS was measured as previously described[10,13] and expressed as percentage of C3H/HeN response.

lymphocyte proliferative response to LPS are both defective in A/J mice, and comparable to the defective responses of C3H/HeJ mice. In contrast to C3H/HeJ, however, A/J macrophages are very sensitive to the toxic effect of LPS *in vitro*, response also controlled by the *Lps* gene[19]. Conversely, P/J mice are completely sensitive to LPS. Their macrophage defect appears therefore distinct from that of C3H/HeJ also genetically.

By genetic analysis, it was possible to demonstrate that C3H/HeJ and A/J macrophage defects (phenotypically indistinguishable) had different genetic basis. In fact, hybrids between the two defective strains (A/JxC3H/HeJ)F_1 mice, were able to develop tumoricidal macrophages following *in vivo* stimulation with BCG[20].

Table 7. Genetic control of tumoricidal and proliferative defects in A/J mice

Mouse strain	Macrophage cytotoxicity[a]		Lymphocyte proliferation to LPS[b]	
	% responsive	% non-responsive	% responsive	% non-responsive
C3H/HeN	100	0	100	0
A/J	0	100	0	100
(A/JxC3H/HeN)F_1	100	0	100	0
A/JxF_1	80	20	48	52

[a] 30 to 57 mice/strain were individually tested for BCG-induced macrophage cytotoxicity,as described in Table 2.

[b] 11 to 16 mice/strain were individually tested for spleen lymphocyte proliferative response to LPS, as described in Table 6.

The genetic control of the two defects in A/J mice (i.e.macrophage cytotoxicity and proliferative response to LPS) was studied by the use of backcross mice (Table 7).

The macrophage tumoricidal activity of 20% of the tested A/Jx(A/JxC3H/HeN) backcross mice was defective and comparable to that of the A/J parental strain, while 80% of the mice were able to develop cytotoxic macrophages,similarly to the hybrid parent. The expected percentage of backcross mice with the parental trait is equal to $(\frac{1}{2})^n$,where n is the number of genes controlling the character[21]. It is therefore possible to conclude that the A/J tumoricidal defect is controlled by at least two genes (the predicted percentage would be 25%). Conversely,the lymphocyte proliferation defect in response to LPS appears under single gene-control. In fact, 52% of backcross mice are unresponsive like A/J mice(the expected percentage for a one gene-controlled trait is 50%). By backcross linkage analysis, a clear dissociation between the two defects was then demonstrated[20].

In conclusion,the identification of mouse strains defective for macrophage cytotoxicity has provided new useful tools for the analysis of intermediary reactions on macrophage activation and of the genetic control of macrophage cytotoxic activity. The A/J macrophage defect is genetically controlled by two autosomal genes [20] distinct from the *Lps* gene, which is responsible for the C3H/HeJ defect. Hence, the identification of the phenotypical difference between the two strains would provide important information on the intermediary reactions of macrophage activation.

P/J macrophage tumoricidal defect is of particular interest. In fact, such defect is both phenotypically and genetically[15] different from that of C3H/HeJ mice and involves different steps of the macrophage activation pathway: development of normal inflammatory response, production of macrophage activating factors, normal macrophage reactivity to activating stimuli. A very powerful tool for the dissection of the genetic control of P/J defects and of their relative importance in the development of activated macrophages would be provided by recombinant inbred (RI) strains[22]. RI strains derived from P/J mice (work in progress) would allow to analyze each defect separately, to study their linkage and finally to understand their *in vivo* importance in susceptibility to tumors.

REFERENCES

1. D.S. Nelson ed., "Immunobiology of the Macrophage", Academic Press, New York (1976).
2. J.B. Hibbs Jr., R.R. Taintor, H.A. Chapman Jr. and J.B. Weinberg, Macrophage tumor killing : Influence of the local environment, Science 197: 279 (1977).
3. S.W. Russell, W.F. Doe, and A.T. McIntosh, Functional characterization of a stable, noncytolytic stage of macrophage activation in tumors, J. Exp.Med. 146: 1511(1977).
4. L.P. Ruco, and M.S. Meltzer, Macrophage activation for tumor cytotoxicity : development of macrophage cytotoxic activity requires completion of a sequence of short-lived intermediary reactions, J. Immunol. 121: 2035 (1978).
5. J. Watson and R. Riblet, Genetic control of responses to bacterial lipopolysaccharides in mice. I. Evidence for a single gene that influences mitogenic and immunogenic responses to lipopolysaccharides, J. Exp. Med. 140:1147 (1974).
6. L.P. Ruco, and M.S. Meltzer, Defective tumoricidal capacity of macrophages from C3H/HeJ mice, J. Immunol. 120: 329 (1978).
7. L.P. Ruco, and M.S. Meltzer, Macrophage activation for tumor cytotoxicity : Tumoricidal activity by macrophages from C3H/HeJ mice requires at least two activation stimuli, Cell. Immunol. 41: 35 (1978).
8. D. Boraschi, and M.S. Meltzer, Macrophage activation for tumor cytotoxicity: genetic variation in macrophage tumoricidal capacity among mouse strains, Cell. Immunol. 45: 188 (1979).
9. D. Boraschi, and M.S. Meltzer, Defective tumoricidal capacity of macrophages from A/J mice. I. Characterization of the macrophage cytotoxic defect after in vivo and in vitro activation stimuli, J.Immunol. 122: 1587 (1979).
10. D. Boraschi, and M.S. Meltzer, Defective tumoricidal capacity of macrophages from P/J mice : Characterization of the macrophage cytotoxic defect after in vivo and in vitro activation stimuli, J. Immunol. 125: 771 (1980).

11. L.P. Ruco, and M.S. Meltzer, Macrophage activation for tumor cytotoxicity : Increased lymphokine responsiveness of peritoneal macrophages during acute inflammation,J.Immunol. 120: 1054 (1978).
12. L.S. Kaplow, Simplified myeloperoxidase stain using benzidine dihydrochloride, Blood 26: 215 (1965).
13. D. Boraschi, and M.S. Meltzer, Defective tumoricidal capacity of macrophages from A/J mice. II. Comparison of the macrophage cytotoxic defect of A/J mice with that of lipid A-unresponsive C3H/HeJ mice, J. Immunol. 122:1592 (1979).
14. R. Van Furth, M.M.C. Dusselhoff-Den Dulk, and H. Mattie, Quantitative study on the production and kinetics of mononuclear phagocytes during an acute inflammatory reaction, J. Exp. Med. 138: 6 (1973).
15. D. Boraschi, and M.S. Meltzer, Defective tumoricidal capacity of macrophages from P/J mice : Tumoricidal defect involves abnormalities in lymphokine-derived activation stimuli and in mononuclear phagocyte responsiveness, J. Immunol. 125: 777 (1980).
16. R. Evans, and P. Alexander, Rendering macrophages specifically cytotoxic by a factor released from immune lymphoid cells, Transplantation 12: 227 (1971).
17. D. Boraschi, and A. Tagliabue, Type I interferon-induced enhancement of macrophage-mediated tumor cytolysis and its difference from activation by lymphokines, Eur.J.Immunol. in press.
18. L. P. Ruco, M.S. Meltzer, and D.L. Rosenstreich, Macrophage activation for tumor cytotoxicity : Control of macrophage tumoricidal capacity by the LPS gene, J. Immunol. 121: 543 (1978).
19. L.M. Glode, A. Jacques, S.E. Mergenhagen, and D.L. Rosenstreich, Resistance of macrophages from C3H/HeJ mice to the in vitro cytotoxic effects of endotoxin, J. Immunol. 119: 162(1977).
20. D. Boraschi, and M.S. Meltzer, Defective tumoricidal capacity of macrophages from A/J mice. III. Genetic analysis of the macrophage defect, J. Immunol. 124: 1050 (1980).
21. C.C. Little, A possible mendelian explanation for a type of inheritance apparently non-Mendelian in nature, Science 40: 904 (1914).
22. D. W. Bailey, Recombinant-inbred strains. An aid to finding identity, linkage, and function of histocompatibility and other genes, Transplantation 11: 325 (1971).

DISCUSSION
Edited by A. Mantovani

Tedesco

Is there any difference in the binding sites for LPS between the low responder and the normally responder strains of mice?

Boraschi

There are several reports in the literature demonstrating that cells of non-responder mice bind LPS as well as cells from responder mice. The defect of unresponsive mice, therefore, would not be in the ability of binding LPS but at a later step of the activation chain.

O_2-INDEPENDENT KILLING OF GRAM-NEGATIVE BACTERIA BY INTACT GRANULOCYTES. THE ROLE OF A POTENT BACTERICIDAL MEMBRANE-PERTURBING PROTEIN

J. Weiss, O. Stendahl and P. Elsbach

Department of Medicine
New York University School of Medicine
New York, N.Y. 10016

During the past two decades many students of the microbicidal activity of polymorphonuclear leukocytes (PMN) have become increasingly convinced that the important role of the PMN in host defense against bacterial infection depends predominantly on the cell's ability to generate cytotoxic systems that utilize oxygen[1,2]. This view has been strongly supported by the observations on patients with chronic granulomatous disease, whose leukocytes cannot generate these systems[1,2].

The release during phagocytosis of preformed granule-associated antimicrobial substances (mainly proteins) into the phagocytic vacuole is an established phenomenon. However, it has not been settled whether these O_2-independent factors serve an essential antibacterial function, or, as suggested by Klebanoff[1], merely as a back-up system for O_2-dependent antibmicrobial mechanisms, normally operating in the phagocytizing PMN.

The work in our laboratory concerns the study of a potent bactericidal protein, isolated from human and rabbit PMN, that is specific for gram-negative bacteria. The properties of these two proteins which have been purified to near homogeneity will be summarized and the results will be discussed of experiments which suggest that these proteins play a role in the overall antimicrobial activity of the intact PMN that the O_2-dependent systems do not appear to fulfill.

These investigations had their origin in our earlier studies showing that intact PMN and crude fractions of PMN effectively killed gram-negative bacteria such as Escherichia coli without inflicting major structural and functional disorganization[3]. The evidence of a highly discrete bactericidal action of the PMN on E.coli is both structural and functional[4-9]. Thus, E.coli, although no longer capable of colony formation, show no signs of major envelope or cytoplasmic alterations by electron microscopy[6,10], and continue biosynthesis of all their constituents at near normal rates.

The finding that broken cell preparations from resting PMN are no less potently bactericidal than intact cells suggested that O_2-dependent bactericidal systems were not necessary for the killing and that preformed bactericidal agents produced the subtle lesion resulting in bacterial death.

Conventional fractionation of homogenized human or rabbit PMN showed that a crude granule fraction contained most of the bactericidal activity of the whole homogenate[5]. This activity can be solubilized by extraction with strong acid and is destroyed by various proteases[5], suggesting that the agent(s) is(are) protein in nature.

Purification of each of these acid extracts of human or rabbit PMN led to the isolation of a single protein[11,12]. These two proteins are remarkably similar in their molecular and biological properties[11,12]. Both proteins represent about 1% of total PMN protein. The human protein has an M_r of 59,000 daltons[11], the molecular weight of the rabbit protein is 50,000[12], indicating that these proteins are larger than any previously described antibacterial granule protein. Both proteins are strongly basic (pI > 9.6) and relatively heat resistant [11,12]. The amino acid composition of the two proteins is quite similar[12]. Antisera prepared against each of the two antigens partially cross-react.

Biological properties (Table 1)

The two proteins produce virtually the same dose dependent effects on susceptible gram-negative bacteria: 1) Loss of ability to multiply; 2) A discrete increase in permeability, resulting in the passage of the normally impermeant hydrophobic antibiotics Actinomycin D and Rifampic [4-12], 3) Activation of a bacterial envelope phospholipase A_1[7,12].

The envelope alterations appear restricted to the outer membrane of the gram-negative bacterial envelope. Thus, the permeability barrier to Actinomycin D and Rifampicin is in the outer membrane[13] and the phospholipase A_1 is an outer membrane enzyme[14]. In contrast, the inner (cytoplasmic) membrane retains its integrity as judged by the finding that K^+ fluxes are not affected by bactericidal protein concentrations and that the biochemical machinery remains almost fully active[4-12].

The activation of the bacterial phospholipase A_1 is not needed for the action of the protein on viability and envelope permeability because these effects are also produced in a phospholipase-deficient E.coli mutant[7].

Table 1. Biological Properties of Bactericidal/Permeability Increasing Proteins of PMN

	Human	Rabbit
Bactericidal Specificity	Gram neg.	Gram neg.
Effects on:		
Outer Membrane		
Permeability to Actinomycin D	↑	↑
Activation of bacterial PLA*	+	+
Inner Membrane		
Macromolecular synthesis	-	-
K^+ uptake/leakage	-	-
Entry of ONP-Galactoside	-	-
pH Optimum	7.0	5.5 - 7.0
Binding (to OM)	+	+

*PLA is phospholipase A.

The bactericidal and permeability increasing activities of both proteins are optimal at neutral pH.

Binding

Binding is an absolute requirement for all three biological effects[5,15]. Binding affinity appears to be determined by the structure of the outer membrane of the gram-negative bacterial envelope. Available evidence suggests that the major, if not the sole, determinant of binding affinity is the length of the saccharide chain of the outer membrane lipopolysaccharides. Thus, the longer the saccharide chain, the higher the concentration of the protein in the medium must be to produce a saturating (bactericidal) amount of attached protein[15].

The importance of the outer membrane anatomy as a site of interaction of the protein and the bacterium is indicated by the fact that microbial species, lacking lipopolysaccharides as components of their wall structure, such as a number of gram-positive bacteria and two strains of the fungus Candida are insensitive to high concentrations of the protein[11,12].

Action of the bactericidal/permeability increasing protein in intact PMN

In crude extracts of human and rabbit PMN, the bactericidal/permeability increasing protein accounts for nearly all the bactericidal activity toward numerous strains of E.coli and Salmonella typhimurium[11,12]. Fractions devoid of the protein exhibit little or no activity toward these bacterial species, suggesting that this protein is an important component of the antimicrobial apparatus of the PMN. However, the conditions that prevail within the intact PMN during phagocytosis may either enhance or diminish the antimicrobial action of a given agent. Hence, our observations on the purified bactericidal proteins from human and rabbit PMN, which show a striking potency toward E.coli and S.typhimurium in vitro (approximately five-fold greater than equimolar concentrations of the antibiotic Polymixin B[16]) do not permit a conclusion concerning the contribution of these proteins to killing by the intact PMN of ingested bacteria.

To study the participation of O_2-independent bactericidal mechanisms in the killing by intact PMN of bacteria that are highly susceptible to the action of the purified bactericidal/permeability increasing protein, we compared the killing by rabbit PMN, incubated in room air or in a nitrogen atmosphere, of three types of bacteria[17]: Two strains of S.typhimurium that are sensitive to the purified protein, one a rough strain (395 MR10) that is effectively ingested by PMN without opsonization, and one smooth strain (395 MS) that must be opsonized for ingestion; and gram-positive Staphylococcus epidermidis 160, which is insensitive to the human and rabbit protein.

Anaerobiosis created by N_2-flushing effectively eliminated the typical respiratory burst as measured by hexose monophosphate shunt activity (conversion of ^{14}C-1-glucose to $^{14}CO_2$) or superoxide production. Bacterial ingestion was measured by colony formation in aliquots of whole suspensions of PMN and bacteria (unsonicated). Intracellular bacterial survival was measured in sonicated suspensions. Sonication breaks up PMN, releasing intracellular bacteria without affecting bacterial viability.

The extent and speed of ingestion and killing of both non-opsonized S.typhimurium MR10 and opsonized S.typhimurium MS are the same under aerobic and anaerobic conditions.

Opsonization has no effect on the sensitivity of S.typhimurium MS to killing either by crude PMN fractions or by purified bactericidal/permeability increasing protein, suggesting that opsonization although required for ingestion of this smooth organism, is not needed for effective intracellular killing.

Staphylococcus epidermidis, however, is only effectively killed under aerobic conditions. In the absence of oxygen most of the ingested Staphylococci survive.

We tentatively conclude that the PMN kills S.typhimurium by O_2-independent mechanisms. Furthermore, in the light of the resistance or sensitivity of the test organisms, used in these experiments, to the action of the bactericidal/permeability increasing protein, we consider it likely that this protein is, in fact, the principal bactericidal agent in the intact PMN against these bacteria.

Additional unpublished observations further suggest that the bactericidal activity of the O_2-independent systems toward S.typhimurium 395 is only limited by the ingestive capacity of the PMN.

In summary

The purified bactericidal/permeability increasing protein of human and rabbit PMN is the most potent single bactericidal protein yet described. Preliminary evidence suggests that this agent is primarily involved in the killing of several strains of gram-negative bacteria by intact PMN.

REFERENCES

1. S.J. Klebanoff, Antimicrobial systems of the polymorphonuclear leukocyte, in: "The Phagocytic Cell in Host Resistance," J.A. Bellanti and D.H. Dayton, eds., Raven Press, New York (1975).
2. B.M. Babior, Oxygen-dependent microbial killing by phagocytes, New Engl. J. Med. 298:659;721 (1980).
3. P. Elsbach, P. Pettis, S. Beckerdite and R. Franson, The effect of phagocytosis by rabbit granulocytes on macromolecuar synthesis and degradation in different species of bacteria, J. Bacteriol.115:490 (1973).
4. S. Beckerdite, C. Mooney, J. Weiss, R. Franson, and P. Elsbach, Early and discrete changes in permeability of E.coli and certain other gram negative bacteria during killing by granulocytes, J. Exp. Med.410:396 (1974).
5. J. Weiss, R.C. Franson, S. Beckerdite, K. Schmeidler, and P. Elsbach, Partial characterization and purification of a rabbit granulocyte factor that increases permeability of E.coli, J. Clin. Invest. 55:33 (1975).
6. J. Weiss, K. Schmeidler, S. Beckerdite-Quagliata, R.C. Franson, and P. Elsbach, Reversible envelope effects during and after killing of Escherichia coli by a highly purified rabbit polymorphonuclear leukocyte fraction, Biochim. Biophys. Acta 436:154 (1976).
7. J. Weiss and P. Elsbach, The use of a phospholipase A-less Escherichia coli mutant to establish the action of granulocyte phospholipase A on bacterial phospholipids during killing by a highly purified granulocyte fraction. Biochim. Biophys. Acta 466:23 (1977).
8. C. Mooney and P. Elsbach, Altered phospholipid metabolism in

Escherichia coli accompanying killing by disrupted granulocytes, Infect. Immun. 11:1269 (1975).
9. P. Elsbach, S. Beckerdite, P. Pettis, and R. Franson, Persistence of macromolecular synthesis by Escherichia coli during killing by disrupted granulocytes, Infect. Immun. 9:663 (1974).
10. A.J. Van Houte, P. Elsbach, A. Verkleij, and J. Weiss, Killing of Escherichia coli by a granulocyte fraction occurs without recognizable ultrastructural alterations in the bacterial envelope, as studied by freeze-fracture electron microscopy, Infect. Immun. 15:556 (1977).
11. J. Weiss P. Elsbach, I. Olsson, and H. Odeberg, Purification and characterization of a potent bactericidal and membrane-active protein from the granules of human polymorphonuclear leukocytes, J. Biol. Chem. 253:2664 (1978).
12. P. Elsbach, J. Weiss, R.C. Franson, S. Beckerdite-Quagliata, A. Schneider, and L. Harris, Separation and purification of a potent bactericidal/permeability increasing protein and a closely associated phospholipase A_2 from rabbit polymorphonuclear leukocytes. Observations on their relationship, J. Biol. Chem. 254:11000 (1979).
13. H. Nikaido, Permeability of the outer membrane of bacteria, Angew. Chem. Int. Ed. Engl. 18:337 (1979).
14. M. Vos, J. Op den Kamp, S. Beckerdite-Quagliata, and P. Elsbach, Acylation of monoacylphosphorylethanolamine in the inner and outer membranes of the envelope of an Escherichia coli K12 strain and its phospholipase-A deficient mutant, Biochim. Biophys. Acta 508:165 (1978).
15. J. Weiss, S. Beckerdite-Quagliata, and P. Elsbach, Resistance of gram-negative bacteria to purified bactericidal leukocyte proteins. Relation to binding and bacterial lipopolysaccharide structure, J. Clin. Invest. 65:619 (1980).
16. J. Weiss, S. Beckerdite - Quagliata, and P. Elsbach, Determinants of the action of phospholipases A on the envelope phospholipids of Escherichia coli, J. Biol. Chem. 254:11010 (1979).
17. J. Weiss, O. Stendahl, and P. Elsbach, Killing of gram-negative bacteria by polymorphonuclear leukocytes: The role of an O_2-independent bactericidal system. Clin. Res. 28:A515 (1980).

DISCUSSION
Edited by P. Elsbach

R. Johnston: Salmonellae are a fairly common cause of infection in patients with chronic granulomatous disease (CGD). In view of this, how do you explain your anerobiosis experiments with this organism? Have you tried to kill your strain with CGD leukocytes? Is there something different about your strain?

P. Elsbach: We have not done experiments with CGD leukocytes. I know of unpublished experiments of others with CGD leukocytes indicating that there is no impairment of killing of Pseudomonas aeruginosa, another gram-negative organism that is effectively killed by the purified cationic protein.

D. Romeo: Comment: I would not like the audience to be left with the impression that O_2-independent factors kill only gram-negative organisms. In fact, we have recently found that bovine neutrophils contain O_2-independent bactericidal factors, very likely cationic proteins, which efficiently kill both gram-negative and gram-positive bacteria. The bactericidal activity under physiological conditions of pH and ionic strength is not inhibited by divalent cations.

F. Tedesco: Is there any evidence that the binding of complement components and particularly of C3, confers to the smooth strain of gram-negative bacteria a rough strain-type hydrophobicity, similar to that obtained with IgG binding?

P. Elsbach: I do not know the answer to this question, but Olle Stendahl probably does.

O. Stendahl: Complement components do indeed confer hydrophobic, rough-like, properties on smooth Salmonella typhimurium strains.

R.A. Clark: Have you identified any inhibitors of the bactericidal activity of the purified BPI that might be useful probes in the studies with intact cells? Polyanions are one potentially useful type of probe.

P. Elsbach: Polyanions such as lipopolysaccharides (LPS) and heparin are indeed very effective inhibitors of the cationic bactericidal protein, as one would expect.

Bacterial LPS probably have too many biological effects to be useful.

Heparin has two drawbacks, 1) it probably will not get into the cell; 2) heparin at concentrations used clinically and for collection of blood inhibits phagocytosis and a number of other neutrophil functions (unpublished observations).

We hope to be able to introduce enough anti-bactericidal protein-antibody (we have a potent antiserum against both human and rabbit purified bactericidal protein) into the intact neutrophil, perhaps by coating bacteria with the antibody, to get convincing results.

J. Palmblad: Is the susceptibility of bacteria to antibiotics, in particular aminoglycosides, enhanced by the presence of the permaability increasing factors?

P. Elsbach: As I indicated in the early part of my talk, the native resistance of many gram-negative bacteria to antibiotics such as actinomycin D is abolished by the bactericidal protein. This is also the case of another hydrophobic antibiotic, rifampicin.

The entry or the susceptibility to other antibiotics appears not to be markedly altered. We have not yet carried out a comprehensive study, however.

J. Verhoef: Most gram-negative bacteria of clinical importance are encapsulated e.g. K antigen positive E.coli strains. Did you test the sensitivity of these strains for the cationic protein.

Most of the rough mutants are serum sensitive. How do the kinetics of the activity of the cationic protein against the Rc mutants compare with that of the serum bactericidal activity.

P. Elsbach: We have not yet examined K antigen positive E.coli for their sensitivity to these cationic proteins. Such experiments are planned in collaboration with Olle Stendahl.

Normal serum is very destructive toward rough gram negative bacteria (Infect. Immun. 11:758-766, 1975;J. Lipid Research, 21: 617-624, 1980). The cationic proteins exert their bactericidal effect without major structural or functional disorganization. The kinetics of loss of viability by bacteria exposed to serum or purified bactericidal protein are roughly comparable.

P. Patriarca: What are the relationships of your bactericidal protein with phospholipase A_2 activity?

P. Elsbach: Dr. Patriarca is referring to observations in our laboratory (J. Clin. Invest. 55:33-42, 1975;J. Biol. Chem. 254:11010-11014, 1979) indicating that rabbit bactericidal protein (but not human bactericidal protein) during purification remains closely associated with the leukocyte's phospholipase A_2 (J.Biol. Chem. 254: 11000-11009, 1979).

PMN preparations that contain both the bactericidal protein and the phospholipase A_2 cause phospholipid degradation in an E.coli mutant that lacks phospholipase A.

The leukocyte phospholipase A_2, separated by purification from the bactericidal protein has no effect on E.coli phospholipids, nor any other parameter of structure or function. However, exposure of the E.coli to recombined rabbit PMN phospholipase A_2 and bactericidal protein, again results in phospholipid degradation.

Interestingly, several other phospholipases A_2 (enzymes with remarkably similar properties regardless of source) cannot substitute for the rabbit PMN phospholipase A_2. Human bactericidal protein, even though practically indistinguishable from the rabbit protein in so many respects, has no facilitating effect on the action of rabbit PMN phospholipase A_2, nor of several other phospholipases A_2 (J. Biol. Chem. 254:11010, 1979).

THE EXTRACELLULAR STIMULATION OF INTRACELLULAR KILLING BY PHAGOCYTES

P.C.J. Leijh, Th.L. van Zwet, M.Th. van den Barselaar, and R. van Furth

Department of Infectious Diseases, University Hospital, Leiden, The Netherlands

INTRODUCTION

Phagocytes, granulocytes, and mononuclear phagocytes are the major effector cells in the host defence against invading micro-organisms, which these cells ingest, kill intracellularly, and digest. It has been known since the beginning of the century that serum proteins called opsonins are obligatory for optimal ingestion of micro-organisms by phagocytes[1-4]. However, the effect of not only these opsonic but also other serum proteins on the other stages of the phagocytic process, and in particular killing and digestion is less clear. This is due mainly to the fact that until recently it was almost impossible to assess intracellular killing from the continuous ingestion of new live bacteria affecting this measurement[5-7]. In the present study we used a recently developed method,[8,9] to investigate the effect of extracellular stimuli, serum proteins as well as membrane pertubating agents, on the intracellular killing of micro-organisms by granulocytes and monocytes.

Intracellular killing assay

The method employed to measure the intracellular killing of micro-organisms by granulocytes and monocytes has been described in detail elsewhere[8,9]. In short, a suspension containing 5 x 10^6 phagocytes/ml (granulocytes or monocytes from the blood of healthy donors), prepared according to Bøyum,[10] and 5 x 10^6 bacteria/ml, pre-opsonized with 10% fresh AB serum for 25 min at 37°C under rotation (4 rpm), was incubated for 3 min at 37°C under rotation (4 rpm). The ingestion phase was terminated by cooling the suspension to 4°C in crushed ice, and the extracellular non-ingested bacteria were removed by differential centrifugation (4 min, 110 x g) and two washes. Finally, the phagocytes containing ingested bacteria were re-incubated at 37°C under rotation (4 rpm)and the number of viable intracellular bacteria was determined microbiologically after disruption of the cells. The decrease in the number of viable intracellular bacteria represents the intracellular killing of the micro-organisms.

RESULTS AND DISCUSSION

Effect of extracellular serum on the intracellular killing of micro-organisms by monocytes and granulocytes

As shown previously[8,11], during incubation of monocytes or granulocytes containing ingested *S. aureus* a maximal decrease in the number of viable intracellular bacteria is only observed when incubation is performed in the presence of fresh serum. In the absence of serum no decrease in the number of viable intracellular bacteria was observed when monocytes were used; for granulocytes the decrease in the number of viable intracellular bacteria was sub-optimal compared with the decrease in the presence of fresh serum (Fig. 1). In the presence of heat-inactivated serum both granulocytes and monocytes kill *S. aureus* at levels intermediate between those obtained in the absence and presence of fresh serum.

The stimulation of the killing process by heat-inactivated serum was found to be due to IgG interacting via its Fc part with the Fc receptors in the surface of granulocytes and monocytes. Proof that such interactions occurred was provided by the following observations. Incubation of phagocytes containing ingested bacteria in the presence of purified heterogeneous IgG, pFc fragments of IgG, the subclasses IgG_1 and IgG_3, for which membrane receptors are present on phagocytes,[12,13] resulted in a decrease in the number of viable intracellular bacteria similar to that seen in the presence of heat-inactivated serum (Fig. 1). Incubation of phagocytes containing *S. aureus* in the presence of IgM, $(Fab^1)_2$ fragments of IgG, IgG_2, or IgG_4, for which no membrane receptors are detectable, gave a degree of

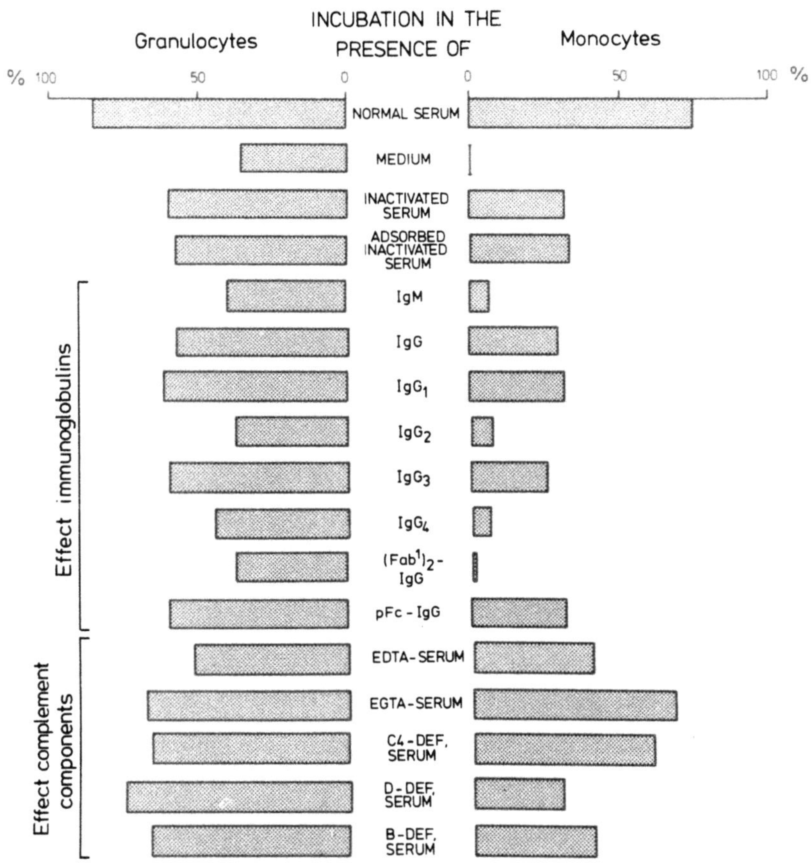

Fig. 1. Effect of various sera and serum proteins on the intracellular killing of S. aureus by monocytes and granulocytes.

killing that was not higher than that observed with medium alone (Fig. 1).

To find out why fresh serum is able to induce a larger decrease in the number of viable intracellular bacteria than is obtained with heat-inactivated serum, we investigated the killing stimulatory activities of various sera in which one or both of the pathways of complement activation were blocked (Fig. 1). Incubation of granulocytes or monocytes containing ingested S. aureus in the presence of sera in which the classical and the alternative pathway of complement activation were blocked, i.e., heat-inactivated serum and fresh serum chelated with EDTA, gave in both cases an intracellular killing similar to that seen in the presence of IgG. Incubation of granulocytes containing S. aureus in the presence of sera in which the classical pathway of complement activation was blocked, i.e., fresh serum chelated with EGTA or serum deficient in C4, resulted in a killing by granulocytes intermediate between the levels observed in the presence of fresh and heat-inactivated serum; monocytes, however, kill bacteria in the presence of these sera to the same degree as in the presence of fresh serum. Incubation of granulocytes or monocytes containing S. aureus in the presence of sera in which the alternative pathway of complement activation is blocked, i.e., sera deficient in B or D, showed that under these conditions monocytes kill bacteria as effectively as in the presence of heat-inactivated serum, whereas for granulocytes the killing is comparable to that in the presence of sera with the classical pathway blocked (Fig. 1). Together these results indicate that in all probability C3b generated by both complement pathways in the case of granulocytes and especially by the alternative pathway in the case of monocytes is able to stimulate intracellular killing.

To find out whether the stimulation by IgG and C3b requires interaction with specific membrane receptors for these proteins, the killing assay was performed after modulation of the number of Fc and C3b receptors on the granulocytes and monocytes by treatment of these cells with pronase, an antiserum against granulocytes, or an antiserum against monocytes. The results indicated that a decrease in the number of Fc and/or C3b receptors was accompanied by a decrease in the intracellular killing in the presence of heat-inactivated or fresh serum, which means that the stimulation of this process occurs via an interaction with Fc and C3b receptors on the cell membrane.[8]

To determine whether extracellular stimulation by serum is required for the intracellular killing of other micro-organisms besides S. aureus by monocytes, we used streptococci, Staphylococcus epidermidis, and Escherichia coli in similar experiments. In the absence of serum S. epidermidis and E. coli, which, like S. aureus, are catalase-positive micro-organisms, remained viable. This is in contrast to the findings for the catalase-negative streptococci, which were killed intracellularly in the absence of serum as well.[14] In the presence of fresh serum both groups of micro-organisms were killed optimally and to the same degree.

This difference between the intracellular killing of catalase-positive and catalase-negative micro-organisms is also known for the killing of micro-

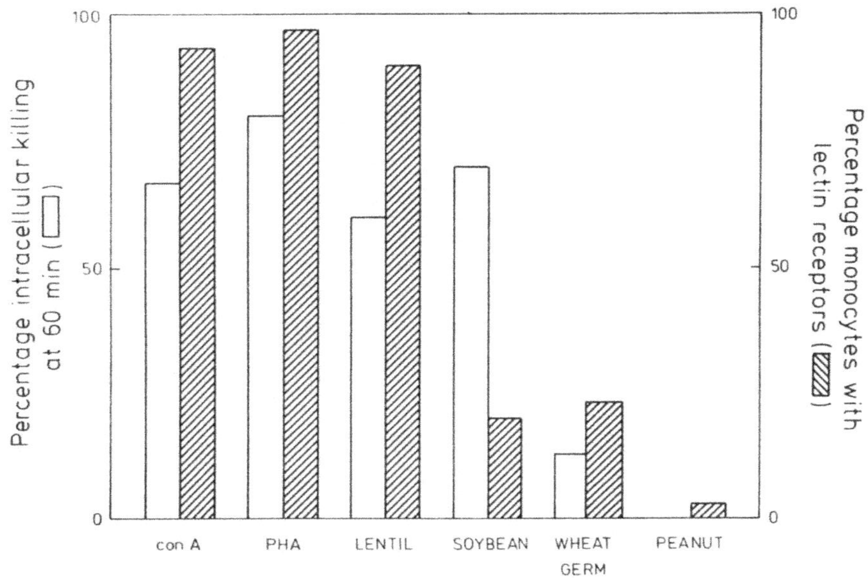

Fig. 2. Effect of various lectins (concanavallin A (con A) , phytohaemagglutinin (PHA), lentil, soybean, wheatgerm and peanut) on intracellular killing by monocytes (□). The number of monocytes carrying specific receptors for these lectins was also determined (□).

organisms by phagocytes of patients suffering from a chronic granulomatous disease.[15] The explanation of this divergence is probably that H_2O_2, which is normally produced by phagocytes and is necessary for optimal functioning of the myeloperoxidase-dependent bactericidal system, is absent in CGD phagocytes.[16,17] However, catalase-negative micro-organisms form H_2O_2 and therefore kill themselves when they enter a phagosome, where myeloperoxidase in combination with H_2O_2 generates bactericidal products.

Effect of other membrane-pertubating agents on the intracellular killing of S. aureus by monocytes

To find out whether other membrane-ligand interactions stimulate intracellular killing as well, the effect of lectins and phorbol myristate acetate (PMA) on the intracellular killing of S. aureus by monocytes was investigated. Incubation of monocytes containing ingested bacteria in the presence of various lectins (soybean, wheatgerm, peanut, lentil, phytohaemagglutinin (PHA), or concanavallin A (con A)) only resulted in a decrease in the number of viable intracellular bacteria when lectin receptors were demonstrable on the monocyte surface (Fig. 2); PMA did not give any decrease. Proof that lectin-receptor interaction is involved in the stimulation of intracellular killing was provided by the following experiment. Incubation of monocytes containing ingested S. aureus in the presence of con A and various concentrations of its specific sugar α-methyl mannoside (αMM), revealed that con A is only able to stimulate killing optimally in the absence of this sugar.

Conclusion

The results of the present studies show that optimum intracellular killing of micro-organisms by human granulocytes and monocytes requires stimulation of these cells by extracellular factors interacting with specific receptors in the membranes of these cells (Fig. 3).

ACKNOWLEDGEMENT

This study was partially supported by the J.A. Cohen Institute of Radiopathology and Radiation Protection.

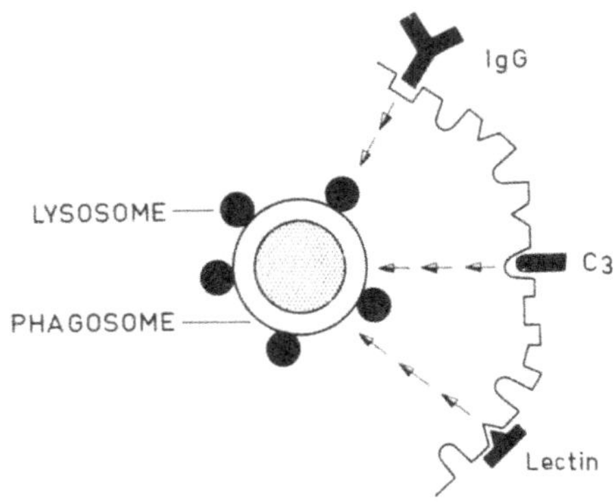

Fig. 3. Schematic representation of the extracellular stimulation of intracellular killing of bacteria by granulocytes and monocytes.

REFERENCES

1. A.E. Wright, and S.R. Douglas, An experimental investigation of the role of the blood fluids in connection with phagocytosis. Proc. Roy. Soc., London s. B., 72:357 (1904).
2. T.M. Saba, F.A. Blumenstock, W.A. Scovill, and H. Bernara, Cryoprecipitate reversal of opsonic α_2-surface binding glycoprotein deficiency in septic surgical and trauma patients. Science 201:622 (1978).
3. F.M. Griffin, J.A. Griffin, J. Leider, and S.C. Silverstein. Studies on the mechanism of phagocytosis. I. Requirements for circumferential attachment of particle-bound ligands to specific receptors in the macrophage plasma membrane.J. Exp. Med. 142:1263 (1975).

4. F.M. Griffin, J.A. Griffin, and S.C. Silverstein, Studies on the mechanisms of phagocytosis. II. The interaction of macrophages with anti-immunoglobulin IgG-coated bone marrow derived lymphocytes. J. Exp. Med. 144:788 (1976).
5. O. Maaløe. On the relation between alexin and opsonin. Copenhagen, Munksgaard (1946).
6. Z.A. Cohn, and S.I. Morse, Interactions between rabbit polymorphonuclear leucocytes and staphylococci. J. Exp. Med. 110:419 (1959).
7. C.P. Craig, and E. Suter, Extracellular factors influencing staphylocidal capacity of human polymorphonuclear leukocytes. J. Immunol. 97:287 (1966).
8. P.C.J. Leijh, M.Th. van den Barselaar, Th.L. van Zwet, M.R. Daha, and R. van Furth, Requirement of extracellular complement and immunoglobulin for intracellular killing of micro-organisms by human monocytes. J. Clin. Invest. 63:772 (1979).
9. R. van Furth, Th.L. van Zwet, and P.C.J. Leijh, In vitro determination of phagocytosis and intracellular killing by polymorphonuclear and mononuclear phagocytes, in Handbook of Experimental Immunology, D.M. Weir, ed., 3rd ed., ch. 32, Blackwell Scientific Publications, Oxford, London, Edinburgh, Melbourne (1978).
10. A. Bøyum, Separation of leucocytes from blood and bone marrow. Scand. J. Lab. Invest. 21 (Suppl.): 97, (1968).
11. P.C.J. Leijh, M.Th. van den Barselaar, and R. van Furth. Manuscript in preparation.
12. W.P. Arend, and M. Mannik, Quantitative studies on IgG receptors on monocytes, in Mononuclear Phagocytes in Immunity, Infection and Pathology, R. van Furth, ed., Blackwell Scientific Publications, Oxford, p. 303 (1975).
13. R.P. Messner, and J. Jelinck, Receptors for human γG on human neutrophils. J. Clin. Invest. 49:2165 (1970).
14. P.C.J. Leijh, Th.L. van Zwet, and R. van Furth, Effect of extracellular serum in the stimulation of intracellular killing of streptococci by human monocytes. Infect. Immun. 30:421 (1980).
15. E.L. Kaplan, T. Laxdall, and P.G. Quie, Studies of polymorphonuclear leucocytes from patients with chronic granulomatous disease of childhood: Bactericidal capacity for streptococci. Pediatrics 41:591 (1968).

16. G.L. Mandell, and E.W. Hook, Leucocyte bactericidal activity in chronic granulomatous disease: correlation of bacteria hydrogen peroxide production and susceptibility to intracellular killing. J. Bacteriol. 100:531 (1969).
17. J. Pitt, and H.P. Bernheimer, Role of peroxide in phagocytic killing of pneumococci. Infect. Immun. 9:48 (1974).

DISCUSSION

Edited by P. Elsbach

R. Gallily: Did you try to increase bactericidal activity of macrophages by adding IgG to the cells which were previously incubated with ConA and bacteria?

P. Leijh: We have not performed that experiment.

B. Seligmann: PMA stimulates the extracellular production of a very large amount of O_2^- and related reactive species. Have you studied the effect of other activators such as the calcium ionophore A23187 or f-met-leu-phe which are different both in extent of O_2^- production that they stimulate and with respect to the basic mechanisms by which they activate granulocytes.

P. Leijh: We have not yet investigated the stimulatory activity of A23187 or f-met-leu-phe on killing, but they are on our list.

D. Roos: From your earlier work I suspected that the killing-increasing effect of serum factors was due to the induction of respiratory burst activity in the monocytes. However, you now told us that PMA is unable to increase killing, although it is one of the most potent stimulators of the oxidative metabolism. Do you have any idea how the killing is actually stimulated in your system?

P. Leijh: We also are puzzled by the fact that PMA is unable to enhance killing of ingested bacteria. At the moment our working hypothesis is that after binding of the stimulant to its receptor, the receptor-stimulus complex is internalized and by a kind of shuttle mechanism transported to the phagosome where it exerts its effect. However, this is only an unproven hypothesis.

P. Elsbach: Your conclusions stand or fall with the conclusiveness of your evidence that the cell-associated bacteria were indeed intracellular. At a ratio of bacteria:leukocyte=1:1, how convincing is your evidence?

If opsonins were limiting before addition of serum factors, what may be promoted is uptake of previously attached bacteria, rather than killing of already intracellular live bacteria.

P. Leijh: I think that we may be quite clear that almost all cell-associated bacteria represent intracellular ones. In a number of control experiments we showed that about 80% of the bacteria are intracellular and only about 20% are extracellular (J.Clin.Invest. 63,772, 1979). Also, the fact that lectins which do not stimulate ingestion are able to stimulate the intracellular killing supports this conclusion.

O. Stendahl: You suggested that the extracellular stimulation of intracellular killing could be due to the internalization of lectin-membrane complex. Does this hold true for the enhancing effect of IgG and C3 as well - are they also internalized or do they induce membrane recycling?

P. Leijh: As I indicated in answer to Dr. Roos's question, our hypothesis is that probably membrane-stimulant complexes are internalized but we have no proof for this speculation. The work of Tulkins (*) as well as of Muller and Steinman (*) has indicated the existence of a membrane-recycling system.

(*) In: Mononuclear Phagocytes, Functional Aspects. R. Van Furth, ed. Martinus Nyhoff Publishers, The Hague, Boston, London.

R.S. Weening: Are you able to induce intracellular killing in GCD neutrophils by the addition of serum or serum components?

P. Leijh: So far we have investigated monocytes from 4 patients, suffering from CGD, for their ability to kill S. aureus. In none of the cases was added serum able to stimulate the killing process, indicating that the stimulation required O_2-dependent mechanisms. However, this conclusion is not supported by the lack of killing upon addition of PMA.

MOLECULAR MECHANISM OF THE BACTERICIDAL ACTION OF MYELOPEROXIDASE-H_2O_2-CHLORIDE

Mic. N. Hamers[1] and Herman J. Sips[2]

[1]Central Laboratory of the Netherlands Red Cross Blood Transfusion Service and Laboratory of Experimental and Clinical Immunology of the University of Amsterdam, Plesmanlaan 125, 1066 CX Amsterdam, and [2]Laboratory of Biochemistry, B.C.P. Jansen Institute, University of Amsterdam, Plantage Muidergracht 12, 1018 TV Amsterdam (The Netherlands)

INTRODUCTION

The enzyme myeloperoxidase (MPO), present in high concentrations in neutrophilic granulocytes of many species, has been shown to have a bactericidal effect when combined with hydrogen peroxidase and chloride as cofactors (for a review see[1]). At present, the exact basis of the myeloperoxidase-mediated killing of bacteria has not yet been elucidated. Both hypochlorous acid (OCl^-) and singlet oxygen ($^1O_2$) have been designated as the microbicidal agents in this system.

In the present study the effect of purified human myeloperoxidase on the permeability properties of the E. coli cell envelope was investigated. A mutant strain of E. coli has been used in which the lactose permease is absent and the cytoplasmic enzyme β-galactosidase is formed constitutively. These properties imply that the substrate for β-galactosidase cannot pass the bacterial cell wall under normal conditions. Thus, this bacterial strain creates an attractive experimental set-up to investigate the permeability properties of the E. coli cell envelope, simply by measuring the accessibility of the intracellular enzyme for its substrate.

MATERIALS AND METHODS

Human myeloperoxidase, purified according to Bakkenist et al.[2] was a gift from Dr. R. Wever, University of Amsterdam. Hydrogen peroxidase (Perhydrol[R], 30% H_2O_2) was diluted to appropriate concentrations just before use. All chemicals used were of analytical grade. E. coli ML 35 lac $i^-z^+y^-$) was supplied by Dr. A. Kepes, University of Paris, France. The bacteria were grown aerobically at 37°C to stationary phase, diluted 1:25 in fresh medium, subsequently harvested in the late logarithmic phase of growth by centrifugation for 10 min at 2500 x g, 4°C, washed twice in PBS (phosphate-buffered saline, consisting of 10 mM phosphate, pH 7.4, 154 mM NaCl), and finally suspended at the desired concentration. Bacterial concentrations were determined spectrophotometrically by measuring the absorbance at 600 nm (A_{600nm}= 1.0 equals 10^9 bacteria/ml). Bacterial homogenates were obtained by sonication of E. coli stock suspensions at 0°C.

Bacteria were incubated with myeloperoxidase and H_2O_2 for various periods in either 100 mM sodium phosphate buffer, pH 7.4, plus 154 mM sodium chloride, or in 50 mM sodium acetate buffer, pH 5.0, plus 154 mM sodium chloride. The H_2O_2 concentration in all experiments was 470 μM, which gives the optimal Cl^-/H_2O_2 ratio for the action of myeloperoxidase at pH 5.0[3]. The reactions were started by the addition of H_2O_2, and stopped by a ten-fold dilution in 100 mM sodium phosphate buffer, pH 7.4, containing 2 mM sodium azide to inhibit the myeloperoxidase activity. Subsequently, the activity of β-galactosidase was assayed by adding O-nitrophenyl-β-galactopyranoside (ONPG) to a final concentration of 1 mM, followed by incubation at 37°C for 10-25 min. The reaction was stopped by a 1:1 dilution with 300 mM glycine-sodium hydroxide buffer, pH 10.6. The amount of O-nitrophenol formed was determined by measuring the absorbance at 420 nm. An absorption coefficient of 7.5 $mM^{-1}\cdot cm^{-1}$ at 420 nm was used[4]. Whenever bacterial viability was measured, a further 10^6-fold dilution in PBS was carried out immediately after stopping the myeloperoxidase reaction to prevent bactericidal effects of sodium azide (control experiments indicated that azide itself has no effect under these conditions). From this solution 100 μl volumes, containing approx. 200-300 bacteria, were plated in triplicate and grown overnight at 37°C.

Neutrophilic granulocytes were isolated from fresh, defibrinated blood according to Weening et al.[5]. Intracellular killing of E. coli ML 35 was measured with a modification of the method described by Roos et al.[6]. Granulocytes (final concentration of 4.3 x 10^6 cells/ml) were incubated for 1 min at 37°C in the presence of human serum (10% w/v) with 12 live bacteria/neutrophil. After 1 min 250 units penicillin and 250 units streptomycin were added per ml mixture to kill the extracellular bacteria. Samples from this micture were incubated for 10 min with phenylbutazone to stop intracellular killing and further endocytosis. The granulocytes were centrifuged for 10 min at 400 x g, 4°C, and washed twice

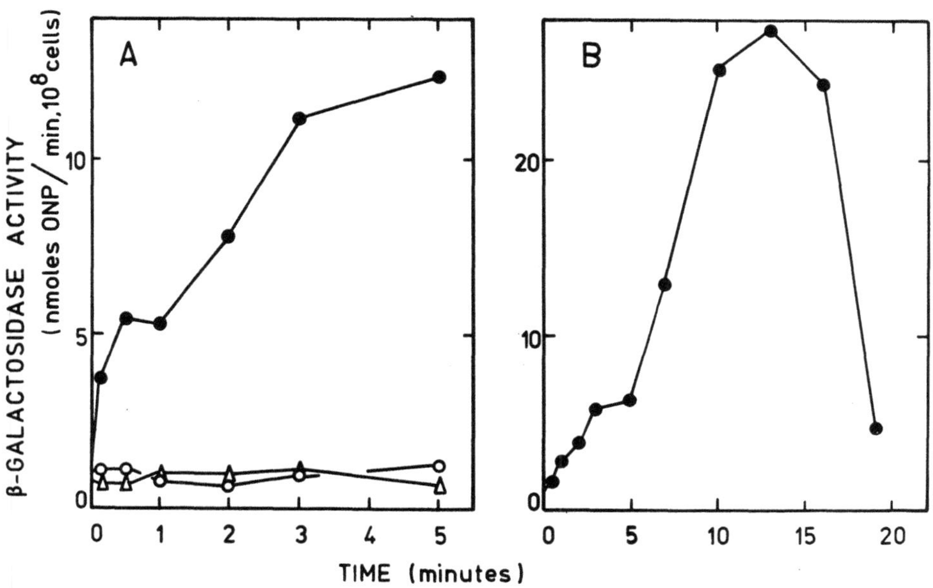

Fig. 1. Effect of the MPO-H_2O_2-Cl^- bactericidal system at pH 7.4 on the β-galactosidase activity. A. Bacteria (final concentration of 2.1×10^8 cells/ml) were incubated with Cl^- and MPO (140 nM) (O-O); Cl and H_2O_2 (Δ-Δ); or the complete system (●-●). B. Bacteria (final concentration of 2×10^9 cells/ml) were incubated with H_2O_2, Cl and MPO (140 nM) for a prolonged period of time.

with ice-cold PBS containing 10% (w/v) human albumin. The lysis of the granulocytes was accomplished with a 30 s hypotonic shock of the pellet in 1 ml ice-cold water, followed by the addition of 9 ml PBS/albumin mixture. Finally, the β-galactosidase activity and the bacterial viability was measured as described above.

RESULTS

When bacteria were incubated at pH = 5.0 with H_2O_2, Cl^- and human myeloperoxidase (7 nM or 14 nM), a rapid increase in β-galactosidase activity occured, followed by a fast decrease. At higher myeloperoxidase concentrations (140 nM) no enzyme activity could be detected, presumably due to an immediate inactivation of the B-galactosidase. We decided to decrease the reaction velocity by altering the pH from 5.0 to 7.4[3]. Fig. 1A demonstrated that the complete system, when applied at pH 7.4, resulted in a drastic increase in

Table 1. Localization of the β-galactosidase activity after exposure of the cells to the MPO-H_2O_2-Cl^- system

Time in minutes	β-galactosidase activity in nmol O-nitrophenol/min/10^8 cells: Before centrifugation	After centrifugation: supernatant	After centrifugation: pellet
5	20.8	0.8	18.6
10	12.9	0.3	9.6
15	0.7	0.03	0.5
20	0.2	0.03	0.2

Bacteria (final concentration 2.1 x 10^8 cells/ml) were incubated with H_2O_2, Cl^- and myeloperoxidase (140 nM) at pH 7.4 for various periods. Part of the treated bacteria was centrifuged, and β-galactosidase activity was measured.

β-galactosidase activity, whereas the incomplete system had no effect. Upon prolonged incubation a decrease in β-galactosidase was observed (Fig. 1B), thus confirming the observations made at pH 5.0.

It was found that the β-galactosidase activity present in bacterial homogenates correlates with the activity as maximally exposed in whole bacteria in the myeloperoxidase system. Furthermore, incubation of the bacterial homogenates with the myeloperoxidase system resulted in an inactivation of the β-galactosidase activity, whereas omission of a component abrogated this effect. It can thus be concluded that all β-galactosidase activity present in the intact bacteria is first exposed to the myeloperoxidase system, followed by an inactivation. The question arises whether in the experiments shown in Fig. 1 β-galactosidase is inactivated inside the bacteria, or, alternatively, the enzyme leaks out of the bacteria as a consequence of the disruption of the envelope and is inactivated in the surrounding medium. Table 1 shows that the β-galactosidase activity remained within the bacterial cells. Thus, it can be concluded that the action of the MPO system on the bacterial cell probably results in a limited permeability, i.e. the cell becomes freely accessible for the low-molecular weight ONPG, but retains its permeability barrier for larger molecules, such as β-galactosidase or myeloperoxidase.

Both OCl^- and $^1O_2$ have been postulated as intermediates in the myeloperoxidase-catalyzes bacterial killing. Fig 2 demonstrates that the effects of the myeloperoxidase system could be obtained with

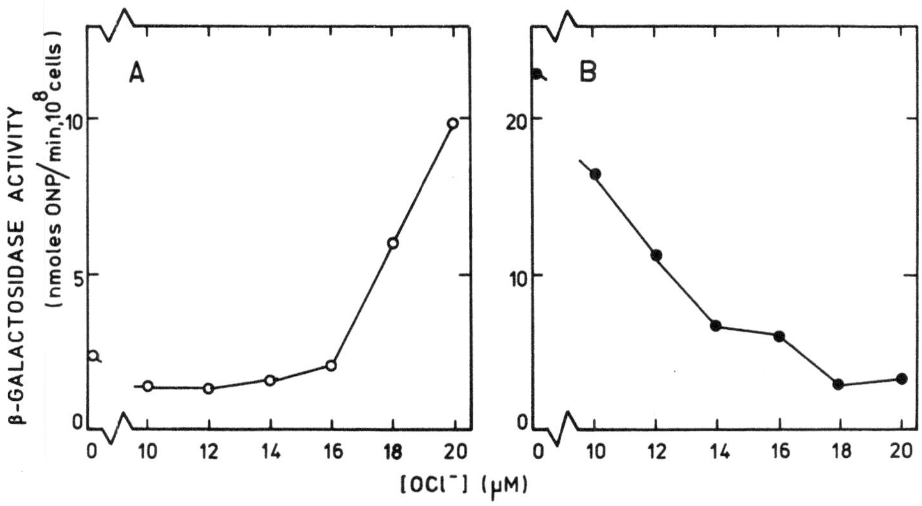

Fig. 2. Effect of OCl^- at pH 7.4 on the β-galactosidase activity in intact and homogenate of E. coli. A. Bacteria (final concentration of 2.2 x 10^8 cells/ml) were incubated with different amounts of OCl^- for 5 min; subsequently, β-galactosidase was assayed as described in Materials and Methods. B. An equal amount of bacteria was homogenized and incubated as described above.

OCl^- alone. Titration with OCl^- revealed both an increase in enzyme activity when OCl^- was incubated with intact bacteria (Fig. 2A), and a decrease in β-galactosidase activity in a bacterial homogenate (Fig. 2B). Addition of H_2O_2 to this reaction mixture, which resulted in the generation of singlet oxygen[7], had no, not even a slight, inhibitory effect on the action of OCl^- (data not shown). These observations support OCl^- rather than $^1O_2$ as the active intermediate. It must be realized that OCl^- is very unstable, especially at neutral pH, and therefore a relatively large variation in the experiments was observed. Nevertheless, qualitatively identical patterns as shown in Fig. 2 were obtained in all experiments. Moreover, addition of OCl^- to the extrabacterial medium may result in substantially different concentrations of OCl^- than those obtained with myeloperoxidase in the direct environment of the bacterial envelope. This may be concluded from the higher bactericidal activity of this system found when bacteria are coated with myeloperoxidase before addition of the cofactors[8].

The correlation between increase in permeability and killing of E. coli via the myeloperoxidase system is shown in Fig. 3. It was found that the survival of bacteria, as indicated by the amount of colony-forming units after myeloperoxidase treatment, correlates

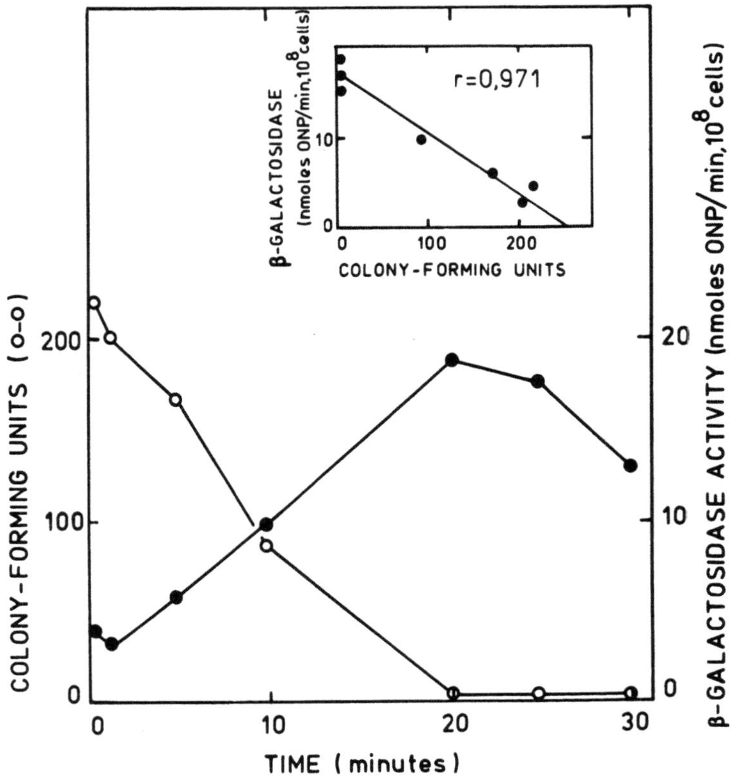

Fig. 3. Correlation between the increase in permeability and killing of E. coli by the MPO-H_2O_2-Cl^- bactericidal system at pH 7.4. Bacteria (final concentration of 2 x 10^9 cells/ml) were incubated with H_2O_2, Cl^- and MPO (140 nM). The reaction of MPO was stopped, and both β-galactosidase activity (●-●) and cell viability (0-0) were measured. For further details see Materials and Methods.

with the increase in permeability of the bacterial cell envelope ($p<0.0005$). Thus, when the myeloperoxidase system had permeabilized every bacterium and caused exposure of all enzyme activity, no viable cells were left in the incubation. These results support the hypothesis that the mechanism of the killing of bacteria via myeloperoxidase proceeds by increasing the permeability of the bacterial cell wall.

Fig. 4 shows that the effects of purified myeloperoxidase could be mimicked by human neutrophils, which contain this enzyme. In the neutrophil system also a correlation between exposure and subsequent inactivation of β-galactosidase and the bacterial viability was found, whereas serum alone had no effect. The β-galactosidase

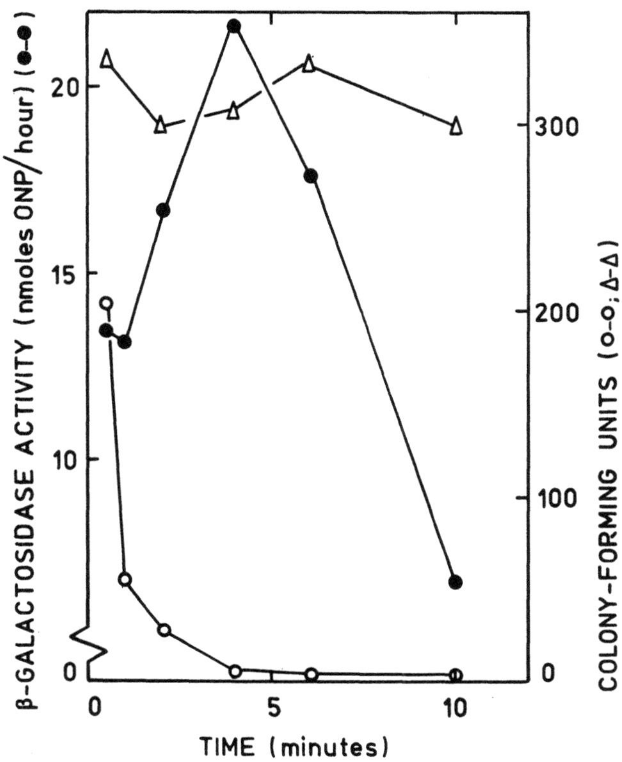

Fig. 4. Correlation between the increase in permeability and killing of E. coli by human neutrophils. Bacteria were incubated with isolated human neutrophils (equivalent to a total amount of 140 nM MPO) in a ratio of 12 bacteria to 1 neutrophil, in the presence of human serum (10%, w/v); ●-●, β-galactosidase activity of endocytosed bacteria; 0-0. viability of bacteria after incubation with human serum.

activity of the neutrophils appeared to be $\leqslant$ 3 nmol/h/incubation.

DISCUSSION

The experiments described in this paper confirm earlier observations that a system consisting of purified myeloperoxidase, H_2O_2 and Cl^- has microbicidal activity[1]. This activity, however, was established after prolonged incubation, i.e. 20 min or longer, and to date no mechanism has been postulated for this killing activity. Evidence is provided for a very rapid increase in permeability of the E. coli envelope; this permeability increase correlates with the killing phenomenon.

Interestingly, both a substantial increase in permeability and killing are observed at pH 7.4, although the Cl^-/H_2O_2 ratio employed is two orders of magnitude smaller than the optimal ratio at pH 7.4[3]. This implies that under physiological conditions the amount of H_2O_2 needed for optimal myeloperoxidase action outside neutrophilic granulocytes is about two orders of magnitude smaller than within the azurophilic granules of the granulocyte. Thus, it is conceivable that neutrophilic granulocytes are able to kill extracellular bacteria very efficiently by releasing myeloperoxidase via degranulation during endocytosis.

Another mechanism of rendering microorganisms permeable has been reported by Elsbach et al.[9,10]. In these studies the bactericidal agent was proved to be a cationic protein that did not contain a detectable activity of myeloperoxidase. The authors concluded that this protein is the principal antibacterial agent at neutral pH towards gram-negative bacteria[10]. Our results, however, show that purified myeloperoxidase is also able to permeabilize and to kill bacteria at neutral pH and within minutes. Therefore, myeloperoxidase may contribute significantly to the antibacterial action of granulocytes at neutral pH.

An interesting application of the correlation between permeability of the cell wall and the killing would be an alternative killing assay for clinical purposes. Fig. 4 illustrates the possibility of such a new assay. It is obvious, however, that further experiments are needed to specify and optimalize this assay. It is noteworthy that both the loss of the permeability barrier and killing, with either myeloperoxidase or intact neutrophils, occur within minutes. Possibly, this β-galactosidase activity assay enables discrimination between the myeloperoxidase-mediated killing and other bactericidal systems known to be present in the neutrophil. Experiments with myeloperoxidase-deficient and chronic granulomatous disease (CGD) granulocytes are in progress.

REFERENCES

1. S.J. Klebanoff and R.A. Clark, "The neutrophil: function and clinical disorders" (pp. 410-434) North-Holland, Amsterdam (1978).
2. A.R.J. Bakkenist, R. Wever, T. Vulsma, H. Plat and B.F. van Gelder, Isolation procedure and some properties of myeloperoxidase from human leucocytes, Biochim. Biophys. Acta 524:45 (1978).
3. A.R.J. Bakkenist, J.E.G. de Boer, H. Plat and R. Wever, The halide complexes of myeloperoxidase and the mechanism of the halogenation reactions, Biochim. Biophys. Acta 613:337 (1980).
4. A.B. Pardee, F. Jacob and J. Monod, The genetic control and cytoplasmic expression of "inducibility" in the synthesis of -galactosidase by E. coli, J. Mol. Biol. 1:165 (1959).
5. R.S. Weening, D. Roos and J.A. Loos, Oxygen consumption of

phagocytizing cells in human leukocyte and granulocyte preparations: a comparative study. J. Lab. Clin. Med. 83:570 (1974).
6. D. Roos, R.S. Weening, A.A. Voetman, M.L.J. van Schaik, A.A.M. Bot,L.J. Meerhof and J.A. Loos, Protection of phagocytic leukocytes by endogenous glutathione: studies in a family with glutathione-reductase deficiency. Blood 53:851 (1979).
7. R.C. Allen, Halide dependence of the myeloperoxidase mediated antimicrobial system of the polymorphonuclear leukocyte in the phenomenon of electronic excitation. Biochem. Biophys. Res. Comm. 63:675 (1975).
8. A.J. Sbarra, R.J. Selvaraj, B.B. Paul, P.K.F. Poskitt, G.W. Mitchell, F. Louis and M.A. Asbell, Granulocyte biochemistry and a hydrogen peroxide-dependent microbicidal system, in: "The Granulocyte: Function and Clinical Utilization", T.J. Greenwalt and G.A. Jamieson, eds., pp. 29-48, Alan R. Liss, Inc., New York (1977).
9. S. Beckerdite, C. Mooney, J. Weiss, R.C. Franson and P. Elsbach, Early and discrete changes in permeability of Escherichia coli and certain other gram-negative bacteria during killing by granulocytes. J. Exp. Med. 140:396 (1974).
10. P. Elsbach, J. Weiss, R.C. Franson, S. Beckerdite-Quagliata, A. Schneider and L. Harris, Separation and purification of a potent bactericidal/permeability-increasing protein and a closely associated phospholipase A_2 from rabbit polymorphonuclear leukocytes. J. Biol. Chem. 254: 11000 (1979).

DISCUSSION
Edited by P. Elsbach

R. Johnston: In your experiments utilizing whole cells, why do you ascribe the effect to the H_2O_2-MPO system alone? Could not other oxidative (as well as non-oxidative) mechanisms be involved?
M. Hamers: You are right, of course. All the killing mechanisms are operating during the experiment with whole cells, but in view of the very fast kinetics of killing which parallels that of the MPO-H_2O_2-Cl^- system as shown in the other experiments, we tend to speculate that it is the effect of the MPO system we have observed in these particular experiments.

P. Elsbach: These are very interesting studies.
I wonder whether your E.coli strain which is constitutive for β-galactosidase is unusually susceptible to damage, because we have noted some years ago (J. Bacteriol.115:490-497, 1973) that killing by intact PMN of an E.coli K12 strain is accompanied by production of β-galactosidase, in response to a suitable inducer, and sustained activity for at least one hour.
M. Hamers: Yes I know that paper and I don't have any other explanation for this fact, than the differences in E.coli strains used in both studies. The ML 35 strain we have used was selected from a list of mutant strains which comprised all known sugar-transport mutations and I don't know whether other characteristics, like membrane type etc. are known.

R.A. Clark: In view of the very rapid time course described, it is important to be certain that the reaction was, in fact, stopped at the early time points. How was this accomplished?
M. Hamers: The incubation mixture, containing E.coli,MPO, H_2O_2 and Cl^- was diluted 10 times in the same buffer containing 2 mM sodium azide. This will inhibit the MPO system immediately (for further details see the manuscript).

THE ROLE OF THE CELL MEMBRANE IN THE KILLING MECHANISM OF POLYMORPHONUCLEAR LEUCOCYTES (PMN)

Jörg Menzel, Klaus-Michael Debatin and Helmut Bannert

Fachklinik Königstuhl, Heidelberg and Institut für Immunologie und Serologie der Universität Heidelberg
6900 Heidelberg

It is generally assumed that the killing of bacteria by leucocytes is performed exclusively intracellularly. However, the observation of a complement augmented cellular bactericidal mechanism raised doubts in that assumption (1). Complement activity is labile and transient and its intracellular action after phagocytosis therefore hard to imagine. Further studies revealed that the subsequent stages of phagocytosis were affected differently by complement activity. The attachment of bacteria to the cell membrane was particularly augmented but, surprisingly, the consecutive ingestion was not (2). This observation raised the question as to whether serum complement can act on the even later intracellular events, i.e. intracellular killing.

Experiments were therefore performed to elucidate the site of complement activity on the bactericidal system(s) of the cell. Experiments were performed in the following way: A monolayer of isolated PMN in tissue culture dishes was used to which opsonized radiolabeled and complement resistent bacteria were added. The cells were allowed to phagocytize for different time intervals. The monolayer was then cooled to inhibit further cell activity. After separation of non cell adherent bacteria the cells were lysed by a solution of heparin-sucrose. The cellular material was removed and separated by differential centrifugation following a method described extensively elsewhere (3). Two fractions were obtained. Fraction 1 (800 x g pellet) contained cell membranes and nuclei and fraction 2 (800 x g supernatant) contained granula and cytoplasm. The effisiency of the separation procedure was monitored by electron microscopy and by testing for characteristic enzymes. Alkaline phosphatase was used as a membrane marker

TABLE 1

Distribution of enzymes after cell fractionation (% of total enzyme activity)

	alk.Ph.[a]	LDH[b]	Lysozyme
800 x g pellet	88	-	5
800 x g supernatant	12	100	95

a: alkaline Phosphatase, b: Lactate Dehydrogenase

and LDH as a marker for the cytoplasm. Lysozyme was selected from the lysosomal enzymes (Table 1.). The total number of bacteria in the fractions was evaluated by the radioactive label and the number of surviving bacteria was enumerated by colony counting following incubation of aliquots on agar plates.

Using a smooth strain of E. coli it became apparent that the distribution of bacteria between the two fractions, i.e. the transport from the membrane into the cell was unaffected by the mode of opsonization. But it became also apparent that most bacteria were killed while still in contact with the outer cell membrane and before ingestion into the cell (Table 2.).
The killing was particularly effective when complement was used for opsonization compared to the opsonization with IgG allone. Therefore it was concluded that phagocytes possess in addition to the intracellular bactericidal system another bactericidal activity in or associated with the outer cell membrane the latter

TABLE 2

% of E. coli in the cell fractions (20 minutes)

bacteria opsonized with	IgG	IgG + Complement
total number of cell bound bacteria	$3{,}5 \times 10^5$	$3{,}7 \times 10^7$
800 x g pellet	15 ± 2	16 ± 1,5
killed*	60	95
800 x g supernatant	85 ± 5	84 ± 5
killed*	95	99

* % killed bacteria in the fraction

TABLE 3

% of killed bacteria in cell fractions (40 minutes)

	E. coli	S. typhim.	Staph. aur.
800 x g pellet	93 ± 2	95 ± 5	81 ± 6
800 x g supernatant	96 ± 3	97 ± 1	92 ± 5

being stimulated by activated serum complement.

Further experiments were performed to see whether this mechanism is engaged in the killing of other strains of bacteria. Identical experiments were undertaken with S. typhimurium and Staph. aureus both of which prooved to be complement resistent (Table 3.) Results indicate that the membrane associated killing mechanism is operative against various strains of bacteria. Obviously there are differences in the affictivity being least with Staph. aureus.

We conclude that phagocytes have at least two, possibely independant bactericidal systems. The membrane associated system may be governed by extracellular factors such as serum complement. Therefore a great number of dead intracellular microorganisms may be killed and then ingested pretending a very effective intracellular killing.

REFERENCES

(1) Menzel, J., Jungfer, H. and Gemsa, D. (1978), Contribution of Immunoglobulin M and G, Complement and Properdin to the intracellular killing of Escherichia coli by Polymorphonuclear Leucocytes. Infect.Immunity; 19, 659 - 665.

(2) Berto, R. and Menzel, J. (1979), The Interaction of Opsonins with human Polymorphonuclear Leucocytes (PMN), I. The Influence of human Complement (C) and IgG on Ingestion and Digestion of C-resistant E. coli, Z. Immun.-Forsch, 155, 189 - 199.

(3) Chodirker, W.B., Bock, G.N. and Vaughan, J.H. (1968), Isolation of human PMN leucocytes and granules: Oberservation on early blood dilution and on heparin. J.Lab.Clin.Med. 71, 9 - 19

DISCUSSION
Edited by P. Elsbach

B. Seligmann: The separation of plasma membrane and plasma membrane associated "extracellular" bacteria from the phagosome granule membranes, and associated "intracellular" bacteria is crucial to this study. You have shown that the appropriate soluble enzymatic markers are well separated but I wonder for instance if it might be possible that dead ingested bacteria in phagosomes might pellet more readily with the plasma membrane fraction, accounting for the results. One might expect that the soluble phagosome enzyme markers might be lost in this case and thus account for the lack of their association with the contaminates of the "plasma membrane" fraction.
J. Menzel: Of course, we were quite aware that adequate separation was crucial for the interpretation of our experiments. Microscopic and electron microscopic observations showed no contamination of the membrane fraction by lysosomes/phagosomes where controls were performed as stated in the answer to Dr. Johnson.

But I do not think that dead bacteria in phagosomes would be able to contaminate the membrane fraction, phagosomes proved to be even lighter or do not settle as readily as bacteria even at much higher g-forces.

R. Johnston: I want to extend Dr. Seligmann's question - Did you do electron micrography on your membrane fraction?

Comment: Perhaps relevant to your findings are those of Newman in our laboratory using erythrocytes as the target particle. If one places C3b on this particle in the complete absence of antibody, there is no surface stimulation of superoxide-release or exocytosis of granule enzymes. If a small concentration of antibody is added to the C3b-coated particles, there is marked enhancement of superoxide and granule enzyme release over that obtained by opsonization by the same quantity of antibody alone.
R. Menzel: Yes we have done some electron microscopy. In fact, it was quite difficult to find bacteria in the membrane fraction. But we found hardly any granules (lysosomes) or phagosomes. So, by this criterion, the separation into an outer membrane fraction and an intracellular membrane fraction was good.

Another type of control experiment was performed: To rule out gross contamination of the plasma membrane fraction by bacteria, which had been intracellular, we made cell preparations, lysed the cells by sucrose-heparin, and then added bacteria (radiolabeled). The subsequent centrifugation at 800 x g revealed a contamination of the membrane fraction by about 2% of added bacteria. So we think that contamination was negligible.

L. De Chatelet: What are the details of membrane isolation? I am amazed at the relative purity you achieve with simple differential centrifugation.
J. Menzel: The separation followed precisely previously published methods, except for the fact that we added some DNAase to the cell material to diminish the viscosity which resulted from the destroyed nuclei.

F. Tedesco: Two questions: 1) I did not quite understand from your presentation whether the E.coli opsonized with either IgG or IgG plus complement were actually washed before being offered to PMN.
2) Did you also use E.coli coated with IgM and complement?
J. Menzel: Answer 1) All bacteria were washed after the opsonization and before being offered for phagocytosis.
2) The killing of E.coli by PMN was also investigated with bacteria coated with IgM and complement. Killing was most efficient with IgM and C as opsonin. - In contrast, IgM alone proved to be very inefficient in promoting phagocytosis or cell-mediated killing (Menzel et al. Infect. Immun. 1978).

LEPTOSPIRES MACROPHAGE INTERACTIONS

Marina Cinco, Elena Banfi & Maria Rosa Soranzo*

Institute of Microbiology, Institute of General Pathology*
University of Trieste, Trieste, Italy

INTRODUCTION

Leptospires are known to be invasive germs: their possibility to spread through the body fluids and tissues could reside in the fact that pathogenic leptospires are able to escape for a time the natural defenses of the host, namely phagocytosis and the antibody-complement system. According to Johnson[1,2] normal serum of various mammalian species exerts its antileptospiral activity primarily against saprophytic, not pathogenic leptospires. This finding suggests that the basis for the virulence of the pathogenic species lay in their capacity to withstand the bactericidal complement action.

The role of phagocytosis in leptospiral infection was scarcely studied. Faine in 1964[4,5] on the basis of microscopic observations performed in vitro, reported that both mouse and guinea pig macrophages engulfed leptospires. However neither the cytotoxic effect of the leptospires on the macrophages nor their intracellular survival within the macrophages was investigated.

The present preliminary study was undertaken to examine the interaction between macrophage population and suspension of either pathogenic or saprophytic leptospires, in order to ascertain whether their invasiveness was indeed associated with their ability to survive within the phagocytic cells. In particular the ability of macrophages to kill leptospires was investigated either in the presence or in the absence of antibodies. To this end an accurate and reproducible method for obtaining viable counts of leptospires was developed. By the same cultural technique and in conjunction with electron microscopy observations, the rate of phagocytosis was also evaluated.

MATERIALS AND METHODS

As bacterial strains, one of virulent leptospires and one of saprophytic ones were used and cultivated as already described[6].

As phagocytic cells guinea pig peritoneal macrophages were employed[6].

Assessment of bactericidal and phagocytic capacity of macrophages on leptospires and electron microscopy were performed as reported[6]. In addition opsonization experiments were carried on leptospires previously opsonized with 50 agglutinating doses of homologous antiserum.

RESULTS

Leptospires/macrophages interactions in absence of immune serum.

No bactericidal activity against either the virulent strain PB-3 or the saprophytic Isola Sacra 1 was exhibited by the macrophages as shown in fig. 1, through the 120' period of incubation. As shown in fig. 2 during the 120' period of incubation of the mixtures of leptospires and macrophages there was a significant increase in the number of colony-forming units/ml mixture associated with the macrophages. This increase was for both the virulent strain PB-3 and the saprophytic strain Isola Sacra 1 significant at the 0.001 level. There was a concomitant decrease with number of c.f.u./ml mixture in the supernatants. These findings suggest that a certain number of both virulent and saprophytic leptospires becomes slowly associated with the macrophages over the 120' period of incubation together.

Electron microscopy observations (photograph 1) performed after 5' and 30' contact between macrophages and leptospires reveal that in general the organisms are outside the macrophage membranes and that, in particular,

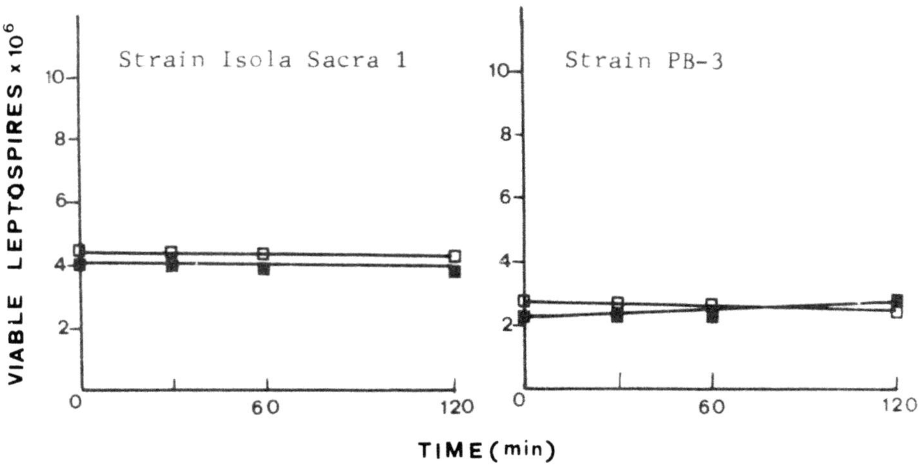

Fig. 1. Survival of strains Isola Sacra 1 and PB-3 with (■) or without (□) guinea pig macrophages.

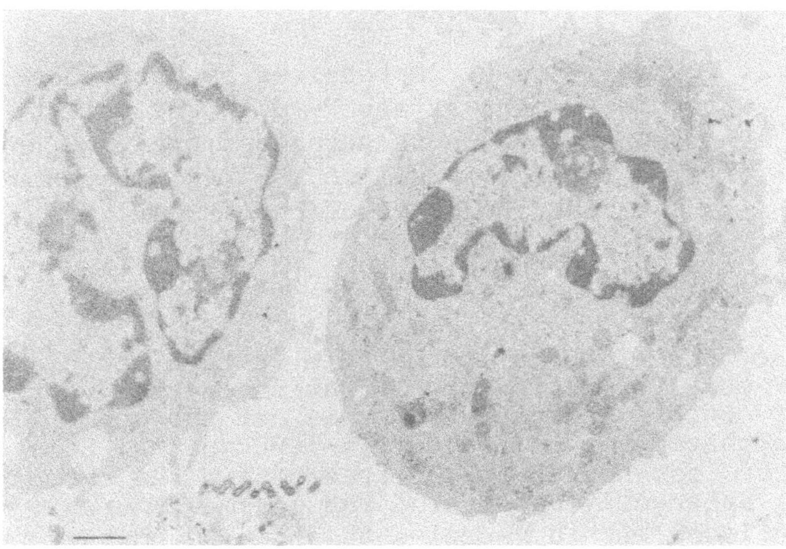

Photograph 1. Leptospira strain PB-3 after 30' incubation with macrophages. Bar = 1 μm. (from Cinco et al., J. Gen. Microbiol., in press).

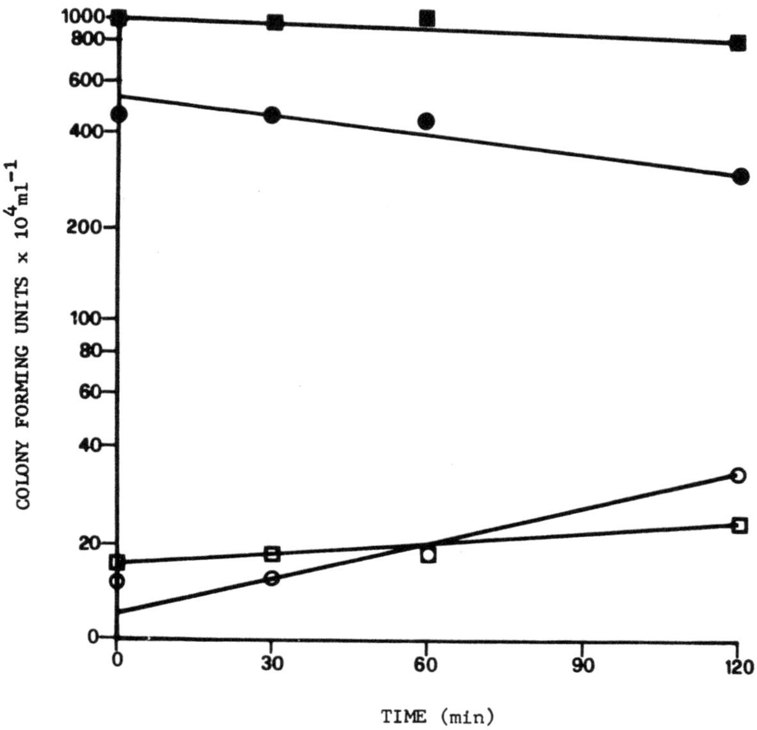

Fig. 2. Phagocytosis of leptospires by macrophages. Number leptospira strain PB-3 in the pellet of macrophages (○),in supernatant (●),and strain Isola Sacra 1 in pellet (□),in supernatant (■) after centrifugation of the mixture.

the macrophages appear to be in a resting state. After 60' or more incubation the situation changes. More macrophages contain leptospires, until at the end of 2h 30% of the phagocytic cells contain leptospires. Some of the ingested leptospires appear free within the cytoplasm, having apparently penetrated the plasma membrane, others are included within vacuoles as shown in photograph 2. Anyway the structure of the bacterial bodies appears undamaged.

Results of opsonization.

Leptospires opsonized by homologous inactivated serum showed to be killed by phagocytic cells, as reported in fig. 3. In both the strains the number of leptospires,

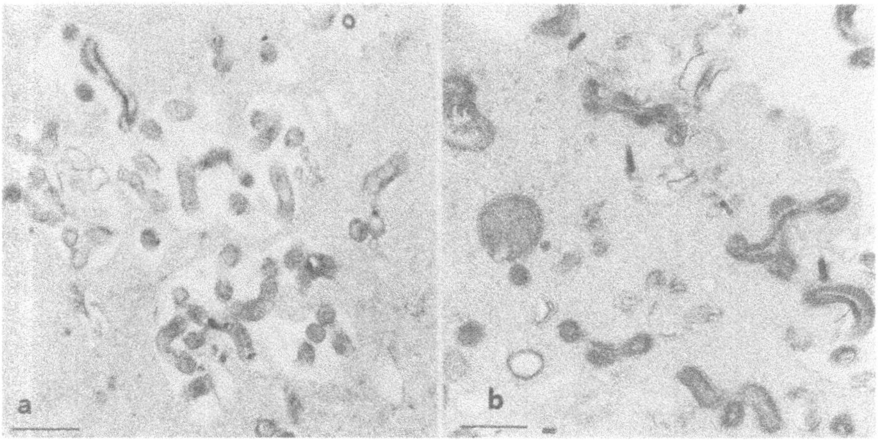

Photograph 2. Leptospira strain Isola Sacra 1 after 60' incubation with macrophages; a: leptospires in vacuoles, b: leptospires penetrating membranes. Bar = 0.5 µm. (from Cinco et al., J. Gen. Microbiol., in press).

as in the supernatant as in the pellet, becomes significatively lower, after 1h of incubation. Electron microscopy observations performed at 5' and 60' show that after 5' there are about 30% of cells ingesting leptospires and this number clearly increases in the time, as reported in fig. 4.

CONCLUSIONS

Data presented in this paper indicate that neither virulent or saprophytic strain of leptospira are sensitive to the bactericidal action of phagocytes during the two hours of the test, in absence of opsonization. In addition, as shown by electron microscopy, no significant phagocytosis takes place during the first 30' of incubation. However, on incubation for longer periods of time, a certain number of leptospires are found, both microscopically and culturally, to be associated with phagocytes and their number increases with time. These findings are in contrast with those obtained with other bacteria such as Staphylococcus aureus, Escherichia coli, Listeria and further more Brucella.
It is interesting to recall that, among Spirochaetales, Treponema pallidum is not phagocytosed unless immune serum is present[7]. Even then, phagocytosis occurs at a

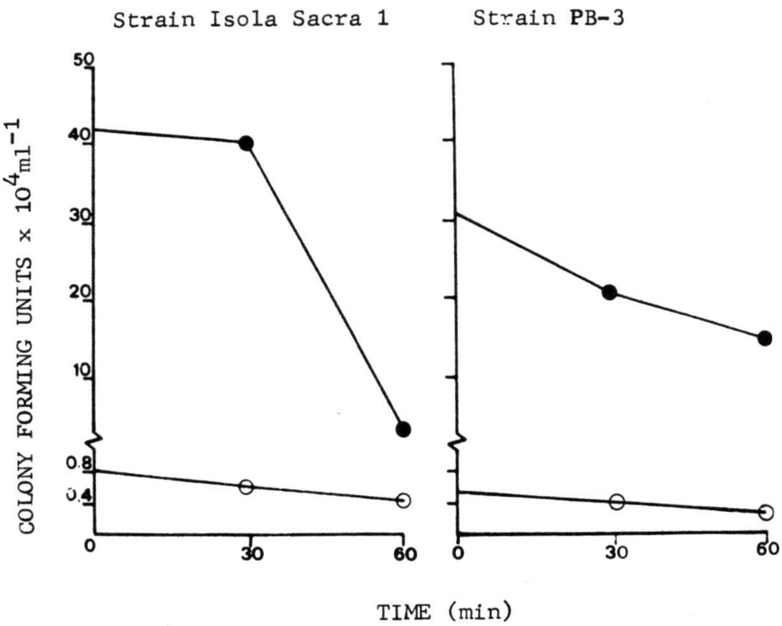

Fig. 3. Survival of the two strains of opsonized leptospires incubated with macrophages in the supernatant (●) and in the pellet (○), after centrifugation of the mixture.

very low rate and becomes evident only after 4h of incubation. The presence of leptospires within phagocytes, in the absence of specific antibodies, may represent the result of either phagocytosis or the active penetration of these microorganisms into macrophages. One is not in the position to decide between the two alternatives, though data in the literature would support the latter hypothesis.

The most significant conclusion that can be drawn from these studies, is that , in absence of opsonizing factors, leptospires both outside and inside the phagocytic cells are still capable of multiplication.
On the contrary the addition of antibodies promote an active phagocytosis associated to an efficient killing of the leptospires.

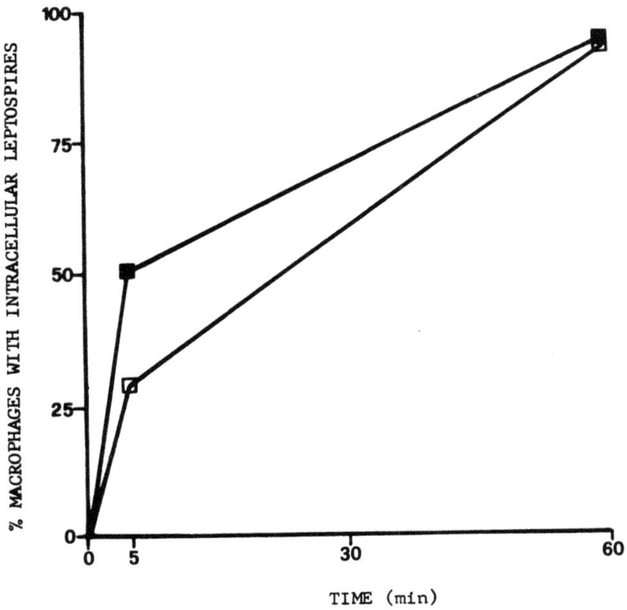

Fig. 4. Phagocytic activity of macrophages against opsonized leptospires, strain Isola Sacra 1 (■), and strain PB-3 (□).

REFERENCES

1. R.C. Johnson and L.H. Muschel, Antileptospiral activity of normal serum, J. Bacteriol. 89:1625 (1965).
2. R.C. Johnson and L.H. Muschel, Antileptospiral activity of serum.I. Normal and immune serum, J. Bacteriol. 91:1403 (1966).
3. R.C. Johnson and V.G. Harris, Antileptospiral activity of serum.II. Leptospiral virulent factors, J. Bacteriol. 93:513 (1967).
4. S. Faine, Reticuloendothelial phagocytosis of virulent leptospires, Am. J. Vet. Res. 25:830 (1964).
5. S. Faine, A. Shahar and M. Aronson, Phagocytosis and its significance in leptospiral infection, Aust. J. Exp. Biol. Med. Sci. 42:579 (1964).
6. M. Cinco, E. Banfi and M. R. Soranzo, Studies on the interaction between macrophages and leptospires.I, J. Gen. Microbiol., in press.
7. S.A. Lukehart and J.N. Miller, Demonstration of the in vitro phagocytosis of T. pallidum by rabbit peritoneal macrophages, Infec. Immunity 121:2014(1978).

DISCUSSION

Edited by P. Elsbach

J. Breton-Gorius: After phagocytosis, have you observed the degranulation into the phagosome and the degradation of Leptospirae?

M. Cinco: We have not investigated this point.

R.J.H. Beelen: How did you elicit your macrophages and did you differentiate into different subpopulations of macrophages on the electron-microscopic level?

M. Cinco: We elicited our macrophages by intraperitoneal of 1% casein. We did not differentiate the subpopulations of macrophages.

RELATIONSHIP BETWEEN THE ENHANCED OXIDATIVE METABOLISM AND THE ENHANCED MICROBICIDAL ACTIVITY OF ACTIVATED MACROPHAGES

Richard B. Johnston, Jr., Masataka Sasada, Lindsay A. Guthrie, and Michael J. Pabst

National Jewish Hospital and Research Center and Departments of Pediatrics and Biochemistry, University of Colorado School of Medicine, Denver, Colorado 80206, USA

Macrophages harvested from animals recently infected by intracellular parasites such as mycobacteria exhibit an increase in size and spreading and other manifestations of a state of activation.[1,2] Macrophages elicited by injection of inflammatory agents, e.g., lipopolysaccharide (LPS), exhibit most of these properties. Of most direct importance to host defense, infection-activated macrophages have been reported to have an increased capacity to kill certain bacteria, viruses, and protozoa. The interrelationships between macrophage activation, phagocytosis-associated oxidative metabolism, and microbicidal activity have been reviewed recently.[3] Certain aspects of these interrelationships will be described in this and the accompanying report.[4]

MACROPHAGE ACTIVATION INCLUDES PRIMING FOR AN ENHANCED OXIDATIVE RESPONSE

Mouse peritoneal macrophages elicited by injection of LPS or obtained from mice infected with bacillus Calmette-Guérin (BCG) are primed compared with normal (resident) macrophages to release up to 17 times more superoxide anion (O_2^-) when stimulated by phagocytosis or contact with phorbol myristate acetate (PMA).[5] Similar results have been reported with H_2O_2 release from BCG-activated macrophages.[6] In fact, with respect to all aspects of the phagocytosis-associated respiratory burst that have been studied, elicited and infection-activated macrophages have responded more vigorously during phagocytosis or plasma membrane perturbation than have resident cells.[3]

The molecular basis of this priming of activated macrophages for an enhanced oxidative response has not been defined. Possibilities include modification of the plasma membrane so that contact with or binding of the stimulus is greater, increased efficiency of the "triggering" mechanism that lies functionally between plasma membrane perturbation and stimulation of the enzyme (oxidase) responsible for O_2^- production, or an increased content or efficiency of the oxidase. We have explored this question by comparing O_2^- production by subcellular fractions of resident and LPS-elicited macrophages. Peritoneal macrophages were obtained from mice injected intraperitoneally with LPS 4 days earlier or from control mice.[5] After culture overnight, the cells from both sources were washed, stimulated by exposure to PMA, then disrupted by sonication and centrifuged.[7] A plasma membrane-rich pellet was obtained by removing the supernate from centrifugation at 9,500g and centrifuging it again at 27,000g. Various concentrations of NADPH were added as electron donor, and O_2^- was quantitated as superoxide dismutase (SOD)-inhibitable reduction of cytochrome c.[7]

In our initial five paired experiments using PMA-stimulated cells we have found that the K_m for NADPH in the particulate fraction (PF) from LPS-elicited cells was 0.07 mM, compared to 0.12 mM for the PF from resident cells. The V_{max} was approximately doubled in the PF from LPS-elicited cells. At physiologic concentrations of NADPH, the velocity of the oxidase from LPS-elicited macrophages was calculated to be 3.3 times higher than that in the PF from resident cells. (In these experiments, the LPS-elicited cells were only mildly "activated"--the intact cells produced approximately 3 times more O_2^- than did resident cells when stimulated by PMA for 60 min.) Thus, it appears that the efficiency of the NADPH oxidase is increased in LPS-elicited cells compared to resident cells and, therefore, that the enzyme has been modified by the process of "activation." Content of the enzyme might also be increased in the LPS-elicited cells.

ENHANCED OXIDATIVE METABOLISM PLAYS AN IMPORTANT PART IN THE INCREASED MICROBICIDAL ACTIVITY OF ACTIVATED MACROPHAGES

Chronic granulomatous disease, as an experiment of nature, indicates that a normal oxidative burst during phagocytosis is required for microbicidal activity. The specific involvement of O_2^-, H_2O_2, and hydroxyl radical (•OH) in the killing of bacteria by neutrophils and monocytes has been indicated by the marked inhibition of such killing by SOD, catalase, and scavengers of •OH.[8,9] Recent work on the role of oxygen metabolites in the killing of microorganisms by macrophages has been reviewed.[3]

We have explored the role of oxidative metabolism in macrophage microbicidal activity using *Candida albicans* and *Candida parapsilosis*.[10] Our results indicate that efficient macrophage candidacidal activity depends on the generation of oxygen metabolites by the macrophage: (i) BCG-activated and LPS-elicited peritoneal macrophages, which release greater amounts of O_2^- than do resident macrophages when stimulated by PMA, opsonized zymosan, or candida, killed candida more effectively than did resident cells. (ii) Cells of the murine macrophage-like line J774.1, which produced < 10% as much O_2^- as resident macrophages did, could ingest the candida normally but could not kill them. (iii) Killing of both candida species by resident, elicited, and activated macrophages was inhibited effectively by SOD, catalase, and agents believed to scavenge •OH (benzoate) or singlet oxygen (azide). (iv) Resident, elicited, and activated macrophages killed *C. parapsilosis* much more effectively than *C. albicans*. *C. parapsilosis* stimulated a more prompt and vigorous consumption of oxygen and release of O_2^- than did *C. albicans*. Neither candida species scavenged O_2^- or H_2O_2 effectively when intact, and both species were equally susceptible to killing by oxygen metabolites. Thus the species that was more difficult to kill avoided full stimulation of the oxidative burst.[10]

Acknowledgments. This work was supported by USPHS grants AI 14148 and DE 05494 from the National Institutes of Health.

REFERENCES

1. M. L. Karnovsky and J. K. Lazdins, Biochemical criteria for activated macrophages, *J. Immunol.* 121:809 (1978).
2. Z. A. Cohn, The activation of mononuclear phagocytes: fact, fancy, and future, *J. Immunol.* 121:813 (1978).
3. R. B. Johnston, Jr., Enhancement of phagocytosis-associated oxidative metabolism as a manifestation of macrophage activation, *Lymphokine Reports*, in press (1981).
4. N. P. Cummings, M. J. Pabst, and R. B. Johnston, Jr., This volume.
5. R. B. Johnston, Jr., C. A. Godzik, and Z. A. Cohn, Increased superoxide anion production by immunologically activated and chemically elicited macrophages, *J. Exp. Med.* 148:115 (1978).
6. C. F. Nathan and R. K. Root, Hydrogen peroxide release from mouse peritoneal macrophages: dependence on sequential activation and triggering, *J. Exp. Med.* 146:1648 (1977).
7. L. C. McPhail, L. R. DeChatelet, and R. B. Johnston, Jr., Generation of chemiluminescence by a particulate fraction isolated from human neutrophils, *J. Clin. Invest.* 63:648 (1979).
8. R. B. Johnston, Jr., B. B. Keele, Jr., H. P. Misra, J. E. Lehmeyer, L. S. Webb, R. L. Baehner, and K. V. Rajagopalan, The role of superoxide anion generation in phagocytic bactericidal activity, *J. Clin. Invest.* 55:1357 (1975).
9. A. L. Sagone, Jr., G. W. King, and E. N. Metz, A comparison of the metabolic response to phagocytosis in human granulocytes and monocytes, *J. Clin. Invest.* 57:1352 (1976).
10. M. Sasada and R. B. Johnston, Jr., Macrophage microbicidal activity: correlation between phagocytosis-associated oxidative metabolism and the killing of candida by macrophages, *J. Exp. Med.* 152:85 (1980).

RELEASE OF SUPEROXIDE ANION AND ENHANCED CANDIDACIDAL ACTIVITY AS A MANIFESTATION OF MACROPHAGE ACTIVATION: STUDIES WITH MURAMYL DIPEPTIDE

Nancy P. Cummings, Michael J. Pabst, and Richard B. Johnston, Jr.

National Jewish Hospital and Research Center and Departments of Pediatrics and Biochemistry, University of Colorado School of Medicine, Denver, Colorado 80206, USA

N-acetylmuramyl-L-alanyl-D-isoglutamine, or muramyl dipeptide (MDP), is a synthetic glycopeptide which has been formulated to correspond to a component found in a water-soluble extract of the cell wall of mycobacteria. It is the minimal structure that is capable of duplicating Freund's complete adjuvant in the primary antibody response.[1,2] MDP added to in vitro systems stimulates a number of functions not necessarily related to the antibody response to specific antigens, such as release of pyrogens and lymphocyte activating factor from blood monocytes, enhanced collagenase and prostaglandin production from blood monocytes and mouse peritoneal cells, and inhibition of growth of mastocytoma cells.[2,3]

Macrophages obtained from the peritoneal cavity of mice treated with BCG or lipopolysaccharide (LPS), agents that act as adjuvants, display an enhanced ability to generate superoxide anion (O_2^-) in response to stimulation by opsonized zymosan or membrane-reactive agents such as phorbol myristate acetate (PMA).[4] These "activated" macrophages are better able to kill microorganisms and tumor cells.[5,6] Since MDP has been shown in in vitro systems to alter macrophage functions, we have explored the effect of MDP on macrophage generation of O_2^- and killing of candida. Studies have been performed after direct contact of MDP with cultured peritoneal cells[7] and with macrophages obtained after injection of MDP into the intact animal.[3]

ADDITION OF MDP TO MACROPHAGES CULTURED IN VITRO

Peritoneal cells were harvested from Swiss-Webster mice. Cells were adhered for 2 hr to tissue culture dishes in Dulbecco's modified Eagle's medium (DMEM) containing penicillin, streptomycin and 20%

heat-inactivated fetal calf serum (FCS). After vigorous washing, cells were greater than 97% phagocytic and appeared to be macrophages. When macrophages were cultured overnight with MDP, 37% of the cells showed spreading (defined as a length:width ratio > 1.5), compared to 25% of control cells. These cells were examined for their ability to produce O_2^- in response to stimulation with PMA (0.5 µg/ml), using as assay superoxide dismutase-inhibitable reduction of ferricytochrome _c_.[3] Results were expressed as nanomoles O_2^- released per mg cell protein. MDP was effective at priming macrophages for enhanced O_2^- generation over a wide range of concentrations, with the greatest effect at concentrations greater than 0.5 µg/ml, although activity persisted down to 0.05 µg/ml (Fig. 1). This capacity of MDP to prime macrophages was obtained within 4 hr after addition. Incubation with the D-D stereoisomer MDP, which is inactive as an adjuvant, did not prime the macrophages for enhanced O_2^- response.

The question of whether MDP acted directly on macrophages or interacted with lymphocytes to cause production of lymphokines was explored in several experiments.[6] For example, cells that were non-adherent after a 2-hr incubation were removed, washed, and added back to macrophage cultures. The presence of these non-adherent cells did not enhance the response of the macrophages to MDP. In addition, macrophages maintained in culture for 24 to 48 hr to further reduce the possibility of lymphocyte contamination, responded as well to subsequent priming by MDP as did cells whose priming began after the initial 2-hr adherence. Macrophages from nude mice, which lack mature T-cell function, responded as well to MDP as did control cells, suggesting that mature T-cell function is not essential for the MDP-mediated enhancement of O_2^- release.

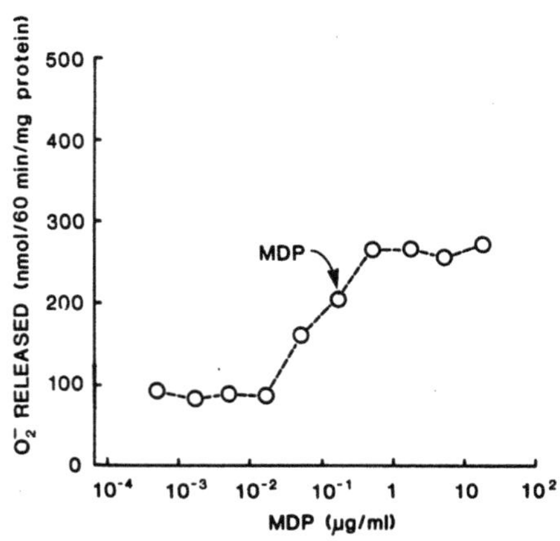

Fig. 1. Effect of concentration of MDP added to cultured macrophages on PMA-stimulated release of O_2^-. Means ± SEM of 6 experiments. (From reference 7, with permission)

STUDIES OF MACROPHAGES FROM ANIMALS INJECTED WITH MDP

When MDP was given subcutaneously to mice and peritoneal macrophages were harvested 18 hr later, macrophage spreading after a 2-hr adherence was present in 12% of cells from mice treated with MDP compared to 4% of cells from non-treated controls. Protein content of macrophages from animals treated with MDP was higher (140 μg protein/10^6 adherent cells) than the protein content of control macrophages (95 μg). Macrophages from mice treated with MDP were primed to release 4 to 5 times more O_2^- than macrophages from control mice (Fig. 2). By 1 hr after injection PMA-stimulated O_2^- generation was already enhanced, and the peak of the enhanced O_2^- response was present by 3 hr after MDP injection. This priming was more prompt than would be expected for a lymphokine-mediated effect. Macrophages from nude mice injected with MDP responded in a similar fashion to macrophages from their normal littermates, suggesting that mature T-cell function is not essential for the MDP-mediated enhancement of O_2^- release.

Macrophages from mice injected subcutaneously with MDP were examined for specific activities of the lysosomal hydrolase β-glucosaminidase and the cytoplasmic enzyme lactate dehydrogenase (LDH). Enhanced enzymatic activity has previously been described in macrophages elicited by thioglycollate, streptococcus A cell wall, and proteose peptone medium.[8] Specific activity of β-glucosaminidase was increased for macrophages from mice treated with MDP (173 compared to 116 nmol/

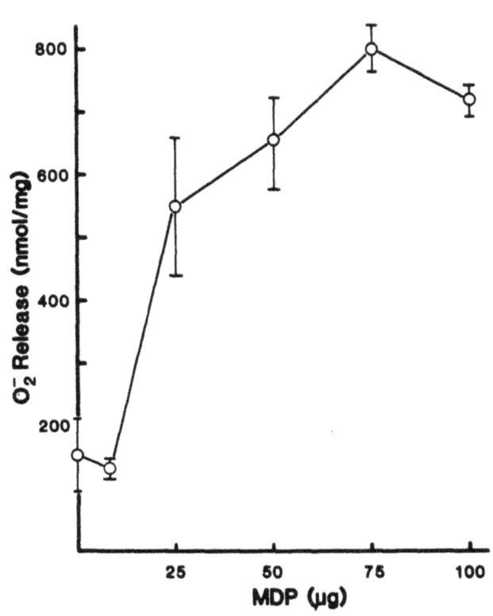

Fig. 2. Release of O_2^- by PMA-stimulated macrophages plotted as a function of the amount of MDP injected. Means ± SEM (bars) of results with macrophages from 4-6 mice are shown. (From reference 3, with permission)

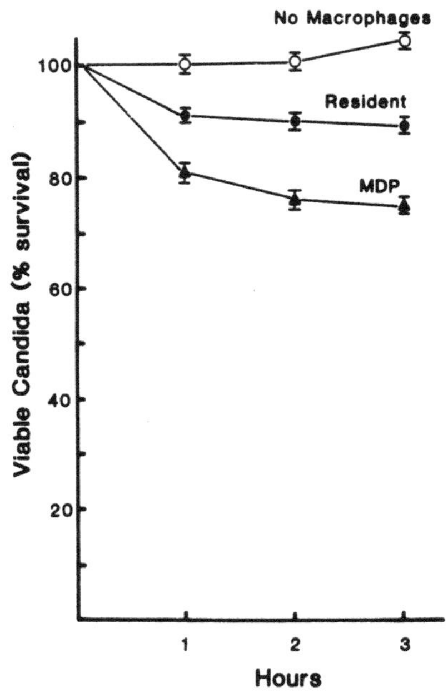

Fig. 3. Killing of C. albicans by macrophages from control or MDP-treated mice. Means ± SEM of 6 matched experiments. (From reference 3, with permission)

min/mg). Similarly, LDH activity was increased (1791 and 1143 units/mg for MDP and control, respectively). When macrophages were cultured for 24 or 48 hr in DMEM, penicillin and streptomycin, and 20% FCS, this relationship continued, with macrophages from mice treated with MDP always showing increased enzyme activity. Release of β-glucosaminidase into the medium was not significantly different for macrophages from control and MDP-treated animals. Release of LDH into the medium, as an indication of cell death, was also similar for MDP macrophages and control macrophages.

The capacity of macrophages from MDP-treated mice to phagocytize Candida albicans was compared to control macrophages.[6] This assay was performed on macrophages which had been cultured overnight. The candida-to-macrophage ratio was 3:1. At 10, 20, 30 and 60 min only a slightly higher percentage of macrophages from MDP-treated mice had ingested one or more candida.

Traditionally, macrophage "activation" has been defined as an enhanced ability to kill microorganisms or tumor cells. We were able to demonstrate enhanced candidacidal activity in macrophages from mice treated with MDP (Fig. 3). Neither control macrophages nor macrophages from mice treated with MDP killed candida easily; however, in our studies, macrophages from treated animals killed more than two times more candida than did control cells. These results are identical to those we have obtained with macrophages activated by infection of mice with BCG.[6]

As an in vivo correlate to the finding of enhanced macrophage killing ability in vitro, we were able to demonstrate protection by MDP against lethal infection by Candida albicans (Fig. 4). Mice were given 50 µg of MDP subcutaneously either 18 hr before, immediately after, or 24 hr after being infected intravenously with 2 x 10^6 C. albicans. Mortality was recorded for 16 days. There were 10 to 15 mice in each MDP-treated and each control group. As shown in Figure 4, top, MDP given 18 hr prior to candida challenge resulted in a 60% survival of the mice by day 7, when all control mice had died. In Figure 4, middle, survival was also 60% by day 12 in mice treated with MDP at the time of injection of candida. Some enhancement of survival was observed even when MDP was given one day following the candida challenge, with a 25% survival in 15 MDP-treated mice at day 12, compared to no survival of the 15 control mice (Fig. 4, bottom).

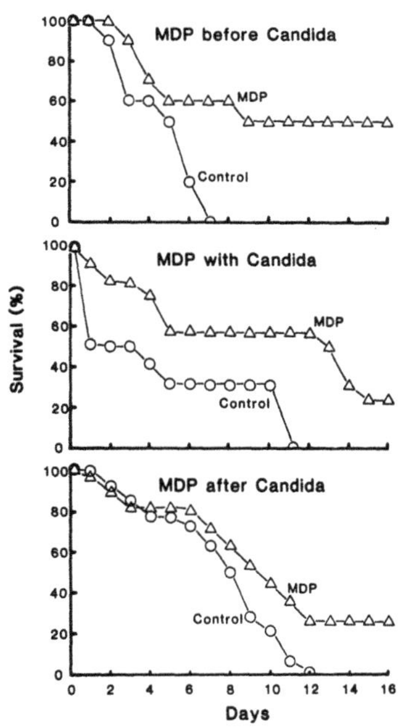

Fig. 4. Effect of injection of MDP into mice on their survival from infection with C. albicans. Top: MDP, 50 µg, or saline as control was administered subcutaneously to 10 mice 18 hr prior to administration of candida intravenously. Middle: MDP was administered to 12 mice within 1 min after candida challenge. Bottom: MDP was administered to 15 mice 24 hr after candida challenge. (From reference 3, with permission)

CONCLUSIONS

When cultured mouse peritoneal macrophages are incubated with MDP, they are primed to release several times more O_2^- after stimulation by PMA. This priming of peritoneal macrophages can be achieved by giving mice MDP subcutaneously. The priming process both in vivo and in vitro occurs within a matter of hours after exposure to MDP and is concentration-dependent. Manifestations of macrophage activation, such as increased spreading, are present with both in vivo and in vitro treatment with MDP. Total cellular content of the lysosomal hydrolase β-glucosaminidase and of the cytoplasmic enzyme LDH is increased when MDP is given to the intact mouse. Evidence suggests that MDP added in vitro achieves its effect through direct interaction with the macrophage.

To confirm macrophage activation by MDP, macrophages from mice pretreated with MDP consistently killed more *C. albicans* than did resident macrophages. The molecular basis for this enhanced killing capacity of activated macrophages has not been completely defined. However, it appears likely that it is related to the capacity of the macrophage to produce more O_2^- when stimulated.[6] The higher content of macrophage hydrolytic enzymes, as represented by β-glucosaminidase, may also be relevant to enhanced candidacidal activity. When injected subcutaneously, MDP protected animals against death from *Candida albicans* infection. Improved survival in mice pretreated with MDP has also been shown with lethal *Klebsiella pneumoniae*[9] and *Trypanosoma cruzi*[10] infections. Enhanced tumoricidal activity has been reported with certain MDP analogs when given in combination with trehalose dimycolate.[11]

MDP should serve as an important probe of macrophage physiology. Side effects noted thus far have been limited to mild pyrogenicity in rabbits.[12] The adverse side effects of BCG and larger cell wall extracts,[13] such as granuloma formation and adjuvant polyarthritis, are lacking when MDP is given in saline. Since MDP can induce macrophage activation even in the absence of mature T cells, MDP stimulation may be particularly useful in the immunocompromised host.

Acknowledgments. This work was supported by USPHS grants AI 14148 and DE 05494 from the National Institutes of Health.

REFERENCES

1. Kotani, S., Watanabe, Y., Kinoshita, F., Shimono, T., Morisaki, I, Shiba, T., Kusumoto, S., Tarumi, Y., and Ikenaka, K., 1975, Immunoadjuvant activities of synthetic N-acetyl-muramyl-peptides or -amino acids, *Biken J.* 18:105.
2. Chedid, L., Audibert, F., and Johnson, A. G., 1978, Biological activities of muramyl dipeptide, a synthetic glycopeptide analogous to bacterial immunoregulating agents, *Prog. Allergy* 25:63.

3. Cummings, N. P., Pabst, M. J., and Johnston, R. B., Jr., Activation of macrophages for enhanced release of superoxide anion and greater killing of Candida albicans by injection of muramyl dipeptide, J. Exp. Med., in press.
4. Johnston, R. B., Jr., Godzik, C. A., and Cohn, Z. A., 1978, Increased superoxide anion production by immunologically activated and chemically elicited macrophages, J. Exp. Med. 148:115.
5. Nathan, C. F., Brukner, L. H., Silverstein, S.C., and Cohn, Z. A., 1979, Extracellular cytolysis by activated macrophages and granulocytes. I. Pharmacologic triggering of effector cells and the release of hydrogen peroxide, J. Exp. Med. 149:84.
6. Sasada, M. and Johnston, R. B., Jr., 1980, Macrophage microbicidal activity: correlation between phagocytosis-associated oxidative metabolism and the killing of candida by macrophages, J. Exp. Med. 152:85.
7. Pabst, M. J. and Johnston, R. B., Jr., 1980, Increased production of superoxide anion by macrophages exposed in vitro to muramyl dipeptide or lipopolysaccharide, J. Exp. Med. 151:101.
8. Schnyder, J. and Baggiolini, M., 1978, Secretion of lysosomal hydrolases by stimulated and nonstimulated macrophages, J. Exp. Med. 148:435.
9. Chedid, L., Parant, M., Parant, F., Audibert, F., Lefrancier, F., Choay, J., and Sela, M., 1979, Enhancement of certain biological activities of muramyl dipeptide derivatives after conjugation to a multi-poly (DL-alanine)-poly (L-lysine) carrier, Proc. Natl. Acad. Sci. U.S.A. 76:6557.
10. Kierszenbaum, F. and Ferraresi, R. W., 1979, Enhancement of host resistance against Trypanosoma cruzi infection by the immunoregulatory agent muramyl dipeptide, Infect. Immun. 25:273.
11. McLaughlin, C. A. and Schwartzman, S. M., 1980, Regression of tumors in guinea pigs after treatment with synthetic muramyl dipeptides and trehalose dimycolate, Science 208:416.
12. Dinarello, C. A., Elin, R. J., Chedid, L., and Wolff, S. M., The pyrogenicity of the synthetic adjuvant myramyl dipeptide and two structural analogues, J. Infect. Dis. 138:760.
13. Kohashi, O., Kotani, S., Shiba, T., and Ozawa, A., 1979, Synergistic effect of polyriboinosinic acid: polyribocytidylic acid and either bacterial peptidoglycans or synthetic N-acetylmuramyl peptides on production of adjuvant-induced arthritis in rats, Infect. Immun. 26:690.

DISCUSSION
Edited by P. Elsbach

P. Rossi: 1) Don't you think that part of the increase in superoxide production in activated macrophages is due to an increase in release? In fact, you show that NADPH oxidase increases 3 times and superoxide recovery at least 6 times.
2) Secondly we have found no great difference in the degree of the respiratory response to PMA between peritoneal resident and elicited macrophages in guinea pig. If there is some difference, it is in favor of the resident macrophage.
3) Comment: We use the term elicited because the term activated might generate confusion. In my opinion activated means a different state of the same type of cells, induced by a particular treatment or a set of conditions. The elicited macrophages are a new population coming into the peritoneal cavity from the blood. They are not necessarily activated. They might be more active or less active! The macrophages that we raised seem to be less active compared to the resident cells, at least as far as the degree of the respiratory burst is concerned.
R. Johnston: 1) There certainly could be a difference in superoxide release. However, the whole cell superoxide release, determined on the same batches of cells used to obtain particulate fractions showed about a 30-fold difference, since the LPS cells were not very "activated". We have begun experiments in which we are attempting to relate the difference in whole cell superoxide release to that by particulate fractions made from the same cells.
2) I do not understand your findings with guinea pig cells. There is a great deal of data now from many laboratories indicating that elicited and activated macrophages have an increased respiratory burst (reviewed in Johnston, Lymphokine Reports, 1981).
3) Your point is a good one. There is no general agreement, as you know, in most of these situations as to whether original cells have been changed or new cells arrived. Our data, showing modification of resident cells *in vitro* with respect to at least some aspects of "activation", suggest that the same cell can differentiate into a more active cell (J. Exp. Med. Jan.,1980). The rapid changes in intraperitoneal cells produced by MDP injection, shown by Dr. Cummings, also favors such a differentiation.

R. Gallily: I believe that in order to obtain conclusive evidence that MDP potentiates directly bactericidal activity of the macrophage one should add this agent *in vitro* to macrophage cultures and thereafter examine the bactericidal activity of the cells.

In our study (published in Infect. Immun. 1978) we demonstrated that when MER (methanol extracted residue of BCG) was added to macrophages *in vitro*, an increase in hydrolytic enzyme occurred whereas no effect could be detected on phagocytic or bacteriostatic capacity of the cells.

Potentiation of phagocytic and bacteriostatic activity of

macrophages was obtained only following incubation of macrophages with supernatants of lymphocytes preincubated with MER.
N. Cummings: We have added MDP to in vitro macrophage cultures and found the macrophages were primed to release increased amounts of superoxide when stimulated by phorbal myristate acetate. We have not, however, conclusively demonstrated in vitro macrophage activation by microbicidal criteria, as we have with MDP given to the intact animal.

P. Elsbach: 1) What does MPD do to other metabolic activities than the respiratory burst?

For example, does MDP increase biosynthesis of protein (bactericidal factors, including candidacidal protein (Lehrer))?

2) Different kinetics of oxidases. Does this reflect appearance of another enzyme with different properties? In other words are your results attributable to the presence of 2 enzymes, either in one population, or in two populations of cells?
R. Johnston: 1) It is highly likely that non-oxidative microbicidal properties are also increased in cells from MDP-treated mice. The increased in vitro killing and protection against death could easily also be due to increases in these factors, whatever they might be. 2) Our data suggest only that there is a more efficient enzyme present in particulate fractions from LPS-elicited cells. We have done preliminary studies with fractions from cells stimulated in vitro by LPS, but we do not have sufficiently firm data yet. Therefore, our increased activity could be that of a second population of cells. It is still an increased activity (efficiency), however.

V. Esmann: The kinetic constants, Km and Vmax, of the NADPH-oxidase from activated macrophages were both changed compared to the oxidase from resident cells. I wonder how well the experimental data fit hyperbolic kinetics - or whether you perhaps have a more complicated system.
R. Johnston: I apologize for not showing you the plots, which we deleted because of time. If we obtained peritoneal cells from about 90 mice, we had enough material to study 4 or 5 concentrations of NADPH in duplicate in each experiment and to develop fairly decent and consistent curves. There certainly may be more enzyme in the activated cells, as indicated by the increased Vmax, but there also appears to be a change in the efficiency of the enzyme.

KILLING OF *LEISHMANIA DONOVANI* AMASTIGOTES BY MURINE MACROPHAGES

Peter F. Bonventre and Constantine G. Haidaris

Department of Microbiology
University of Cincinnati College of Medicine
Cincinnati, Ohio USA

INTRODUCTION

Activated tissue macrophages are the effector cells responsible for containment of a variety of intracellular, bacterial, mycotic and protozoan parasites. The term "activated" macrophage employed originally to describe enhanced bactericidal capability (1) has been extended in recent years to tumoricidal activity (2,3). An activated macrophage system arises as a consequence of stimuli provided by one or more soluble mediators of T lymphocyte origin (4). The mechanisms responsible for destruction of intracellular parasites and killing of tumor cells by activated macrophages may be related (5,6).

Leishmania donovani, the agent of visceral leishmaniasis, parasitizes host tissue macrophages, particularly those of the liver and spleen. This and other Leishmania species exist as two distinct morphological entities. The flagellated promastigotes develop in the gut of the phlebotomine arthropod vector and convert to the amastigote form in tissue macrophages after introduction of the parasite into a suitable vertebrate host. Since promastigotes can be readily propagated in cell free culture media, considerable information regarding physiology and biochemistry of this leishmanial form is available (7,8,9). The amastigote, an obligate intracellular parasite has been little studied from a biochemical standpoint, and thus virtually nothing of its physiology and metabolism is published. Together with our colleague A.J. Mukkada, we are initiating studies toward this end by utilizing macrophage tumor cell lines as well as infected tissues as a source of *L. donovani* amastigote forms.

The data presented here provide convincing evidence that activated macrophages can effectively eliminate intracellular amastigotes under appropriate experimental conditions. A requirement for prolonged maintenance of the activated state suggests that killing of L. donovani is a more rigorous functional criterion of macrophage activation than is tumoricidal activity. As far as we know, this is the first report of direct macrophage killing of this important human pathogen.

The experimental model chosen to study the leishmania-macrophage relationship was based on previous studies conducted by Mauel et al. (10) who noted that mouse peritoneal macrophages serve as excellent host cells for Leishmania enrietti amastigotes. It was noted in those studies that mouse macrophages are non-per missive; i.e., they do not support intracellular multiplication of amastigotes. Similarly, we found that L. donovani retain viability within the parasitophorous vacuole of macrophages during several weeks of in vitro culture, but no intracellular multiplication occurs. Our study utilized peritoneal exudate cells harvested from C57Bl/6 mice and established the adherent cell population as monolayer cultures. The Syrian Golden hamster which is susceptible to L. donovani infection, develops an enormous parasite burden in the liver and spleen during late stages of disseminated leishmaniasis. The infected spleen served as the source of amastigotes used to infect macrophages in vitro.

MATERIALS AND METHODS

In all cases peritoneal macrophages from female C57Bl/6 mice were used. The state of macrophage activation was assessed by tumor cell cytostasis, tumor cell killing and intracellular destruction of L. donovani amastigotes. A comparison of the macrophage's capacity to be cytotoxic for tumor cell targets and ability to kill ingested amastigotes provided a convenient model system to assess pertinent functional characteristics associated with the activated macrophage.

Tumor cell cytostasis

Inhibition of growth of tumor cells was measured by ^{3}H-thymidine incorporation (11). EMT 6, a Balb/c mammary carcinoma cell line was the tumor cell target. Peritoneal exudate cells suspended in 0.5 ml Minimal Eagle's Medium (MEM) + 10% fetal calf serum (FCS) were added at a concentration of 2 X 10^6 per ml to 16 mm tissue culture wells for 1 hr at 37 C in 5% CO_2. Nonadherent cells were removed and 5 X 10^4 EMT 6 tumor cells added to each well. After 48 hr incubation in 2.0 ml MEM, the cell mixture was washed thoroughly with MEM and the cells were pulsed for 1 hr with

^{3}H-thymidine (New England Nuclear; sp. act. 20 Ci/mmole) and 10^{-6} M fluorodexyuridine (Sigma Chemical Co.). Cells were solubilized with 0.05% Triton X and radioactivity measured by liquid scintillation counting.

Tumor cell killing

The cytotoxicity expressed by macrophages for EMT 6 tumor cells was measured by a qualitative "spot test" (12,13) and a quantitative clonogenic assay developed by Thomossen and Stewart (14). The spot test is based on a visible clearing of tumor cells in an area of a monolayer in contact with tumoricidal adherence macrophages. One tenth ml of peritoneal exudate cells (PEC) containing 4 X 10^5 cells was placed in the middle of a 35 mm plastic dish (Falcon). Nonadherent cells were removed after 1 hr and EMT 6 cells were added at a concentration so as to form a confluent monolayer after three days. After washing, the cells were fixed with methanol and Giemsa stained. A zone of clearing over the area of macrophages indicative of tumor cell killing was considered a positive tumoricidal test. The quantitative clonogenic assay measures the number of EMT 6 tumor cells which retain viability after interaction with macrophages for 48 hr. The assay determines the number of tumor cells able to form discrete colonies before and after interaction with macrophages and is analagous to determination of colony forming units of bacteria by dilution.

Leishmanicidal Activity of Macrophages

L. donovani Sudan strain 1S was maintained by serial passage in hamsters. Every two months, amastigotes were obtained from infected spleens by the method of Brun et al. (15). On average, a yield of 10^9 amastigotes per spleen was obtained. Dilutions were made in MEM + 10% FCS and coverslip cultures of macrophages were infected for 4 hr over a range of multiplicity of infection between 10:1 and 100:1. Amastigotes not phagocytized after this time were removed by washing of the monolayers and the infected macrophages reincubated in MEM + 10% FCS. The course of infection was followed by microscopic examination of Wright stained coverslip cultures. At least 500 infected macrophages were scored for each sample taken in duplicate for each time point. Estimation of amastigote killing by macrophages was based on the physical elimination of ingested parasites. Killing appears to precede digestion and disappearance of the amastigote within phagocytic vacuoles by approximately 1 day. Mauel et al. (11) established this relationship with *L. enrietti* by correlating conversion of amastigotes to promastigotes in vitro with estimates of destruction by microscopic examination of infected mouse macrophages. Our data would suggest that the same relationship holds for *L. donovani*.

Types of Macrophages

Peritoneal macrophages were elicited and/or activated in several ways. Thioglycollate (3% Brewer's) in 2.0 ml amounts intraperitoneally 4 days prior to harvest was used to elicit PEC. These were activated in vitro with concanavalin A (Con A) stimulated spleen cell culture supernatants according to standard techniques (16). Preactivated macrophages were obtained from mice injected i.p. 7 days previously with 700 µg Corynebacterium parvum (Burroughs Wellcome, Durham N.C.). Alternatively, activated peritoneal macrophages were obtained from 21 day Mycobacterium tuberculosis ($H_{37}R_a$) infected mice stimulated with 50 ug purified protein derivative (PPD) 2 days prior to harvest of PEC.

RESULTS

Our preliminary efforts were directed toward establishing conditions required to demonstrate tumor cell cytostasis and tumoricidal activities of peritoneal macrophages. This was a necessary prerequisite so that we could then correlate killing of tumor cell targets with leishmanicidal activity.

Cytostatic and Cytotoxic Activities of Macrophages in vitro

Peritoneal macrophages elicited with thioglycollate and activated with lymphokine containing spleen cell supernatants were tested for ability to inhibit the growth or EMT 6 cells by the ^{3}H-thymidine assay and also tested for tumoricidal activity by the spot test and clonogenic assays. Table 1 shows that elicited cells (Thio Mϕ) are neither tumoristatic or tumoricidal. Addition of lymphokine (30%) or normal spleen cells (NSC) and conconavalin A (Con A; 5 µg/ml) for 48 hr however, resulted in highly activated macrophage populations as measured by killing of the tumor cell target. All three assays for anti-tumor cell function correlated well.

Killing of L. donovani by thioglycollate elicited macrophages

Figure 1 shows that the amastigotes survive within inflammatory peritoneal macrophages for the 12 day period of the experiment. This shows the similarity in behaviour of the L. donovani mouse macrophage relationship with that of the guinea pig pathogen L. enritti (11).

Table 1. TUMOR CELL KILLING BY ACTIVATED MACROPHAGES

Macrophage Population	Cytostasis ^{3}H-Thymidine Uptake (% Inhibition)	Cytotoxicity Spot Test	Cytotoxicity Clonogenic Assay (% Killing)
1. Thio Mϕ	0	Negative	18
2. Thio Mϕ + NSC + Con A	99	++++	92
3. Thio Mϕ + NSC	22	Negative	14
4. Thio Mϕ Lymphokine	96	++++	90

NSC = Normal Spleen Cells.
Mϕ = Macrophages

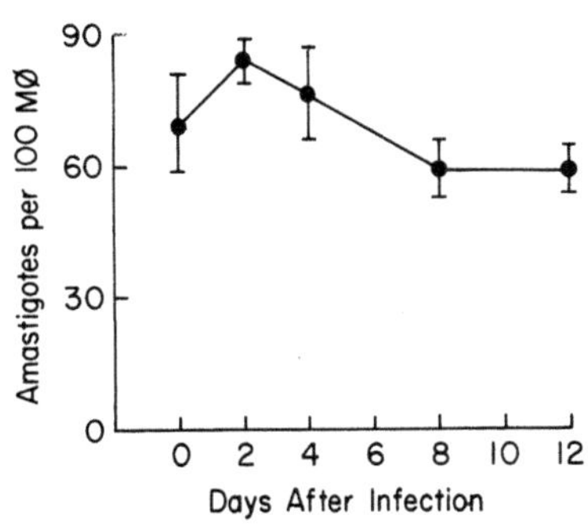

Fig.1 Survivial in elicited macrophages.

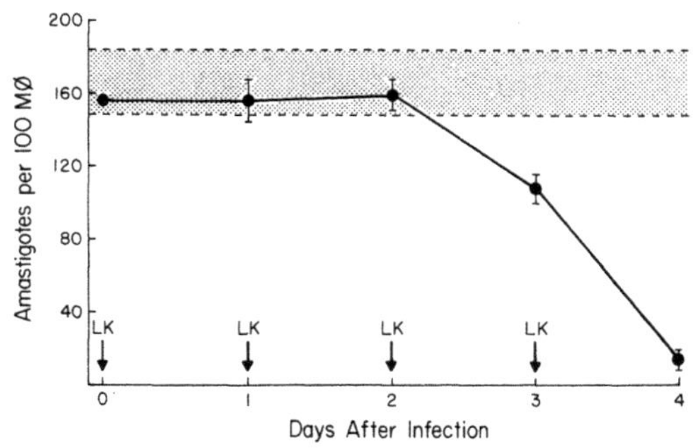

Fig. 4.

When elicited macrophage monolayers are treated with lymphokine containing spleen cell supernatants for 24-48 hours the cells become tumoricidal (data not shown). In contrast, the capacity to kill L. donovani is acquired only after maintenance of the activated state by daily replenishment of the lymphokine containing medium for 4 days. Figure 2 shows that some amastigotes are eliminated with 3 pulses of lymphokine (LK) but efficient destruction of the parasite requires all 4 daily pulses with LK. The shaded area in Figure 2 represents numbers of amastigotes observed in infected macrophages incubated with culture medium containing normal spleen cell supernatants at the same concentration as the LK containing preparations (i.e., 30%). It was found that killing of L. donovani was dependent upon the concentration of lymphokine used since leishmanicidal activity diminished considerably when used at 15% (v/v) and disappeared at a concentration of between 5 and 10%. The data allow us to conclude that tumoricidal activity of macrophages is acquired more readily than leishmanicidal potential. Indeed, these two functions of the fully activated macrophage appear to be dissociated and acquired sequentially.

Comparison of Tumoricidal and Leishmanicidal Activities

In order to determine how stringent a criterion of macrophage activation elimination of intracellular L. donovani represents, we compared EMT 6 tumor cell killing and parasite elimination by several different macrophage populations which had been preactivated in vivo or and/or stimulated in vitro with lymphokines. The data in Table 2 compared these activities by macrophages derived from thioglycollate elicited and C. parvum stimulated mice.

Table 2. TUMORICIDAL AND LEISHMANICIDAL ACTIVITIES OF MACROPHAGES

Macrophage Population	Tumoricidal Activity: Spot Test	Tumoricidal Activity: ^{3}H-Thymidine uptake (% Inhibition)	Killing of Leishmania (Amastigotes/100 Mϕ): 24 hr	Killing of Leishmania (Amastigotes/100 Mϕ): 96 hr
1. Thio Mϕ	Negative	None	123	177
2. C. parvum Mϕ	++++	90	278	239
3. Thio Mϕ + Lymphokine	++++	99	151	2
4. C. parvum Mϕ + Lymphokine	++++	99	219	30

As noted by others (17) the inflammatory macrophages are neither cytostatic nor cytotoxic for tumor cells. Macrophages from C. parvum treated mice, in contrast are activated by the criterion of tumor cell killing immediately after explant from the peritoneal cavity. Both cell populations, however fail to express leishmanicidal activity after 4 days of infection. When either of these macrophage populations was treated with the lymphokine during the 4 day period, elimination of intracellular amastigotes occurred. Thus, this emphasizes again the fact that tumoricidal macrophages are insufficiently activated to kill L. donovani in vitro unless maintained in the activated state for several days. In another set of experiments utilizing macrophages preactivated by Myco. tuberculosis infection and PPD stimulation before collection of PEC, similar results were obtained (data not shown). Although the preactivated macrophages were tumoricidal and phagocytized considerably more amastigotes than did inflammatory macrophages, both cell types required the lymphokine stimulus for 3-4 days before the parasites were killed and eliminated.

DISCUSSION

The criteria of tumor cell killing and enhanced microbial capabilities of murine peritoneal macrophages modulated in several ways were compared. In addition to establishing for the first time that mononuclear phagocytes can be stimulated sufficiently to kill the intracellular form of human visceral leishmaniasis, our observations allow several other conclusions to be drawn:
1) Murine peritoneal macrophages serve as suitable host cells for L. donovani amastigotes but do not permit intracellular multiplication. Whether the non-permissive nature of the host parasite relationship, also reflected by the natural resistance of mice to disseminated infection, can be ascribed to nutritional or immunological factors remains to be determined. 2) Macrophages activated by appropriate stimuli (lymphokine, C. parvum or mycobacterial infection) to become tumoricidal are not able to kill the tissue form of L. donovani. This demonstrates the high resistance of this human pathogen to the considerable antimicrobial mechanisms possessed by "activated" macrophages. The data show unequivocally that the tumoricidal macrophage is insufficiently activated to kill L. donovani. 3) Macrophages activated in vivo or in vitro and stimulated with lymphokine containing preparations for a sufficiently long period (72-96 hr) kill intracellular amastigotes. Indeed, parasite burdens as high as 50-100 organisms per cell were eliminated by macrophages stimulated with lymphokine for 4 days. Thus, the unusually high resistance of L. donovani to the microbicidal mechanisms of mononuclear phagocytes can be overcome provided the activating stimulus is maintained for several days. 4) Lastly, it may be concluded that killing of L. donovani is a much more rigorous functional test of macrophage activation than is tumor cell killing. It will be instructive to test several other tumor cell targets to establish if this relationship can be generalized.

Preliminary experiments suggest that specific opsonizing antibody against L. donovani does not modify the resistance of the parasite to intracellular killing as is the case with Toxoplasma gondii (18). We are now beginning experiments to ascertain if killing by activated macrophages of this protozoan parasite is dependent upon hydrogen peroxide and other intermediates of molecular oxygen (19,20).

REFERENCES

1. G.B. Mackaness, Cellular resistance to infection, J. Exp. Med. 116:381 (1962).
2. R. Evans and P. Alexander, Mechanisms of extracellular killing of nucleated mammalian cells by macrophages, in "Immunobiology of the Macrophage", D.S. Nelson, ed., Academic Press, New York, (1976).
3. J.B. Hibbs, Jr., L.H. Lambert, Jr., and J.S. Remington, Control of carcinogenesis: a possible role for the activated macrophage, Science, 177:998 (1972).
4. H.B. Simon and J.N. Sheagren, Enhancement of macrophage bactericidal capacity by antigenically stimulated immune lymphocytes, Cell. Immunol., 4:163 (1972).
5. C. Nathan and Z. Cohn, Role of oxygen-dependent mechanisms in antibody-induced lysis of tumor cells by activated macrophages, J. Exp. Med., 152:198 (1980).
6. H.W. Murray and Z.A. Cohn, Macrophage oxygen-dependent antimicrobial activity. I.Susceptibility of Toxoplasma gondii to oxygen intermediates, J. Exp. Med., 150-938 (1979).
7. L. Simpson, The leishmania-leptomonad transformation of Leishmania donovani: Nutritional requirements, respiration changes and antigenic changes, J. Protozool., 15:201 (1968).
8. E. Martin and A.J. Mukkada, Respiratory chain components of Leishmania tropica promastigotes, J. Protozool., 26:138 (1979).
9. M.W. Simon, E. Martin, and A.J. Mukkada, Evidence for a functional glycoxylate cycle in the Leishmaniae, J. Bact., 135:895 (1978).
10. J. Mauel, Y. Buchmuller and R. Behin, Studies on the mechanisms of macrophage activation. I Destruction of intracellular Leishmania enriettii in macrophages activated by cocultivation with stimulated lymphocytes, J. Exp. Med., 148:393 (1978).
11. A.M. Kaplan, P.L. Walker, and P.S. Morahan, Tumor cell cytotoxicity versus cytostasis of pyran activated macrophages, in "Modulation of Host Immune Resistance in the Prevention or Treatment of Induced Neoplasias", M.A. Chirigos, ed., DHEW Publication No. (NIH) 77-893, Washington, D.C. (1974).
12. D.L. Thomasson and C.C. Stewart, Strain differences of tumoricidal actions by mouse peritoneal macrophages, Abstracts of Fifteenth National Meeting of the Reticuloendothelial Society (1978).
13. J.B. Hibbs, Jr., R.R. Taintor, H.A. Chapman, Jr., and J.B. Weinberg, Macrophage Tumor Killing: Influence of the Local Environment, Science, 197:279 (1977).
14. D.L. Thomasson and C.C. Stewart, J. Immunol Methods (in press), (1981).

15. R. Brun, R.L. Berens, S.M. Krassner, Inhibition of L. donovani transformation by hamster spleen homogenates and active human lymphocytes, Nature, 262:689 (1976).
16. Y. Buchmüller and J. Mauel, Studies on the mechanisms of macrophage activation. II Parasite destruction in macrophages activated by supernatants from concanavalin A stimulated lymphocytes, J. Exp. Med., 150-359 (1979).
17. Z.A. Cohn, The activation of mononuclear phagocytes: fact, fancy and future, J. Immunol., 121:813 (1978).
18. J.C. Jones, L. Len, and J.G. Hirsch, Assessment of in vitro immunity against Toxoplasma gondii, J. Exp. Med. 141:466 (1975).
19. H.W. Murray and Z.A. Cohn, Macrophage oxygen-dependent antimicrobial activity. III Enhanced oxidative metabolism as an expression of macrophage activation, J. Exp. Med., 152:1596 (1980).
20. M. Sasada and R.B. Johnston, Jr., Macrophage microbicidal activity. Correlation between phagocytosis-associated metabolism and the killing of Candida by macrophages, J. Exp. Med., 152:85 (1980).

SELECTIVE DEPRESSION OF PHAGOCYTES INTRACELLULAR KILLING ACTIVITY

Daniela Iannello, Letterio Bonina, and Demetrio Delfino

Institute of Microbiology, University of Messina, Italy

INTRODUCTION

Valid resistance to infections is the result of a number of complex interactions between infectious agents and host's immune response. Virulence of microorganisms, genetic factors, host's immune mechanisms may in different ways influence the infectious agent host relationship. Phagocytes, including polymorphonucleates and macrophages, are the final effector cells in the elimination of a variety of bacteria and fungi.[1]

Recovery from infections due to extracellular bacteria occurs mainly through specific antibodies and polymorphonucleates, while infections due to intracellular parasites are efficiently circumscribed by mechanisms of cellular immunity, including cooperation between T lymphocytes and macrophages.[2] In some cases, however, a naturally occurring activation of macrophages can lead by itself to an efficient host defense against bacteria, as shown by Nickol[3] in congenitally athymic nude mice.

Defense against Listeria monocytogenes at least in the early phases of infection appears to depend mainly on macrophages, while protection against Pseudomonas aeruginosa appears to depend mainly on polymorphonucleates phagocytes, as recently shown by Tatsukawa.[4]

Infectious agents by themselves may influence host's immune response. Systemic candidiasis is associated with suppression of mitogen or antigen-induced blastogenesis in mice;[5] many bacterial products on the other hand are able to inhibit immune system cells functions.[6]

Conditions as hypogammaglobulinemia, granulocytopenia, depressed cellular immunity are often associated with opportunistic infections sustained by different pathogens.

Such infections are frequent in neoplastic patients in which many authors demonstrated a depression in monocyte-macrophages activity. In these patients and in experimental tumor-bearing animals depressed phagocytic activity, macrophage migration and chemotactic response were described.[7,8]

In previous experiments we demonstrated, in an experimental model, a marked depression of phagocytosis of latex particles and microorganisms by peritoneal macrophages obtained from tumor-bearing rats. The extent and time-course of this defect was, however, different, as previously reported,[9,10] when latex particles or microorganisms of different types were incubated with macrophages.

MATERIALS AND METHODS

In all experiments Norvegicus strain rats of either sex, three months old, were used. Atypic transplantable epithelioma T8 of Guerin was employed. The tumor was transplanted in the rat's dorsal subcutaneous region, and tumor-bearing rats were sacrificed five, ten, and twenty days after tumor graft.

Peritoneal cells were harvested three days after i.p. injection of proteose peptone in saline, repeatedly washed and resuspended in MEM to obtain a suspension containing 1×10^6 peritoneal cells/ ml.

Bacterial suspensions containing 1×10^6 colony forming units (cfu) /ml were prepared from 12 hours cultures of Escherichia coli and Staphylococcus aureus in broth. Candida albicans was grown 48 hours in Sabouraud, then resuspended in MEM to obtain 1×10^6 cfu/ml.

A micromethod to assess phagocytosis and intracellular killing by macrophages was employed, originally described by Smith and Rommel[11] and modified. Peritoneal cells after adhesion to sterile coverglasses in Petri dishes were incubated with microorganisms for 45 min. in 5 % CO_2 incubator.

Slides were then stained for 1 min. with acridine orange solution containing 1.44 mg/100 ml and observed in U.V. light. A differential count was carried out of cells containing viable or killed microorganisms, respectively assuming green or red-orange staining. The results were expressed as percentage of positive macrophages containing viable or killed microorganisms.

RESULTS AND DISCUSSION

Figure 1 shows the phagocytic activity of peritoneal macrophages obtained from tumor-bearing rats, five, ten, or twenty days after tumor graft respectively, determined by staining with acridine orange. The 81 %, 84 %, and 86 % of macrophages from control rats were able to phagocytize different microorganisms after 45 min. incubation.

Five days after graft, only 50 % of macrophages were able to phagocytize E. coli, 56 % S. aureus and 52 % C. albicans. Ten days after tumor graft, the percentage of positive cells containing different microorganisms was 66 %, 73 %, and 49 % of the total counted macrophages.

Twenty days after graft, 98 % and 73 % of macrophages were able to phagocytize E. coli or S. Aureus, while 84 % contained C. albicans.

Table I represents the percentage of macrophages containing viable or killed E. coli in control or tumor-bearing rats at different times from tumor graft. Although there was a depression of phagocytic activity, intracellular killing of E. coli was similar to controls.

In Table 2 the percentage of cells containing killed S. aureus in control and tumor-bearing rats is reported. The percentage of macrophages containing killed bacteria after 45 min. incubation was not significantly different from controls.

Table 3 shows the results obtained after incubation of macrophages from tumor-bearing rats, at different times from graft, with a suspension of C. albicans.

Macrophages obtained 5 days after graft showed a depression both in phagocytic activity and in intracellular killing. The 32.5 % of positive cells in fact contained living C. albicans. Cells obtained ten days after graft contained viable Candida, while twenty days after

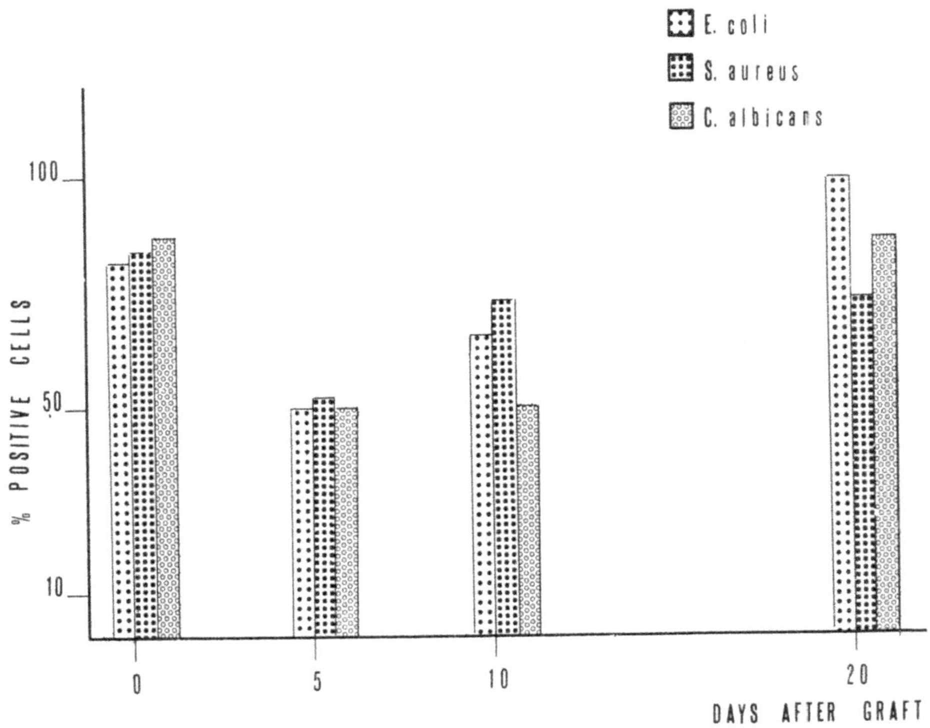

Fig. 1 Phagocytic activity of peritoneal macrophages from tumor-bearing rats at different times from graft.

graft the percentage of cells containing viable microorganisms was similar to controls.

In previous experiments[9] we showed a depression in the phagocytic activity of peritoneal macrophages from tumor-bearing rats; this reached the maximum ten days after graft when latex particles were used to assess phagocytosis.

However, tests employing living microorganisms are more reliable in the evaluation of complex interactions between phagocytes and the phagocytized particles. Such processes can in different ways be influenced by many intrinsic or extrinsic conditions.

Preliminary data obtained in our experimental model show a different behaviour of phagocytes from tumor-bearing rats , when incubated with microorganisms taxonomically different. Depression of phagocytosis of E. coli

Table 1 - Phagocytosis and intracellular killing determined by acridine orange staining

Macrophages source	Phagocytosis and killing[a] Escherichia coli			
	N	P	K	L
Control rats	19	81	93	7
Tumor-bearing rats				
- 5 days after graft	50	50	93	7
- 10 days after graft	34	66	96	4
- 20 days after graft	2	98	97	3

Table 2 - Phagocytosis and intracellular killing determined by acridine orange staining

Macrophages source	Phagocytosis and killing[a] Staphylococcus aureus			
	N	P	K	L
Control rats	15.5	84.5	99	1
Tumor-bearing				
- 5 days after graft	44	56	98	2
- 10 days after graft	27	73	93	7
- 20 days after graft	16	84	98	2

[a] N=% negative cells; P= % of positive cells; K= % of cells containing killed microrganisms; L= % of cells containing viable microrganisms.

and S. aureus in fact was evident at different times from graft, while intracellular killing was not affected.

Macrophages incubated with C. albicans on the other hand showed a marked depression of phagocytosis and intracellular killing. Ten days after graft both functions were depressed and returned to normal values only twenty days after graft.

The observed depression of phagocytic function was longer lasting when cells were incubated with fungi than with bacteria. In previous experiments carried out on the same experimental model we showed[12] a marked depression of phagocytosis and killing of Listeria monocytogenes.

Many authors described depression of monocyte-macrophages and polymorphonuclear phagocytes in neoplastic patients and in experimental tumors-bearing animals[13,14].

Table 3 - Phagocytosis and intracellular killing determined by acridine orange staining

Macrophages source	Phagocytosis and killing [a] Candida albicans			
	N	P	K	L
Control rats	13.75	86.25	93	7
Tumor-bearing rats				
- 5 days after graft	47.5	52.5	67.5	32.5
- 10 days after graft	51.5	49.5	77	23
- 20 days after graft	16	84	98	2

[a] N= % negative cells; P= % of positive cells; K= % of cells containing killed microorganisms; L= % of cells containing viable microorganisms.

On the other hand, some cases of selective deficiency of intracellular killing of C. albicans or S. aureus were described. In our experimental model, however, independently from the microorganisms employed to assess phagocytosis and intracellular killing, the observed depression of both functions is a transitory phenomenon. In fact, values obtained in tumor-bearing rats twenty days after graft were not significantly different from controls.

These observations suggest the need for a further investigation of these particular aspects of monocytes-macrophages functions in tumor-bearing animals and in neoplastic patients, in order to better understand the biology of complex tumor-host's immune system interactions.

REFERENCES

1. D.S. Nelson, "Immunobiology of the Macrophage," Academic Press, New York (1976).
2. R. Van Furth, "Mononuclear Phagocytes in Immunity, Infection and Pathology: Conference on mononuclear Phagocytes," Blackwell, Oxford, (1975).
3. A.D. Nickol and P.F. Bonventre, Anomalous high resistance of athymic nude mice to bacterial pathogens. Infect. Immun. 18/3: 636 (1977).
4. K. Tatsukawa, M. Mitsuyama, K. Takeja, and K. Nomoto. Differing contribution of polymorphonuclear cells and macrophages to protection of mice against Listeria monocytogenes and Pseudomonas aeruginosa, J. Gen. Microbiol. 115: 161 (1979).
5. T.J. Rogers and E. Balish, Suppression of lymphocyte blastogenesis by Candida albicans, Clin. Immunol. Immunopathol. 10/3: 298 (1978).
6. R.J. Smialovica and J.H. Schwab, Inhibition of macrophage phagocytic activity by group A streptococcal cell walls, Infect. Immun. 20/1: 258 (1978).
7. D.A. Boetcher and E.J. Leonard, Abnormal monocyte chemotactic response in cancer patients, J. Natl. Cancer Inst. 52: 1091 (1974).
8. R. Snyderman, M.C. Pike, B.L. Blaylock and P. Weinstein, Effect of neoplasms on inflammation: depression of macrophage accumulation after tumor implantation, J. Immunol. 116: 585 (1976).
9. L. Bonina, D. Iannello, and P. Mastroeni, Macrophage functions in rats treated with bacterial adjuvants, Giorn. Batt. Virol. Immun. LXXII:

278 (1979).

10. D. Iannello, L. Bonina, and M. Carbone, Comparazione di due differenti metodi per la valutazione della attivita fagocitaria e di killing intracellulare "in vitro" dei macrofagi di ratto, in press on Abstracts XIX Congr. Naz. Soc. Ital. Microbiol., Catania (1980).
11. D.L. Smith and F.J. Rommel, A rapid micromethod for the simultaneous determination of phagocytic-microbiocidal activity of human peripheral blood leukocytes in vitro, J. Immunol. Meth. 17: 241 (1977).
12. L. Bonina, A. Focà, D. Iannello, R. Merendino, and D. Delfino, Valutazione di alcuni parametri di funzionalità macrofagica in ratti trattati con adiuvanti batterici, in press on Giorn. Mal. Infett. Parass.
13. R. Snyderman and C. Stahl, Defective immune effector function in patients with neoplastic and immune deficiency diseases, in "The Phagocytic Cell in host resistance," J.A. Bellanti and D.H. Dayton, ed. Raven Press, New York, (1975).
14. R. Snyderman and M.C. Pike, Macrophage migratory disfunction in cancer, a mechanism for subversion of surveillance, Am. J. Pathol. 88: 727 (1977).

MODULATION OF THE INFLAMMATORY RESPONSE BY THE NEUTROPHIL MYELOPEROXIDASE SYSTEM*

Robert A. Clark

Evans Memorial Department of Clinical Research
and Department of Medicine
Boston University Medical Center
Boston, MA. 02118
USA

INTRODUCTION

The polymorphonuclear leukocyte (PMN)† is a highly specialized cell geared to the destruction of ingested microorganisms through the concentration of toxic agents within phagocytic vacuoles. Among the microbicidal systems of the PMN is one comprised of the granule enzyme, myeloperoxidase (MPO), the product of oxygen metabolism hydrogen peroxide, and a halide cofactor such as chloride (1,2). During phagocytosis or on exposure to certain soluble activating agents, the components of the MPO system are secreted into the extracellular fluid where they exert such biologic functions as the lysis of mammalian cells (3-5), the killing of hyphal forms of invasive fungal pathogens (6), and the oxidative inactivation of humoral mediators of inflammation (7,8). I will review some of our work on the inactivation of chemotactic factors and protease inhibitors by the isolated MPO system by stimulated human PMN.

*Supported by Grant CA24353 and Research Career Development Award CA00441 from the National Cancer Institute.

†Abbreviations: DABCO-diazobicyclo(2,2,2)octane; MPO-myeloperoxidase; α_1-PI-α_1-protease inhibitor; PMA-phorbol myristate acetate; PMN-polymorphonuclear leukocyte; SOD-superoxide dismutase.

INACTIVATION OF CHEMOTACTIC FACTORS

Effect of the MPO System and Intact PMN

The synthetic peptide chemoattractant, f-met-leu-phe, was incubated with the components of the MPO system or with PMN and then tested for residual chemotactic activity (Table 1). Exposure to MPO, H_2O_2, and chloride resulted in nearly complete inactivation of the peptide. This required each component of the system and was inhibited by azide or catalase, but not by heated catalase. H_2O_2 could be replaced by glucose oxidase plus glucose (data not shown, see 7) or by xanthine oxidase plus hypoxanthine. In the latter case, inactivation of f-met-leu-phe was inhibited by azide or catalase, but not by superoxide dismutase (SOD) or heated catalase. Under the conditions used, xanthine oxidase plus hypoxanthine had no activity in the absence of either MPO or chloride. In the xanthine oxidase system, a single-electron reduction results in the initial formation of superoxide anion which may then undergo dismutation to form H_2O_2 (9). The foregoing data indicate that in our system the reactivity of superoxide is low relative to that of the H_2O_2 formed from it, at least in the presence of MPO and chloride. Thus, superoxide anion serves primarily as an intermediate in the formation of H_2O_2.

Table 1. Inactivation of f-met-leu-phe by the myeloperoxidase system and by neutrophils[a]

Supplements	Inactivation (percent) $MPO-H_2O_2-Cl^-$	$MPO-XO-HX-Cl^-$	$PMN-PMA-Cl^-$
Complete system	98.3	94.6	94.3
MPO omitted	4.1	8.4	-
PMN omitted	-	-	4.4
H_2O_2 omitted	4.8	-	-
XO omitted	-	21.0	-
HX omitted	-	4.2	-
PMA omitted	-	-	12.7
Cl^- omitted	3.2	3.4	10.5
Azide added	13.4	5.8	12.1
Catalase added	6.7	3.0	8.9
SOD added	-	86.2	88.8

[a]The reactions were carried out as decribed (7) in 0.5 ml of sodium phosphate buffer pH 7.0 with 50 pmol f-met-leu-phe and supplements as follows: MPO, 16 mU/ml; H_2O_2, 10 μM; xanthine oxidase (XO), 0.02 mU/ml; hypoxanthine (HX), 10 μM; NaCl, 0.1M ($MPO-H_2O_2-Cl^-$ system) or 10 μM (XO and PMN systems); PMN, 10^5/ml; PMA, 1 ng/ml; NaN_3, 0.1 mM; catalase, 500 U/ml; SOD, 50 μg/ml.

Intact human PMN, in the presence of chloride and phorbol myristate acetate (PMA), a soluble activator of oxidative metabolism and enzyme secretion (10,11), have the ability to inactivate f-met-leu-phe (Table 1). This requires each of the system components and is inhibited by azide or catalse, but not by SOD or heated catalase. These data suggest that secreted MPO and H_2O_2 combine with chloride

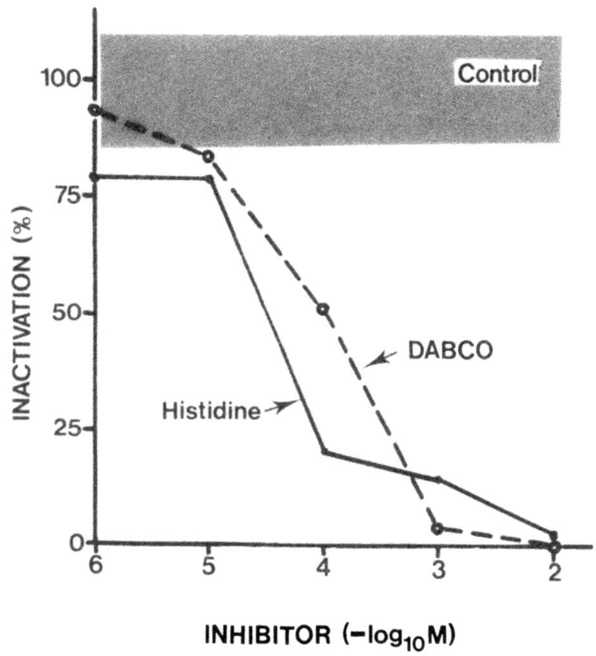

Fig. 1. Effect of histidine and DABCO on inactivation of f-met-leu-phe by the MPO-H_2O_2-Cl^- system. The conditions were as described in Table 1 except for the addition of varying concentrations of the inhibitors as indicated. The shaded area shows the degree of chemoattractant inactivation in the absence of inhibitors.

to form a system that inactivates the chemotactic factor. As in the xanthine oxidase system, superoxide anion displays no direct reactivity but is presumably an intermediate in H_2O_2 formation. Additional evidence for involvement of MPO and H_2O_2 is the failure of PMN from patients with hereditary MPO deficiency or chronic granulomatous disease to inactivate the chemoattractant (data not shown).

The proximate oxidant generated by the MPO system in these experiments cannot be stated with certainty, although hypochlorous acid and singlet oxygen should each be considered candidates (1,2). The finding of dose-related inhibition of the inactivation process by histidine and DABCO (Figure 1) is consistent with a singlet oxygen mechanism. However, the lack of absolute specificity of these agents as scavengers of singlet oxygen (6) precludes a definite conclusion regarding the involvement of this reactive species.

Requirement for Methionine Oxidation

Several approaches have been taken to determine the biochemical change in the MPO- and PMN-treated chemoattractant that is responsible for its inactivation. It is clear that certain chemotactic factors are very sensitive to inactivation, whereas others are almost totally resistant (Table 2). One of the structural differences between these two groups of mediators is the presence of a methionine residue in the MPO-sensitive agents and its absence from those that resist inactivation (see 8). We have further shown that MPO-treated f-met-leu-phe has a markedly decreased affinity for PMN surface receptors, whereas similar treatment of a methionine-free agent, f-norleu-leu-phe, had no effect on its binding affinity (8). The addition of low concentrations of free methionine markedly inhibits chemoattractant inactivation (Table 3, Figure 2). This was observed for both C5a and f-met-leu-phe and for the isolated MPO system as well as the intact PMN system. Oxidized methionine had no effect. Nonspecific reducing agents - 2-mercaptoethanol and ascorbic acid - also inhibited inactivation. In these experiments, methionine may well be acting as a competitive substrate for oxidants generated by the MPO system. The results thus support methionine oxidation as the basis for chemotactic factor inactivation. We have presented more direct proof of this hypothesis based on thin layer chromatographic analysis of control and MPO-treated chemoattractants (8).

INACTIVATION OF α_1-PROTEASE INHIBITOR

On the basis of the foregoing observations, we have been interested in the possibility that other methionine-containing biologically active agents might be sensitive to oxidative inactivation by the MPO system and by stimulated PMN. The plasma protein, α_1-protease inhibitor (α_1-PI), contains a methionine residue at the critical protease binding site and is sensitive to inactivation by oxidants of chemical or biological origin (12-15). Since α_1-PI is important in defense against elastolytic tissue injury, for example in pulmonary emphysema, we have examined its sensitivity to PMN oxidant systems.

Table 2. Inactivation of various chemoattractants by the myeloperoxidase system and by neutrophils[a]

Chemoattractant	Inactivation (percent)	
	$MPO-H_2O_2-Cl^-$	$PMN-PMA-Cl^-$
C5a	94.4	82.7
f-met-leu-phe	98.3	94.3
f-met-leu-phe-lys	75.6	-
f-phe-leu-phe-leu-phe	10.9	2.6
f-norleu-leu-phe	0.1	18.3
Pepstatin	16.7	0.7

[a]The reactions were carried out as in Table 1 except that the chemoattractant was varied as indicated. Maximally effective concentrations of each attractant were used (see 8).

Purified α_1-PI was incubated with the components of the MPO system or with human PMN and then assessed for its ability to bind ^3H-labeled porcine pancreatic elastase by a gel filtration assay (16). In the presence of MPO, H_2O_2, and chloride, there was nearly complete loss of elastase-binding ability (Table 4). This depended on each system component and was inhibited by azide or catalase, but not by heated catalase. Human PMN in the presence of PMA and chloride also abolished elastase-binding activity of α_1-PI. This effect was blocked by azide or catalase, but not by SOD or heated catalase. These data suggest that the inactivation of α_1-PI occurs by a mechanism similar to that observed with chemotactic factor inactivation, i.e., secretion of MPO and H_2O_2.

Table 3. Inhibitors of chemoattractant inactivation by the myeloperoxidase system and neutrophils[a]

Inhibitor	Inactivation (percent)		
	$MPO-H_2O_2-Cl^-$		$PMN-PMA-Cl^-$
	C5a	f-met-leu-phe	f-met-leu-phe
None	97.4	109.6	93.1
Methionine 0.1mM	0.4	8.6	-8.7
Methionine sulfoxide 0.1mM	90.2	96.0	101.9
2-mercaptoethanol 1mM	24.2	21.8	12.9
Ascorbic acid 1mM	-	6.0	18.9

[a]The reactions were carried out as in Table 1 except for the addition of various inhibitors as indicated.

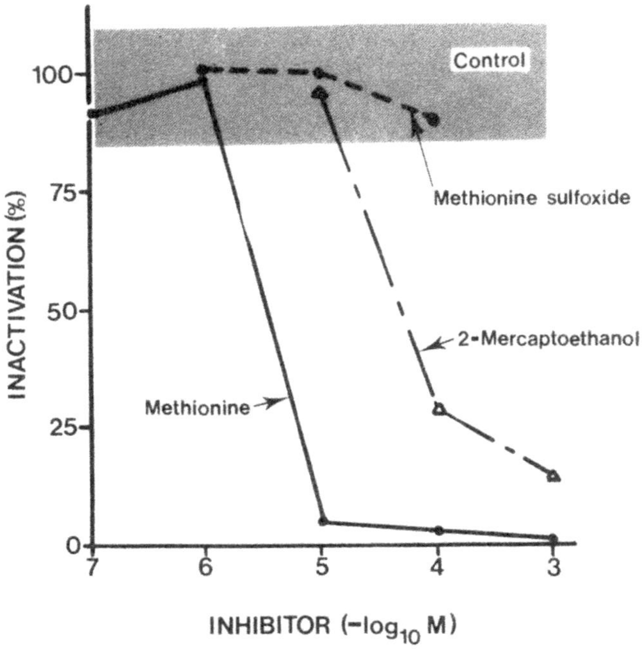

Fig. 2. Effect of methionine, methionine sulfoxide, and 2-mercaptoethanol on inactivation of f-met-leu-phe by the MPO-H_2O_2-Cl^- system. The conditions were as decribed in Table 1 except for the addition of varying concentrations of the inhibitors as indicated. The shaded area is as decribed in Figure 1.

SUMMARY AND CONCLUSIONS

The MPO-H_2O_2-halide system is capable of oxidatively inactivating chemotactic factors and α_1-PI. The mechanism of inactivation is by oxidation of methionine residues critical to biological function. A similar effect is observed when these mediators are incubated with activated human PMN. The results indicate that secreted MPO and H_2O_2 combine with a halide cofactor to oxidatively inactivate chemotactic factors or protease inhibitors. The potential significance of chemoattractant inactivation relates to the modulation of the acute inflammatory response. The observations on α_1-PI inactivation have implications for the pathogenesis of elastolytic lung injury and suggest that PMN might provide both elastase and agents that

Table 4. Inactivation of α_1-protease inhibitor by the myeloperoxidase system and by neutrophils[a]

Supplements	Elastase Binding (percent of control)	
	$MPO-H_2O_2-Cl^-$	$PMN-PMA-Cl^-$
None (control)	100	100
Complete system	6.7	2.4
MPO omitted	111.5	-
PMN omitted	-	111.9
H_2O_2 omitted	113.6	-
PMA omitted	-	98.7
Cl^- omitted	90.2	79.0
Azide added	94.8	96.7
Catalase added	100.3	87.6
SOD added	-	2.6

[a]The reactions were carried out as in Table 1 except that the chemoattractant was replaced by 50 μg of α_1-PI and the supplements were changed as follows: NaCl, 0.1 M; PMN, 10^6/ml; PMA, 10 ng/ml.

eliminate a major anti-elastase defense. Finally, it may be that secretion of the components of the MPO system and other leukocytic oxidants by activated PMN constitutes a general mechanism for the modification of oxygen-sensitive humoral mediators.

REFERENCES

1. S.J. Klebanoff, Antimicrobial mechanisms in neutrophilic polymorphonuclear leukocytes, Semin. Hematol. 12:117 (1975).
2. B.M. Babior, Oxygen-dependent microbial killing by phagocytes, N. Engl. J. Med. 298:659 and 721 (1978).
3. R.A. Clark and S.J. Klebanoff, Neutrophil-mediated tumor cell cytotoxicity: role of the peroxidase system, J. Exp. Med. 141:1442 (1975).
4. R.A. Clark and S.J. Klebanoff, Role of the myeloperoxidase-H_2O_2-halide system in concanavalin A-induced tumor cell killing by human neutrophils, J. Immunol. 122:2605 (1979).
5. R.A. Clark and S.J. Klebanoff, Neutrophil-platelet interaction mediated by myeloperoxidase and hydrogen peroxide, J. Immunol. 124:399 (1980).
6. R.D. Diamond, R.A. Clark, and C.C. Haudenschild, Damage to Candida albicans hyphae and pseudohyphae by the myeloperoxidase system and oxidative products of neutrophil metabolism in vitro, J. Clin. Invest. 66:908 (1980).

7. R.A. Clark and S.J. Klebanoff, Chemotactic factor inactivation by the myeloperoxidase-hydrogen peroxide-halide system: an inflammatory control mechanism, J. Clin. Invest. 64:913 (1979).
8. R.A. Clark, S. Szot, K. Venkatasubramanian, and E. Schiffmann, Chemotactic factor inactivation by myeloperoxidase-mediated oxidation of methionine, J. Immunol. 124:2020 (1980).
9. I. Fridovich, Superoxide dismutases, Ann. Rev. Biochem. 44:147 (1975).
10. J.E. Repine, J.G. White, C.C. Clawson, and B.M. Holmes, The influence of phorbol myristate acetate on oxygen consumption by polymorphonuclear leukocytes, J. Lab. Clin. Med. 83:911 (1974).
11. I.M. Goldstein, S.T. Hoffstein, and G. Weissmann, Mechanisms of lysosomal enzyme release from human polymorphonuclear leukocytes - Effects of phorbol myristate acetate, J. Cell. Biol. 66:647 (1975).
12. D. Johnson and J. Travis, The oxidative inactivation of human α-1-proteinase inhibitor, J. Biol. Chem 254:4022 (1979).
13. A.B. Cohen, The effects in vivo and in vitro of oxidative damage to purified α_1-antitrypsin and to the enzyme-inhibiting activity of plasma, Amer. Rev. Resp. Dis. 119:953 (1979).
14. N.R. Matheson, P.S. Wong, and J. Travis, Enzymatic inactivation of alpha-1-proteinase inhibitor by neutrophil myeloperoxidase. Biochem. Biophys. Res. Commun. 88:402 (1979).
15. H. Carp and A. Janoff, In vitro suppression of serum elastase-inhibitory capacity by reactive oxygen species generated by phagocytosing polymorphonuclear leukocytes. J. Clin. Invest. 63:793 (1979).
16. P.J. Stone, J.D. Calore, G.L. Snider, and C. Franzblau, The dose-dependent fate of enzymatically active and inactivated tritiated methylated pancreatic elastase administered intratracheally in the hamster. Amer. Rev. Resp. Dis. 120:577 (1979).

Discussion
Edited by R.A. Clark

Dr. A. Segal: If you consider this system to be important in the regulation of the inflammatory reaction, do you find any evidence of disordered inflammatory responses in patients with myeloperoxidase deficiency?

Dr. R. Clark: The inflammatory response has not been systematically evaluated in these patients. In a few patients, there are clinical associations with infection and certain inflammatory conditions, although their basis is uncertain. Some MPO-deficient subjects appear to have normal health, but the lack of detailed studies precludes any statement about the status of the inflammatory response in MPO deficiency. In addition, the increased oxygen-dependent MPO-independent systems in these cells may compensate, at least in part, for the missing enzyme. Finally, it seems likely that multiple regulatory systems are operative in vivo and that the absence of one system may not result in obvious clinical manifestations.

Dr. A. Siccardi: Have you tried adding methionine to a regular chemotaxis assay? If inactivation of methionine-containing chemoattractants is a regulatory mechanism, you might observe enhanced migration.

Dr. R. Clark: We have not yet done this, but it is a potentially interesting experiment.

Dr. L. DeChatelet: Have you looked at α_1-PI levels in patients with hereditary MPO deficiency as a means of examining possible in vivo significance? Alternatively, you might examine α_1-PI in infected patients with CGD as opposed to patients with no active infection.

Dr. R. Clark: I doubt that differences in plasma levels of the protease inhibitor would be demonstrable. It might be more rewarding to examine the events in a local site of inflammation or tissue destruction. We intend to do this in an animal model of elastase-induced pulmonary emphysema.

Dr. J. Hakim: Have you measured myeloperoxidase release in these experiments?

Dr. R. Clark: We have not measured myeloperoxidase release by enzymatic methods, although we have demonstrated it indirectly by determining the peroxidase-catalyzed iodination of extracellular protein. Other workers have detected MPO release from activated neutrophils by enzymatic assay. In general, small amounts of MPO are detectable in the extracellular space, although these amounts appear to be sufficient to catalyze the oxidation of the mediators.

Dr. D. Roos: We have exactly the same experience. After 30 minutes of incubation of human neutrophils with serum-opsonized zymosan, we found about 5-10% of the total MPO activity in the extracellular medium but a decrease in the cell-bound activity of about 40-50%. Thus, there is a substantial loss of activity during phagocytosis.

ELASTASE SECRETION BY MOUSE PERITONEAL MACROPHAGES:

EFFECTS OF RIFAMPIN AND CORTICOSTEROIDS*

J. Bernard L. Gee, Cynthia A. Stevens and Lois M. Hinman

Department of Internal Medicine
Yale University School of Medicine
New Haven, Ct. 06510

INTRODUCTION

The traditional focus on macrophages as phagocytic cells has recently been enlarged to include biosynthetic and secretory processes. Among such secretory products are the neutral proteases, plasminogen activator, collagenase and elastase. The latter two are particularly important in tissue injury and repair since they can attack their native insoluble substrates at neutral pH in the extracellular fluid. Such enzymes are not stored in macrophages, nor do they exist as proenzymes. Thus, their secretion directly depends on continuous synthesis which is known to be inhibitable by such protein synthesis inhibitors as cycloheximide.

A number of factors are known to initiate and regulate neutral protease secretion in mouse peritoneal macrophages (MPM). They will be briefly indicated. First, such secretion is greatly enhanced in the activated macrophage.[1] Activation can be initiated in vivo with such nonspecific inflammatory agents as intraperitoneal thioglycollate injections. In vitro activation has also been demonstrated with lymphocyte-derived factors,[2] water soluble materials from mycobacteria,[3] synthetic muramyl dipeptide and endotoxin.[4] Additionally, sustained in vitro enhancement of collagenase secretion by normal resident MPM occurs with short exposures to the Fc fragment of human IgG.[5] Similar effects also occur with concanavalin A.

A second factor is prostaglandin metabolism. Activation and collagenase secretion produced by endotoxin are associated with

*Supported by USPHS grant HL 19237.

heightened release of PGE_2[4] and all three responses to endotoxin are blocked by indomethacin. However, the latter drug has no effect on collagenase secretion when added 24 hours after endotoxin. Similarly, neither aspirin nor indomethacin affects elastase secretion by MPM cultured after in vivo activation with thioglycollate.[6]

The third factor enhancing collagenase secretion also occurrs with agents elevating 3'-5' cyclic AMP.[7] Such agents include a biphasic response to varying concentrations of dibutyryl cyclic AMP and additionally to both cholera toxin and phorbol myristate.

Thus, the present evidence suggests that there is a primary need for an initiating agent, probably acting on the macrophage cell membrane, with secondary regulation of protease secretion by both prostaglandins and cyclic nucleotides. Our studies have focussed on agents of potential therapeutic value in man, i.e. agents that might diminish tissue injury by depressing protease secretion in already activated macrophages. Two agents were considered, namely, corticosteroids and an agent known to inhibit protein synthesis, the antituberculous drug, rifampin. This latter drug affects protein synthesis in both bacteria and mainly in lymphocytes.[8] Details of our methods are described elsewhere.[9,10]

RESULTS

Dexamethasone

Effects of 10^{-7} dexamethasone on secretion of elastase and on the total system content of both lysozyme and β-glucuronidase are indicated in Table 1. It should be noted that all the elastase is extracellular and approximately 85% of the lysozyme is to be found in the extracellular fluid; the remainder being detected in the cell pellet. By contrast, only 24% of the β-glucuronidase activity is to be found in the extracellular fluid. These data demonstrate minimal effects on lysozyme and β-glucuronidase in contrast with the approximate 50% inhibition of the elastase secretion rates by mouse peritoneal macrophages.

Table 1. The Effects of Dexamethasone (% control)

Elastase	55 ± 3.0
Lysozyme	89 ± 8.2
β-glucuronidase	94 ± 6.1

Total enzyme activities are indicated. Representative experiment performed in triplicate. Mean and range are indicated.

Table 2. Effects of Rifampin[a]

	Control		Rifampin			
			25 μg/ml		50 μg/ml	
	Total Activity (units)	Percent Activity in medium	Total Activity (% control)	Percent Activity in medium	Total Activity (% control)	Percent Activity in medium
Elastase	1.2 ± 0.40	100	80.0 ± 15.0[b]	100	44.1 ± 10.4[c]	100
Lysozyme	97 ± 7.8	86 ± 4	82.0 ± 6.1[b]	83 ± 3	62.1 ± 7.5[c]	88 ± 4
β-glucuronidase	86.1 ± 15.2	23 ± 3	97.1 ± 4.1	24 ± 5	94.1 ± 4.6	24 ± 3
LDH % Release		8.2 ± 1.2		9.2 ± 1.9		12 ± 2

[a]Control data (8 experiments) for elastase, lysozyme and β-glucuronidase are total cell + medium enzyme activities expressed per mg cell protein in 48 hour cultures. The units are, respectively, 1 μg porcine pancreatic elastase equivalent, 1 μg egg white lysozyme equivalent, and Δ O.D. units, where 1 unit activity releases 1 μg phenolphthalein/hr. Note: in order to compare directly the effects of rifampin on elastase, lysozyme and β-glucuronidase with the appropriate control experiments, the rifampin data are expressed as the % of the matched control experiments. Data (mean ± 1 S.D.) for rifampin at 25 and 50 μg/ml were respectively derived from 5 and 6 separate experiments performed in triplicate.

[b]$p < 0.05$
[c]$p < 0.01$

Rifampin

The effects of rifampin on these same enzyme activities, namely, elastase, lysozyme and β-glucuronidase, are summarized in Table 2. Additionally, it should be noted that the percent LDH release into the medium was little affected by the addition of 25 μg or 50 μg/ml of rifampin to the system. Thus, evidence of cell cytotoxicity is minimal or absent. The data indicate the activity of these three enzymes together with the percent found in the medium. Additionally, the effects of rifampin at 25 and 50 μg/ml are also indicated. To summarize this data, it is apparent that approximately 50% inhibition of elastase secretion occurrs with 50 μg/ml of rifampin. Lysozyme secretion is inhibited to about 62% of control, whereas there are no effects upon the β-glucuronidase. Thus, these data indicate that two enzymes, whose secretion is inherently dependent upon protein synthesis, namely elastase and lysozyme, are significantly depressed by the presence of rifampin in the extracellular medium.

DISCUSSION

These data indicate the two drugs currently in standard therapeutic use for other purposes in man, are capable of producing significant depression of elastase secretion in our in vitro system. As regards dexamethasone, doses of 10^{-7} M can be produced in tissue fluids following standard therapeutic doses of this drug. Likewise, with regards to rifampin, it should be noted that doses of 600 mg are currently used therapeutically in the management of tuberculosis. This dose corresponds to 7.5 mg/kg in an 80 kg man. It should be noted that doses of 10 mg/kg body weight in animals yield tissue concentrations in the range of 64 and 9 μg/ml in liver and lung tissue, respectively.[11] Thus, both drugs can produce in vitro changes in elastase secretion at levels approximately similar to those that might be expected in therapeutic situations in man.

As regards rifampin, it should be noted that this drug can also impair protein synthesis by lymphocytes.[8] Thus, rifampin can be shown, at least in vitro, to act upon two critical cells in the immunological system, namely lymphocytes and macrophages, in such a way as to damp down immunologically determined inflammatory mechanisms, whose final effector arm is a tissue injuring neutral protease such as elastase. This in vitro evidence is supported by the known effects of rifampin on delayed skin hypersensitivity reactions to PPD in man.

Our particular interest has been in pulmonary disease and, in that regard, unregulated elastinolysis has been proposed as a major mechanism for the development of pulmonary emphysema. In mice, cigarette smoke has been reported to increase intra-alveolar macrophage accumulation and also to enhance secretion of elastase in vitro by such cells.[12] Whether these particular mechanisms which focus on alveolar macrophages operate in man is not clear, since there are differing views on the presence or absence of a true elastase in the secretions of cultured human alveolar macrophages.[13] In spite of that uncertainty, it is of potential interest to consider the possibility that rifampin, by damping down both lymphocyte and macrophage mediated immune responses, may still remain a useful adjunct in the management of immunologically determined lung injury. This point is particularly enhanced if one considers the fact that human alveolar macrophages are certainly a rich source of human polymorphonuclear leukocyte chemotactic materials. Thus, modification of macrophage activity by impairment of protein synthesis may not only damp down the secretion of enzymes by that cell but may also diminish the production of chemotactic factors, thereby also diminishing the accumulation of leukocytes and leukocyte elastase in lung tissues.

REFERENCES

1. Z. Werb and S. Gordon, Elastase secretion by stimulated macrophages: characterization and regulation, J. Exp. Med. 142:361 (1975).
2. L.M. Wahl, S.M. Wahl, S.E. Mergenhagen, and G.R. Martin, Collagenase production by lymphokine-activated macrophages, Science 187:261 (1975).
3. S.M. Wahl, L.M. Wahl, J.B. McCarthy, L. Chedid, and S.E. Mergenhagen, Macrophage activation by mycobacterial water soluble compounds and synthetic muramyl dipeptide, J. Immunol. 122: 2226 (1979).
4. L.M. Wahl, C.E. Olsen, A.L. Sandberg, and S.E. Mergenhagen, Prostaglandin regulation of macrophage collagenase production, Proc. Natl. Acad. Sci. 74:4955 (1977).
5. J.H. Passwell, J.-M. Dayer, K. Gass, and P.J. Edelson, Regulation by Fc fragments of the secretion of collagenase, PGE_2, and lysozyme by mouse peritoneal macrophages, J. Immunol. 125:910 (1980).
6. J.B.L. Gee, C.A. Stevens, and L.M. Hinman, Therapeutic modulation of elastase secretion by macrophages, Fed. Proc. 39(part 1): 320 (1980).
7. J.B. McCarthy, S.M. Wahl, J.C. Rees, C.E. Olsen, A.L. Sandberg, and L.M. Wahl, Mediation of macrophage collagenase production by 3'-5'cyclic adenosine monophosphate, J. Immunol. 124:2405 (1980).
8. W.C. Buss, R. Morgan, J. Guttmann, T. Barela, and K. Stalter, Rifampin inhibition of protein synthesis in mammalian cells, Science 200:432 (1978).
9. J.B.L. Gee, C.A. Stevens, and L.M. Hinman, Inhibition by rifampin of elastase and lysozyme secretion in mouse peritoneal macrophages, Biochem. Pharmacol. 29:3037 (1980).
10. J.B.L. Gee, C.A. Stevens, and L.M. Hinman, Elastase and lysozyme secretion by peritoneal macrophages: effects of dexamethasone and rifampin, J. Physiol. Path. Respiratoire. (in press).

11. G. Binda, E. Domenishini, A. Gottardi, B. Orlandi, E. Ortelli, B. Pacini, and G. Fowst, Rifampicin: a general review, Arzneim. Forsch (Drug Res.) 21:1907 (1971).
12. R. White, J. White, and A. Janoff, Effects of cigarette smoke on elastase secretion by murine macrophages, J. Lab. Clin. Med. 94:489 (1979).
13. L.M. Hinman, C.A. Stevens, R.A. Matthay, and J.B.L. Gee, Elastase and lysozyme activities in human alveolar macrophages, Am. Rev. Respir. Dis. 121:263 (1980).

DISCUSSION

Edited by: R.A.Harkness

SICCARDI: Rifampin has some detergent action, could this account for your results rather than affects on protein synthesis?

GEE: We have not measured total protein synthesis but major depression is lethal to cultured macrophages. Low dose cycloheximide does inhibit elastase secretion and the most likely mechanism to explain our results is impaired elastase synthesis.

GALLILY: Did you assay plasminogen activator?

GEE: No, we have not studied other neutral proteases.

SYNTHESIS AND RELEASE OF FACTOR INCREASING MONOCYTOPOIESIS (FIM) BY MACROPHAGES

W. Sluiter, Elly Hulsing-Hesselink, and
R. van Furth

Department of Infectious Diseases
University Hospital
Leiden, The Netherlands

INTRODUCTION

In the steady-state condition more than 90% of the cells in the peritoneal cavity of CRF Swiss mice are resident macrophages.[1] These resident macrophages originate from monocytes in the circulation, which are produced in the bone marrow.[2] During an acute inflammation a short-lasting increase of the monocyte production in the bone marrow gives rise to an increase of the number of macrophages at the site of the lesion. It has been shown that this increase of the monocyte production is due to the action of factor increasing monocytopoiesis (FIM) occurring in the circulation during the initial phase of the inflammatory reaction.[1] A recent study reveals that FIM also occurs in the circulation of rabbits during the first phase of an acute inflammation.[3]

FIM is a protein without essential carbohydrates, has no colony-stimulating activity, is cell-line specific and stimulates the mitotic activity of the promonocytes and probably also the proliferation of the monoblasts.[4]

The aim of the present study was to localize the origin of FIM and to investigate the mechanism of its production.

Release of FIM in vitro

Previously, it has been shown that extracts of normal peritoneal cells of CRF Swiss mice contain FIM.[1]

To find out whether these cells can be stimulated to secrete FIM in vitro, normal peritoneal cells were incubated on Teflon-film dishes[5] in the presence of latex particles pre-opsonized with fetal calf serum (latex-to-cell ratio about 100:1). After various intervals the amount of FIM activity in the supernatants of these cultures was measured in vivo by intravenous injection into untreated mice followed by determination of the increase in the number of blood monocytes[1].

The results, expressed as percentage increase relative to the number of blood monocytes in normal CRF mice, show that after the addition of latex to the peritoneal cell suspension, FIM activity occurs in the supernatant with peak values at 3 min, 6 hr and 18 hr (Fig. 1). In the supernatants of peritoneal cells incubated without latex and the supernatant of pre-opsonized latex particles alone no FIM activity was demonstrable.

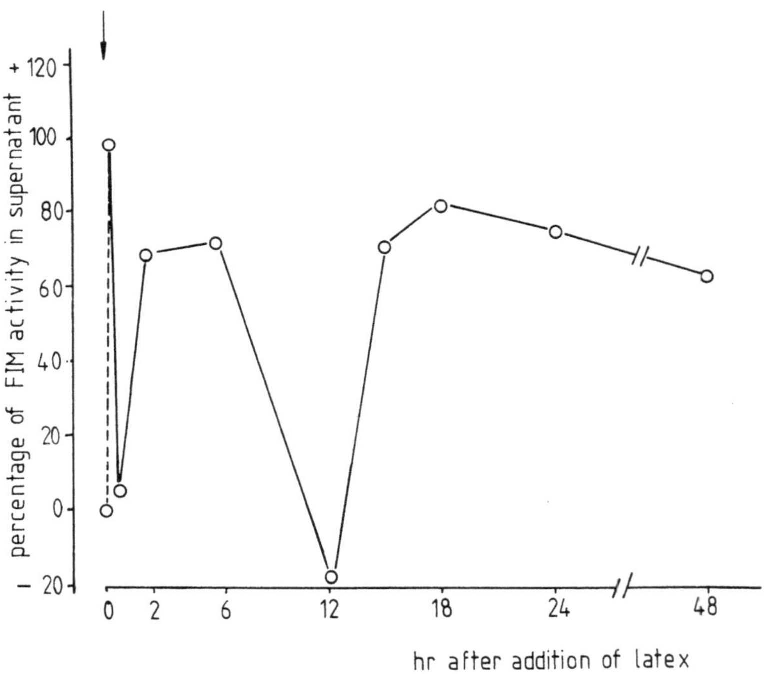

Fig. 1. Course of FIM activity in the supernatant of 2×10^6 normal peritoneal cells after addition of polystyrene latex particles preopsonized with fetal calf serum (latex-to-cell ratio about 100:1).

Origin of FIM

Since the occurrence of FIM is correlated with the presence of a phagocytic stimulus, it seems likely that FIM is directly released by macrophages and/or granulocytes. To find out which kind of phagocytes release FIM, peritoneal cell suspensions of various composition were adjusted to equal numbers of granulocytes or macrophages and incubated in the presence of pre-opsonized latex particles for 18 hr, after which the FIM activity in the supernatants was determined.

The results show that when adjusted to the same number (about 4.0×10^4) of granulocytes FIM activity was only found in the supernatant of the suspension with about 2×10^6 macrophages (Fig. 2). However, at the same number (about 2×10^6) of macrophages FIM activity was equal in the supernatants containing 4×10^4 and 6×10^6 granulocytes (Fig. 2).

These results indicate that the amount of FIM activity in the supernatant was correlated with the number of macrophages but not with the number of granulocytes. It is therefore most likely that macrophages are the origin of FIM.

Inhibition of FIM release

To obtain insight in the mechanism of FIM release, the relation with protein synthesis was studied. By culturing in the presence of cycloheximide, to prevent protein synthesis, neither the phagocytosis of latex, nor the viability of the peritoneal cells was affected. However, the results given in Table I show that FIM activity in the supernatants collected after 3 min and 6 hr is as high as in the cultures without cycloheximide, but in the supernatant of an 18-hr culture is greatly diminished.

Membrane stimulation and FIM release

To find out whether the entire process of phagocytosis is needed to induce FIM release, the effect of membrane stimulation alone by phorbol myristate acetate (PMA) was studied. In the presence of PMA the supernatant of normal peritoneal cells incubated for 3 min showed moderate FIM activity. However, after an 18-hr incubation a high FIM activity was found (Table I). PMA injected intravenously into mice did not evoke monocytosis.

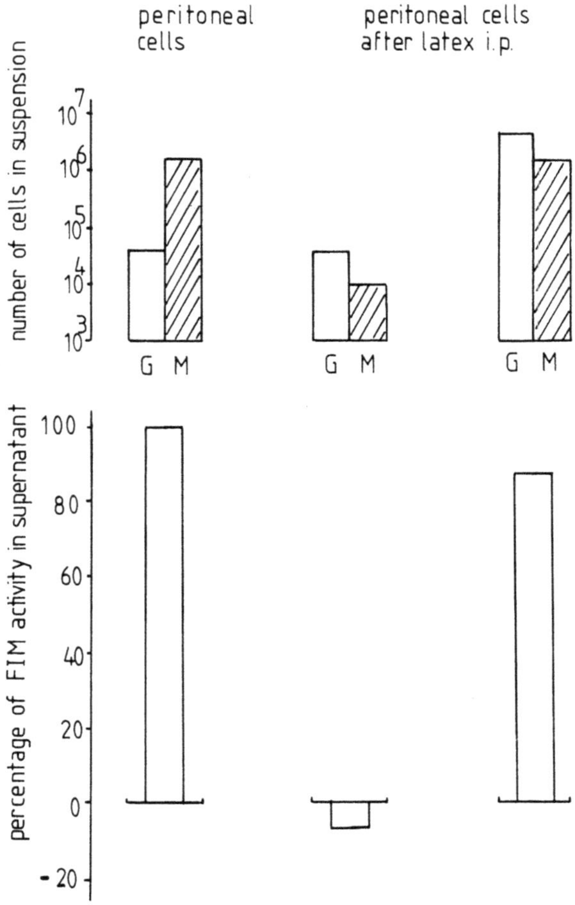

Fig. 2. Percentage of FIM activity in the supernatant of suspensions of normal peritoneal cells and peritoneal cells harvested 6 hr after latex i.p., as function of the number of granulocytes (G) and macrophages (M).

DISCUSSION

The present results show that FIM is released in vitro from resident peritoneal macrophages upon triggering with a phagocytic stimulus. Granulocytes do not respond.

After addition of latex to a peritoneal cell

Table I. Effect of various conditions on FIM release by normal peritoneal macrophages

Latex added to cultures*	Incubation conditions	Relative FIM activity** in culture medium after incubation for		
		3 min	6 hr	18 hr
		%	%	%
+	$37^{o}C$	+98	+72	+82
+	$37^{o}C$ 0.5 μg/ml cycloheximide	+95	+77	+15
-	$37^{o}C$ 20 ng/ml PMA	+40		+78

*Pre-opsonized polystyrene latex particles (0.81 μ) added to about 1×10^{7} normal peritoneal cells per ml (latex-to-cell ratio about 100:1).

**Expressed as percentage and calculated according to the formula $\frac{Ne-Nc}{Ne} \times 100\%$, in which Ne is the number of blood monocytes in test mice 24 hr after injection of supernatant, and Nc is the number of blood monocytes in control mice.

suspension, three peaks of FIM activity were found, which differs from the course of FIM activity in the circulation of mice during an acute inflammation.[1] This difference could be due to the fluctuation in the number and composition of peritoneal cells in vivo. However, since the characteristics of FIM in the supernatant[6] are in many respects similar to those of FIM in serum,[4] it is highly probable that FIM, present in serum during the early phase of an inflammatory reaction, originates from macrophages at the site of inflammation.

Based upon the experiments with an inhibited protein synthesis the release of FIM in vitro can be divided in two phases: the first phase, comprising of the first two peaks of the curve, does not depend on protein synthesis, which makes it likely that FIM stored in the macrophages is released during this phase. This is in full agreement with the results of van Waarde,[1] who found that extracts of normal peritoneal cells have a high level of FIM activity. The second phase of FIM release is affected by the inhibition of protein synthesis and therefore FIM released during this phase is synthesized apparently <u>de novo</u>.

It seems not very likely that phagocytosis is needed for induction of the release of FIM, since, after addition of a membrane-active agent, like PMA, FIM activity could also be dectected in the supernatant of macrophages. However, the kinetics of FIM release induced by latex and PMA are different.

ACKNOWLEDGEMENTS

This study was supported by the Foundation for Medical Research FUNGO, which is subsidized by the Netherlands Organization for the Advancement of Pure Research (ZWO).

REFERENCES

1. D. van Waarde, E. Hulsing-Hesselink, L.A. Sandkuyl, and R. van Furth. Humoral regulation of monocytopoiesis during the early phase of an inflammatory reaction caused by particulate substances. <u>Blood</u> 50:141 (1977).
2. R. van Furth, and Z.A. Cohn. The origin and kinetics of mononuclear phagocytes. <u>J. Exp. Med.</u> 128:415 (1968).

3. W. Sluiter, D. van Waarde, E. Hulsing-Hesselink, I. Elzenga-Claasen, and R. van Furth. Humoral control of monocyte production during inflammation. in: Mononuclear Phagocytes - Functional Aspects. R. van Furth, ed., Martinus Nijhoff Publishers, The Hague, Boston, London (1980).
4. D. van Waarde, E. Hulsing-Hesselink, and R. van Furth. Properties of a factor increasing monocytopoiesis (FIM) occurring in serum during the early phase of an inflammatory reaction. Blood 50:727 (1977).
5. J.W.M. van der Meer, D. Bulterman, Th.L. van Zwet, I.L. Elzenga-Claasen, and R. van Furth. Culture of mononuclear phagocytes on a Teflon surface to prevent adherence. J. Exp. Med. 197:271 (1978).
6. W. Sluiter, E. Hulsing-Hesselink, I. Elzenga-Claasen, and R. van Furth. Manuscript in preparation.

Discussion
Edited by R.A. Clark

Dr. L. DeChatelet: I think this is a very nice study, but one facet puzzles me. The experiments employing phagocytizing or PMA-treated cells involved incubation for 16 hours at 37°C, following which FIM activity was measured. Yet earlier you demonstrated that FIM is thermolabile with a half-life of about 20 minutes. How do you explain this discrepancy?

Dr. P. Lejih: We suppose that a portion of the synthesized and secreted FIM is inactivated and that the actual production is probably much greater than detected. Furthermore, we know that the addition of soybean trypsin inhibitor prevents the inactivation of FIM.

Dr. L. Ruco: Did you test the supernatant from your macrophage cultures for colony-stimulating activity?

Dr. P. Lejih: We have tested the purified fraction containing FIM for colony-stimulating factor (CSF), but this fraction doesn't induce colony formation by normal bone marrow cells. Furthermore, CSF produced by peritoneal macrophages has a molecular weight of 37,000 (Cell Tissue Kinet. 7:9, 1979) and will, therefore, be separated from FIM by ultracentrifugation through a membrane with a cut-off level of 25,000 daltons.

Dr. M. Ka'vai: Is FIM a specific early signal of macrophage activation?

Dr. P. Lejih: I think it is interesting to speculate about FIM as a marker for macrophage activation, but so far we haven't evaluated this possibility.

Dr. J. Breton-Gorius: Is the monocyte able to synthesize FIM?

Dr. P. Lejih: We have not tested the ability of monocytes to synthesize FIM but, as shown, macrophages which are derived from monocytes are able to do so.

BIOCHEMICAL AND BIOLOGICAL CHARACTERISTICS

OF LEUCOCYTE PROTEINASE INHIBITORS

M. Kopitar, M. Drobnič-Košorok, J. Babnik, J. Brzin, V. Turk, M. Korbelik[1], U. Batista[1], S. Svetina[1], J.Škrk[2], T. Giraldi[3] and G. Sava[3]

Department of Biochemistry, J. Stefan Institute; [1]Department of Biophysics, Medical Faculty, E.Kardelj University; [2]Oncological Institute, Ljubljana, Yugoslavia [3]Istituto di Farmacologia, Università di Trieste, Italia

In last decade there has been an intensifying interest in naturally occurring protease inhibitors. Inhibitors which are present in the same cells as the enzymes they inhibit are of great physiological significance. A vast amount of literature[1] reports that leucocytes are cells which contain different types of proteinases: neutral, thiol and acid ones. Only for the latter group of enzymes has the endogenous inhibitor not yet been determined, whereas for all other types the specific endogenous inhibitors have already been detected. Many groups of investigators have studied the isolation[2-6], molecular characterization[7-9] and mode of inhibition of these proteins[10].

The present review will emphasize studies of our laboratory and our coworkers which have delinated the biochemical and biological properties of endogenous proteinase inhibitors.

SUBCELLULAR DISTRIBUTION OF INHIBITORS

Leucocytes were isolated[3] from 6400 ml of pig peripheral blood and the amount of leucocytes obtained was 4 - 5 . 10^{10} cells. According to the evidence of previous investigations, leucocyte endogenous

TABLE I

Subcellular pattern of leucocyte proteinase inhibitors

Cell homogenate (0.34 M sucrose)

Differential centrifugation

Nuclear pellet	PGS	Granula
1st extraction	I_1, I_2, I_3	
0.9% NaCl, pH 7.0		

Centrifugation (5000 g/15 min)

S_1-supernatant	Sediment
Concentrated (UM-10)	2nd extraction 0.029 M NaCl pH 7.0
Dialysis (H_2O, pH 6.0)	Centrifugation (5000 g/15 min)

Dialysate (discarded)	Dialysand	S_3-supernatant	Sediment
	Centrifugation (5000 g/15 min)		

S_2-supernatant	Sd (sediment)

---

inhibitors could be isolated only from the soluble phase of disrupted pig leucocytes, freed of nuclei and granules, by differential centrifugation, named the post-granular supernatant (PGS)[2,3,5]. But so far no work has been published dealing with the extraction of these inhibitors from nuclei.

As shown in Table I., the active inhibitors can be obtained directly only from the post-granular supernatant, whereas in the case of nuclei, selective conditions have to be found. Successful extraction was achieved on the basis of the different solubility of proteinases and inhibitors at different ionic strength and pH of salt solutions. A two step extraction of the inhibitors from the nuclear pellet was performed. The first was made with 0.9% NaCl, pH 7.0. The extract (S_1)

showed no inhibitory activity against the tested enzymes (urokinase, elastase, chymotrypsin-like neutral protease, papain and trypsin). The inhibitory activity was generated after dialysis of the concentrated supernatant - S_1, against distilled water, pH 6.0. Dialysis caused precipitation (Sd) of some proteins, whereas the inhibitors remain in solution (S_2). The sediment (Sd) obtained during dialysis of S_1 was extracted with 20 mM Na_2HPO_4 containing 1.5 M NaCl and tested for proteolytic activity. It showed a small neutral proteolytic activity, tested on casein, pH 7.5 and an acid proteolytic activity, tested on hemoglobin, pH 3.5. From these results it can be concluded that generation of nuclear inhibitory activity under the chosen conditions is a time dependent process, presumably based on dissociation of the enzyme - inhibitor complex and on precipitation of neutral proteinases mainly inactivated under the given conditions.

In the second extraction with 0.029 M NaCl, pH 7.0, followed by 4-times repeated freezing in liquid air released proteins (S_3) showed high inhibitory activity against the tested enzymes. Extract S_2 and S_3 and post-granule supernatant fraction were further examined for their contents of proteinase inhibitors.

PURIFICATION AND PROPERTIES OF LEUCOCYTE INTRACELLULAR INHIBITORS

For purification of the inhibitors, the post-granular supernatant fraction and the nuclear extracts were used. Subsequent purification was achieved by cation exchange chromatography and by gel chromatography on Sephadex G-100, G-150 and G-50[6,7,9]. In the case of the urokinase inhibitor, affinity chromatography of inhibitor bound to urokinase Sepharose was also used. Through all the isolation procedures the eluted fractions were tested for inhibitory activity against the neutral proteinases (elastases, chymotrypsin-like neutral proteinases, papain, urokinase and trypsin). So far we have succeeded in isolating 3 types of proteinase inhibitors from the post-granular supernatant, and two of them could also be obtained from the nuclear extracts.

Disc gel electrophoresis of inhibitor samples showed a high homogeneity (Fig. 1), and this is also evident from isoelectric point analysis determined by gel analytical isoelectric focusing, as shown in Fig. 2.

Table II summarizes the main characteristics of the three types of proteinase inhibitors isolated from leucocytes. The inhibitors are acidic proteins, with isoelectric points between pH 4.4 to 5.7. They differ also in molecular weight, location and thermal stability. Only inhibitor I_1 showed high thermal stability, even at 80 ^oC (10 min) retaining its full inhibitory activity. Another important characteristic of these inhibitors was also determined, namely their pH-stability. It was found that highly purified inhibitors are quite stable in buffer solutions from pH 3 - 5. The loss of inhibitory activity, at

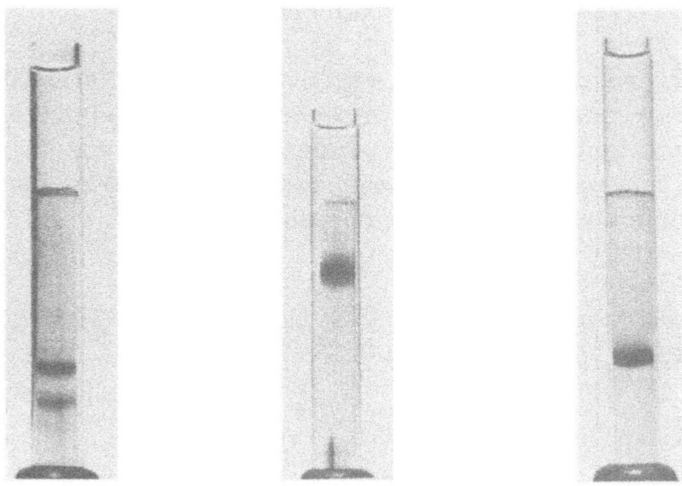

Fig. 1. Polyacrylamide gel electrophoresis of inhibitors: I_1, I_2, I_3.

acid pH, of post-granular supernatant or nuclear extracts is connected with a specific interaction with acid proteinases. The mechanism is clarified in a subsequent section of this paper.

The immunological properties were determined using antisera obtained by immunization of rabbits with an inhibitor I_1 and I_3 mixture, and a second mixture of I_2 and I_3. By double immunodiffusion inhibitor I_1 and I_3 as well I_2 and I_3, showed cross reactivity of precipitin lines (Fig. 3), which confirms that the inhibitors are unrelated proteins.

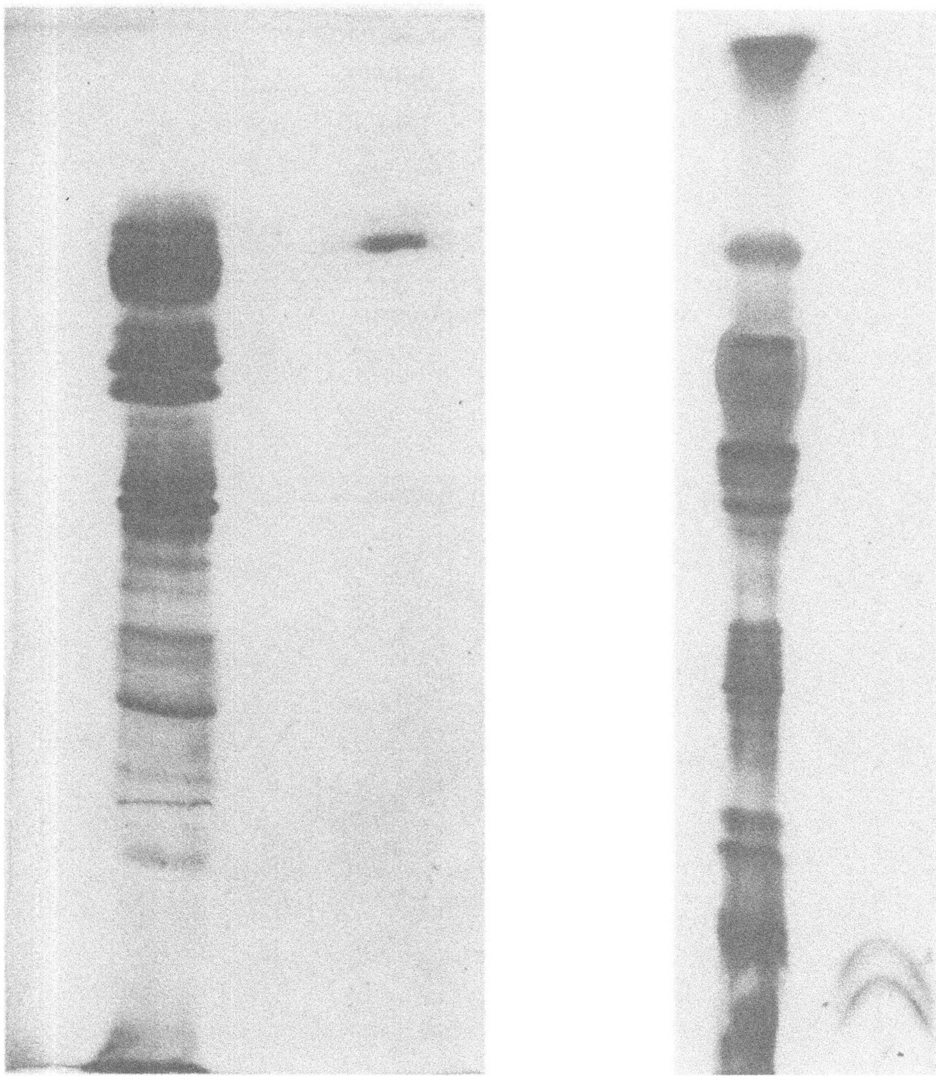

Fig. 2. Gel isoelectric focusing of inhibitor I_3 and standards and standards and inhibitor I_1; determined on an apparatus used for analytical isoelectric focusing (Desaga), using 5% polyacrylamide plates, standards with Ip 3 - 10.

TABLE II

Biochemical characteristics of leucocyte inhibitors

Inhibitor	Source N	Source C	M.W.x10^3 approx.	Ip
I_1	+	+	13 - 15	4.6 - 4.8
I_2		+	40	5.6 - 5.8
I_3	+	+	68	4.4 - 4.5

N - nuclea, C - cytoplasm

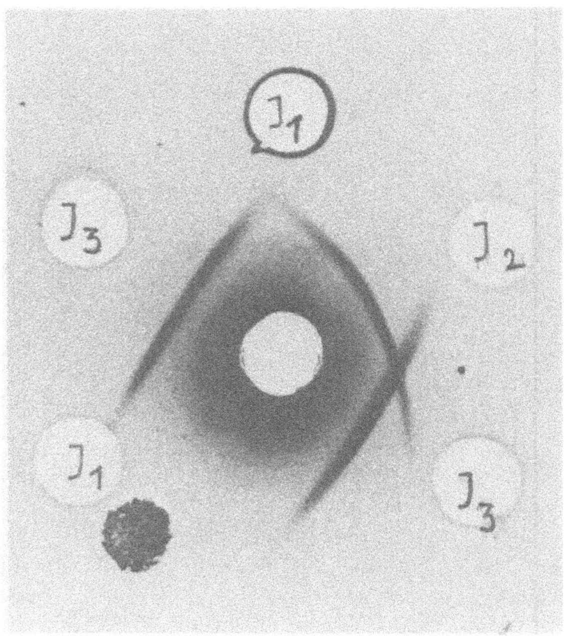

Fig. 3. Immunodiffusion plate in 3% agar, buffer PBS, pH 7.1. In outher well I_3, I_2, I_1, I_3, I_1. In center well 10 μl of antibody. Developing plate 72 hrs, staining Comassie blue. Antiserum was prepared with the mixture of I_2 and I_3.

INHIBITION STUDIES OF INTRACELLULAR INHIBITORS ON PROTEINASES

As shown in Table III inhibitor I_1 inhibits chymotrypsin-like neutral proteinases and thiol proteinases (papain, cathepsin B and H), but does not inhibit elastase and plasminogen activator (urokinase). Inhibitor I_2 is a specific inhibitor for leucocyte as well as for spleen elastase, and is also a weak inhibitor of neutral proteinases of the chymotrypsin type. Inhibitor I_3 is a specific inhibitor of plasminogen activator. These inhibitors belong to the type of so called fast reacting inhibitors of proteinases. That is, they reach their full inhibitory ability toward the tested proteinases in a very short time of preincubation (1 min) and the complexes formed were stable over the tested time intervals, of 1 - 240 min.

TABLE III

Inhibition effect of inhibitors on proteinases

Inhibition of						
elastases	chymotryp. neutr.prot.	urokinase	papain	Cathepsin B	Cathepsin H	Inhibitor
-	++	-	++	+	+	I_1
++	+	-	-	-	-	I_2
-	-	++	-	-	-	I_3

++ strong, + weak, - no inhibition

INHIBITION EFFECTS IN SYSTEMS IN VIVO AND IN VITRO

The regressive effect of leucocyte intracellular inhibitors on tumour growth and metastases formation has already been reported. It was found that leucocyte inhibitors significantly reduce the growth of the primary tumour in mice bearing Lewis lung carcinoma. The number of lung metastases found at sacrifice was also significantly reduced to about 64% with respect to the control and their mass was also significantly reduced. A similar regressive effect on metastase formation was also obtained when animals were treated with spleen intracellular inhibitors[11,12]

The in vitro inhibitory effect of pig leucocyte inhibitor I_1 and a trypsin(plasmin) inhibitor T_1 from bovine spleen was first tested on colony forming ability of V-79-379A cells (diploid lung fibroblasts of Chinese hamster, Gray Laboratory, Mount Veron Hospital, Nothwood, England). The cells were seeded in plastic Petri dishes (200-300 cells in each), using Eagle MEM culture medium supplemented with 15% foetal calf serum. After 3 hrs the medium was changed with medium containing different concentrations of inhibitor. The cells were then

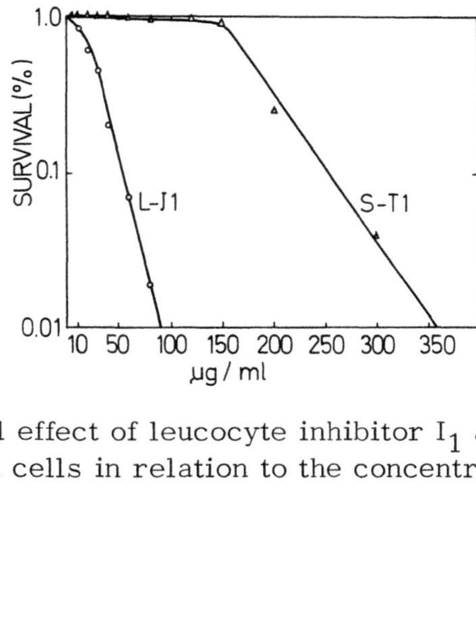

Fig. 4. Cytocidal effect of leucocyte inhibitor I_1 and spleen inhibitor T_1 on V-79-379A cells in relation to the concentrations of inhibitors.

further maintained in CO_2 in incubator for 7 days. After 6 days the grown colonies were stained with 10% Giemsa solution and counted. The cytotoxic effect was expressed as survival fractions of cells (proportion of numbers of colonies grown out of cells treated with inhibitors and numbers of control colonies without inhibitor) against tested concentrations of inhibitors. Our results showed (Fig. 4) very pronounced difference between two curves. Inhibitor I_1 revealed stronger cytocydal effect in comparison to trypsin inhibitor T_1.

INACTIVATION-DEGRADATION OF INTRACELLULAR INHIBITORS BY CATHEPSIN D

In our first study[3] on leucocyte intracellular inhibitors we found that the inhibition ability of leucocyte cytoplasm was highly diminished at acid pH from 3 - 5. In further investigations[8,13], the mechanism of the acid inactivation of cytoplasm inhibitor ability was clarified. Highly purified, isolated inhibitors are quite stable over the pH range from 3 - 8, but two of them, I_2 and I_3 could be inactivated when treated with cathepsin D at acid pH. However, when I_2 and I_3 are already complexed with enzymes, cathepsin D is without effect, namely, no reactivation of enzyme activity was noted. We also observed that cathepsin D does not inactivate I_1. In our last study[14] we determined that cathepsin D inactivates intracellular inhibitors by hydrolysis of the inhibitor molecule. The conversion of active inhibitor into inactive protein proceeds catalytically (Fig. 5). The inactivation

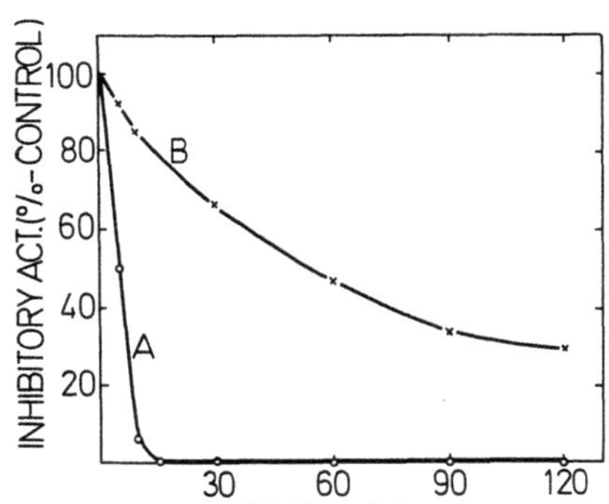

Fig. 5. Inactivation of I_3 by cathepsin D. Samples of the inhibitor and cathepsin D in molar ratio 2:1 (A) and 100:1 (B) were incubated at 37 °C, at pH 4.0, and samples were removed at the time points indicated and assayed for residual inhibitory activity toward urokinase.

as well as the degradation of the inhibitor depends on pH and the concentration of cathepsin D. These experiments suggest a new pathway by which the active intracellular proteinase inhibitors in leucocyte cells may be inactivated, so that endogenous proteolysis may occur. Thus cathepsin D could readily decrease the protective concentration of these inhibitors in cells.

CONCLUSIONS

Pig leucocytes contain inhibitors of neutral and thiol proteinases. These proteins could be isolated from post-granule supernatant fraction as well as from nuclear extract using ion exchange chromatography, gel chromatography and affinity chromatography. Inhibitors differ in molecular weight, isoelectric point, immunologically and their inhibition ability against tested enzymes.

Another important properties of intracellular inhibitors were observed, namely, they inhibit the growth of metastase formation (in vivo) as well as the growth of cells in colonies (in vitro).

The purified inhibitors are stable in acid as well as in neutral pH, but it was found that inhibitor I_2 and I_3 could be inactivated - degraded by the action of cathepsin D.

The role of these inhibitors in leucocyte cells is presently unknown, it is only known that they are located in cytoplasm and nuclea and that its activity could be regulated by cathepsin D.

The latest studies on neutral proteinases: elastases, chymotrypsin-like neutral proteinases and plasminogen activators showed that these enzymes are connected with many processes of fibrinolytic enzyme system[15], cancer metabolism[16,17] emphysema[18] and other diseased. During last two years appear evidence that chymotrypsin-like neutral proteinases of leucocyte cells, are involved in superoxide production[19]. With reference to these possible roles of intracellular neutral proteinases, its intracellular inhibitors are of great interest.

ACKNOWLEDGEMENT. The excellent technical assistance of Mrs. M. Božič and Mr. K. Lindič is greatfully acknowledged. Supported by the research grant from the Research Council of Slovenia.

REFERENCES

1. A.J.Barrett, ed., "Proteinases in Mammalian Cells and Tissues", North Holland Publishing Co., Amsterdam, New York (1977).

2. A. Janoff and J. Blondin, Inhibition of elastase-like esterase in human leucocyte granules by human leucocyte cell soup, Proc. Soc.Exp.Biol.Med. 136: 1050 (1971).

3. M. Kopitar and D. Lebez, Intracellular distribution of neutral proteinases and inhibitors in pig leucocytes, Europ.J.Biochem. 56: 571 (1975).

4. M. Kopitar, J. Babnik, I. Kregar and A. Suhar, Neutral proteinases and inhibitors of leucocyte cells, in "Movement, Metabolism and Bactericidal Mechanisms of Phagocytes, F. Rossi, P. Patriarca, D. Romeo, eds., Piccin Medical Books, Padua, London (1977).

5. A. Dubin, Polyvalent proteinase inhibitor from horse blood leucocyte cytosol, Europ.J.Biochem. 73: 429 (1977).

6. M. Kopitar, A. Suhar, T. Giraldi and V. Turk, Biochemical and Biological properties of cell and tissue neutral proteinases and inhibitors, Acta Biol.Med.Germ. 36: 1863 (1978).

7. M. Kopitar, J. Brzin, T. Zvonar, P. Ločnikar, I. Kregar and V. Turk, Inhibition studies of an intracellular inhibitor on thiol proteinases, FEBS Letters 91: 355 (1978).

8. M. Kopitar, T. Giraldi, P. Ločnikar and V. Turk, Biochemical and biol ogical properties of leucocyte intracellular inhibitors of proteinases, in "Macrophages and Lymphocytes", Part A, M.C. Escobar and H. Friedman, eds, Plenum Publishing Co., New York and London (1980).

9. M. Kopitar, Isolation and characterization of urokinase inhibitors isolated from pig leucocytes, Haemostasis, in press.

10. F.S. Steven, D.W. Milsom and J.A.D. Hunter, Human polymorphonuclear leucocyte neutral protease and its inhibitor, Europ.J. Biochem. 67: 165 (1976).

11. T. Giraldi, M. Kopitar and G. Sava, Antimetastatic effects of a leucocyte intracellular inhibitor of neutral proteases, Cancer Res. 37: 3834 (1977).

12. T. Giraldi, G. Sava, M. Kopitar, J. Brzin and V. Turk, Neutral proteinase inhibitors and antimetastatic effects in mice, Europ.J.

Cancer 16: 449 (1980)

13. M. Kopitar, J. Brzin, J. Babnik, V. Turk and A. Suhar, Intracellular neutral proteinases and their inhibitors, in: "Enzyme regulation and mechanism of action" Vol. 60, P. Mildner, B.Ries, eds., Pergamon Press, Oxford, New York (1980).

14. M. Drobnič-Košorok, M. Kopitar, J. Babnik and V. Turk, Inactivation studies of an endogenous inhibitor of urokinase by cathepsin D, Mol.Cell.Biochem. in press.

15. B. Astedt, E. Lundgren, G. Ross and G. Abu Sinna, Release of various molecular forms of plasminogen activators during culture of human ovarian tumours, Thrombosis Res. 13: 1031 (1978).

16. J.C. Unkeless, K. Dano, G.M. Kellerman and E. Reich, Fibrinolysis associated with oncogenic transformation, J.Biol.Chem. 249: 4295 (1974).

17. L.A. Liotta, J. Kleinerman and G.M. Seidel, The significance of hematogenous tumour cell clumps in the metastatic process, Cancer Res. 36: 889 (1976).

18. A. Janoff, B. Sloan, G. Weinbaum, V. Damiano, R.A.Sandhaus, J.Elias and P. Kimbel, Experimental emphysema induced with purified human neutrophil elastases: tissue localization of the instilled protease, Am.Rev. Respir.Dis. 115: 461 (1977).

19. S. Kitagawa, F. Takalen, and S. Sakamoto, Evidence that proteases are involved in superoxide production by human polymorphonuclear leucocytes and monocytes, J.Clin.Invest. 65: 74 (1980).

Discussion
Edited by R.A. Clark

Dr. P. Venge: Is there any antigenic similarity between the intracellular protease inhibitors and those found in plasma?

Dr. M. Kopitar: Until the present experiments, no similarity could be detected between extracellular (serum) and intracellular inhibitors.

OXIDATIVE DAMAGE TO LYSOSOMAL ENZYMES IN HUMAN PHAGOCYTOSING NEUTROPHILS[1]

Ron S. Weening, Alwin A. Voetman, Mic N. Hamers, Louis J. Meerhof, Annet A.A.M. Bot and Dirk Roos

Central Laboratory of the Netherlands Red Cross Blood Transfusion Service and Laboratory for Experimental and Clinical Immunology, University of Amsterdam; and Department of Pediatrics, Binnengasthuis, Academic Hospital, University of Amsterdam
P.O. Box 9190, 1006 AD Amsterdam, The Netherlands

SUMMARY

During phagocytosis neutrophils from 8 patients with chronic granulomatous disease released 2-3 times more activity of lysozyme and beta-glucuronidase than did normal neutrophils. This difference was caused by the partial inactivation of these enzymes by normal neutrophils. The inactivation of granular enzymes depends on oxidative products and takes place mainly in the phagolysosomes. Myeloperoxidase is involved in this phenomenon.

Abbreviations: CGD, chronic granulomatous disease; STZ, serum-treated zymosan; LDH, lactate dehydrogenase; MPO, myeloperoxidase.

INTRODUCTION

Neutrophils exhibit their microbicidal activity by releasing oxidative products (such as hydrogen peroxide) and granular enzymes into phagocytic vacuoles (1). Neutrophils from patients with chronic granulomatous disease (CGD) are unable to produce the oxidative bactericidal agents (2). These neutrophils, therefore, fail to kill bacteria that do not themselves produce hydrogen peroxide.

The release of lysosomal enzymes by CGD neutrophils was studied extensively by morphological and biochemical means. Morphological examinations suggested either normal or defective degranulation by CGD cells (3,4). However, it is very difficult to quanti-

[1]Part of this study will be published in the J. Clin. Invest.

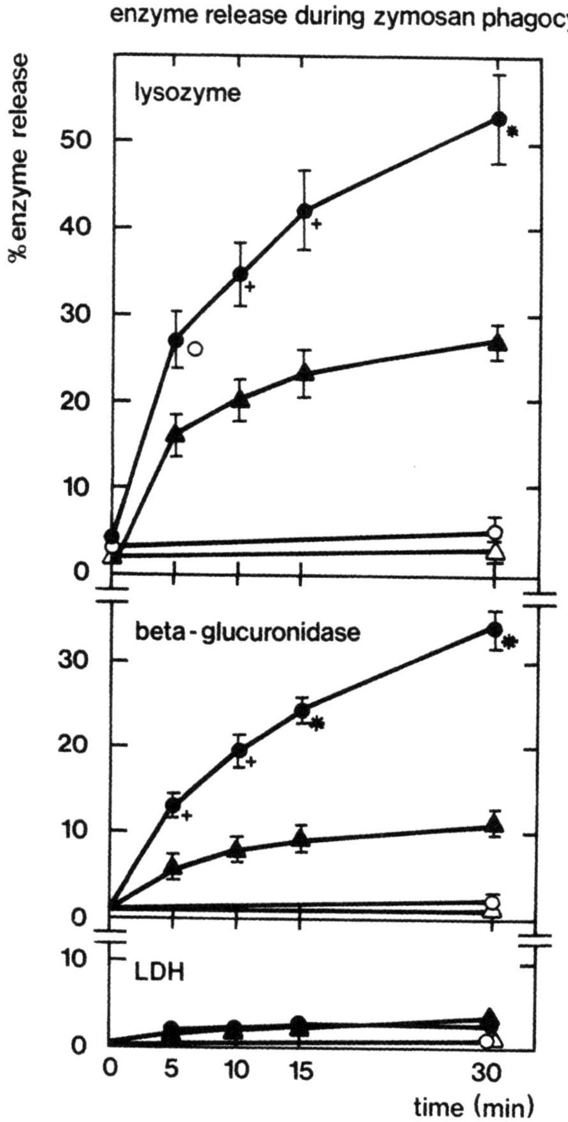

Fig. 1 - Release of enzymes during phagocytosis. The results are given as percentage of the enzymic activities in the cells at time zero. (▲), enzymic activity in the supernatant of normal cells incubated with serum-opsonized zymosan (1 mg/ml); (Δ), ibid. of normal cells incubated with an equal volume of 154 mM NaCl; (●), ibid. of CGD cells incubated with serum-opsonized zymosan (1 mg/ml); (o), ibid. of CGD cells incubated with an equal volume of 154 mM NaCl. Results are the means of 8 paired experiments ± SEM. Statistical significance between results obtained with phagocytosing normal neutrophils and CGD neutrophils (t-test for paired observations): *, $p<0.001$; +, $p<0.005$; o, $p<0.01$. For methods of cell isolation and enzyme assays, see ref. 10.

tate degranulation in morphological studies. Measuring the activities of granular enzymes in cell pellets or phagocytic vacuoles, most investigators were unable to detect significant differences in the release of these enzymes by CGD or normal cells (5,6). Other authors, however, reported normal (7), decreased (8) or even enhanced (9) rates of degranulation when CGD neutrophils were exposed to IgG-coated surfaces.

In the present study we measured degranulation from neutrophils of 8 CGD patients after challenge with serum-treated zymosan (STZ). In each experiment, CGD neutrophils released 2-3 times more activity of granular enzymes than did normal neutrophils. These differences, however, were found to be caused by partial inactivation of granular enzymes in normal, but not in CGD neutrophils.

RESULTS AND DISCUSSION

Inactivation of granular enzymes

The activities of lysozyme (N-acetylmuramide glycanohydrolase, E.C. 3.2.1.17) and beta-glucuronidase (beta-D-glucuronide glucurono-hydrolase, E.C. 3.2.1.31) released by resting and phagocytosing, normal and CGD neutrophils into the medium is shown in Fig. 1. CGD cells released twice as much lysozyme and three times as much beta-glucuronidase as did normal cells. Resting cells released only minimal activities of these enzymes. No differences were found in the enzymic activities in the cells prior to the incubations (not shown). The loss of the cytoplasmic enzyme lactate dehydrogenase (L-lactate: NAD oxido-reductase, E.C. 1.1.1.27) was always less than 5%, indicating that the cells remained intact during the incubation. Therefore, it seems that phagocytosing neutrophils from CGD patients release significantly more enzymic activities into the medium than normal cells. In principle, this difference might be due to a higher rate of phagocytosis by CGD cells. However, we did not detect differences in this respect (not shown).

On the other hand, the following experiments indicate that the release of granular enzymes by CGD cells only seem to be enhanced, because normal cells partly inactivate their lysosomal enzymes. Fig. 2 shows both the release of enzymic activities in the cell-free supernatants and the concomitant decrease of these activities in the cell pellets of normal and CGD neutrophils. In phagocytosing CGD neutrophils, the enzymic activities that disappeared from the cells were completely recovered in the cell-free supernatants. In normal cells, however, the decrease in enzymic activities in the cell pellets was not compensated by an equal increase of activities in the supernatants. Thus, in normal cells, but not in CGD cells, about 20 per cent of the total lysozyme activity was lost. Fig. 3 shows that about the same loss of total myeloperoxidase (donor: H_2O_2 oxidoreductase, E.C. 1.11.1.7) activity was observed. Similar results were obtained with beta-glucuronidase (not shown). Cyto-

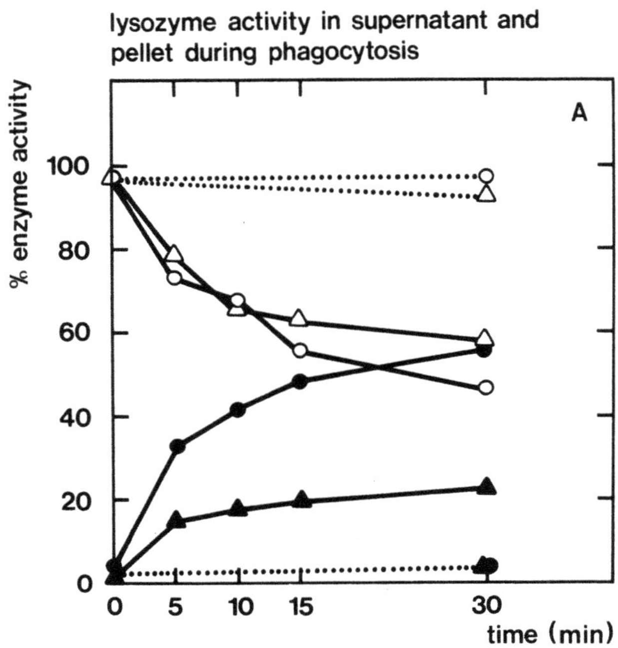

Fig. 2 - Enzymic activities in supernatant and in neutrophils during phagocytosis. The results are given as percentage of the enzymic activities in the cells at time zero. At various times, samples were taken and centrifuged. The enzymic activity was measured in the cell-free supernatant and in the pellet.
(Δ), enzymic activity in the pellet of normal neutrophils.
(▲), ibid. in the supernatant of normal neutrophils.
(o), ibid. in pellet of CGD neutrophils.
(•), ibid. in supernatant of CGD neutrophils.
Solid line, cells incubated with serum-opsonized zymosan (1 mg/ml).
Broken line, cells incubated with an equal volume of 154 mM NaCl.
Results are the means of 2 experiments.

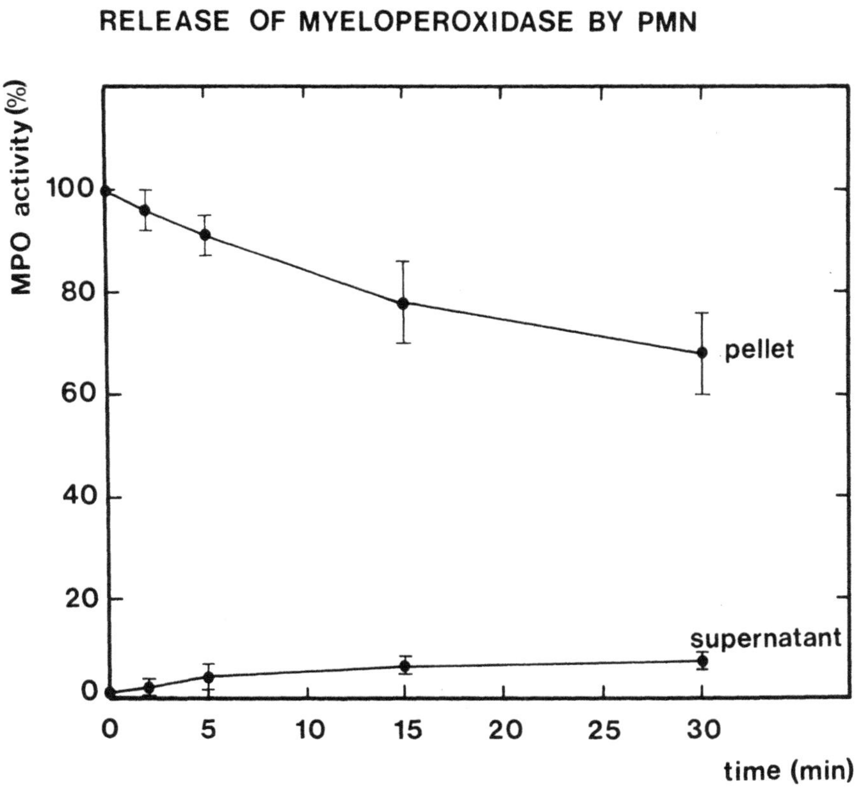

Fig. 3 - Myeloperoxidase activity in supernatant and cells during phagocytosis. Comparison of the residual activities of myeloperoxidase (MPO) in normal phagocytosing neutrophils (PMN) (pellet) with the activities measured in the cell-free supernatants at various times (abscissa). The activities of MPO are expressed as percentage of the total amount present in neutrophils before incubation (ordinate). Results are the means of 9 experiments ± SEM.

plasmic enzymes, such as LDH (Table 1) and catalase, were not affected.

This loss of activity was not an artefact of the separation procedure of the cell pellets from the supernatants, because suspensions of phagocytosing cells treated with Triton X-100 showed the same decrease in total enzymic activities. Furthermore, mixing experiments excluded the generation of inhibitors in normal phagocytosing neutrophils. With phorbol myristate acetate instead of STZ as a cell stimulant, similar results were obtained (not shown).

We hypothesized that in normal cells the loss of enzymic acti-

Table 1. Granular enzyme activities in supernatant and cell pellet from MPO-deficient neutrophils

	enzyme activities*		
	supernatant	pellet	total
MPO-deficient cells			
lysozyme	75	31	106
beta-glucuronidase	30	69	99
LDH	7	96	103
Normal neutrophils			
lysozyme	26	59	85
beta-glucuronidase	8	78	86
LDH	5	93	98

*Values are expressed as percentage of the original activities, after 30 min of incubation with STZ (1 mg/ml) (means of duplicate samples in one experiment).

vities was due to inactivation of the enzymes by the reactive oxygen compounds produced during phagocytosis. Indeed, when normal cells were allowed to phagocytose particles for 30 min under anaerobic conditions, the total enzymic activities recovered were 99% of the original lysozyme activity and 95% of the beta-glucuronidase activity (means of three experiments). The values obtained under aerobic conditions were 84 and 80%, respectively, of the original activities. These differences are statistically significant (t-test for paired observations, $p<0.005$). From these studies we conclude that inactivation of granular enzymes is due to oxidative damage, because neither phagocytosing CGD cells (Fig. 2) nor normal cells under anaerobic conditions lose enzymic activity.

Site of inactivation

The site of inactivation is probably the phagolysosome, where the reactive oxygen metabolites and the granular enzymes meet in high concentrations. This hypothesis was tested by measuring the specific activity of the lysosomal enzyme acid alpha-glucosidase (alpha-D-glucoside glucohydrolase, E.C. 3.2.1.20) in normal phagocytosing neutrophils (11). Cells were incubated with STZ and samples were taken after 0, 10 and 30 minutes. After centrifugation, the supernants were discarded and the cell pellets were lysed with 0.2% Triton X-100. In the cell pellets, the activity of acid alpha-glucosidase was measured. Furthermore, by adding various amounts of Sepharose-coupled antibody to this enzyme, the amount of

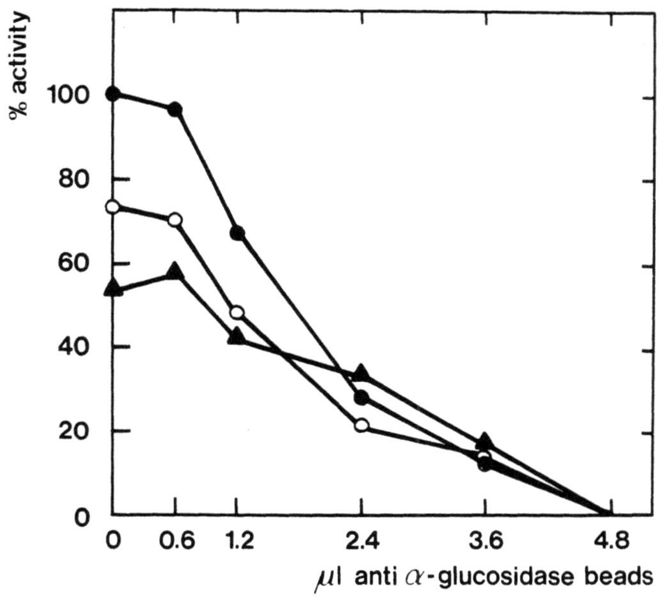

Fig. 4 - Precipitation of acid alpha-glucosidase with Sepharose-coupled antibody to acid alpha-glucosidase. Values are given as percentage of the acid alpha-glucosidase activity in the cells at time zero with no beads present.
(●), pellet at time zero.
(o), pellet after 10 min incubation with serum-treated zymosan.
(▲), pellet after 30 min incubation with serum-treated zymosan.
Values are means of duplicate determinations in one experiment.

acid alpha-glucosidase protein was determined. Fig. 4 shows that in the cell pellets obtained after 0, 10 and 30 min of incubation, the same amount (4.8 μl) of Sepharose-coupled antibody was needed to remove the activity of acid alpha-glucosidase from the cell pellets. This indicates that the amount of acid alpha-glucosidase protein in the cells remained constant during the time of incubation. The activity of this enzyme, however, decreased during the incubation to about half the original value. Thus, the specific activity of acid alpha-glucosidase in the cells decreased by about 50 per cent. Because hardly any activity was released into the cell-free supernatant (not shown), these findings indicate that the oxidative damage to lysosomal enzymes takes place mainly intracellularly, before leakage into the medium.

An intracellular site of inactivation also explains why the activity of purified acid alpha-glucosidase or lysozyme was not affected when added and incubated for 30 min either in the presence of STZ-stimulated neutrophils or in the presence of a system that generates hydrogen peroxide and superoxide (acetaldehyde plus xanthine oxidase).

Addition of catalase, superoxide dismutase or a combination of these two enzymes, to normal phagocytosing neutrophils, did not prevent inactivation of granular enzymes, although these enzymes remained active. Therefore, we have to consider the possibility that these enzymes did not reach the site of inactivation or did not work at that site. As indicated before, the phagolysosomes are probably the sites of inactivation; therefore, the low pH in these vacuoles might be suboptimal for these enzymes. Moreover, it might be impossible to mimick in vitro the environmental conditions that exist in phagolysosomes.

The addition of reduced glutathione (GSH) to a final concentration of 5 mM did give partial protection (the release of lysozyme during 30 min of incubation with STZ was enhanced from 22 $\pm$ 1% to 28 $\pm$ 1% (means $\pm$ SEM of 5 paired experiments)). Probably, these small molecules were able to reach the site of inactivation.

Mechanism of inactivation

As to the mechanism of inactivation, we concluded already that it depends on the production of oxygen compounds. Furthermore, in myeloperoxidase-deficient neutrophils, no inactivation of lysozyme and beta-glucuronidase was observed during phagocytosis (Table 1). Because it is known that MPO-deficient neutrophils release normal or enhanced amounts of hydrogen peroxide (12), our results indicate that both H_2O_2 and myeloperoxidase are required for the inactivation of granular enzymes. Which MPO-mediated oxidative products are responsible for the inactivation of granular enzymes is still unknown.

Conclusion

We conclude from these studies that granular enzymes are partly inactivated by MPO-dependent oxidative products generated by phagocytosing normal neutrophils. The site of inactivation is probably the phagolysosomes. As a consequence, the activities of enzymes released from normal cells are lower than those released from CGD neutrophils. Therefore, the release of granular enzymes from CGD neutrophils only seems to be enhanced. In fact, the decrease of granular enzyme activities from the cell pellets shows that the degranulation of CGD neutrophils is most probably comparable to that of normal cells. Furthermore, these studies stress the point that kinetic studies on release of lysosomal enzymes can only be properly interpreted if the residual enzymic activities in the cells are taken into account.

REFERENCES

1. S.J. Klebanoff, Antimicrobial mechanisms in neutrophilic polymorphonuclear leukocytes. Semin. Hematol. 12:117 (1975).
2. B. Holmes, A.R. Page, and R.A. Good, Studies of the metabolic activity of leukocytes from patients with a genetic abnormality of phagocytic function. J. Clin. Invest. 46:1422 (1967).
3. E. Kauder, L.L. Kahle, H. Moreno, and J.C. Partin, Leukocyte degranulation and vacuole formation in patients with chronic granulomatous disease of childhood. J. Clin. Invest. 47:1753 (1968).
4. P.G. Quie, J.G. White, B. Holmes, and R.A. Good, In vitro bactericidal capacity of human polymorphonuclear leukocytes: diminished activity in chronic granulomatous disease of childhood. J. Clin. Invest. 46:668 (1967).
5. R.L. Baehner, M.J. Karnovsky, and M.L. Karnovsky, Degranulation of leukocytes in chronic granulomatous disease. J. Clin. Invest. 47:187 (1968).
6. T.P. Stossel, R.K. Root, and M. Vaughan, Phagocytosis in chronic granulomatous disease and the Chediak-Higashi syndrome. New Engl. J. Med. 286:120 (1972).
7. G.L. Mandell, and E.W. Hook, Leukocyte function in chronic granulomatous disease of childhood. Am. J. Med. 47:473 (1969).
8. S.B. Gold, D.M. Hanes, D.P. Stites, and H.H. Fudenberg, Abnormal kinetics of degranulation in chronic granulomatous disease. New Engl. J. Med. 291:332 (1974).
9. R.J. Ulevitch, P. Henson, B. Holmes, and R.A. Good, An in vitro study of exocytosis of neutrophil granule enzymes in chronic granulomatous disease neutrophils. J. Immunol. 112: 1383 (1974).
10. D. Roos, R.S. Weening, A.A. Voetman, M.L.J. van Schaik, A.A.M. Bot, L.J. Meerhof, and J.A. Loos, Protection of phagocytic leukocytes by endogenous glutathione: studies in a family with glutathione reductase deficiency. Blood 53:851 (1979).
11. A.W. Schram, B. Brouwer-Kelder, W.E. Donker-Koopman, C. Loonen, M.N. Hamers, and J.M. Tager, Use of immobilized antibodies in investigating acid α-glucosidase in urine in relation to Pompe's disease. Biochim. Biophys. Acta 567:370 (1979).
12. H. Rosen, and S.J. Klebanoff, Chemiluminescence and superoxide production by myeloperoxidase-deficient leukocytes. J. Clin. Invest. 58:50 (1976).

Discussion
Edited by R.A. Clark

Dr. E. Schell-Frederick: Do your results suggest that enzymes already released into the medium are relatively protected from inactivation? What, then, is the best way to represent results of granular enzyme release?

Dr. R. Weening: I think our data indeed suggest that once the enzymes are released, little additional oxidative damage will occur. In our opinion, the best way of determining enzyme release is the measurement of specific enzyme activities in cell-free supernatants. Since this is not always very easy, I would also suggest taking the decrease in enzyme activity in the cell pellets into account.

Dr. D. Bass: Have you had the opportunity to examine these events in myeloperoxidase-deficient granulocytes?

Dr. R. Weening: Not yet, but it would be very useful to do so.

Dr. R. Clark: Could the loss of enzymatic activity from the cell pellets be the result of inactivation of enzymes in remaining intact granules as well as in phagolysosomes?

Dr. R. Weening: Certainly, since we have no evidence whether the inactivation by oxygen compounds takes place in the phagolysosomes, in the lysosomes, or in both. On theoretical grounds, however, I would expect that the site of inactivation is the phagolysosome, rather than the lysosome.

Dr. N. Borregaard: Did you look for a potential protective effect of mannitol on the inactivation of enzymes?

Dr. R. Weening: Indeed we did. With both benzoate and mannitol, the inactivation of enzymes by oxidative compounds could be partly prevented. However, we also observed that 25mM benzoate strongly inhibited oxygen consumption. Apart from the question whether these scavengers are specific enough to draw any conclusions, it is quite obvious from these experiments that with the concentration of scavengers used the cells were damaged.

Dr. D. Romeo: A few years ago, we published data showing that MPO associated with granules derived from phagocytizing PMN exhibits an increased affinity for the substrate H_2O_2 and a decreased Vmax. These changes in MPO kinetic properties could be reproduced by exposing granules of resting cells to either glucose-glucose oxidase or xanthine-xanthine oxidase. These data suggest that products of O_2 reduction can cause a change in the kinetic properties of granule-bound enzymes. Have you checked whether the decrease in specific activity of the granule enzymes you have studied is a function of substrate concentration or, generally speaking, of assay conditions?

Dr. R. Weening: We have not investigated enzyme kinetics. In principle, it might be possible that changes in the enzyme characteristics occur. This is, however, an intrinsic feature of the damage.

Dr. B. Seligmann: You suggest that the loss of lysosomal enzyme activity is directly mediated by an O_2-dependent reaction. However, you mention that purified acid alpha-glucosidase added to phagocytosing normal neutrophils was not inactivated. This seems to suggest that more is involved than the direct action of reactive species on the lysosomal enzymes. To pursue this, have you determined whether purified alpha-glucosidase or crude lysosomal enzyme activity is reduced when incubated in the presence of a cell-free system such as xanthine plus xanthine oxidase with or without MPO and halide?

Dr. R. Weening: Purified acid alpha-glucosidase or lysozyme added to phagocytosing PMN were indeed not inactivated. Purified lysozyme incubated in a cell-free system with either glucose-glucose oxidase or xanthine-xanthine oxidase, both in the presence and absence of purified myeloperoxidase, was unaffected, too. In our opinion, this only means that the intracellular situation during phagocytosis was not efficiently mimicked by these cell-free systems. Anyway, it is beyond any doubt that the inactivation of granular enzymes depends on oxidative damage since this phenomenon is not observed in CGD cells and under anaerobic conditions.

Dr. P. Elsbach: With regard to Dr. Seligmann's question, experts on peroxidative reactions tell me that peroxidative damage to macromolecules often takes place in a sequential manner. It may, therefore, be possible that the presence of a mixture of macromolecules (in the cell) enhances such a propagation of peroxidative damage, which is not observed in the presence of a single compound (enzyme).

Dr. R. Weening: I fully agree. You cannot reproduce the intracellular situation in cell-free systems.

THE RELEASE OF PLATELET-ACTIVATING FACTOR DURING PHAGOCYTOSIS BY POLYMORPHONUCLEAR NEUTROPHILS AND MONOCYTES

G. Camussi, F. Bussolino, ^M. Aglietta, ^W. Piacibello,
C. Tetta and G. Segoloni
Laboratorio di Immunopatologia, Cattedra di Nefrologia,
^Cattedra di Patologia Medica I
10126 Torino - Italy

INTRODUCTION

Platelet-activating factor (PAF) is a soluble factor that induces platelet aggregation and release reaction (1-4), independently from ADP- and Arachidonic Acid/Thromboxane A_2 -mediated pathways (2, 4). PAF is a low molecular weight phospholipid (1-0-alkyl-2-acetyl--sn-glyceryl-3-phosphorylcholine)(5,6). PAF was initially obtained from antigen-stimulated, IgE-sensitized rabbit basophils (2). In man PAF release from basophils still awaits confirmatory evidence, mainly because in man basophils constitute a much smaller fraction of the total leukocytes than in the rabbit. Lewis recovered a PAF-like activity from purified leukemic basophils in the presence of Calcium Ionophore (7). Later, we have shown that immune complexes (IC), the specific antigen, anaphylatoxin and neutrophil cationic proteins stimulate PAF release from human leukocytes (8). Although basophil degranulation was coincident with PAF release, the cell populations used in those experiments were relatively impure. More recently, ultrastructural evidence of platelet interaction only with degranulated basophils but not with either neutrophils or monocytes was obtained when basophils were degranulated by the specific antigen, anti-IgE or Synachten (9). However, PAF release occurred independently from basophil degranulation during phagocytosis by polymorphonuclear neutrophils (PMN) and monocytes (9,10). Lotner et al. demonstrated that purified human PMN are capable of releasing PAF (11). The purpose of the present report was to comparatively study the mechanism of PAF release from PMN and monocytes under phagocytic stimuli.

MATERIALS AND METHODS

Human PMN and plastic-adherent monocytes were obtained as previously described (9). PMN preparations were more than 90% pure. Monocytes (i.e. naphtol-AS-D-chloroacetate-positive cells) were more than 95% pure. One x10^6 PMN or 1x10^5 monocytes per ml of Tris-buffered Tyrode, pH 7.4, supplemented with 0.25% bovine serum albumin (TTBSA) were challenged with the following stimuli: a) Ionophore A 23187, b) Baker's Yeast Spores (BYS) - 10-times more than the cells -, activated with C3b (C3b-BYS) or C3d (C3d-BYS) prepared as already described (9), c) unopsonized or opsonized Zymosan (Z)(1.5 mg/ml), d) C5a anaphylatoxin (0.2 μg/ml)generated and purified as described in (12). PAF passive release was assessed by incubating the cells overnight at pH 10.6 at 20°C in TTBSA. PAF was detected by aggregation of washed rabbit platelets prepared as previously described (2,8) in an aggregometer (Elvi 840). Two to five x10^7 platelets were stirred at 900 rpm in 300 μl of TT supplemented with 0.25% gelatin in presence of Indomethacin (10^{-5}M) which inhibits the cyclooxygenase, and of the Creatine Phosphate (CrP) Creatine Phosphokinase (CPK) enzymatic system (CrP/CPK)(31.25 μg/ml^{-1} of CrP and 15.25 μg/ml^{-1} of CPK), which converts AMP and ADP to ATP, thus blocking both Arachidonic Acid and ADP dependent platelet aggregation. PAF activity was expressed in arbitrary units per milliliter as defined in (8). The amount of PAF in ng/ml was calculated using as reference substance semi-synthetic PAF (Dr. Benveniste's gift) with a specific activity of 6 arbitrary units/ng. PAF from different cell populations was purified as previously described (13,14). Briefly, the crude PAF sample in TTBSA, BSA serving as PAF carrier, was extracted with 70% ethanol and phased into chloroform. The lipid fraction thus obtained was chromatographed on a silicic acid column and developed with chloroform, chloroform-methanol mixtures and pure methanol. PAF--active fraction was further chromatographed on silica gel plates and developped with chloroform-methanol-water (65:35:4, v/v). The biologically active material was characterized as PAF and compared to semi--synthetic PAF according to the following criteria: a) the ability to induce platelet aggregation by a pathway independent from both ADP-and Arachidonic Acid/Thromboxane A_2-mediated pathways; b)treatment with lipases: inactivation by phospholipase A_2, C and D and resistance to lipase A1; 3) same Rf on thin layer chromatography as the semi-synthetic PAF and migration between sphyngomyelin and lyso-phosphatidylcholine (13,14). Histamine was measured by a spectrofluorimetric procedure (15). Lactate dehydrogenase was assayed with NADH (16), β-glucuronidase according to (17), lysozyme according to (18).

RESULTS

As shown in Fig.1, purified neutrophils released PAF after challenge with non- and opsonized Z, opsonized BYS, Ionophore A 23187 or by incubation in an alkaline environment (pH 10.6). The release of PAF from PMN was compared to that from plastic-adherent monocytes(Table I). Plastic-adherent monocytes were capable of releasing PAF under several stimuli: IgG-coated and C3b-,C3d- BYS, Ionophore A 23187 and at pH 10.6. PMN released PAF in response to IgG-coated and C3b- BYS, C5a anaphylatoxin, Ionophore A 23187 and at pH 10.6. The release of PAF from PMN and monocytes differed in that monocytes were unable to release

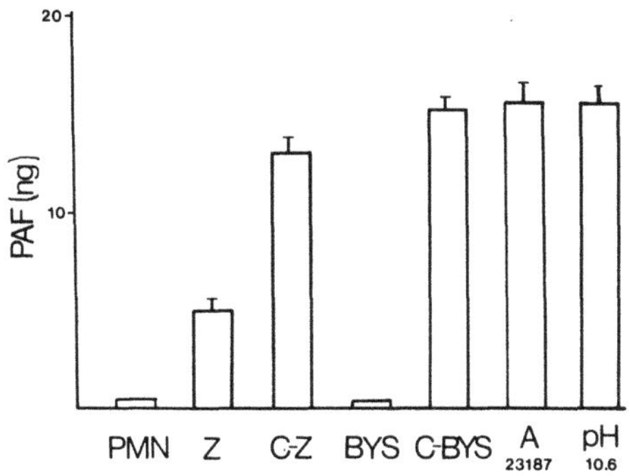

Fig. 1: Release of PAF from purified PMN ($1x10^6$ cells/ml of TTBSA with Ca^{++} and Mg^{++}) either at pH 10.6 or challenged (15 min at 37°C) with 1 µg/ml of Ionophore A23187 or with non- and opsonized phagocytable substrates such as Z (and C-Z) or BYS (and C-BYS). The figure is representative of ten experiments.

PAF after challenge with C5a anaphylatoxin, whereby PMN, which release PAF in response to C5a, were unable to release PAF when challenged with C3d-BYS. Both in PMN and monocyte preparations, metachromatically staining basophils were absent or less than 0.1% and the histamine amount was negligible. Increasing the percentage of contaminating cells by addition of basophils or lymphocytes, purified on Ficoll gradient, to PMN or monocyte prepations did not affect PAF release during phagocytosis. However, 10% monocytes or 15% PMN contaminating purified PMN and monocyte preparations, respectively, significantly increased

TABLE I. PAF Release from PMN and Monocytes

	PMN δ ($1x10^6$/ml)	PLASTIC-ADHERENT MONOCYTES($1x10^5$/ml)δ
IgG-BYS	11.8	12.6
C3b-BYS	12.7	13.3
C3d-BYS	0	11.8
C5a	5.4	0
Ionophore A23187	14.1	20.2
pH 10.6	14.1	20.2

δ Comparative release of PAF (expressed in ng) from purified PMN and plastic-adherent monocytes under various phagocytic and soluble stimuli. The table is representative of ten experiments.

($p < 0.01$) the amount of PAF released during phagocytosis of C3b-BYS. Furthermore, the release of PAF was well correlated with the percentage of phagocytic cells (Fig. 2). The temporal correlation of PAF release from PMN and monocytes with phagocytosis of C3b-BYS and lysosomal enzyme secretion is shown in Figure 3. Phagocytosis started rapidly (1-2 min) to finally reach the maximal percentage of phagocytic cells at 10 min for PMN and 15 min for monocytes. PAF release shows a profile superimposable to that of the percentage of phagocytic cells. The number of particles ingested per cell, which increased until the 30th to the 60th minute, did not affect the release of PAF. In contrast, lysosomal enzyme release continued to rise until the 60th min incubation. Lactate dehydrogenase was not released in significant amount

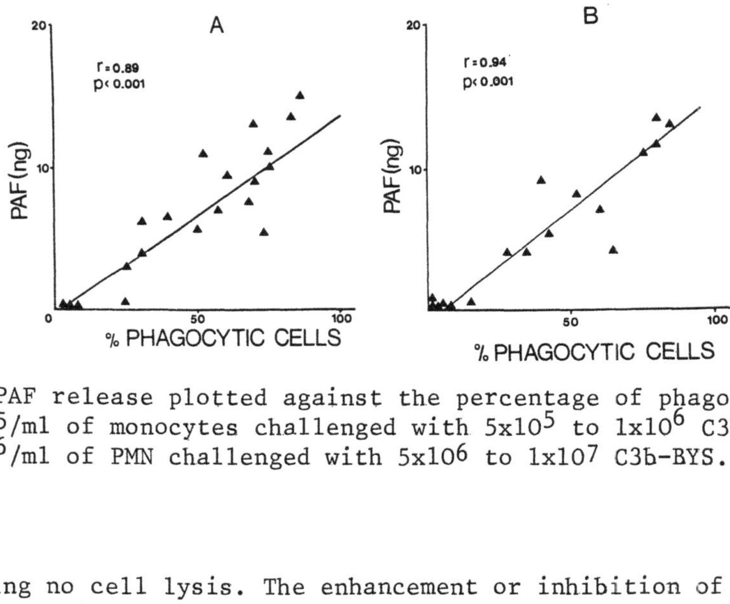

Fig. 2: PAF release plotted against the percentage of phagocytic cells. A = $1x10^5$/ml of monocytes challenged with $5x10^5$ to $1x10^6$ C3b-BYS, B = $1x10^6$/ml of PMN challenged with $5x10^6$ to $1x10^7$ C3b-BYS.

indicating no cell lysis. The enhancement or inhibition of PAF release from phagocytic cells was investigated by using several chemicals having selective biochemical effects (Table II). Bromo-diphenacyl-bromide, an inhibitor of phospholipase A_2, and Ethylendiaminetetraacetate (EDTA), a cation ligand, inhibit PAF release from PMN and plastic-adherent monocytes. Eicotetrayenoic (EDTA), an inhibitor of both cyclo- and lypo-oxygenase, enhances the release of PAF, but indomethacin, an inhibitor of cyclooxygenase, is uneffective. Prostacyclin (PGI_2) and N^6, O^2 dibutirryl Adenosine-3',5'-cyclic phosphoric acid (dib-cAMP), that increase the cellular level of cAMP, inhibit PAF release. Cytochalasin B, that damages the cell subplasmalemmal microfilaments, enhances the release of PAF from C5a-stimulated PMN, but blocks PAF release from PMN and monocytes under phagocytic stimuli. The addition of Sodium acetate markedly enhances the release of PAF.

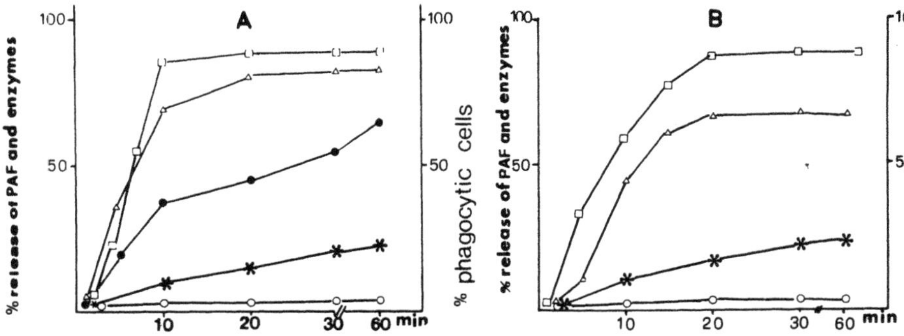

Fig. 3: Kinetics of the release of PAF△—△, β-glucoronidase *—*, lysozime ●—●, lactate dehydrogenase ○—○ from phagocytic monocytes (B) and PMN (A). PAF and lysosomal enzyme release is expressed as the percentage of the total amount recovered after incubation at pH 10.6 or with Triton X 100, respectively. The figure is representative of eight experiments. Phagocytosis □—□

TABLE II. EFFECT OF VARIOUS CHEMICALS ON PAF RELEASE FROM PMN AND MONOCYTES

	MONOCYTES + C3B-BYS	PMN + C3b-BYS	PMN + C5a
	% VARIATION		
BROMO-DIPHENACYL- -BROMIDE ($1x10^{-6}M$)	-67 [δ]	-60	-78
EDTA ($5x10^{-4}M$)	-100	-100	-100
ETYA ($1x10^{-5}M$)	+34	+37	+30
INDOMETHACIN ($5x10^{-6}M$)	0	0	0
dib-cAMP ($1x10^{-3}M$)	-53	-50	-59
CYTOCHALASIN B ($2.5x10^{-5}M$)	-68	-60	+38
PGI_2 ($1x10^{-7}$ M)	-71	-77	-68
SODIUM ACETATE ($1x10^{-5}M$)	+145	+168	+149

[δ] These numbers are percentages of variations either as enhancement or inhibition of PAF release from phagocytic monocytes and PMN or from C5a-stimulated PMN in the presence of various chemicals as compared with control values (see Table I).

DISCUSSION

PAF released from leukocytes under several stimuli is an extraordinarly potent soluble mediator of platelet aggregation and secretion of their granular constituents. During recent years, the chemical nature of PAF has been the subject of intensive research. PAF is a small lipid as it is entirely recovered in the chloroform phase of chloroform-methanol-water mixtures (3,13) with an estimated molecular weight between 700 and 1,000 daltons on Sephadex LH 20 in 80% ethanol (13). Its solubility in chloroform alone, together with a low migration between sphyngomyelin and lyso-phosphatidylcholine on silica gel thin layer chromatography, suggested a neutral highly polar lipid (3,13,19). Given these data, possible candidates were glycosylacylglycerydes, neutral phosphoglycerides and neutral sphyngolipids. Lipids having known hormone-like activity (steroids and prostaglandins) were eliminated along with glycosylacylglycerides and glycosphyngolipids as PAF is resistant to sodium nitrite, which cleaves glycosidic linkages, and to reagents for free or vicinal hydroxyl and amino groups (19). PAF resists boiling in water for 5 min; it resists 0.5 N ethanolic HCl at 37°C for 3 hr and 0.03 N NaOH at 37°C for 20 min in water, but its activity disappears after the latter treatment when in ethanol (13,19). Therefore, PAF is not a regular glycophosphate, a sphyngomyelin or a plasmalogen. Analysis of PAF by lipases further contributes to the interpretation of its structure (14). PAF is inactivated by phospholipase A2 from either porcine pancreas or snake venom, by phospholipase C from B. _cereus_ and D from cabbage. However, a phospholipase C specific for sphyngomyelin is totally uneffective thus ruling out the possibility of PAF being a sphyngomyelin. Lipase A1 from _Rhizopus arrhizus_, under conditions when 100% of phosphatidylcholine is converted into lyso phosphatidylcholine, has no effect on PAF. Finally, PAF activity is not affected by diazomethane, indicating the presence of a choline head group. Thus, PAF is a glycerophospholipid devoid of an ester linkage at position 1. Initially, the bulk of this evidence led to the proposal that the backbone of the molecule were a 2-acyl (or 1-lyso)-glyceryl-phosphorylcholine (13,14). However, 1-lyso-phosphatidylcholine fails to mimick PAF activity(14). More recently, Benveniste et al and Demopoulos et al. independently recognized a 1-fatty ether glycerol-phosphatidylcholine, which, following acetylation at position 2, generated a compound identical in biological activity and chromatographic behaviour to PAF (5,6). The relevance of an acetyl group in the molecule was noticed as acetic anhydride treatment increased the PAF-like activity of crude but not of purified PAF preparations (5). The chemical structure of PAF which well accords with its insensitivity to lipase A1 but inactivation by phospholipase A2, C and D is as follows:

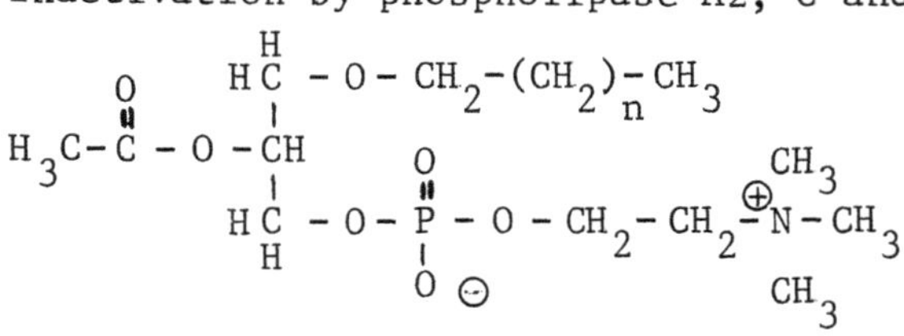

More recently, the aliphatic chain in position 1 has been elucidated by Godfroid et al., who synthesized a 1-O-octadecyl-2-O-acetyl-sn-glycero-3-phosphorylcholine displaying functional and physicochemical characteristics identical to PAF (20) and by Hanahan et al. (21), who demonstrated that the alkyl side chain of rabbit-derived PAF were composed of octadecyl (90%) and hexadecyl (10%) residues.

Several reports have indicated that human leukocytes release PAF (3,7,8). PAF has been recovered from human mixed leukocyte suspensions concomitantly with basophil degranulation (8) and, more recently, from monocytes and PMN (9,10,11). The stimuli capable of releasing PAF in-vitro from these cells show some specificity for the cell target(9,10). Monocytes and PMN release PAF during phagocytosis. Neutrophils but not monocytes release PAF in response to C5a anaphylatoxin, neutrophil cationic proteins and their Carboxypeptidase B-derived desarginated fragments (9).

In the present paper, purified neutrophils and plastic-adherent, sterase-positive mononuclear cells are shown to release PAF under phagocytic stimuli. Nevertheless, the mode by which PMN and mononuclear cells release PAF seems to be different. Phagocytosis-associated release of PAF from PMN and monocytes is observed in the presence of IgG- or C3b-BYS. However, monocytes but not PMN release PAF during phagocytosis of C3d-BYS. In contrast, PMN but not monocytes release PAF when challenged with soluble stimuli such as purified C5a anaphylatoxin. During phagocytosis, monocytes release a partial amount (50%) or their total PAF (releasable at pH 10.6 or after challenge with Ionophore A23187), whereby PMN liberate almost the totality of their dismissable PAF. The kinetics of PAF release is temporally closely associated with phagocytosis. The release of PAF from both the two cell types is well related with the percentage of phagocytic cells but is independent from the number of ingested particles. The release of PAF during phagocytosis is an active process requiring Ca^{++} and Mg^{++} ions and metabolic energy. Our studies on the effect on PAF release from PMN and monocytes by several chemical compounds, known to have selective biochemical activities, suggest the activation of a membrane phospholipase A_2 since the inhibitory effect of bromo-diphenacyl-bromide, the modulating role of arachidonate/lipooxygenase and of cyclic AMP since the enhancing effect of ETYA, an inhibitor of both cyclooxygenase and lypooxygenase, and the inhibition by increased intracellular c-AMP levels by PGI_2 and dib-cAMP. Of critical importance is the integrity of subplasmalemmal microfilaments as,in the presence of Cytochalasin B, PAF release is blocked during phagocytosis but is enhanced when PMN are challenged with soluble stimuli.The increase of PAF yield by Sodium acetate suggests the improtance of an acetylation process, the availability of the cellular acetyl donor during the synthesis and/or release of PAF, and the critical importance of the acetyl group ester linked at position 2 from PMN and monocyte -derived PAF to be active. This observation coupled with Benveniste's report of increased PAF-like activity of crude but not of purified PAF preparations by treatment with acetic anhydride (5) could indicate that a 2-lyso-PAF release, possibly following activation of membrane phospholipase A2, undergoes acetylation in position 2 and acquires biological activity.

The increasing awareness that PAF may be a membrane phospholipid belonging to various cells capable of degranulating under several stimuli suggests that PAF may have a relevant role in inflammatory and hypersensitivity reactions.

ACKNOWLEDGEMENT

This work was supported by C.N.R., Rome, Contract no.79.01998.04.

REFERENCES

1) P.M. HENSON and C.G. COCHRANE: Immune complex disease in rabbits. The role of complement and of a leukocyte-dependent release of vasoactive amines from platelets. J.Exp.Med. 133: 554 (1971).
2) J.BENVENISTE, P.M. HENSON and C.G. COCHRANE: Leukocyte-dependent, histamine release from rabbit platelets: the role of IgE, basophils and platelet-activating factor. J.Exp.Med. 136: 1356 (1972).
3) J.BENVENISTE: Platelet-activating factor, a new mediator of anaphylaxis and immune complex deposition from rabbit and human basophils. Nature (Lond.) 249: 581 (1974).
4) M.CHIGNARD, J.P. LE COUEDIC, M. TENCE, B.B. VARGAFTIG and J. BENVENISTE: The role of platelet-activating factor in platelet aggregation. Nature (Lond.) 279: 799 (1979).
5) J.BENVENISTE, M. TENCE, J. BIDAULT, C. BOULLET and J. POLONSKY: Sémi-synthèse et structure proposée du facteur activant les plaquettes (PAF): PAF-acether, un alkyl ether analogue de le lysophosphatidylcholine. C.R.Acad.Sci. (Paris), 289: 1037 (1979).
6) C.A.DEMOPOULOS, R.N. PINCKARD and D.J. HANAHAN: Platelet-activating factor. Evidence for a 1-0-alkyl-2-acetyl-sn-glyceryl-3-phosphorylcholine as the active component (a new class of lipid chemical mediators). J.Biol.Chem. 254, 9355 (1979).
7) R.A.LEWIS, E.J. GOETZL, S.J. WASSERMAN, S.I. VALLONE, R.H. RUBIN and K.F AUSTEN: The release of four mediators of immediate hypersensitivity from human leukemic basophils. J.Immunol. 114: 87 (1975).
8) G.CAMUSSI, J.M. MENCIA-HUERTA and J. BENVENISTE: Release of platelet activating factor and histamine. I. Effect of immune complexes, complement and neutrophils on human and rabbit mastocytes and basophils Immunology 33: 523 (1977).
9) G.CAMUSSI, M. AGLIETTA, R. CODA, F. BUSSOLINO, W. PIACIBELLO and C. TETTA: Release of platelet activating factor (PAF) and histamine. II. The cellular origin of human PAF: monocytes, polymorphonuclear neutrophils and basophils. Immunology (1980) in press.
10) G.CAMUSSI, F. BUSSOLINO, C.TETTA, J. BENVENISTE and A. VERCELLONE: Platelet-activating factor and glomerulonephritis. In "Hemostasis, Prostaglandins and Renal Disease" Eds. G.Remuzzi et al., Raven Press New York, 133, 1980.
11) G.Z.LOTNER, J.M. LYNCH, S.J. BETZ and P.M. HENSON: Human neutrophil derived platelet activating factor. J.Immunol. 124: 676 (1980).
12) G.CAMUSSI, C. TETTA, F. BUSSOLINO, F. CALIGARIS CAPPIO, R. CODA, C. MASERA and G. SEGOLONI: Mediators of immune complex-induced aggrega-

tion of polymorphonuclear neutrophils. I.C5a Anaphylatoxin, Neutrophil Cationic Proteins and Their Cleavage Fragments. Int.Archs Allergy appl.Immun. 62: 1-15 (1980).

13) J.BENVENISTE, G.CAMUSSI, J.POLONSKY: PLatelet-activating factor. Monogr.Allergy 12: 138 (1977).

14) J.BENVENISTE, J.P. LE COUEDIC, J. POLONSKY and M. TENCE. Structural analysis of purified platelet activating factor by lipases. Nature (Lond.) 269: 180 (1977).

15) F.RUFF, A.SANDELLE, E. DUTRIPAN and J.L. PARROT: Continous automatic fluorimetric evaluation of total blood histamine. Nature (Lond.) 214: 279 (1967).

16) W.E.WACKER, D.D.ULMER and B. VOLEE. Metalloenzymes and myocardial infarction. New Engl.J.Med. 225: 449 (1956).

17) G.BRITTINGER, R.HIRSCHORN, S.D. DOUGLAS and G. WEISSMAN: Studies on lysosomes. XI. Characterization of hydrolase-rich fraction from human lymphocytes. J.Cell.Biol. 37: 394 (1968).

18) D.J.PROCKOP and W.P. DAVIDSON: A study of urinary and serum lysozyme in patients with renal disease. New Engl.J.Med. 270: 269,1964

19) R.M.PINCKARD, R.S. FARR and D.S. HANAHAN: Physicochemical and functional identity of rabbit platelet-activating factor (PAF) released in vivo during IgE anaphylaxis with PAF released in vitro from IgE-sensitized basophils. J.Immunol. 123, 1847 (1979).

20) J.J. GODFROID, F. HEYMANS, E. MICHEL, C. REDEVILH, E. STEINER and J.BENVENISTE: Platelet-activating factor (PAF-acether): total synthesis of 1-O-octadecyl-2- O -acetyl-sn-glycero-3-phosphorylcholine FEBS Letters 116: 161 (1980).

21) D.J. HANAHAN, C.A. DEMOPOULOS, J. LIEHR and R.N. PINCKARD: Identification of PF isolated from rabbit basophils as acetyl glyceryl ether phosphorylcholine. J.Biol.Chem. 25: 5514 (1980).

Discussion
Edited by R.A. Clark

Dr. P. Elsbach: Would you indicate how you assayed for PAF? Since sodium acetate stimulates PAF release (production?), this should permit you to label PAF and make your assay easier. Do you know the localization of PAF and its production in leukocytes?

Dr. G. Camussi: The PAF assay is performed on washed rabbit platelets, preincubated with creatine phosphate/creatine phosphokinase enzymatic system and indomethacin so as to inhibit ADP- and arachidonic acid-induced platelet aggregation. We examined aggregometry and release of ^{3}H-serotonin. Studies are in progress in our laboratory to obtain labeled PAF. The localization of PAF in leukocytes is not known. It is not a preformed mediator in a strict sense, but it is probably a membrane phospholipid or component generated by a process probably enzymatic in nature in stimulated leukocytes.

THE SECRETION OF LYSOSOMAL ENZYMES BY HUMAN POLYMORPHONUCLEAR LEUCOCYTES (PMN) AND ITS MODULATION BY SERUM COMPLEMENT

Klaus-Michael Debatin[1] and Jörg Menzel

Institut für Immunologie und Serologie der Universität
Heidelberg and Fachklinik Königstuhl
D-6900 Heidelberg

When engaged in phagocytosis PMN release considerable amounts of their lysosomal constituents. This "regurgitation during feeding" is considered to be caused by passive loss out of incompletely closed phagosomes (1). However, there is an obvious discrepancy between the assumption, that the enzymes are lost during the phagocytic process and the important intraphagosomal function they are thought to have in the destruction of ingested microbes. In this study we investigated the release of lysozyme, myeloperoxidase (MPO) and protease activity at pH=5,5 in relation to the ingestion of E.coli. These enzymes are considered to have important bactericidal as well as digestive functions.

A standard phagocytosis system was used consisting of a PMN-monolayer ($5x10^6$/ml) to which ^{3}H-thymidine labelled E.coli (complement-resistant) in Hank's balanced salt solution (HBSS) were added. The bacteria ($2x10^8$/ml) were opsonized with complement (human serum) and/or specific IgG antibody (rabbit). Phagocytosis was quantified by measurement of PMN associated radioactivity calculating the number of bacteria from the specific label. In some experiments the PMN were fractionated using sucrose-heparin according to a procedure described in detail elsewhere (2). Lysosomal enzymes were measured in the supernatants after centrifugation to sediment cells and bacteria: MPO according to Pütter (3), lysozyme with micrococcus lysodeikticus (4), and acid protease using casein as substrate (5). As a marker of cell viability lactic dehydrogenase, which was also determined, was constantly found with no more than 5-10% of total activity in the supernatants.

[1]supported by Deutsche Forschungsgemeinschaft

The first experiments investigating the kinetic of enzyme release showed that obviously differences exist for each enzyme (Fig. 1): While lysozyme was maximally released within the first 20 minutes of incubation, the release of acid proteases was delayed and only a little MPO activity was detectable in the medium.

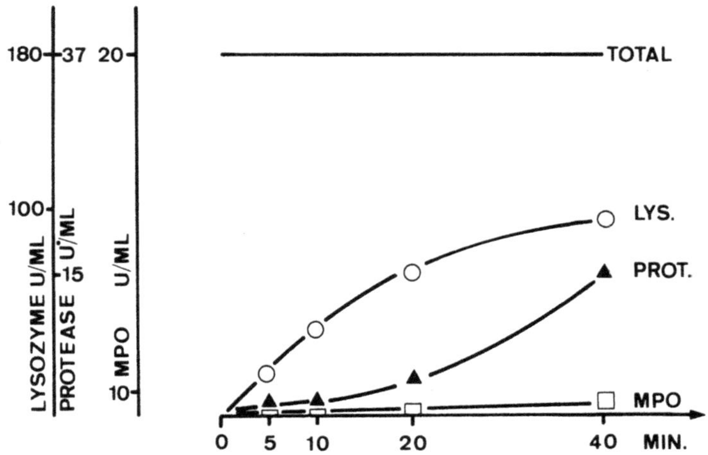

Fig. 1 Kinetic of enzyme release during phagocytosis of E.coli opsonized with IgG plus complement.

The kinetic of phagocytosis (Fig. 2) suggested that the lysozyme release closely paralled the increase in ingested bacteria (i.e. cell bound ^{3}H-thymidine).

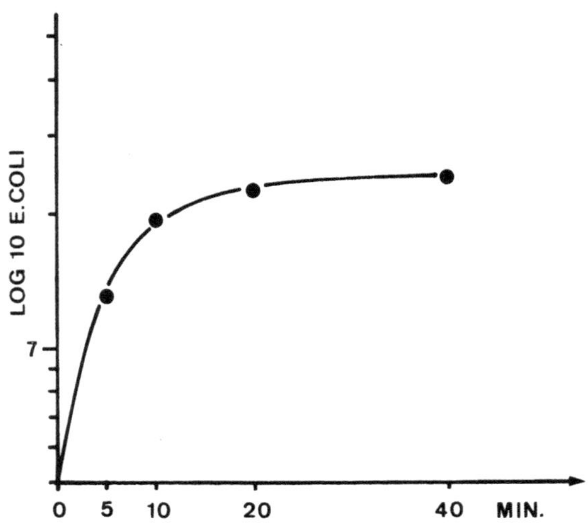

Fig. 2 Kinetic of phagocytosis

In order to study the relationship between enzyme release and the different steps of ingestion, i.e. the attachment of bacteria to the cell membrane and the subsequent formation of phagolysosomes, the experimental protocol was changed. The PMN monolayer were incubated with bacteria for a short period of time, washed thereafter and incubated again in HBSS. After different times the cells were treated with sucrose-heparin followed by centrifugation to result in a sediment containing the membrane fraction. The initial incubation period (5 minutes) was expected to result in a considerable amount of bacteria bound to the cell membrane but yet not being internalized. The following incubation in medium should allow the observation of enzyme release and internalization of bacteria simultaneously.
As shown in Fig. 3, under these conditions the pattern of enzyme release was the same as before (Fig. 1).

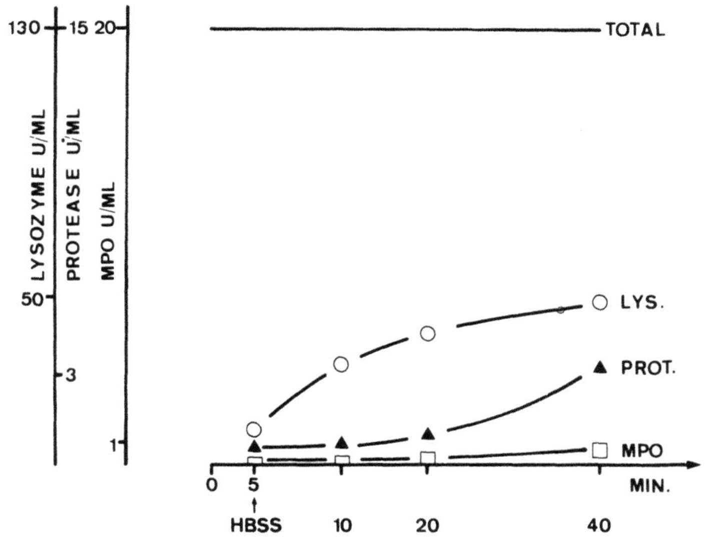

Fig.3 Kinetic of enzyme release after removal of unbound bacteria.The values for HBSS incubation represent activities released during the 5 min. period plus additional release during incubation in medium.

Corresponding to the release of lysozyme an increase in intracellular bacteria, as measured by ^{3}H-thymidine label, was observed (Fig. 4). However at 40 minutes the intracellular label was reduced and increased again in the membrane fraction. This was paralled by the loss of about 10% of the total cell associated radioactivity (data not shown).

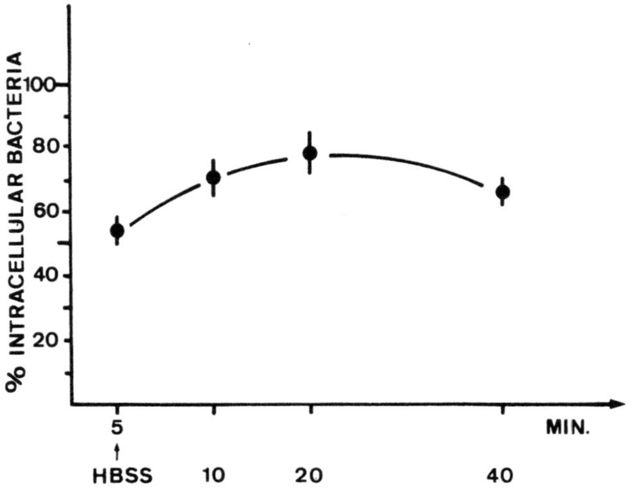

Fig.4 Percentage of bacteria (^{3}H-thymidine label) not associated with the membrane fraction.

This unexpected redistribution of bacteria and the loss of total radiolabel may be the consequence of exocytic processes. This was suggested by the appearence of free ^{3}H-thymidine in the cell- and bacteria free supernatant (Fig. 5). The amount released quantitatively correlated with the loss in total cellular label (10%).

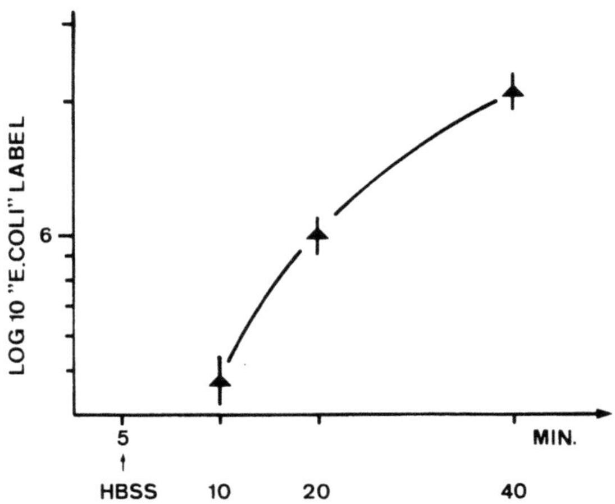

Fig.5 Release of free ^{3}H-thymidine expressed as the number of corresponding E.coli calculated from the specific label.

Since it has been demonstrated that bacteria opsonized with IgG plus complement were not only ingested to a higher extend but also killed more efficiently when compared to those coated with IgG alone (7), we were interested to see whether opsonizing complement affects the amount of enzyme activity released during phagocytosis. The PMN monolayer were incubated for 30 minutes with E.coli opsonized by different procedures (IgG alone or IgG plus complement), after which the supernatant was removed and tested for lysozyme activity. The ratio between lysozyme release and amount of bacteria ingested was calculated for both groups of opsonized bacteria. As shown in Table 1 in the presence of complement the lysozyme activity released per one ingested E.coli was dramatically reduced, the number of bacteria per granulocyte being enhanced as expected.

Table 1. Influence of opsonizing complement upon the release of lysozyme

Exp.No.	Opsonization	Lysozyme[1]/E.coli	E.coli/PMN
2	IgG	3,3	0.25
	IgG+C	0,4	9
4	IgG	3,5	0,25
	IgG+C	0,5	11
5	IgG	22,0	0,10
	IgG+C	1,7	7
6	IgG	12,0	0,13
	IgG+C	1,0	9

[1]Lysozyme activity: U/ml x 10^{-6}

To summarize our results, the data indicate that the release of lysosomal enzymes from PMN during phagocytosis is a more complex process rather than a passive "regurgitation during feeding". This mechanism originally considered to be responsible for the degranulation may nevertheless account for the release of lysozyme, which seems to be in close correlation with the internalization of bacteria. In this context the lack of MPO release may reflect the observation, that azurophil granules, which contain all MPO activity degranulate at later stages of phagolysosomal formation than specific granules (8), which contain half of the lysozyme activity. The kinetic of the release of acid proteases however clearly shows that other mechanisms, possibly secretory processes exist, by which granule material is released by the phagocyte. This is confirmed by the finding that extrusion of radiolabel occured, which indicates exocytosis of degrated bacterial compounds.

In addition to the heterogeneity of release mechanisms complement components present on the bacterial membrane seem to be able to modify the degranulation process. Since opsonizing complement only enhances the number of bacteria bound to the phagocyte membrane but leaves unaffected the rate of ingestion (2), the reduction of enzyme release as a consequence of changes in the phagocytic process itself, a more rapid internalization for example, seems to be unlikely. Although not tested so far, the reduced release of lysosomal enzymes may be a direct consequence of C3b receptor stimulation thus enhancing the bactericidal capacity of the phagocyte.

REFERENCES

(1) G.Weissmann, J.E. Smolen, S.Hoffstein, Polymorphonuclear leukocytes as secretory organs of inflammation. J.invest.Dermatol. 71: 95 (1978).

(2) R.Berto, J.Menzel, The interaction of opsonins with human polymorphonuclear leukocytes (PMN). I. The influence of human complement (C) and IgG on ingestion and digestion of C-resistant E.coli. Z. Immun.-Forsch. 155: 189 (1979).

(3) J.Pütter, Peroxidasen. in: "Methoden der enzymatischen Analyse" H.U.Bergmayer ed. Verlag Chemie Weinheim/Bergstr. (1974).

(4) S.Gordon, J.Todd, Z.A.Cohn, In vitro synthesis and secretion of lysozyme by mononuclear phagocytes. J.exp.Med.139: 1228 (1974).

(5) Y.Lin, G.E.Means, R.E.Feeny, The action of proteolytic enzymes on N,N-dimethyl-proteins. J.Biol.Chem. 244: 789 (1969).

(6) F.Wroblewski, J.S.LaDue, Lactic dehydrogenase activity in blood. Proc.Soc.exp.Biol.Med. 90: 210 (1955).

(7) J.Menzel, H.Jungfer, D.Gemsa, Contribution of IgG,IgM, complement and properdin to the intracellular killing of Escherichia coli by polymorphonuclear leucocytes. Infect.Immun.19: 659 (1978).

(8) D.F.Bainton, Sequential degranulation of the two types of polymorphonuclear leukocyte granules during phagocytosis of microorganisms. J.Cell.Biol. 58: 249 (1973).

Discussion
Edited by R.A. Clark

Dr. R. Weening: May I make two comments. First, myeloperoxidase is very susceptible to intracellular inactivation by oxygen compounds generated during phagocytosis. As a result, human neutrophils release only 5-10% MPO activity during 30 minutes of incubation with serum-opsonized zymosan, whereas the activity in the cell pellets decrease about 40-50%. Second, we have to be extremely careful in interpreting kinetics of enzyme release since the susceptibility to inactivation by oxygen products is different for the several lysosomal enzymes.

Dr. E. Schell-Frederick: In order to conclude that the different kinetics of lysozyme and acid protease release are specifically related to bacterial phagocytosis, would it not be useful to test a nonparticulate stimulus, for example, the ionophore A23187?

Dr. K. Debatin: Yes, this would be useful. We did not use the ionophore, but zymosan-activated serum, as a source of C5a, a nonparticulate stimulant. In this case, we found that lysozyme, acid protease and elastase were released with the same kinetics and dose response characteristics, although the amount of lysozyme released was greater than that of the others.

Dr. D. Roos: We have done similar experiments with human neutrophils and zymosan particles. We used particles coated with either IgG alone or both IgG and C3b. We found that the presence of C3 increased both phagocytosis and the release of lysosomal enzymes by a factor of about two. Therefore, the ratio between these two parameters stayed about the same. In other words, the change in this ratio that you observed with bacteria may not be a general phenomenon.

Dr. K. Debatin: I believe that. But, again, I want to state that the phagocytosis of large particles may be completely different from that of bacteria. Our results refer, of course, only to the latter situation.

Dr. P. Elsbach: I have three questions. 1] Did you use a smooth or a rough E. coli strain? 2] Did you determine the effect of serum components on release of label? The extent of release of thymidine into the supernatant is very much greater than we have encountered in experiments with nonopsonized ingested bacteria. 3] With respect to the different release rate of lysozyme and protease, did you examine the possibility that lysozyme is adherent to either intra- or extracellular E. coli?

Dr. K. Debatin: 1] The E. coli strain employed was a smooth, complement-resistant stain, 08 K 27. 2] Because the release of radiolabel was unexpected, we looked at the label in the supernatant of centrifuged bacteria after opsonization. The label in the supernatant was less than 1% of that incorporated by the bacteria. 3] We looked at the possibility that the enzyme activities released bind to bacteria or are even able to destroy bacteria which possibly had received sublethal damage by complement. We could not find detectable binding of lysozyme to the bacteria nor could we find killing of opsonized bacteria by incubation with supernatants.

THE BOVINE NEUTROPHIL: SEPARATION AND PARTIAL CHARACTERIZATION OF PLASMA MEMBRANE AND CYTOPLASMIC GRANULES

Renato Gennaro[1], Domenico Romeo[1], Beatrice Dewald[2] and Marco Baggiolini[2]

Istituto di Chimica Biologica , Università di Trieste Italy and Wander Research Institute , a Sandoz Research Unit, Wander Ltd., P.O.Box 2747, CH-3001 Berne, Switzerland

Bovine neutrophils are a convenient material for biochemical studies since they can be obtained easily in large quantities. We report here some of the biochemical and functional properties of these cells. In contrast to the neutrophils thus far described, bovine neutrophils contain three distinct types of cytoplasmic storage granules and exhibit powerful oxygen-independent antibacterial properties.

Purification of the Cells

The cells are isolated from fresh cattle blood by differential centrifugation and hypotonic lysis of the erythrocytes (1,2). The main contaminants are eosinophils which in some cases may account for up to 30% of the total granulocyte count. Eosinophils are resolved from the neutrophils by centrifugation in gradients of colloidal silica (Percoll), as already described (3). With or without the latter preparation step, large batches of several billions of neutrophils, more than 95% pure, can be obtained within 2 hours.

Oxidative Metabolism

Bovine neutrophils respond with a typical respiratory burst to challenge with opsonized bacteria. Oxygen consumption increases about 10-fold and both O_2^- and H_2O_2 are generated (2,4). They contain less than 1/10 of the catalase and twice as much glutathione peroxidase and glutathione reductase than human neutrophils. Accordingly, the glutathione cycle appears to be the main pathway of H_2O_2 disposal in bovine neutrophils as suggested by the fact that H_2O_2 is readily scavenged in the presence of azide concentrations sufficient to block catalase (2).

The Vacuolar Apparatus

As in other neutrophils, the major cytoplasmic structure of the bovine cells are the granules (2). In thin sections of specimens processed by standard procedures (5), large numbers of round and very light staining granule profiles are seen. In addition, smaller, round or elongated peroxidase-positive and peroxidase-negative profiles are also found. The incidence of peroxidase-positive profiles is much lower than in human neutrophils.

Subcellular fractionation by either zonal sedimentation or isopycnic equilibration (6,7) leads to the resolution of three granule populations. The lightest population, with modal equilibrium density of 1.20 g/ml, contains myeloperoxidase and neutral proteinase activities. A second population of peroxidase-negative granules is slightly resolved from the above one. Its average sedimentation velocity is somewhat higher and its modal equilibrium density is about 1.21 g/ml. These granules contain vitamin B_{12}-binding proteins and about half of the total cellular content of lactoferrin. On the basis of their appearance and biochemical properties, the two above organelles qualify as azurophil and specific granules (6,7). Granules of the third type are visibly larger and sediment much faster than azurophils and specifics. Their modal equilibrium density is 1.23 g/ml. These granules contain the remainder of lactoferrin, but are devoid of other enzymes known to occur in neutrophil granules (8). Most remarkably, however, these large granules contain virtually all of the oxygen-independent antibacterial activity of the bovine neutrophil. These antimicrobial factors are released from intact cells exposed to opsonized zymosan, phorbol-12-myristate-13-acetate or the ionophore A 23187. This indicates that the large granules share with the azurophil and specific granules the ability to discharge their content into phagocytic vacuoles or into the extracellular space.

Oxygen-independent Antibacterial Activity

Lysates of bovine neutrophils have antibacterial effects against *E.coli* and *S.aureus* which are not observed with similar preparations of human neutrophils (2). Some of the properties of these antibacterial factors, which are associated with the novel population of large granules, were studied. Their activity is maximal in hypotonic saline media at pH 7 to 8. In contrast to the "cationic proteins" of human neutrophils (9), however, the large granule proteins still retain considerable activity in isotonic media and acid pH, i.e. at the conditions which are likely to occur within phagocytic vacuoles.

Gel electrophoresis and dialysis through small-pore membranes indicate that they are highly cationic and have a molecular weight below 20,000. Their activity is heat-resistant.

Plasma Membrane

A discontinuous sucrose density gradient was used to resolve a plasma membrane fraction starting from a cytoplasmic extract (3). A fraction highly-enriched in plasma membrane fragments which accounted for about 1% of the total protein was recovered at the interface between layers of 20 and 30 percent (w/w) sucrose. On the basis of the relative specific activities of alkaline phosphatase and Ca^{2+}-ATPase, which were enriched 25 to 35 fold, and of the granule and mitochondrial markers myeloperoxidase, β-glucuronidase and succinate: cytochrome c oxidase it was calculated that the membrane fraction contained more than 80% of plasma membrane - derived material. The polypeptide components of this membrane fraction were analyzed by high-resolution SDS polyacrylamide gel electrophoresis which resolved approximately 40 bands. Nine out of 10 major polypeptides (with apparent molecular weights 34,000, 36,000, 39,000, 55,000, 57,000, 61,000, 70,000, 78,000 and 81,000) were enriched markedly in the membrane fraction as compared with the cytoplasmic extract. A further major band with apparent molecular weight 43,000, which was prominent both in the membrane fraction and in the cytoplasmic extract, is likely to correspond to monomeric actin.

REFERENCES

1. Carlson, G.P. and J.J.Kaneko. 1973. Isolation of leukocytes from bovine peripheral blood. Proc. Soc. Exp. Biol. Med. 142: 853.
2. Gennaro, R., C.Schneider, G.de Nicola, F.Cian, and D.Romeo. 1978. Biochemical properties of bovine granulocytes, Proc. Soc. Exp. Biol. Med. 157: 342.
3. Mottola, C., R.Gennaro, A.Marzullo and D.Romeo. 1980. Isolation and partial characterization of the plasma membrane of purified bovine neutrophils. Eur. J. Biochem. 111: 341.
4. Gennaro, R. and D.Romeo. 1979. The release of superoxide anion from granulocytes: Effect of inhibitors of anion permeability. Biochem. Biophys. Res. Commun. 88: 1979.
5. Bainton, D.F., J.L.Ullyot, and M.G.Farquhar. 1971. The development of neutrophilic polymorphonuclear leukocytes in human bone marrow. Origin and content of azurophil and specific granules. J. Exp. Med. 134: 907.
6. Baggiolini, M., J.G.Hirsch and C.de Duve. 1969. Resolution of granules from rabbit heterophil leukocytes into distinct populations by zonal sedimentation. J. Cell Biol. 40: 529.
7. Bretz, U. and M.Baggiolini. 1974. Biochemical and morphological characterization of azurophil and specific granules of human neu-

trophilic polymorphonuclear leukocytes. J. Cell Biol. 63: 251.
8. Baggiolini, M. 1972. The enzyme of the granules of polymorphonuclear leukocytes and their functions. Enzyme 13: 132.
9. Odeberg, H. and I.Olsson. 1975. Antibacterial activity of cationic proteins from human granulocytes. J. Clin. Invest. 56: 1118.

Discussion
Edited by R.A. Clark

Dr. L. DeChatelet: I think this is a very elegant study. Do bovine PMN exhibit a respiratory burst? Have you looked for the localization of the oxidase in your gradients?

Dr. R. Gennaro: The bovine neutrophils exhibit a stimulation of the oxidative metabolism similar to that shown by neutrophils of other species (R. Gennaro et al, Proc. Soc. Exp. Biol. Med. 157: 342, 1978). The respiratory burst can be elicited either by exposure of the cells to particulate stimuli, such as opsonized bacteria and serum-treated zymosan, or soluble stimuli, such as PMA. This suggests a membrane location of the primary oxidase.

Dr. P. Brederoo: As you know, azurophil granules are formed in promyelocytes and specific granules in myelocytes. Did you study the bovine bone marrow to determine whether the tertiary granules are formed in an early or in a late stage of development? I ask this question because I will present my findings in guinea pig and rat bone marrow in which I have found a third type of granule which is peroxidase-positive.

Dr. R. Gennaro: Experiments concerning the sequence of appearance of granules in the maturing bovine neutrophil are in progress.

THE RESPIRATORY BURST OF PHAGOCYTIC CELLS: FACTS AND PROBLEMS

Filippo Rossi, Paolo Bellavite, Giorgio Berton, Pietro Dri[+] and Giuliano Zabucchi[+]

Istituto di Patologia Generale dell'Università di Padova, sede di Verona, Strada Le Grazie, Verona (Italy) and [+]Istituto di Patologia dell'Università di Trieste

INTRODUCTION

The phagocytic cells have impressive capacities to respond to a variety of external stimuli with activation of random and oriented movement, with secretion of mediators and enzymes stored in cytoplasmic organelles, with production of new compounds and mediators, with changes of the quality and of the intensity of oxidative metabolism. The most impressive response is the production, in a very short time, of an enormous amount of weapons and projectiles that are used against viruses, bacteria, protozoa, tumor cells etc. These weapons are O_2^-, H_2O_2, $OH^{\cdot}$ and singlet oxygen and are intermediate products of a particular type of respiration, called "respiratory burst", which is induced by a perturbation of the plasma membrane of the phagocytes following the interaction with particulate matter and with a number of soluble factors[1-7].

In spite of a great number of investigations performed during the last twenty years, many aspects of the mechanisms responsible for the respiratory burst remain to be clarified, such as the role of the plasma membrane, the relationship with other responses, the nature of the enzymatic system responsible for the activation of the respiration and of the production of the oxygen radicals, the mechanisms of oxygen reduction, the nature of the substrate of the primary oxidase, the mechanism of activation and of deactivation, the relationship between the activity of the primary oxidase and the net oxygen balance in whole cells, the fate of the intermediate products of oxygen reduction etc.

We shall analyse in this presentation some facts and problems concerning a) the enzymatic basis of the respiratory burst and b) the relationship between the activity of the primary oxidase and the oxygen balance of whole cells.

THE OXIDASE RESPONSIBLE FOR THE RESPIRATORY BURST

Table 1 presents the main hypotheses, advanced during the last twenty years, concerning the enzymatic basis of the activation of the respiratory metabolism of phagocytes.

There is a general agreement on the criteria for the recognition of an enzyme as responsible for the respiratory burst of the phagocytes (table 2). On the basis of these criteria the hypotheses concerning the involvement of myeloperoxidase[2], of aminoacid oxidase,[8] of ascorbate oxidase[9] have been ruled out.

Table 1. Enzymes or enzymatic systems proposed as responsible for the respiratory burst of phagocytes

MYELOPEROXIDASE	(Quastel's group)
"SOLUBLE" NADH OXIDASE	(Karnovsky's group)
MEMBRANE-BOUND NADH OXIDASE	(Segal)
MEMBRANE-BOUND NADPH OXIDASE	(Rossi's group Sbarra's group Hohn and Lehrer Babior's group De Chatelet's group Kakinuma's group Minakami's group)
AMINOACID OXIDASE	(Cline and Lehrer)
ASCORBATE OXIDASE	(De Chatelet)
NAD(P)H DEHYDROGENASE - CYT. b	(Segal)

Table 2. Requirements to be fulfilled for an enzymatic activity to be considered responsible for the respiratory burst

1. The enzyme must consume oxygen and produce O_2^- and H_2O_2
2. The activity must be increased in homogenates or in cell fractions derived from stimulated cells
3. The degree of activation of the enzyme should correlate with the intensity of the respiratory burst of intact cells
4. The Km for the substrate should fall in the range of the physiologic concentration of that substrate
5. The inhibition profile of the enzyme should be similar to that of the respiratory burst of intact cells
6. The activity of the enzyme should account for the net oxygen consumption by intact cells
7. CGD cells are expected to lack the enzyme or the enzymatic activation

During the last twenty years evidences have been presented in our and in other laboratories that the burst is linked to the activation of a membrane bound oxidase that preferentially uses NADPH as substrate[4,5,6,10-15]. In our opinion these evidences are so strong that the controversy concerning the identification of the oxidase should be closed and the research should deal with the problems that remain to be clarified, such as the nature of the enzyme, the involvement of cytochrome b and of other cofactors, the solubilization and purification of the oxidase, the mechanism and the control of its activation.

In spite of this conviction we have been induced to reinvestigate the matter by two events. The first is the new emphasis that recently has been given to the involvement of a "soluble oxidase", specific for NADH, and measurable in the supernatants of homogenates of guinea pig and human granulocytes[16-20]. The second event is the following: about one year ago we submitted for publication a paper showing that in guinea pig granulocytes treated with phorbol-myristate acetate (PMA) the activated NADPH oxidase is located on the plasma membrane. The paper was rejected on the basis of the following comment of one referee: "The choice of guinea pig as a source of neutrophils is very unfortunate. In this animal, more than in rabbit and human, the NADH oxidase contributes to the O_2^- formation. Therefore it is impossible to limit the study of guinea pig neutrophils to the NADPH oxidase".

The activity of the soluble and of membrane bound oxidase in guinea pig granulocytes

It is likely that the disagreements in the identification and characterization of the oxidase derive from the methods and the conditions employed for the assay of the enzymatic activity and from the sources of the enzymatic preparations. For these reasons we have measured the activity of the oxidase following the same experimental conditions used by Badwey and Karnovsky[34], that is: granulocytes (PMN) of guinea pig peritoneal exudates; supernatant and cell free particles of PMN homogenized in isotonic alkaline KCl; measurement of the oxidase as O_2^- formation (SOD sensitive cytochrome c reduction) in HEPES medium pH 7.0, in presence of NADH and NADPH.

The results reported in table 3 show that:

1. The supernatants of resting PMN contain a very low O_2^- forming activity both in presence of NADH and NADPH. The oxidase has the following characteristics: pH optimum between 7.0 and 7.5 (data not shown); a Km for NADPH one order of magnitude lower than that for NADH; an activity and kinetic properties that do not vary in supernatants of PMA activated PMN.
2. The cell free particles of resting cells have a very low oxidase activity both with NADH and NADPH as substrates. The values of the activity are around the lower limit of sensitivity of the assay method. The quantitation has been hard to be performed with accuracy due also to the turbidity of the assay system and hence to the wide scatter of the experimental data. For these reasons the range of the values is given in the table 3 and the Km has not been measured. From these results it can be concluded that the oxidase activity is substantially inactive in PMN in resting state. On the contrary the oxidase activity of cell free particles from PMA treated PMN is very high both with NADH and

Table 3. Activity and kinetic properties of the NAD(P)H-dependent O_2^- production by cell free particles and supernatants from resting and PMA-activated guinea pig PMN

	nmoles O_2^-/min/mg prot.			
	Resting cells		PMA-activated	
	Supernatant	Cell free particles+	Supernatant	Cell free particles
0.15 mM NADH	0.77 ± 0.22	0-1.47	1.31 ± 0.17	18.79 ± 6.4
1.0 mM NADH	1.25 ± 0.45	0-1.55	1.88 ± 0.13	53.40 ± 16.3
0.15 mM NADPH	0.66 ± 0.29	0-0.45	0.74 ± 0.12	45.70 ± 8.4
1.0 mM NADPH	0.74 ± 0.36	0-0.69	0.92 ± 0.23	65.25 ± 15.6
Vmax NADH	1.46 ± 0.28	——	1.57 ± 0.57	62.5 ± 18.7
Vmax NADPH	0.73 ± 0.38	——	1.32 ± 0.6	54.0 ± 18.0
Km NADH	0.178 ± 0.07 mM	——	0.116 ± 0.03 mM	0.64 ± 0.11 mM
Km NADPH	0.016 ± 0.016 mM	——	0.017 ± 0.004 mM	0.057 ± 0.01 mM

The data are means ± SD of 5 experiments.

+ Due to the high variability, the data are expressed as range of activity.

NADPH. This oxidase has the following characteristics: a pH optimum between 7.0 and 7.5; a Km for NADPH one order of magnitude lower than that for NADH; a Vmax with NADPH similar to that with NADH.

The comparison of the values of Km with the values of the intracellular concentrations of NADPH and NADH in guinea pig PMN (table 4) indicates that at the concentration of the substrates within the cells, the activity with NADPH is greatly favoured with respect that with NADH.

The activation of the oxidase of cell free particles parallels the degree of the activation of the respiratory metabolism of whole cells induced by stimulation with different doses of PMA. Fig. 1 reports the data on the activity in presence of NADPH.

We have also investigated the sensitivity to various inhibitors of the oxidase present in supernatants and in cell free particles of PMA-activated PMN. The data presented in table 5 show that:
1. The oxidase of supernatants and of cell free particles, both in

Table 4. Km for NADPH and NADH of the oxidase activity of cell free fractions from PMA activated guinea pig PMN compared with the intracellular concentrations of reduced nicotinamide adenine dinucleotides

	NADPH (mM)	NADH (mM)
Km of supernatant	0.017	0.116
Km of cell free particles	0.057	0.64
Intracellular concentration		
Sbarra's group 21	0.136	0.172
Rossi's group 22	0.273	0.076
Frei's group 23+	0.1	0.16

+ Values calculated by Badwey and Karnovsky 34.

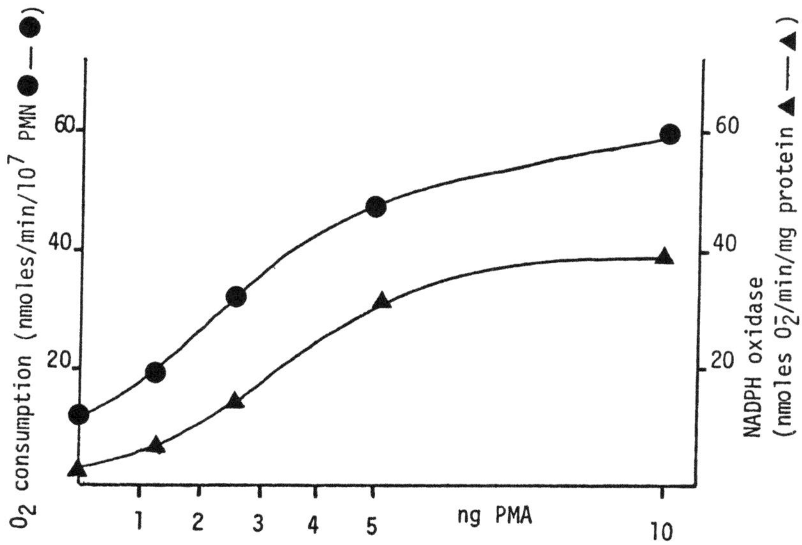

Fig. 1. Relationship between the activation of O_2 consumption by intact guinea pig PMN and the activation of the NADPH oxidase activity. O_2 consumption was measured in presence of different amounts of PMA and NADPH oxidase was measured as O_2^- forming activity of cell free particles obtained from the same cell suspensions.

presence of NADH and NADPH, are insensitive to cyanide, rotenone and NaN_3. 2. The oxidase of supernatants and of cell free particles are inhibited by ATP when NADH is the substrate and not inhibited when NADPH is the substrate. 3. The NADPH oxidase activity of supernatants and of cell free particles and the NADH oxidase activity of cell free particles are inhibited by appropriate concentrations of sulfhydrylic reagents N-ethylmaleimide (NEM) and p-chloromercurybenzoate (pCMB). On the contrary the NADH oxidase activity of supernatants is stimulated by NEM and also by pCMB 0.1 mM.

Fig. 2 shows the effect of various inhibitors on the respiratory activity of guinea pig PMN stimulated by PMA. It can be seen that the respiratory burst is insensitive to cyanide, rotenone and NaN_3, and inhibited by the penetrating sulfhydrylic reagent NEM and by high concentration of non-penetrating pCMB. This behaviour parallels that of the oxidase activity of cell free particles of PMA activated cells, while differs from that of the oxidase of supernatants which in presence of NADH is activated by sulfhydrylic reagents.

Table 5. Effect of inhibitors on the NAD(P)H-dependent O_2^- production by cell free particles and supernatant of PMA-activated guinea pig PMN

Additions		SUPERNATANT		CELL FREE PARTICLES	
		NADH	NADPH	NADH	NADPH
None		100	100	100	100
NaN_3	2.0 mM	80	93	100	104
KCN	0.5 mM	85	91	90	87
Rotenone	0.025 mM	88	111	96	100
ATP	0.5 mM	36	81	23	85
NEM	0.1 mM	188	71	--	--
NEM	1.0 mM	179	36	17	23
pCMB	0.1 mM	151	33	3	3.5
pCMB	1.0 mM	77	0	0	0

The results are expressed as percentage of O_2^- production in the control mixture which contained no inhibitor. NADH 1 mM, NADPH 0.15 mM.

A brief comment.

1. We do not know whether the activities found in the supernatants are due to enzymes solubilized from the membranes. Some differences in Km and in the sensitivity to inhibitors might be indications against this possibility. It is likely, however, that the changes of these properties are due to the solubulization.

We do not know also whether the oxidations of NADPH and of NADH by supernatants are performed by a single or by different enzymes. The different effect of some inhibitor might favour the second possibility.

In any case the results seems to us a strong indication that the oxidase (or the oxidases) of supernatants of guinea pig PMN is not involved in the respiratory burst of intact cells. In fact the activity a) is very low, b) is not modified upon activation of PMN

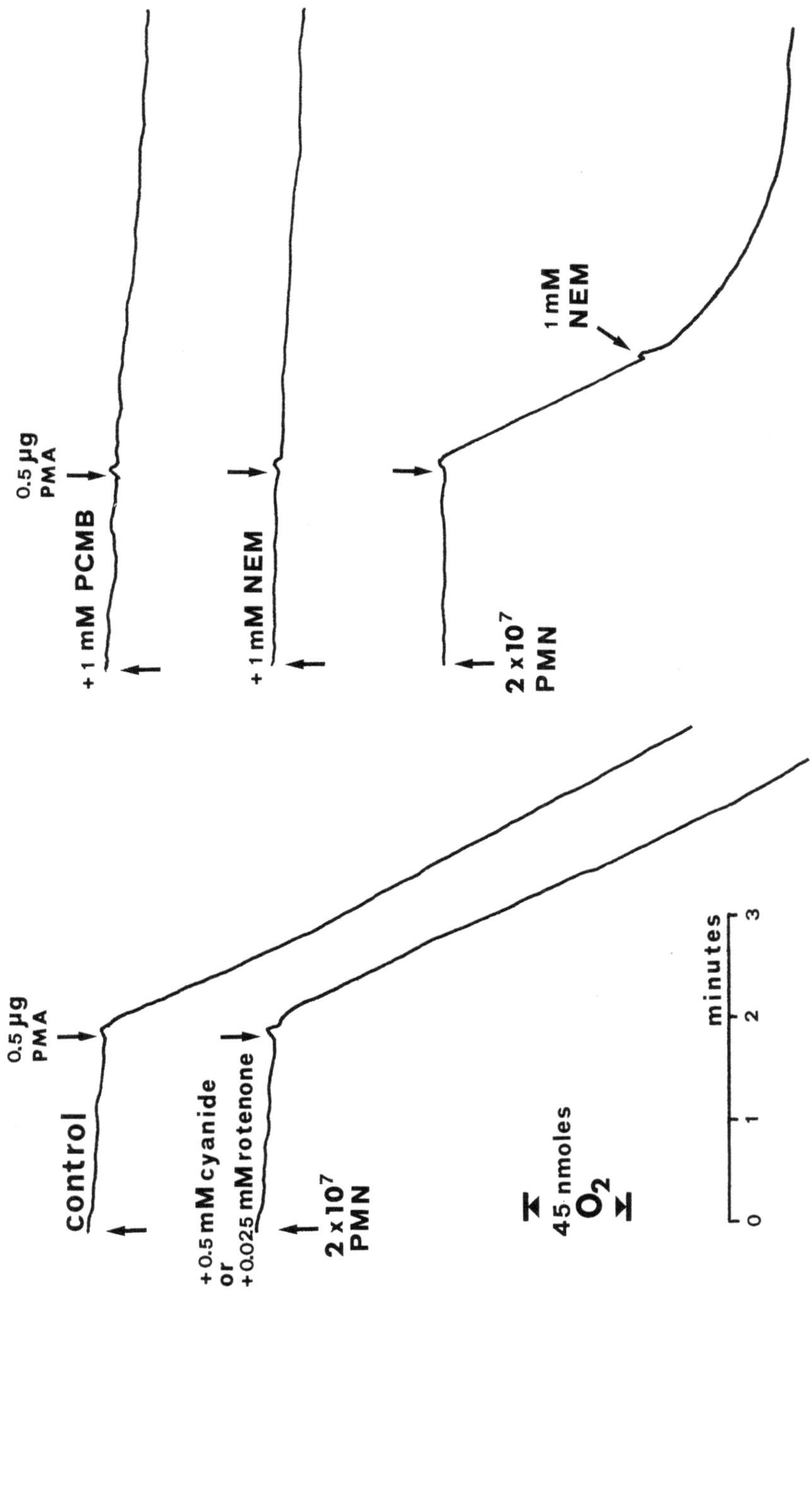

Fig. 2. Effect of inhibitors on the oxygen consumption by guinea pig PMN. Assay medium: 2 ml KRP containing 5 mM glucose and 0.5 mM $CaCl_2$.

by PMA, c) is inadequate to account for the net oxygen consumption by intact cells, d) in presence of NADH is activated by sulfhydrylic reagents whereas the respiration of the cells is inhibited.

2. The data presented here seems to us a strong confirmation of the results that we[4,6,10,11,24,35] and others[5,12,13,14,15] have published in the last twenty years. In other words we have shown that the O_2^- forming activity bound to some membranous structures of the granulocytes is involved in the respiratory burst. This enzyme fulfills the requirements listed above (table 2), and has more affinity for NADPH than for NADH. The comparison between the values of Km and the concentration of NADPH and NADH inside the cells (table 4) clearly indicates that the physiological substrate is NADPH. A more complete survey of the evidences in favour of the NADPH as substrate of the primary oxidase has been published elsewhere by us[6] and by others[5].

The primary oxidase in human PMN and guinea pig macrophages

We have investigated the activity of the oxidase present in supernatants and in cell free particles of resting and PMA stimulated human blood neutrophils, guinea pig peritoneal resident and casein elicited macrophages (data not shown). The results[24,25] clearly indicate that also in these cells the oxidase activity of supernatants both with NADH and NADPH is very low and does not change upon stimulation of intact cells with PMA. On the contrary the oxidase activity of cell free particles of PMA stimulated cells is greatly activated. In cell free particles of PMA-treated human gra ulocytes and of guinea pig peritoneal resident macrophages the degree of activation with NADPH and NADH, and the Km for NADH and for NADPH are similar to those of cell free particles of guinea pig granulocytes. On the contrary only the oxidase activity with NADPH is activated in cell free particles of PMA stimulated guinea pig peritoneal elicited macrophages.

The localization of the primary oxidase

In the last few years the number of indirect and direct evidence on the localization on the plasma membrane of the primary oxidase is increasing[26,27,28,29]. Data have been also published on intracellular localization of NADPH oxidase[30,31,32]. It is likely that the discrepancies depend on the methods of fractionation of cell homogenates, on the different state and degree of activation of the cells, on the methods of assay of the oxidase.

We have reinvestigated the matter[33] by using PMA activated guinea pig granulocytes and macrophages and different methods of fractionation of cell homogenates (isopycnic equilibration, rate zonal centrifugation in continuous sucrose gradient, rate zonal sedimentation in discontinuous sucrose gradient). The results (table 6) show that the distribution of the activated oxidase follows that of 5'-nucleotidase and acid pNPPase, enzymes marker of plasma membrane. These data confirm those obtained by Dewald et al.[29] on human granulocytes activated by PMA. We are now measuring the oxidase activity in the membrane of phagosomes isolated from phagocytosing guinea pig granulocytes and peritoneal elicited macrophages. Preliminary results, whose one example is reported in table 7, seems to indicate that during phagocytosis only the oxidase of the portion of the membrane that forms the phagocytic vacuole is activated. In fact the enrichment of the oxidase in the phagosomal fraction, expressed by the relative specific activity, is much higher than that of 5' nucleotidase, the typical marker of the plasma membrane. This localization of the activated O_2^- forming enzyme is the more strategical one for the intravacuolar use of the oxygen radical and of hydrogen peroxide.

RELATIONSHIP BETWEEN THE ACTIVITY OF THE PRIMARY OXIDASE AND THE RESPIRATORY METABOLISM OF WHOLE CELLS

One of the requirements to be fulfilled for the recognition of an enzyme responsible for the respiratory burst is that its activity should account for the extra oxygen consumption of whole cells. When we deal with this argument we are faced by a number of problems. The main are the mechanisms of the activity of the oxidase on one side, and the complexity of the reactions involved in the oxidative metabolism of phagocytes, that is in the O_2 balance of intact cells, on the other. We shall try to analyze these problems.

The reaction of NAD(P)H oxidation by cell free particles of leucocytes

A. The NAD(P)H oxidase is measured in different conditions of pH and by using different procedures, the main being O_2 consumption, NAD(P)H disappearance, $NAD(P)^+$ formed, O_2^- production, H_2O_2 production. By comparing the activity of the oxidase expressed as O_2^- formation, O_2 consumption and disappearance of the substrate we have been able to show that more than one mechanism is operative

Table 6. Percent distribution of marker enzymes in the four fractions obtained by centrifugation on discontinuous sucrose gradient of postnuclear supernatants of guinea pig elicited peritoneal macrophages

Enzymes	sucrose density (% w/w)			PELLET	RECOVERIES
	20/34	34/40	40/45		
5'-nucleotidase	43.0 ± 13.0	15.6 ± 2.7	7.8 ± 2.2	9.2 ± 1.2	73.2 ± 14.0
Glucose-6 phosphatase	59.3 ± 18.0	46.5 ± 12.0	24.5 ± 7.5	5.0 ± 2.0	130.4 ± 35.0
NADPH-cyt.c reductase	27.0 ± 8.0	23.5 ± 4.0	13.8 ± 5.2	15.8 ± 1.0	76.2 ± 11.5
Succinic-dehydrogenase	13.3 ± 2.4	52.4 ± 18.5	15.8 ± 3.9	5.9 ± 2.8	85.5 ± 20.0
β-glucuronidase	6.8 ± 3.5	18.8 ± 4.2	22.8 ± 1.2	18.2 ± 2.1	62.0 ± 5.0
NADPH-oxidase	36.7 ± 5.3	24.0 ± 2.8	14.2 ± 5.3	12.1 ± 1.9	84.0 ± 7.5
Protein	20.6 ± 5.5	22.8 ± 3.4	17.6 ± 3.3	30.2 ± 6.2	84.0 ± 7.5

Percent distribution was calculated by taking 100,000g pellet of postnuclear supernatant as 100%. The average values of four experiments ± SEM are reported.

Table 7. Relative specific activities of marker enzymes and of NADPH oxidase in fractions obtained from guinea pig PMN that phagocytosed PMA-coated latex

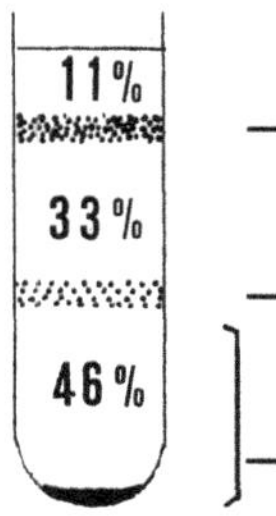

	Acid pNPPase	5'nucleotidase	Myeloperoxidase	β-glucuronidase	NADPH oxidase
PHAGOSOMES	2.9	3.0	0.7	1.4	9.1
FRACTION 33/46	1.8	1.8	1.0	1.4	3.4
PELLET	1.5	1.1	0.8	1.1	1.5

in the oxidation of NAD(P)H and that these mechanisms vary by modifying the conditions of the assay[35]. Table 8 shows in fact that: 1) When measured as O_2 consumption or as NADPH disappearance, the enzymatic activity appears higher at pH 5.5 than at pH 7.0, particularly at 1 mM substrate. 2) In all the conditions the ratio be-

Table 8. O_2 consumption, NADPH oxidation and O_2^- production by cell free particles from phagocytosing guinea pig PMN

	NADPH (mM)	O_2 consumed	NADPH oxidized	O_2^- produced
pH 5.5	1.0	72.7 $\pm$ 9.7	84.3 $\pm$ 6.4	9.4 $\pm$ 2.0
" "	0.15	22.9 $\pm$ 1.2	23.6 $\pm$ 0.5	7.6 $\pm$ 1.7
pH 7.0	1.0	42.0 $\pm$ 4.0	45.5 $\pm$ 4.0	22.1 $\pm$ 4.5
" "	0.15	16.0 $\pm$ 2.9	19.7 $\pm$ 5.4	12.4 $\pm$ 3.4

The data are means of 4 experiments $\pm$ SEM. Values are nmoles/2.5 min/mg protein.

tween O_2 consumed and NADPH oxidized is near 1.0. 3) when assayed as O_2^- generation the oxidase activity is higher at neutral than at acid pH, at both NADPH concentrations. In any case it is always lower than when measured as NADPH oxidized or as O_2 consumed. The amount of O_2^- produced versus NADPH oxidized is much less than that theoretically expected that is 2 versus 1. Similar results have been obtained by using NADH as substrate.

The indications that in these conditions more than one mechaism is responsible for NAD(P) H oxidation have been verified by studying the effect of superoxide dismutase (SOD) and of cytochrome c (cyt. c) on the O_2 consumption and on NAD(P)H oxidation. The rationale of this approach is based on their ability to remove O_2^-, SOD by catalyzing the dismutation and cyt. c by oxidizing the superoxide to molecular oxygen. Table 9 reports the results of four experiments at pH 5.5 and 1 mM NADPH. It can be seen that the addition of SOD and of cyt. c causes a strong decrease in O_2 consumption and in NADPH oxidation. This means that O_2^- produced is subsequently involved in other reactions that oxidize NAD(P)H and consume oxygen and that the these reactions (chain reaction) are hindered when scavenger of O_2^- (in this case cyt. c) are present. By applying appropriate calculations to the data of table 9, at least three components of the NADPH oxidizing activity can be identified. 1. The first corresponds to the SOD inhibitable cyt. c reduction and represents the O_2^- forming activity, that is the reaction catalyzing the "univalent reduction of O_2". 2. The second corresponds to the amount of O_2 consumption that remains after the addition of cyt. c. This aliquot is referred as "residual O_2 consumed" and might represent either a divalent reduction of O_2 with direct formation of H_2O_2, or a formation of O_2^- in a space not accessible to cyt. c. 3. The third component is a "chain reaction" initiated by O_2^- and corresponds to the aliquot of O_2 consumption and of NADPH oxidation inhibited by SOD. It can be seen that at pH 5.5 with 1 mM NADPH most of the activity is contributed by the chain reaction.

Table 10 and table 11 report the data showing that the contributions of each of the three components vary depending on pH and on NADPH and NADH concentrations.

The rate of the "univalent reduction of O_2" increases by increasing the pH and decreases by lowering the substrate concentration. The so-called "residual O_2 consumption" is not significantly affec-

Table 9. Components of the NADPH - oxidizing activity of particles from phagocytosing polymorphonuclear leucocytes calculated on the basis of the effect of superoxide dismutase and of cytochrome c.

Additions	A			B	C	
	O_2 consumed	NADPH oxidized	cyto-chrome c reduced	Calculations	Results	% of total
1. None	72.7 ± 9.7	84.3 ± 6.4		Total activity (as O_2 or NADPH) = line 1	O_2 = 72.7 NADPH = 84.3	100 100
2. Superoxide dismutase	14.1 ± 1.9	14.8 ± 3.9		Chain reaction (as O_2 or NADPH) = line 1 — line 2	O_2 = 58.6 NADPH = 69.5	81 82
3. Cytochrome c	9.2 ± 2.6	24.7 ± 7.8	22.6 ± 2.2	Residual O_2 consumption (as O_2 consumed) = line 3	O_2 = 9.2	13
4. Cytochrome c + superoxide dismutase	13.9 ± 1.9	28.2 ± 7.2	13.2 ± 1.2	Univalent reduction (as O_2^- detected) =cyt c reduced on line 3 — line 4 (as O_2 consumed) $= 1/2\ O_2^-$ detected	O_2^- = 9.4 O_2 = 4.7	 6

The values are given as nmol/2.5 min/mg protein. The mean of four experiments $\pm$ S.E.M. are reported.

Table 10. Components of the NADPH-oxidizing activity of particles from phagocytosing guinea pig PMN at different pH and substrate concentration

ASSAY CONDITIONS		ACTIVITIES (nmoles O_2/2.5 min/mg protein)				% OF TOTAL ACTIVITY		
pH	NADPH mM	Total activity	Univalent reduction	Residual oxygen consumption	Chain reaction	Univalent	Residual	Chain
5.5	1.0	72.7	4.7	9.2	58.6	6.5	12.6	80.6
7.0	1.0	42.0	11.0	11.8	22.1	26.3	28.1	52.6
5.5	0.15	22.9	3.8	7.2	10.1	16.6	31.4	44.1
7.0	0.15	16.0	6.2	6.9	0.7	38.7	43.1	4.4

O_2 consumption by the corresponding intact PMN during 2.5 min of phagocytosis was 99.0 $\pm$ 12.1 (SEM) nmoles O_2.

ted by pH changes and the concentration of the substrate. The "chain reaction" decreases by increasing the pH and by decreasing the substrate concentration. At pH 7.0 with 0.15 mM NAD(P)H the contribution of the chain reaction is negligible.

These findings are worthy of some comments.

1. Some of the discrepancies on the kinetic properties of the NAD(P)H oxidase can be explained by the difference in the conditions and methods of the assay used. For example, a pH optimum of 5.5 has been reported for the activity measured as O_2 consumption or NADPH oxidation[11,17,18] and a pH optimum of 7.0 when the activity has been measured as O_2^- generation[13,24,38]. From the data presented above it is clear that at acid pH the chain reaction accounts for most of the oxygen consumed and of NADPH disappearance. Thus the assay of the oxidase at this pH might give erroneous informations.

2. When the kinetic properties of the oxidase are to be studied or when its activity must be compared with the respiration of intact cells, it is advisable to choose conditions in which the chain reaction is not operative.

3. When the experimental goal is only the detection of the activity of the enzyme, as for example in different cell fractions separated with various methods of fractionation, anyone of the procedures can be used. In some cases it might be quite useful to favour the chain reaction. For example, when the activities of the true enzymatic reaction (the univalent reduction of O_2) are low, the detection of a difference in the activity between preparations from resting and from activated cells might be facilitated by using conditions in which the rate of the chain reaction is high. In some cases the use of manganese, which is known to greatly potentiate the chain reaction[36,37], might be very useful.

B. On the basis of the overall reaction of the oxidase ($2O_2 + NAD(P)H \longrightarrow 2O_2^- + NAD(P)^+$), 2 nanomoles of O_2^- should be produced for one nanomole of NAD(P)H oxidized. In the experimental conditions employed in the experiments reported here we have found 0.4 - 0.8 nmoles of O_2^- produced for 1 nanomole of NAD(P)H oxidized (table 12).

Two hypotheses can be advanced to explain the defective stoichiometry: 1. There are two enzymatic pathways of oxygen reduction: an "univalent" with formation of O_2^- and a "divalent" with direct for

Table 11. Components of the NADH-oxidizing activity of particles from phagocytosing guinea pig PMN at different pH and substrate concentration

ASSAY CONDITIONS		ACTIVITIES (nmoles O_2/2.5 min/mg protein)				% OF TOTAL ACTIVITY		
pH	NADH (mM)	Total activity	Univalent reduction	Residual oxygen consumption	Chain reaction	Univalent	Residual	Chain
5.5	1.0	38.7	4.7	13.3	26.2	12.3	34.4	67.8
7.0	1.0	37.0	4.6	12.9	19.3	12.5	35.0	52.1
5.5	0.15	2.1	0.14	1.8	0.32	6.7	85.7	15.3
7.0	0.15	5.2	0.68	3.6	0.32	13.1	68.7	6.2

See note of table 10.

Table 12. Stoichiometric relationships between NAD(P)H oxidation and O_2^- production by cell free particles from PMA-activated guinea pig PMN

	NAD(P)H oxidation	O_2^- production
THEORETICAL	1	2
EXPERIMENTAL		
with NADH	1	0.4 - 0.8
with NADPH	1	0.5 - 0.8

Assay conditions: NAD(P)H 0.15 mM, pH 7.0.

mation of H_2O_2; 2. Only the univalent pathway is operative but part of O_2^- is formed in a site not accessible to cyt.c.

We shall present evidences against the first hypothesis at least as far as the oxidation of NADPH is concerned. The results are reported in fig. 3 show in fact that the addition of deoxycholate (DOC) increases 2-3 folds the production of O_2^- when cell free particles from activated PMN oxidize NADPH. The detergent increases only slightly the amount of NADPH oxidized. Therefore, in these conditions, the value of ratio O_2^- produced versus NADPH oxidized approximate the theoretical one. These results indicate that all the oxidation of NADPH by cell free particles involves mainly an univalent reduction of O_2.

Thus, the NADPH oxidase activity should be assayed at pH 7.0, in presence of 0.15 mM substrate and in presence of 0.05% DOC. Otherwise the values in absence of DOC should be corrected by a factor calculated on the basis of the insufficient recovery of O_2^- with respect to the theoretical. In our case the factor varies between 5.0 and 2.5.

THE OXYGEN BALANCE OF WHOLE CELLS

The discussion of this argument can be divided in two parts. The first concerns whether or not all the extrarespiration of intact cells involves a univalent reduction of oxygen, as we have

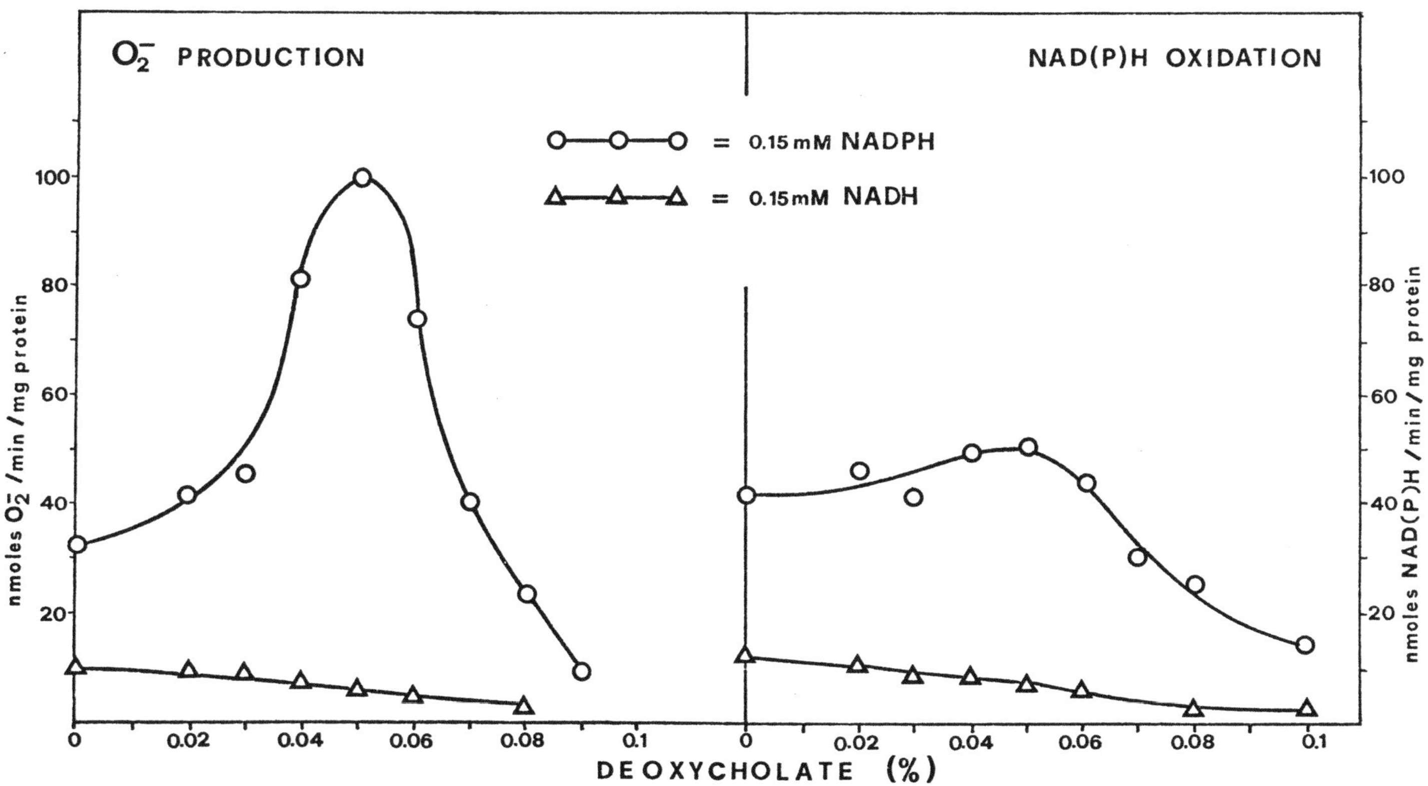

Fig. 3. Effect of deoxycholate on the NAD(P)H oxidation by cell free particles from PMA-activated PMN. O_2^- production was measured as SOD-inhibitable cytochrome c reduction NAD(P)H oxidation was measured following the decrease in absorbance at 340 mM.

shown to occur in cell free particles oxidizing NADPH at pH 7.0 and in presence of DOC. The second regards the complexity of the reactions underlying the net oxygen consumption of intact cells during the burst.

A. We shall present now data showing that, at least in appropriate conditions, all oxygen reduced during the burst passes through the formation of O_2^-.

It is known that during the respiratory burst only the O_2^- released from the cells is measurable. This measurement is based on the ability of O_2^- to reduce a suitable electron acceptor, such as cyt.c, that does not penetrate inside the cell (O_2^- + cyt.c ox. $\longrightarrow$ O_2 + cyt.c red.). Since in this reaction one molecule of O_2 is liberated for everyone molecule of O_2^- that reacts with cyt.c, the addition of this electron acceptor to a leucocyte suspension, in which O_2 is being consumed with formation of O_2^-, should lead to a decrease in the net oxygen consumption and this decrease should correspond to the amount of O_2^- that is released outside the cell. It has been shown in many laboratories that during the early stages of the respiratory burst induced by different stimuli the O_2^- release is only an aliquot of the O_2 consumed and, as a consequence, the inhibition of the respiration by cyt.c varies from 20 to 30%[39,40-42]. Many explanations have been given of this low recovery of O_2^- during the burst: the existence of more than one pathway for oxygen reduction; the production of O_2^- in a site not accessible; the discharge of O_2^- in the endocytic vacuoles where the amount of oxidized cyt.c becomes rapidly insufficient; the discharge of O_2^- inside the cell.

We have found that in some experimental conditions the addition of cyt.c completely inhibits the activation of respiration in termsof oxygen consumption. The fig. 4 shows the recording of the respiration of leucocytes stimulated by the ionophore A23187 in absence and in presence of cyt.c, and the data of table 13 reports the results of 8 experiments. That the inhibitory effect of cyt.c is due to the oxidation of O_2^- is demonstrated by the fact that the inhibition by cyt.c is removed by the addition of SOD that prevents O_2^- from reacting with cyt.c.

Results similar to those presented in fig. 4 have been obtained in our laboratory by using guinea pig PMN and Con A as stimulatory agents in presence of CB (data not shown). These data seem to us a strong evidence that all O_2 consumption passes through an univalent

Table 13. Effect of cytochrome c on the respiratory burst induced by A23187

	1	2	3
ADDITIONS	A23187	A23187 + cyt c	A23187 + cyt c + SOD
Extra O_2 consumed[+]	30.5 ± 4.7	3.0 ± 0.8	34.0 ± 5.5
Extra O_2^- detected[+]		77.4 ±17.0	
O_2 taken off by cytochrome c	A) 1 minus 2 = 27.5		B) 3 minus 2 = 31.0
Ratio $\frac{O_2^- \text{ detected}}{O_2 \text{ taken off}}$	A) $\frac{77.4}{27.5} = 2.8$		B) $\frac{77.4}{31.0} = 2.5$

The data are means ± SEM of 5 experiments. [+]The values are given as nmoles/1.5 x 10^7 PMN/4 min.

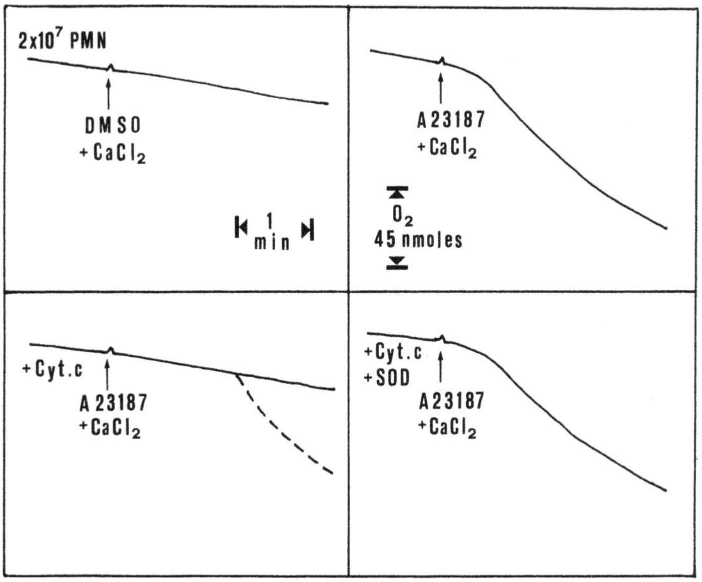

Fig. 4. Effect of cytochrome c on the oxygen consumption induced by the calcium ionophore A23187.

reduction and that all O_2^- formed is discharged outside the plasma membrane, at least in the early stages of the burst.

B. The complexity of the oxygen balance in whole cells has been extensively discussed by us[39] and by others[44]. The measurement of the extrarespiration in whole cells refers to net O_2 uptake, which is the result of several reactions: the reduction of O_2^- catalyzed by the oxidase on one side and reaction of dismutation of O_2^- and of degradation of H_2O_2 on the other (table 14). Since in the reaction of dismutation and in the catalatic degradation of H_2O_2 part of O_2 is recovered, the amount of oxygen univalently reduced by the oxidase is greater than that measured as net O_2 consumed. The ratio between the extent of the reaction 1 and the net O_2 consumed varies depending on the extent of the O_2 given back, that is mainly on the mechanisms of H_2O_2 degradation. The reaction 6 is very low as it is shown by the very low amount of hydroxyl radical measured during the burst[45].

Table 14. Reactions involved in oxygen consumption in whole cells

1.	$2O_2 + NADPH$	$\xrightarrow{\text{oxidase}}$	$2O_2^- + NADP^+ + H^+$
2.	$2O_2^- + 2H^+$	$\xrightarrow{\text{dismutation}}$	$H_2O_2 + O_2$
3.	H_2O_2	$\xrightarrow{\text{catalase}}$	$H_2O + 0.5\ O_2$
4.	$H_2O_2 + RH_2$	$\xrightarrow{\text{peroxidase}}$	$2H_2O + O_2$
5.	$O_2^- + Xoxid.$	$\longrightarrow$	$O_2 + Xred.$
6.	$O_2^- + H_2O_2$	$\longrightarrow$	$OH^\cdot + OH^- + O_2$

$1 + 2 + 3 = 2O_2^-$ formed, $\frac{1}{2}\ O_2$ consumed = Ratio 4 : 1

$1 + 2 + 4 = 2O_2$ formed, $1\ O_2$ consumed = Ratio 2 : 1.

The data of the table 13 allow us to derive some indications concerning the complexity of the reactions involved in the net oxygen consumption during the burst. We can infact calculate how much O_2^- is formed versus the net O_2 consumed in these particular experimental conditions. Furthermore, with this information we can understand whether or not in intact cell, beside the reactions reported in the table 14, some sort of chain reaction is operative.

Before to enter into the details of these problems it is worthy to point out two facts. The data of table 13 show that:

1. the ratio between the nanomoles of O_2^- oxidized by cyt.c and the nanomoles of O_2 not consumed due to the presence of cyt.c is 2.5 - 2.8. This means that in absence of this scavenger for one nanomole of O_2^- produced and released, only 0.4 nanomoles of O_2 are actually consumed.
2. During the respiratory burst, H_2O_2 that accumulates is usually much less than that expected, since the phagocytic cells are equipped with very efficient mechanisms for its degradation. The accumulation of the peroxide increases only when the main mechanism of degradation is inhibited by the addition of NaN_3 or KCN

or by the presence of a system that traps the H_2O_2 released as it is formed[39,43]. The calculations that we are presenting are valid whether, in absence of inhibitor and of scavenger or of trapping systems, all the intermediates of O_2 reduction are degraded. In the experimental conditions employed in this and in other circumstances[39,43], this has been verified: during the respiratory burst "in vitro" O_2^- and H_2O_2 originally discharged outside the cell or in the phagocytic vacuole are subsequently degraded by intracellular mechanisms since they equilibrate with the intracellular space.

We shall discuss now various hypotheses concerning the oxygen balance in whole cells during the burst. The hypotheses consist in different combinations of the reactions of table 14 and of chain reactions originated by O_2^- or by H_2O_2.

HYPOTHESIS n. 1. The oxidase forms O_2^-; O_2^- originates a chain reaction followed by a dismutation and by the degradation of H_2O_2 through catalatic and/or peroxidatic mechanisms

1. $NADPH + H^+ + 2O_2 \longrightarrow NADP^+ + 2O_2^- + 2H^+$

2. (chain reaction)
$$2O_2^- + 2RH_2 \xrightarrow{+\,2H^+} 2RH^\cdot + 2H_2O_2$$
$$2RH^\cdot + 2O_2 \longrightarrow 2R^+ + 2O_2^- + 2H^+$$

3. $2O_2^- + 2H^+ \xrightarrow{\text{dismutation}} H_2O_2 + O_2$

4. $3H_2O_2 \xrightarrow{\text{catalase}} 3H_2O + 1.5\ O_2$

5. $3H_2O_2 + 3RH_2 \xrightarrow{\text{peroxidase}} 6H_2O + 3R$

In the case that these reactions occur in different combinations in whole cells (1+2+3+4; 1+2+3+5; 1+2+3 and 4 and 5 in different proportions), between 3.0 and 1.5 nanomoles of O_2 are actually consumed for 2 nanomoles of O_2^- initially generated by the oxidase or at the end of the reaction 1+2, with ratios O_2^-/O_2 between 0.66 and 1.33. These sequences do not account for the experimental ratio 2.5 - 2.8 (table 13). By increasing the lenght of the chain reaction the net O_2 consumed for the same initial amount of O_2^- generated would be higher. Thus, these sequences of reactions, as indicated in this hypothesis, do not occur in whole cells.

The experimental ratio O_2^-/O_2 does not change appreciably in presence of SOD (data not shown). This is a further evidence that a chain reaction, originated by O_2^- released, does not take place.

HYPOTHESIS n. 2. The oxidase forms O_2^-; O_2^- dismutates to H_2O_2; H_2O_2 initiates a chain reaction followed by the dismutation and by the degradation of H_2O_2 through catalatic or peroxidatic mechanisms

1. $NADPH + H^+ + 2O_2 \xrightarrow{\text{oxidase}} NADP^+ + 2O_2^- + 2H^+$

2. $2O_2^- + 2H^+ \xrightarrow{\text{dismutation}} H_2O_2 + O_2$

3. $2RH_2 + H_2O_2 \longrightarrow 2RH^{\cdot} + 2H_2O$ (chain reaction)
   $2RH^{\cdot} + 2O_2 \longrightarrow 2R^+ + 2O_2^- + 2H^+$

4. $2O_2^- + 2H^+ \xrightarrow{\text{dismutation}} H_2O_2 + O_2$

5. $H_2O_2 \xrightarrow{\text{catalase}} H_2O + 0.5\ O_2$

6. $H_2O_2 + RH_2 \xrightarrow{\text{peroxidase}} 2H_2O + R$

In the case that these reactions occur in different combinations in whole cells (1+2+3+4+5; 1+2+3+4+6; 1+2+3+4 and 5 and 6 in different proportions), between 2.0 and 1.5 nmoles of O_2 are consumed for 2 nmoles of O_2^- initially generated by the oxidase or at the end of reactions 1+2+3, with a ratio O_2^-/O_2 between 1.0 and 1.33. These sequences do not account for the experimental ratio 2.5 - 2.8 (table 13) and therefore do not occur in whole cells.

When the respiratory burst of guinea pig granulocytes takes place in presence of azide, the ratio O_2 consumed versus H_2O_2 accumulated is near to the unity[39]. If a chain reaction initiated by H_2O_2 and independent of MPO is operative, the recovery of the peroxide should be much less than the oxygen consumed. Furthermore, assuming that the chain reaction is operative and mediated by MPO, the inhibition of this enzyme by the presence of NaN_3 or KCN should decrease the O_2 consumption. It is widely known that this is not the case. These are further evidences that a chain reaction initiated by H_2O_2 is not involved in the respiratory burst.

HYPOTHESIS n. 3. The oxidase forms O_2^-; O_2^- initiates a chain reaction with oxygen consumption and H_2O_2 formation; also H_2O_2 initiates a chain reaction followed by dismutation and by degradation of H_2O_2 through catalatic or peroxidatic mechanisms

1. $NADPH + H^+ + 2O_2 \longrightarrow NADP^+ + 2O_2^- + 2H^+$

2. [chain reaction]
$2O_2^- + 2RH_2 \xrightarrow{+\ 2H^+} 2RH^{\cdot} + 2H_2O_2$
$2RH^{\cdot} + 2O_2 \longrightarrow 2R^+ + 2O_2^- + 2H^+$

3. [chain reaction]
$4RH_2 + 2H_2O_2 \longrightarrow 4RH^{\cdot} + 4H_2O$
$4RH^{\cdot} + 4O_2 \longrightarrow 4R^+ + 4O_2^- + 4H^+$

4. $6O_2^- + 6H^+ \xrightarrow{\text{dismutation}} 3H_2O_2 + 3O_2$

5. $3H_2O_2 \xrightarrow{\text{catalase}} 3H_2O + 1.5\ O_2$

6. $3H_2O_2 + 3RH_2 \xrightarrow{\text{peroxidase}} 6H_2O + 3R$

In the case that these reactions occur in different combinations in whole cells (1+2+3+4+5; 1+2+3+4+6; 1+2+3+4 and 5 and 6 in different proportions) between 5.0 and 3.5 nmoles of O_2 are consumed for 2 nmoles of O_2^- initially generated by the oxidase (with a ratio O_2^-/O_2 between 0.4 and 0.57) or for 6 nanomole of O_2^- generated by the reactions 1+2+3 (with a ratio O_2^-/O_2 between 1.2 and 1.7). These sequences do not account for the experimental ratio 2.5 - 2.8 and therefore do not occur in whole cells.

HYPOTHESIS n. 4. The oxidase forms O_2^-; O_2^- dismutates to H_2O_2; H_2O_2 initiates a chain reaction followed by a chain reaction initiated by O_2^-; then O_2^- dismutates and H_2O_2 is degraded

1. $NADPH + H^+ + 2O_2 \xrightarrow{\text{oxidase}} NADP^+ + 2O_2^- + 2H^+$

2. $2O_2^- + 2H^+ \xrightarrow{\text{dismutation}} H_2O_2 + O_2$

3. [chain reaction]
$2RH_2 + H_2O_2 \longrightarrow 2RH^{\cdot} + 2H_2O$
$2RH^{\cdot} + 2O_2 \longrightarrow 2R^+ + 2O_2^- + 2H^+$

4.	$2O_2^- + 2RH_2$	$\xrightarrow{+\ 2H^+}$	$2RH^{\cdot} + 2H_2O_2$	chain
	$2RH^{\cdot} + 2O_2$	$\longrightarrow$	$2R^+ + 2O_2^- + 2H^+$	reaction

$$5.\quad 2O_2^- + 2H^+ \xrightarrow{\text{dismutation}} H_2O_2 + O_2$$

$$6.\quad 3H_2O_2 \xrightarrow{\text{catalase}} 3H_2O + 1.5\ O_2$$

$$7.\quad 3H_2O_2 + 3RH_2 \xrightarrow{\text{peroxidase}} 6H_2O + 3R$$

In the case that these reactions occur in different combinations in whole cells (1+2+3+4+5+6; 1+2+3+4+5+7; 1+2+3+4+5 and 6 and 7 in different proportions), between 4.0 and 2.5 nmoles of O_2 are consumed for 2 nmoles of O_2^- initially generated by the oxidase or at the end of the reactions 1+2+3+4 with a ratio O_2^-/O_2 0.5 - 0.8. These sequences do not account for the experimental ratio 2.5 - 2.8 and therefore do not occur in whole cells.

HYPOTHESIS n. 5. The oxidase forms O_2^-; O_2^- dismutates to H_2O_2; H_2O_2 is degraded by catalatic and/or by peroxidatic mechanism

$$1.\quad NADPH + H^+ + 2O_2 \xrightarrow{\text{oxidase}} NADP^+ + 2O_2^- + 2H^+$$

$$2.\quad 2O_2^- + 2H^+ \xrightarrow{\text{dismutation}} H_2O_2 + O_2$$

$$3.\quad H_2O_2 \xrightarrow{\text{catalase}} H_2O + 0.5\ O_2$$

$$4.\quad H_2O_2 + RH_2 \xrightarrow{\text{peroxidase}} 2H_2O + R$$

In the case that these reactions occur in different combinations in whole cells (1+2+3; 1+2+4; 1+2 and 3 and 4 in different proportions, 1.0 or 0.5 nmoles of O_2 are consumed for 2 nmoles of O_2^- initially generated by the oxidase with a ratio O_2^-/O_2 between 2.0 - 4.0. This sequence could account for the experimental ratio 2.5 - 2.8.

Thus, on the basis of the data and of the calculation presented, one can conclude that the reactions involved in the oxygen metabolism are mainly those indicated in the last hypothesis. We have excluded from the calculation the reaction of oxidation of the O_2^- released (O_2^- + X-oxid. $\longrightarrow$ O_2 + X-red.). Assuming that this reaction takes place, the amount of O_2^- oxidized must be rather low. This is shown by the finding (not shown here) that when the calculation is made by using the values of the respiration in presence of SOD the ratio O_2^-/O_2 does not appreciably vary with respect that calculated using the values of the respiration in absence of SOD (table 13). We have also excluded the reaction between O_2^- and H_2O_2 with forma= tion of $OH^{\cdot}$. Infact the data published by others[45] show that the production of this radical, probably through the Haber-Weiss reac tion ($O_2^- + H_2O_2 \longrightarrow OH^{\cdot} + OH^- + O_2$), is very low with respect to that of O_2^- and H_2O_2.

ADEQUACY OF THE OXIDASE TO ACCOUNT FOR THE RESPIRATORY BURST

On the basis 1) of the reaction involved in oxygen balance in whole cells and 2) of the imperfect stoichiometry between O_2^- measured and NADPH oxidized (or O_2 consumed) when the oxidase activity of particles is measured, it is clear that the estimation of the adequacy of the primary oxidase runs the risk of being rather rough (or even wrong).

The data of table 15 show the comparison between the respiratory burst in intact guinea pig PMN and the NADPH oxidase activity of cell free particles. The values of column "experimental" clearly dem- onstrate that the activity of the oxidase measured as O_2^- formation is inadequate to account for the respiratory burst of intact cells. We have shown that this is partially due to an imperfect capture of O_2^- formed during the oxidation of NAD(P)H by cell free particles and that the measurement of the oxidase in presence of DOC increases the O_2^- measured with a stoichiometry O_2^-/NADPH oxidized near to the theoretical. The data of column "corrected" represent the values of O_2^- formation in intact cells and of O_2^- formation by the oxidase of cell free particles, obtained after homogenization and centrifuga- tion of homogenates of the corresponding cells, corrected on the ba- sis of the defective stoichiometry O_2^-/NADPH shown in table 12.

In spite of these corrections the activity of the oxidase, both in presence of NADPH and of NADH, remains inadequate to account for the rate of the extrarespiration of intact cells. This inadequacy could be explained in different ways. One is that, besides NADPH oxi- dase, other enzymes or other respiratory systems not involving reduced nicotinamide adenine nucleotides are responsible for the burst. Another reason is that we underestimate the oxidase activity because the enzyme is deactivated or inactivated during the procedures of cell disruption and homogenization. Thus, if the oxidase is mea-

Table 15. Comparison between the respiratory burst of intact cells by PMA (expressed as O_2 uptake and O_2^- release) and the activity of NADPH oxidase of cell free particles

		Experimental	Corrected
Whole cells	Net O_2 uptake^x	4.7 ± 1.35	12.2 (a)
	O_2^- production^x	6.5 ± 1.05	
NADPH oxidase of cell^o free particles as O_2^-		0.766 ± 0.140	3.8 (b)
			1.9 (c)

x = nmoles/min/10^6 cells

o = nmoles O_2^-/min/cell free particles from 10^6 cells

a. Experimental (4.7) x 2.6, that is the ratio O_2^- formed/ O_2 consumed (see reference 39). 12.2 are nmoles of O_2^- actually formed for a net O_2 consumption of 4.7 nmoles.

b. Experimental (0.766) x 5 (stoichiometry O_2^-/NADPH 0.4 instead of 2.0).

c. Experimental (0.766) x 2.5 (stoichiometry O_2^-/NADPH 0.8 instead of 2.0).

ured soon after the activation has taken place and by using a procedure that allows a complete accessibility of the substrates to the enzymatic site and of cytochrome c to the O_2^- generating site, one would expect that the actual activity of the enzyme can be fully revealed. This expectation has been verified by using the experimental model described in fig. 5.

The respiratory activity of intact cells stimulated by PMA was measured as O_2 consumption with Clark oxygen electrode. During the linear part of the respiration, the cells were lysed and their respiratory activity was abolished by adding deoxycholate (DOC). At this point NADH 1 mM or NADPH 0.15 mM were added. The addition of NADPH induced a instantaneous resumption of O_2 consumption at a rate similar or even higher than that of the stimulated cells. After

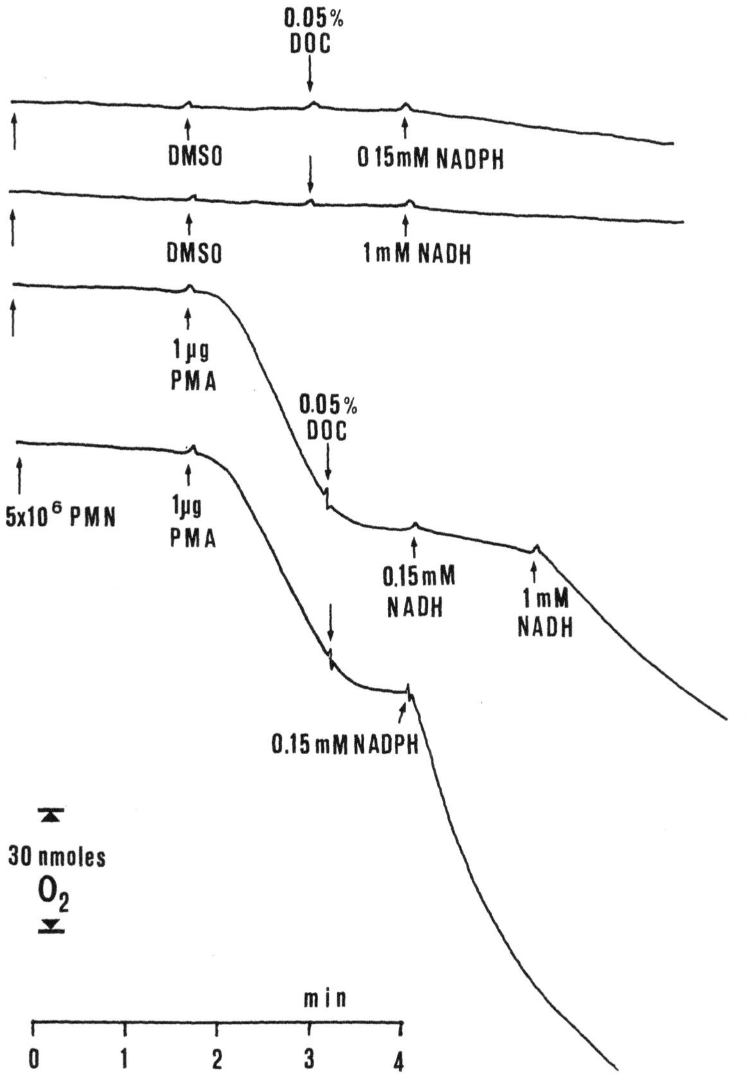

Fig. 5. Polarographyc traces of oxygen consumption by guinea pig PMN and by NAD(P)H oxidase after addition of deoxycholate (DOC). Assay medium: 2 ml KRP containing 5 mM glucose, 0.5 mM $CaCl_2$ and 2 mM NaN_3.

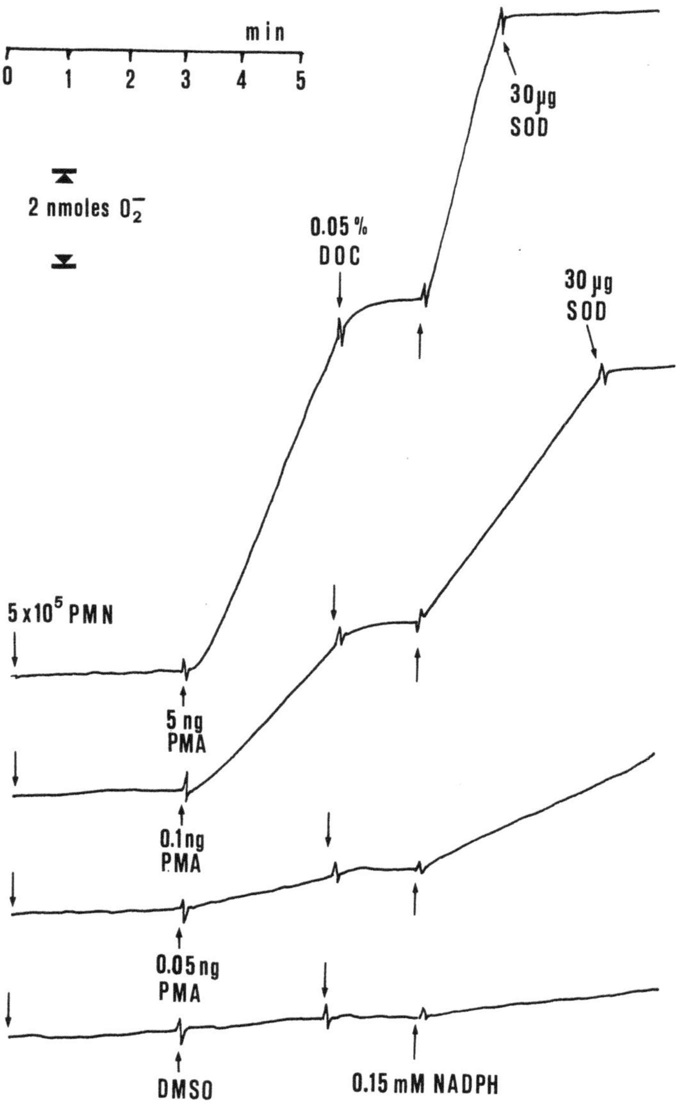

Fig. 6. Relation between the O_2^- production by guinea pig PMN stimulated with PMA and the corresponding NADPH oxidase activity after the addition of deoxycholate (DOC). Assay medium: 1 ml KRP con= taining 5 mM glucose, 0.5 mM $CaCl_2$, 2 mM NaN_3, 100 µM cytochrome c. 37°C.

the addition of NADH the O_2 consumption also resumed but at a rate much lower with respect that of the stimulated cells. DOC has not stimulatory activity on NAD(P)H oxidase activity of resting cells.

Similar results have been obtained by measuring as O_2^- formation both the respiratory burst of intact cells before the addition of DOC and the oxidase activity after the addition of DOC and NAD(P)H.

The validity of this experimental approach, that at the same time avoids the inactivation of the oxidase and allows a good substrate-enzyme interaction and a physiological fate of the intermediates, is also shown by the data presented in fig. 6. It can be seen that the rate of the respiration resumed after the addition of NADPH strictly parallels the extent of the activation of the respiratory metabolism of intact cells induced by different doses of PMA.

These results indicate that the activity of the primary oxidase that oxidizes NADPH, when measured in appropriate conditions,is adequate to account for the respiration of whole cells.

CONCLUSION AND SUMMARY

1. The so called "soluble" oxidase(s) are not involved in the respiratory burst of guinea pig and human granulocytes and of guinea pig peritoneal resident and elicited macrophages.

2. The activation of the oxidation of NADPH by a membrane bound NAD(P)H oxidase is the main mechanism responsible for the activation of the respiration of phagocytes.

3. The oxidase is inactive in resting cells and the activated form works on the plasma membrane.

4. More than one mechanism is operative in the oxidation of NAD(P)H by cell free particles in vitro. These mechanisms vary in relation to the conditions of assay (pH and concentration of substrate).

5. Under optimal conditions in vitro the enzymatic oxidation of NADPH practically involves the univalent pathway of oxygen reduction with stoichiometry of two nanomoles of O_2^- formed for one nanomole of NADPH oxidized.

6. Also in intact cells all O_2 is first univalently reduced to O_2^- and then discharged outside the cell or in the phagocytic vacuoles.

7. The main reactions involved in the O_2 balance in intact cells are the univalent reduction of O_2, the dismutation of O_2^- to H_2O_2 and the degradation of the peroxide through catalatic and peroxidatic mechanisms.

8. The total oxygen univalently reduced by the activated oxidase is 2-4 folds the net oxygen consumed by the cells, depending on the mechanism of H_2O_2 degradation.

9. All the rate of extrarespiration is accounted for by the rate of oxidation of physiological concentration of NADPH by the membrane-bound enzyme. This adequacy can be observed only under appropriate experimental conditions, because the high activity of the oxidase is not a permanent state.

REFERENCES

1. C. W. Baldridge, R. W. Gerard, The extrarespiration of phagocytosis, Amer.J.Physiol. 103:235 (1933).
2. G. Y. N. Iyer, M. F. Islam and H. J. Quastel, Biochemical aspects of phagocytosis, Nature (London) 192:535 (1961).
3. M. L. Karnovsky, Metabolic basis of phagocytic activity, Physiol. Rev. 42:143 (1962).
4. F. Rossi and M. Zatti, Changes in the metabolic pattern of polymorphonuclear leukocytes during phagocytosis, Brit.J.Exp. Pathol. 45:548 (1964).
5. B. M. Babior, Oxygen-dependent microbial killing by phagocytes, N.Engl.J.Med. 298:659 (1978).
6. F. Rossi, D. Romeo and P. Patriarca, Mechanisms of phagocytosis-associated oxidative metabolism in polymorphonuclear leucocytes and macrophages, J.Reticuloendothel.Soc. 12:127 (1972).
7. J. A. Badwey and M. L. Karnovsky, Active oxygen species and the functions of phagocytic leukocytes, Ann.Rev.Biochem. 49:695 (1980).
8. L. R. DeChatelet, C. E. McCall and M. R. Cooper, Aminoacid oxidase in leukocytes: evidence against a major role in phagocytosis, Infect.Immun. 5:632 (1971).
9. L. R. DeChatelet, M. R. Cooper and C. E. McCall. Stimulation of the hexose monophosphate shunt in human neutrophils by ascorbic acid: mechanism of action. Antimicrob. Agents Chemother. 1:12 (1972).

10. F. Rossi and M. Zatti, Mechanism of the respiratory stimulation in saponine-treated leukocytes. The KCN insensitive oxidation of NADPH, Biochim.Biophys.Acta 153:296 (1968).
11. P. Patriarca, R. Cramer, S. Moncalvo, F. Rossi and D. Romeo, Enzymatic basis of metabolic stimulation in leukocytes during phagocytosis: The role of activated NADPH oxidase, Arch.Biochem.Biophys. 145:255 (1971).
12. B. B. Paul, R. R. Strauss, A. A. Jacobs and A. J. Sbarra, Direct involvement of NADPH oxidase with the stimulated respiratory and hexose phosphate shunt activity in phagocytizing leukocytes, Exp.Cell Res. 73:456 (1972).
13. H. J. Cohen, M. E. Chovaniec and W. A. Davies, Activation of the guinea pig granulocytes NAD(P)H dependent superoxide generating enzyme: localization in a plasmamembrane enriched particle and kinetics of activation, Blood 55:355 (1980).
14. L. R. DeChatelet, L. C. McPhail, D. Mullikin and C. E. McCall, An isotopic assay for NADPH oxidase activity and some characteristics of the enzyme from human polymorphonuclear leukocytes, J. Clin. Invest. 55:714 (1975).
15. K. Kakinuma and M. Kaneda, Kinetic studies on the H_2O_2 (O_2^-)-forming enzyme in guinea pig leukocytes, FEBS Letters 111:90 (1980).
16. W. H. Evans and M. L. Karnovsky, A possible mechanism for the stimulation of some metabolic functions during phagocytosis, J. Biol. Chem. 236:Pc30 (1961).
17. R. H. Cagan and M. L. Karnovsky, Enzymatic basis of the respiratory stimulation during phagocytosis, Nature (London) 204-255 (1964).
18. M. L. Karnovsky, Chronic granulomatous disease. Pieces of a cellular and molecular puzzle, Fed. Proc. 32:1527 (1973).
19. G. L. Mandell and G. W. Sullivan, Pyridine nucleotide oxidation by intact human polymorphonuclear neutrophils, Biochim. Biophys. Acta 234:43 (1971).
20. J. A. Badwey and M. L. Karnovsky, Active oxygen species and the functions of phagocytic leukocytes, Ann.Rev.Biochem. 49:695 (1980).
21. R. J. Selvaraj and A. J. Sbarra, The role of the phagocyte in host-parasite interaction. VII. Di and triphosphopyridine nucleotide kinetics during phagocytosis. Biochim.Biophys.Acta 141:243 (1967).

22. R. Rossi, D. Romeo and P. Patriarca, Mechanism of phagocytosis-associated oxidative metabolism in polymorphonuclear leucocytes and macrophages, J. Reticuloendothel. Soc. 12:127 (1972).
23. A. Aellig, M. Maillard, A. Phavorin and J. Frei, The energy metabolism of the leukocyte. IX. Changes in the concentrations of the coenzymes NAD, NADH, NADP, NADPH in polymorphonuclear leucocytes during phagocytosis of Staphylococcus albus and due to the action of phospholipase C, Enzyme 22:207 (1977).
24. P. Bellavite, G. Berton, P. Dri, M. R. Soranzo, Enzymatic basis of the respiratory burst of guinea pig peritoneal resident macrophages. J. Reticuloendothel. Soc. 29:47 (1981).
25. G. Berton, P. Bellavite, P. Dri, F. Rossi, Metabolism of inflammatory macrophages. I. The enzyme responsible for the respiratory burst in elicited peritoneal macrophages of guinea pig, J.Pathology (submitted).
26. R. T. Briggs, D. B. Drath, M. L. Karnovsky and M. J. Karnovsky, Localization of NADH oxidase on the surface of human polymorphonuclear leukocytes by a new cytochemical method. J.Cell Biol. 67:566 (1975).
27. K. Takanaka and P. J. O'Brien, Mechanism of H_2O_2 formation by leukocytes. Evidence for a plasmamembrane location, Arch. Biochem. 169:428 (1975).
28. J. M. Goldstein, M. Cerqueira, S. Lund and H. B. Kaplan, Evidence that the superoxide generating system of human leukocytes is associated with the cell surface, J. Clin. Invest. 59:249 (1977).
29. B. Dewald, M. Baggiolini, J. T. Curnutte and B. M. Babior, Subcellular localization of the superoxide forming enzyme in human neutrophils, J. Clin. Invest. 63:21 (1979).
30. D. B. Iverson, P. Wang-Iverson, J. K. Spitznagel and L. R. De Chatelet, Subcellular localization of NAD(P)H oxidases in human neutrophilic polymorphonuclear leukocytes, Biochem. J. 176:175 (1978).
31. A. I. Tauber and E. J. Goetzl, Subcellular localization and solubilization of the superoxide generating activities (SGA) of human neutrophils, Blood 52 (Suppl. 1):128 (1978).
32. P. Patriarca, R. Cramer, P. Dri, L. Fant, R. R. Basford and F. Rossi, NADPH oxidizing activity in rabbit polymorphonuclear leukocytes: localization in azurophilic granules, Biochem.Biophys.Res.Commun. 53:830 (1973).
33. F. Rossi, P. Patriarca, G. Berton and G. De Nicola, Subcellular localization of the enzyme responsible for the respiratory

burst in resting and phorbol myristate acetate activated leucocytes, in: "Biological and Clinical Aspects of Superoxide and Superoxide Dismutase" (W.H.Bannister and J.V.Bannister eds.) pp. 193-200. Elsevier/North Holland, New York.

34. J. A. Badwey and M. L. Karnovsky, Production of superoxide anion hydrogen peroxide by an NADH-oxidase in guinea pig polymorphonuclear leukocytes, J. Biol. Chem. 254: 11530 (1979).

35. P. Bellavite, G. Berton, P. Dri, Studies on the NADPH oxidation by subcellular particles from phagocytosing polymorphonuclear leukocytes. Evidence for the involvement of three mechanisms, Biochim. Biophys. Acta 591:434 (1980).

36. P. Patriarca, P. Dri, K. Kakinuma, F. Tedesco and F. Rossi, Stu dies on the mechanism of metabolic stimulation in polymorphonuclear leukocytes during phagocytosis. I. Evidence for superoxide anion involvement in the oxidation of $NADPH_2$, Biochim. Biophys. Acta 385:380 (1975).

37. J. T. Curnutte, M. L. Karnovsky and B. M. Babior, Manganese dependent NAD(P)H oxidation by granulocyte particles. The role of superoxide and the nonphysiological nature of the manganese requirement, J. Clin. Invest. 57:1059 (1976).

38. T. G. Gabig and B. M. Babior, The O_2^--forming oxidase responsible for the respiratory burst in human neutrophils. Properties of the solubilized enzyme, J. Biol.Chem. 254:9070 (1979).

39. P. Dri, P. Bellavite, G. Berton and F. Rossi, Interrelationship between oxygen consumption, superoxide anion and hydrogen peroxide formation in phagocytosing guinea pig polymorphonuclear leukocytes, Mol.Cell.Biochem. 23:109 (1979).

40. R. K. Root and J. A. Metcalf, H_2O_2 Release from Human Granulocytes during Phagocytosis, J. Clin. Invest. 60:1266 (1977).

41. R. S. Weening, R. Wever and D. Roos, Quantitative aspects of the production of superoxide radicals by phagocytizing human granulocytes, J. Lab.Clin.Med. 85:245 (1975).

42. B. M. Babior, Superoxide production by phagocytes, Biochem. Biophys.Res.Commun. 91:222 (1979).

43. F. Rossi, G. Zabucchi, P. Dri, P. Bellavite and G. Berton, O_2^- and H_2O_2 production during the respiratory burst in alveolar macrophages, in: "Macrophages and lymphocytes. Natu re, Functions and Interaction" Part A (M.R.Escobar and H. Friedman, eds.) pp. 53-74, Plenum Press, New York (1980).

44. J. A. Badwey, J. T. Curnutte and M. L. Karnovsky, The enzyme of granulocytes that produces superoxide and peroxide: An elusive pimpernel. N.Engl.J.Med. 300: 1157 (1979).

45. A. I. Tauber and B. M. Babior, Evidence for hydroxyl radical production by human neutrophils, J.Clin.Invest. 60:374 (1977).

Acknowledgements.The research was supported by the grant n° 79.03271.04 from the National Research Council of Italy (CNR).

DISCUSSION
Edited by: R.E. Basford

Dr. Minakami:
I would like to congratulate Prof. Rossi for his beautiful accomplishments. Our data, which will be presented tomorrow morning, agree quite well with your statement.

Dr. Rossi:
Thank you.

Dr. De Chatelet:
First I think I'll retract my talk scheduled for tomorrow. The stoichiometry demonstrated is surely outstanding. I still have an unpopular stance in believing the oxidase is present in a granule of the cell. I think there are several other interpretations of our data.

a) The enzyme may be located in a granule in the resting cell and translocated to the membrane by phagocytosis.
b) You are actually isolating phagosomes rather than plasma membrane. It is possible that there is very rapid degranulation into the phagocytic vacuole.

Would you care to comment on these possibilities?

Dr. Rossi:
Yes, it is possible that the presence of the enzyme in phagosomes is due to the translocation from the granules to the membrane of the vacuole or inside the vacuole. However, in this case, MPO or other enzymes bound to granules also should concentrate in the phagosomes. As you have seen from the results I have presented, this is not the case.

Furthermore, I say to you that we found that in resting PMN the oxidase is located in the granule fraction. However, it is likely that this activity corresponds to the oxidase activity of MPO. As you know, my old data, published many years ago, supported the hypothesis that the oxidase was located in the azurophilic granules. Now I am convinced that the enzyme is in the plasma membrane. Obviously we need further work in order to settle this matter.

Dr. Kakinuma:
In our recent study, we measured the rate of H_2O_2 generation by the particulate fraction isolated from phagocytosing polymorphonuclear leukocytes in the presence of NADPH by means of a spectrophotometric assay of cytochrome c peroxidase-H_2O_2 complex. We found that one mole of NADPH produces exactly one mole of H_2O_2 in the NADPH oxidation reaction by this fraction. Therefore, I support, one of your proposed several hypothetical reactions (hypothesis No. 5):

$$\text{NADPH} + H^+ + 2\,O_2 \xrightarrow{\text{oxidase}} \text{NADP} + 2\,O_2^- + 2\,H^+ \quad (1)$$

$$2\,O_2^- + 2\,H^+ \xrightarrow[\text{dismutation}]{} H_2O_2 + O_2 \quad (2)$$

Where one mole of NADPH reacts with oxygen to form one mole of H_2O_2.

DISCREPANCIES IN THE OXYGEN BALANCE OF WHOLE HUMAN NEUTROPHILS AND NEUTROPHIL HOMOGENATES

Dirk Roos[1], Rob van Zwieten[1], Margriet L.J. van Schaik[2] and Mic N. Hamers[1]
[1]Central Laboratory of the Netherlands Red Cross Blood Transfusion Service and Laboratory for Experimental and Clinical Immunology, University of Amsterdam, P.O. Box 9190, Amsterdam, The Netherlands
[2]Pediatric Clinic, Binnengasthuis, Academic Hospital University of Amsterdam, Amsterdam, The Netherlands

SUMMARY

The oxygen consumption, superoxide production and hydrogen peroxide generation was studied in human neutrophils phagocytosing zymosan particles. Application of sodium azide, as an inhibitor of catalase, and/or 1,3-bis(chloroethyl)-1-nitrosourea (BCNU), as an inhibitor of glutathione reductase, led to the conclusion that neutrophils convert about half of the oxygen consumed in the respiratory burst to hydrogen peroxide; the other half is used for formation of organic peroxides, disulfide bridges, etc. These products are rapidly degraded to water by catalase and/or the glutathione redox cycle. Reduction of exogenous cytochrome _c_ accounted for only about 15% of the consumed oxygen.

Neutrophil homogenates contain a badly damaged oxidase system, because oxygen consumption and hydrogen peroxide formation were only about one-tenth of that observed with whole cells. In contrast, cytochrome-_c_ reduction was about three times as high as that found with intact cells. Probably, cytochrome _c_ partly reconstitutes damaged oxidase systems, thus artificially increasing the oxidase activity. We conclude that cytochrome-_c_ reduction is not a good parameter to characterize cell-free oxidase preparations.

INTRODUCTION

The oxygen balance of phagocytes during the respiratory burst has been for a long time, and still is, a matter of considerable controversy. One of the unsettled questions is the fate of the oxygen consumed by these cells. While most people agree that these cells generate large amounts of hydrogen peroxide (1-3), and that these cells are capable of reducing exogenous cytochrome c in an SOD-inhibitable reaction (4-6), the dispute has focused on the quantitative importance of this last reaction. Either the cells generate H_2O_2 in a reaction that involves superoxide only as an enzyme-bound intermediate of which some may leak into the medium (7), or the cells generate and release O_2^- that may then be dismuted into H_2O_2, spontaneously or catalyzed by SOD (2,3).

When studying the oxygen balance of phagocytosing cells, several problems are evident; most have been recognized in a recent commentary (8). First, the species difference. It has been amply proven that human and guinea-pig neutrophils differ completely as to the properties of the O_2^--H_2O_2-generating oxidase(s) (9-11). Also, there are pronounced quantitative differences between the H_2O_2- and O_2^--degrading enzymes in phagocytes from different origins (12). Therefore, in this chapter, we restrict ourselves to the human neutrophil.

This leads to the second point: only net reactions can be measured. The oxygen consumed by the cells reflects utilization of O_2 by the oxidase as well as re-formation of O_2 by dismutation of O_2^-, degradation of H_2O_2 in the catalase reaction, etc. Similarly, O_2^- may react with cytochrome c added to the cells for its detection, but may also disappear into side reactions. H_2O_2 may react with catalase, myeloperoxidase, glutathione peroxidase, etc. (see Fig. 1). Therefore, our detection systems will always underestimate the true capacity of the cell. Of course, one can minimize the side reactions by applying inhibitors, such as azide, and by trapping the products as efficiently as possible with high concentrations of detecting agents. The idea that this is "unphysiologic" (7) does not appeal to us: we regard this as legitimate tools to estimate the cell's capacity.

Third, the detection systems are not ideal: they cannot always reach the site of the oxidase (12-17), and they may be affected by other cell products than those that one wants to measure. For instance, reduced cytochrome c can be re-oxidized by H_2O_2 (18); this leads again to an underestimation of the actual situation.

Finally, when studying the oxygen balance of cell homogenates or fractions thereof, one should always be aware of possible damage to the oxidase system due to the fractionation procedure. Therefore, the balance should not only be based on the input of O_2 and the output of O_2^- and H_2O_2 of the cell-free preparation, but also be compared with the capacity of the original, whole cells.

With these points in mind, we have re-studied the oxygen balance of whole human neutrophils and homogenates.

RESULTS AND DISCUSSION

Fate of hydrogen peroxide

As Fig. 1 shows, H_2O_2 has many catabolic pathways. It is not always realized that organic peroxides can act as substrates both for catalase (19) and for the glutathione redox system (20). Both pathways are very active in human neutrophils (21), and should therefore be taken into account when estimating the oxygen balance of these cells. We have applied azide to block catalase and 1,3-bis-(chloroethyl)-1-nitrosourea (BCNU) as a rather specific inhibitor of glutathione reductase (22,23). The results are shown in Table 1.

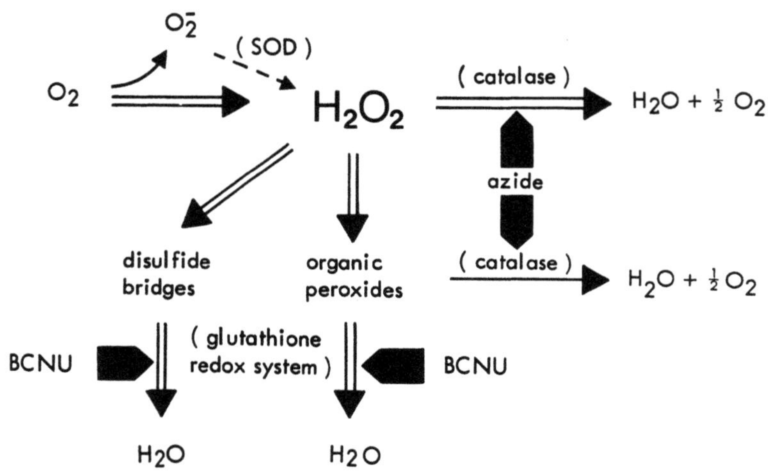

Fig. 1 - Oxygen pathways in phagocytosis.

In the presence of azide (Table 1, line b), only about 50% of the oxygen consumed by phagocytosing neutrophils was recovered as O_2^- + H_2O_2 (H_2O_2 was measured in the presence of SOD). This low recovery of H_2O_2 was not due to the methods used, because the oxygen consumed in the reaction of glucose with glucose oxidase was completely balanced by the H_2O_2 produced (not shown). Our hypothesis is, therefore, that -in human neutrophils- the other 50% of the consumed oxygen is converted into water by the glutathione

Table 1. Oxygen balance of zymosan-stimulated human neutrophils.

Addition	O_2 consumption	$H_2O_2 + O_2^-$ production
	(nmoles per 10^6 PMN per min)	
a. None	7.2 ± 1.1 (89%)	1.9 ± 0.6 (23%)
b. NaN_3 (2 mM)	8.1 ± 2.2 (100%)	3.8 ± 1.1 (47%)
c. BCNU (230 μM)	3.8 ± 1.2 (47%)	1.2 ± 0.1 (15%)
d. BCNU + NaN_3	5.5 ± 0.5 (68%)	2.4 ± 1.2 (30%)

Values are mean ± SEM of 3-8 experiments. Oxygen consumption was measured with an oxygen electrode; assay conditions: neutrophils ($2x10^6$/ml), serum-opsonized zymosan (1 mg/ml). H_2O_2 generation was measured in the presence of superoxide dismutase with the homovanillic acid method (14); assay conditions: neutrophils ($3x10^5$/ml), homovanillic acid (Merck, 1 mM), horse radish peroxidase (Boehringer, 1.44 U/ml), serum-opsonized zymosan (0.1 mg/ml) and superoxide dismutase (Sigma, 7.5 μM).

pathway. In the absence of azide (Table 1, line a), 23% of the consumed oxygen was recovered as $O_2^- + H_2O_2$, while 11% was returned as oxygen. This 11% must have originated from 22% H_2O_2; hence, 45% of the consumed oxygen can be accounted for as H_2O_2 or organic peroxides. Thus, similar to the situation with azide, about half of the H_2O_2 produced is detoxified by catalase and the other half by the glutathione redox cycle.

In the presence of BCNU (Table 1, line c), we had expected to find an accumulation of H_2O_2 due to the inhibition of an important catabolic pathway. Instead, we found less H_2O_2 in the medium. This indicates that, indeed, organic peroxides are formed (either directly or via H_2O_2), and degraded by catalase. We assume that, in the presence of BCNU, of every 100 molecules of organic or hydrogen peroxide formed, 85 are degraded by catalase, returning 43 molecules of oxygen. Thus, we expected to find a net O_2 consumption of 57% of that found with azide. Instead, we found 47%, in reasonable agreement. This again indicates that our hypothesis of Fig. 1 may be correct.

Finally, with both azide and BCNU present (Table 1, line d), 30% of the oxygen was recovered as $O_2^- + H_2O_2$. In this situation, however, the O_2 consumption soon stopped after addition of zymosan, indicative for cell damage by accumulated oxidative products. This is completely analogous to the observation with neutrophils genetically defective in glutathione reductase (24). Nevertheless, we recovered again about half of the consumed oxygen as H_2O_2; the other half must have been converted into organic peroxides.

Thus, our results persistently indicate that human neutrophils convert about half of the oxygen consumed in the respiratory burst to superoxide and hydrogen peroxide, whereas the remainder may have been used for formation of organic peroxides, disulfide bridges, etc. These products may be expected to be rapidly degraded by catalase and/or the glutathione redox cycle. This explains why organic peroxides do not accumulate in the cells (25), even in the presence of azide. It also explains why virtually all oxygen consumed during phagocytosis is recovered in water (25). It may be expected that organic peroxides do accumulate in cells treated with BCNU + azide; this is presently under investigation.

Release of superoxide

In the next set of experiments, O_2 consumption, O_2^- generation and H_2O_2 formation were measured separately. Cytochrome-*c* reduction was assayed in the presence of catalase, to avoid re-oxidation of reduced cytochrome *c*. Indeed, addition of catalase changed the kinetics of cytochrome-*c* reduction from a gradual decrease in the reduction rate to a straight line over 10 min of incubation with either opsonized zymosan (STZ) or phorbol myristate acetate (PMA). Establishment of a steady state in the cytochrome-*c* reduction by phagocytosing neutrophils has been taken to indicate that O_2^- is formed inside phagosomes, beyond reach of cytochrome *c* (15); re-oxidation of reduced cytochrome *c* by H_2O_2 may have contributed to this process, however.

Table 2 shows that only about 15% of the O_2 consumed after stimulation of the cells with STZ could be recovered as O_2^- (footnote 1) and, as described above, about 50% as O_2^- + H_2O_2. Thus, only a small portion of the oxygen is univalently reduced to superoxide, whereas most of the oxygen is directly reduced to H_2O_2 or other peroxides. This explains the rather small effect on the oxygen consumption of cytochrome-*c* addition (7,27,28), which was confirmed by us.

It has been argued that superoxide cannot be measured efficiently in the supernatant of phagocytosing cells, because O_2^- generated in phagosomes will be dismuted into H_2O_2 before leaving the cell (14,15). Indeed, about twice as much cytochrome *c* was reduced when neutrophils were stimulated with PMA as when STZ was used, although the oxygen consumption was about equal (results not shown).

However, it might be that cytochrome *c*, added to the cells "pulls away" electrons from O_2^- when superoxide is still bound to

Footnote 1. Previously (2,5,18,26), we had found higher recoveries of O_2^- and H_2O_2. Due to improvements in the oxygen consumption assay, we now find about two times higher respiration rates and, in consequence, two times lower recoveries of O_2^- and H_2O_2.

Table 2. Oxygen balance of intact stimulated neutrophils and homogenates thereof.

Preparation	O_2 consumption	O_2^- production	H_2O_2+ O_2^- production
10^7 intact stimulated neutrophils (n=3)	78 ± 3	11 ± 2.5	38 ± 4.3
homogenate of 10^7 stimulated neutrophils (n=5)	7 ± 1.2	30 ± 5.3	11 ± 2.2

Values in nmoles/min. Mean ± SEM of (n) paired experiments. Human neutrophils were incubated with 1 mg serum-opsonized zymosan per ml. Oxygen consumption was measured with an oxygen electrode in the presence of 2 mM NaN_3. Superoxide was measured with cytochrome c (150 μM) and catalase (260 U), continuously at 550 nm for about 2 min, followed by addition of SOD (7.5 μM). Hydrogen peroxide was measured discontinuously with the LDADCF method (2) in the presence of NaN_3 (2 mM) and SOD (7.5 μM). Homogenates were prepared from cells ($5x10^7$/ml) treated for 2 min with serum-opsonized zymosan (5 mg/ml) in the presence of 1 mM NaN_3. Immediately thereafter, the cells were homogenized at 0°C and the homogenates were diluted 5 times with ice-cold distilled water. The tests with the homogenates were performed in the presence of 1 mM NADPH.

the oxidase system (7). Without addition of cytochrome c, O_2^- might stay on the oxidase as an intermediate. This idea is strengthened by the next set of experiments, performed with cell homogenates.

Damage to the oxidase system

Table 2 (second line) gives the results obtained with neutrophil homogenates. The problems are evident. First, only 10% of the original respiratory activity was left. The H_2O_2-generating capacity had also decreased drastically. This indicates that the oxidase system of the homogenates is badly damaged. We imagine that many investigators must have observed this phenomenon; nevertheless, it has seldom been described (8,17). In contrast to the decreased O_2 consumption and H_2O_2 production, the cytochrome-c reduction was enhanced when compared to intact cells. The ratio of (O_2^- produced):(O_2 consumed) increased from 0.15 in intact cells to 4 in homogenates. In azide-treated cells, this ratio can maximally be 2 (see Rossi et al., this volume). Thus, in the homogenates there is more output of O_2^- than can be accounted for by O_2 consumed!

What can be the reason of this strange phenomenon? The cytochrome-c assay is not to blame, because:

1) homogenates from resting cells did not reduce cytochrome c (Table 3), so the process is specific for activated cells;
2) the cytochrome-c reduction was inhibitable by SOD, so the process is specific for O_2^-;
3) The cytochrome-c reduction showed normal Michaelis-Menten kinetics with respect to NADPH (Km, 139 µM; Vmax, 17 nmoles/min/mg protein) and NADH (Km, 350 µM; Vmax, 9 nmoles/min/mg protein), and it was proportional to the amount of homogenate used.

Table 3. Effect of an O_2^--generating system on the oxygen balance of neutrophil homogenates.

Incubation mixture	O_2 cons. (exp. 1)	O_2^- prod. (exp. 2)	H_2O_2 prod. (exp. 3)
1. XOD, acetaldehyde, NADPH	17.4	2.0	7.0
2. homogenate of STZ-stimulated PMN, NADPH, acetaldehyde	2.8	5.1	0.1
(1) + (2)	21.4 (106%)	6.8 (96%)	6.7 (95%)
3. homogenate of Z-stimulated PMN, NADPH, acetaldehyde	2.7	0.5	0
(1) + (3)	20.5 (102%)	2.4 (96%)	6.0 (85%)

Values in nmoles/min. Xanthine oxidase (XOD) was used at 20 mU/ml, acetaldehyde at 2 mM, NADPH at 1 mM. STZ, serum-opsonized zymosan; Z, non-opsonized zymosan. For preparation of neutrophil homogenates and for assays, see legend to Table 2.

According to us, two possibilities remain. One is that the cytochrome-c reduction, although dependent on superoxide formation, is not a quantitative measure of this process. For instance, if O_2^- starts a radical chain reaction that reduced cytochrome c, SOD would stop this reaction, but the amount of cytochrome c reduced might be far more than the amount of O_2^- produced. We have tested this possibility by adding acetaldehyde + xanthine oxidase, as an O_2^--generating system, to homogenates of either resting or activated cells. The results are shown in Table 3. Within each experiment, it is evident that the xanthine-oxidase system did not amplify the cellular reactions: we always found the expected sum of the separate values. Therefore, it is unlikely that cytochrome c is reduced by anything else but O_2^-.

The other possible explanation for the high cytochrome-c reduction by the homogenates is the effect of cytochrome c itself on this reaction. It might be, as has been suggested by Segal and Meshulam (7), that addition of cytochrome *c*, especially to a damaged oxidase system, promotes the flow of electrons through this system. For instance, cytochrome *c* might replace, or accept electrons from, cytochrome *b* in the membrane (29,30). As a result, the system will be partially reconstituted. If it were possible to measure O_2 consumption under these conditions, we would expect to also find an increased respiration. In that case, the discrepancy in the oxygen balance would have disappeared.

The consequence of this idea is that cytochrome-*c* reduction is not a proper parameter to characterize cell-free oxidase preparations. Therefore, such preparations should be characterized also by oxygen consumption and H_2O_2 release.

REFERENCES

1. G.Y.N. Iyer, M.F. Islam, and J.H. Quastel, Biochemical aspects of phagocytosis. *Nature* 192:535 (1961).
2. J.W.T. Homan-Müller, R.S. Weening, and D. Roos, Production of hydrogen peroxide by phagocytizing human granulocytes. *J. Lab. Clin. Med.* 85:198 (1975).
3. R.K. Root, J. Metcalf, N. Oshino, and B. Chance, H_2O_2 release from human granulocytes during phagocytosis. I. Documentation, quantification and some regulating factors. *J. Clin. Invest.* 55:945 (1975).
4. B.M. Babior, R.S. Kipnes, and J.T. Curnutte, Biological defense mechanisms: the production by leukocytes of superoxide, a potential bactericidal agent. *J. Clin. Invest.* 52:741 (1973).
5. R.S. Weening, R. Wever, and D. Roos, Quantification of the production of superoxide radicals by phagocytizing human granulocytes. *J. Lab. Clin. Med.* 85:245 (1975).
6. R.B. Johnston Jr., B.B. Keele Jr., H.P. Misra, J.E. Lehmeyer, L.S. Webb, R.L. Baehner, and K.V. Rajagopolan, The role of superoxide anion generation in phagocytic bactericidal activity. Studies with normal and chronic granulomatous disease leukocytes. *J. Clin. Invest.* 55:1357 (1975).
7. A.W. Segal, and T. Meshulam, Production of superoxide by neutrophils: a reappraisal. *FEBS Lett.* 100:27 (1979).
8. J.A. Badwey, J.T. Curnutte, and M.L. Karnovsky, The enzyme of granulocytes that produces superoxide and peroxide. An elusive pimpernel. *New Engl. J. Med.* 300:1157 (1979).
9. B.M. Babior, J.T. Curnutte, and B.J. McMurrich, The particular superoxide-forming system from human neutrophils: properties of the system and further evidence supporting its participation in the respiratory burst. *J. Clin. Invest.* 58:989 (1976).
10. J.A. Badwey, and M.L. Karnovsky, Production of superoxide and hydrogen peroxide by an NADH oxidase in guinea-pig polymorphonuclear leukocytes. Modulation by nucleotides and divalent

cations. J. Biol. Chem. 254:11530 (1979).
11. P.E. Newburger, M.E. Chovaniec, and H.J. Cohen, Activity and activation of the granulocyte superoxide-generating system. Blood 55:85 (1980).
12. F. Rossi, G. Zabucchi, P. Dri, P. Bellavite, and G. Berton, O_2^- and H_2O_2 production during the respiratory burst in alveolar macrophages, in: "Macrophages and Lymphocytes. Nature, Functions and Interaction", part A, M.R. Escobar, and H. Friedman, eds., Plenum Press, New York and London, 53 (1979).
13. D. Roos, J.W.T. Homan-Müller, and R.S. Weening, Effect of cytochalasin B on the oxidative metabolism of human peripheral blood granulocytes. Biochem. Biophys. Res. Commun. 68:43 (1976).
14. R.K. Root, and J.A. Metcalf, H_2O_2 release from human granulocytes during phagocytosis. Relationship to superoxide anion formation and cellular catabolism of H_2O_2: studies with normal and cytochalasin-B-treated cells. J. Clin. Invest. 60:1266 (1977).
15. Z.F. Nabi, K. Takeshige, T. Hatae, and S. Minakami, Hydrogen peroxide formation of polymorphonuclear leukocytes stimulated with cytochalasin D. Exp. Cell. Res. 124:293 (1979).
16. M.S. Cohen, J.A. Metcalf, and R.K. Root, Regulation of oxygen metabolism in human granulocytes: relationship between stimulus binding and oxidative response using plant lectins as probes. Blood 55:1003 (1980).
17. T.R. Green, R.E. Schaefer, and M.T. Makler, Orientation of the NADPH-dependent superoxide-generating oxidoreductase on the outer membrane of human PMN's. Biochem. Biophys. Res. Commun. 94:262 (1980).
18. D. Roos, M.L.J. van Schaik, R.S. Weening, and R. Wever, Superoxide generation in relation to other oxidative reactions in human polymorphonuclear leukocytes, in: "Superoxide and Superoxide Dismutases", A.M. Mickelson, J.M. McCord, and I. Fridovich, eds., Academic Press, London/New York/San Francisco, 307 (1977).
19. K.G. Stern, Spectrography of the reaction of catalase with ethyl hydrogen peroxide. Enzymologia 4:145 (1937).
20. J.M. Oliver, D.F. Albertini, and R.D. Berlin, Effects of glutatione-oxidizing agents on microtubule assembly and microtubule-dependent surface properties of human neutrophils. J. Cell. Biol. 71:921 (1976).
21. A.A. Voetman, and D. Roos, The importance of endogenous catalase in the protection of human blood phagocytes against oxidative damage by extracellularly-generated hydrogen peroxide. Blood 56:846 (1980).
22. H. Frischer, and T. Ahmad, Severe generalized glutathione reductase deficiency after antitumor chemotherapy with BCNU [1,3-bis(chloroethyl)-1-nitrosourea]. J. Lab. Clin. Med. 89:1080 (1977).
23. J.R. Babson, and D.J. Reed, Inactivation of glutathione re-

ductase by 2-chloroethyl nitrosourea-derived isocyanates. Biochem. Biophys. Res. Commun. 83:754 (1978).
24. D. Roos, R.S. Weening, A.A. Voetman, M.L.J. van Schaik, A.A.M. Bot, L.J. Meerhof, and J.A. Loos, Protection of phagocytic leukocytes by endogenous glutathione: studies in a family with glutathione-reductase deficiency. Blood 53:851 (1979).
25. A.W. Segal, J. Clark, and A.C. Allison, Tracing the fate of oxygen consumed during phagocytosis by human neutrophils with $^{15}O_2$. Clin. Sci. Mol. Med. 55:413 (1978).
26. M. Reiss, and D. Roos, Differences in oxygen metabolism of phagocytosing monocytes and neutrophils. J. Clin. Invest. 61: 480 (1978).
27. B.M. Babior, Superoxide production by phagocytes. Another look at the effect of cytochrome c on oxygen uptake by stimulated neutrophils. Biochem. Biophys. Res. Commun. 91:222 (1979).
28. T.R. Green, R.E. Schaeffer, and M.T. Makler, Significance of O_2 availability and cycling on the respiratory burst response of human PMN's exposed to cytochrome c and superoxide dismutase. Biochem. Biophys. Res. Commun. 94:1213 (1980).
29. A.W. Segal, and O.T.G. Jones, Novel cytochrome b system in phagocytic vacuoles of human granulocytes. Nature 276:515 (1978).
30. M.N. Hamers, R. Wever, M.L.J. van Schaik, R.S. Weening, and D. Roos, Cytochrome b and the superoxide-generating system in human neutrophils, in: "Biological and Clinical Aspects of Superoxide and Superoxide Dismutase", W.H. Bannister, and J.V. Bannister, eds., Elsevier/North-Holland, New York/ Amsterdam/Oxford, 242 (1980).

DISCUSSION

Edited by: R.E. Basford

Dr. Markert:

What was the concentration of cytochrome c, as it was shown by Babior that the number of cells and cytochrome c concentration were critical for efficient trapping of superoxide anion?

Dr. Roos:

We used 150 μM cytochrome c and measured for only a few minutes without and a few minutes with SOD. Therefore, cytochrome c certainly cannot have been rate-limiting in our experiments.

Dr. Minakami:

Your assay of O_2^- in whole cells is that of the extracellular O_2^- and not of total O_2^- formation. You cannot discuss the mode of O_2 reduction, univalent or divalent, from your data. I recommend you to add cytochalasin B in your system. It prevents the internalization of the NADPH oxidase, so that the yield of O_2^- may increase.

Dr. Roos:

I agree that this is a possibility. However, we have also studied cells stimulated with PMA, and the amount of cytochrome c reduction was only slightly higher than with opsonized zymosan. Therefore, I do not think that internalization of the oxidase can explain our results.

Dr. De Chatelet:

I don't really feel that your statement about univalent reduction being a relatively unimportant pathway is warranted. Don't you feel that it is reasonable to expect that only a part of the O_2^- would be elaborated outside the cell? The molecule is so reactive that I would not expect to see much outside the phagocytic vacuole.

Dr. Roos:

As I answered to the question of Dr. Minakami, we obtained practically the same results with PMA-stimulated PMN. Therefore, presence or absence of phagocytic vauoles does not make much difference.

Dr. Rossi:

I agree with Dr. Roos in regard to the amount of H_2O_2 degraded by the glutathione cycle in human granulocytes. As far as the production of O_2^- and the consumption of O_2 by human granulocytes, the ratio of 4:1 indicates that in your system, the sequence of events taking place is:

1) univalent reduction of O_2 by the oxidase
2) dismutation
3) degradation of H_2O_2 by catalase

Dr. Roos:

Since the assays of O_2 and H_2O_2 were carried out in the presence of azide, the degradation of H_2O_2 by catalase is impossible. The cytochrome c reduction was done without azide. Therefore, the ratio of O_2 consumed to O_2^- produced should be 2 in the case of univalent reduction to O_2^- followed by dismutation to H_2O_2 and 1 in case of divalent reduction of O_2 to H_2O_2. As I have shown, however, in homogenates we found a 4:1 ratio. To us, this indicates that cytochrome c accelerates the flow of electrons in the damaged oxidase of homogenates.

Dr. Kakinuma:

How did you measure the H_2O_2 produced by leukocyte homogenate?

Dr. Roos:

We measured the H_2O_2 with a fluorometric assay, based on the oxidation of leuko-diacetyl-dichlorofluorescein (LDADCF) by H_2O_2 and horse-radish peroxidase (see Homan-Müller et al., J. Lab. Clin. Med. 85: 198 (1975)).

Dr. Kakinuma:

I wonder whether you measured a real H_2O_2 concentration in the homogenate system. In general, fluorometric assays have been used to measure the H_2O_2 generation by intact cells. Fluometric assays involve the use of a fluorescent dye which acts as hydrogen donor for the horse-radish peroxidase reaction. However, subcellular fractions or cell homogenates contain endogenous hydrogen donors which will react with horse-radish peroxidase-H_2O_2 compound and compete with the fluorescent dye. Furthermore, the H_2O_2 production by subcellular fractions was initiated by the addition of NADPH, which will also be able to compete with indicator dyes. Therefore, the fluorescent assay methods do not provide a precise value of H_2O_2 concentration in the subcellular system.

Dr. Roos:

We dilute the samples 31-fold before mixing with the dye and the peroxidase. therefore, neither endogenous nor exogenous hydrogen donors affect the oxidation of the dye. This was checked with pilot samples of H_2O_2 and with acetaldehyde plus xanthine oxidase, as an H_2O_2-generating system. Moreover, the H_2O_2 generation by cell homogenates is matched by the oxygen consumption (see Table 2).

NEUTROPHIL ACTIVATION STUDIED USING TWO INDIRECT PROBES OF MEMBRANE POTENTIAL WHICH RESPOND BY DIFFERENT FLUORESCENCE MECHANISMS

Bruce Seligmann and John I. Gallin

Bacterial Diseases Section, Laboratory of Clinical Investigation, National Institute of Allergy and Infectious Diseases, National Institutes of Health, Bethesda, M.D. 20205

INTRODUCTION

The involvement of ions in the activation of neutrophils (PMN) has been studied by monitoring cell membrane potential.[1-11] The relationship of membrane potential to ionic gradients and permeabilities is given by the Goldman Equation and is the basis from which changes in membrane potential can be directly related to ion flux changes:

$$\text{Cell membrane potential} = \frac{-RT}{F} \log \frac{P_K[K^+]_i + P_{Na}[Na^+]_i + P_{Cl}[Cl^-]_o + \ldots}{P_K[K^+]_o + P_{Na}[Na^+]_o + P_{Cl}[Cl^-]_i + \ldots}$$

The factor $\frac{RT}{F}$ is a constant consisting of R, the gas constant; T, temperature; and F, a faraday. The intracellular (i) and extracellular (o) concentration of each ion is indicated, weighted by the permeability (P) of each ionic species. Thus, by measuring membrane potential and varying the extracellular concentration of specific ions, one can determine the ionic flux changes giving rise not only to resting membrane potential but also to stimulus-induced changes in membrane potential. Membrane potential of cultured macrophages has been measured by penetrating cells with microelectrodes.[12-16] However, PMN are too small to study in this way with the currently available technology. Therefore, it has been necessary to use indirect molecular probes to study PMN membrane potential.[1-11] The indirect probes currently in use are charged, lipophilic ions which exhibit some property enabling determination of their intracellular concentration. Two classes of such probes are the fluorescent cyanine dyes and radiolabeled probes such as triphenyl methyl phosphonium ion. A number of reports have appeared using these probes to study PMN membrane potential. However, before conclusive interpretations of results can be drawn, it must be shown that the responses are not affected by cellular artifacts independent of membrane potential.

This study demonstrates that two cyanine dyes, 3,3'-dipentyl-oxacarbocyanine [di-O-C_5(3)] and 3,3'-dipropyl-thiocarbocyanine [di-S-C_3(5)] (Fig. 1), can be utilized to study PMN by two different fluorescence mechanisms. We also find that di-S-C_3(5), a cyanine dye used by a number of laboratories to study PMN, is affected adversely by superoxide anion ($O_2^{\cdot}$), peroxide (H_2O_2), or related reactive species consistent with a previous report.[11] However, we demonstrate that this problem can be adequately compensated for by conducting the proper control experiments utilizing cysteamine. Considered together, the results do not support the possibility that cellular effects (unrelated to membrane potential) on probe fluorescence account for the observed responses. Thus, we suggest that the responses stimulated by the chemoattractant N-formyl-methionyl-leucyl-phenylalanine (f-met-leu-phe) and the secretagogue phorbol myristate acetate (PMA) are consistent with changes in membrane potential. However, a definitive characterization of the ionic basis for these responses remains to be achieved, and only with such characterization will it be possible to conclude that the responses accurately reflect membrane potential changes.

MATERIALS AND METHODS

Assays were carried out using phosphate-buffered Hanks' solutions as described previously.[5,6] The croton oil derivative PMA (Consolitated Midland Corp., Brewster, N.Y.), f-met-leu-phe (Sigma Chemical Co., St. Louis, Mo.), and valinomycin (Sigma) were dissolved in dimethyl sulfoxide (DMSO) (Fisher Scientific Co., Pittsburgh, Pa.) to make appropriate stock solutions such that the final concentration of DMSO in the reaction mixture was 0.1%. This amount of DMSO did not have any effect on secretion, $O_2^{\cdot}$ generation, cell viability as measured by trypan blue exclusion and lactate dehydrogenese release, or the fluorescence responses.

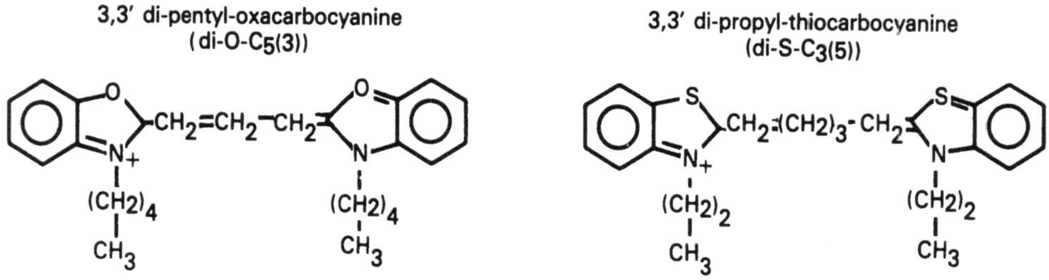

Fig. 1. Molecular structure of di-O-C_5(3) and di-S-C_3(5).

Obtained as follows were Hypaque (Winthrop Laboratories, N.Y.), Ficoll and T250 Dextran (Pharmacia, Uppsala, Sweden). Gramicidin D (Sigma), di-O-C_5(3) and di-S-C_3(5) (provided by Dr. Alan Waggoner, Amherst College, Amherst, Mass.) were dissolved in ethanol to prepare stock solutions from which at least 1000-fold dilutions were made. Superoxide dismutase (SOD), catalase, cysteamine, xanthine, and xanthine oxidase were all obtained from Sigma and dissolved in Hanks' buffer.

Leukocyte Isolation

Peripheral blood PMN from normal human subjects were prepared by Ficoll-Hypaque gradient centrifugation followed by Dextran sedimentation.[5,17] This technique routinely resulted in cell populations containing over 95% PMN. All experiments were completed within 6 hr after obtaining the cells. PMN were kept on ice in calcium- and magnesium-free Hanks' solution to minimize cell aggregation and then were incubated at 37°C for 15 min in divalent cation containing medium with the appropriate probe. The cells were not exposed to the calcium/magnesium containing medium at 37°C for longer than 45 min.

Fluorometric Assay of Membrane Potential

The procedure used has been published previously.[5,6] Unless otherwise indicated, the concentration of di-O-C_5(3) was 2.5×10^{-7} M and of di-S-C_3(5) was 10^{-6} M. Fluorescence was recorded using a modified Farrand Mark I spectrophotometer (Farrand Optical Co., Inc., Valhalla, N.Y.) which enabled automatic sequential determinations of fluorescence from eight cuvettes. The cell suspensions were maintained at 37°C by stirring. The excitation and emission wavelengths used were 460/510 nm for di-O-C_5(3) and 620/660 nm for di-S-C_3(5). Fluorescence changes are reported as a percentage of the total full-scale fluorescence. Cell concentration, viability, and extent of aggregation were determined microscopically at the end of each experiment using trypan blue and phase-contrast microscopy. Aggregation was also monitored continuously during the experiment as a function of the signal noise level; aggregation causes an increase in signal noise. Additions of cells or stimuli to the cuvette required the momentary closing of the photomultiplier tube, causing a spike in the tracing of fluorescence at the point of each addition.

RESULTS

Comparison of di-O-C_5(3) and di-S-C_3(5)

Equilibration of di-O-C_5(3) (2.5×10^{-7}) with neutrophils results in an enhancement of total fluorescence without any change in the absorbance spectra of the dye (Figs. 2A and 3A, respectively). Cyanine dyes form nonfluorescent dimers with a different absorbance maximum when the concentration reaches a sufficiently high critical level. This concentration depends both on the specific structure of the cyanine dye molecule and the nature of the solvent. Since in cells with a negative membrane potential the concentration of

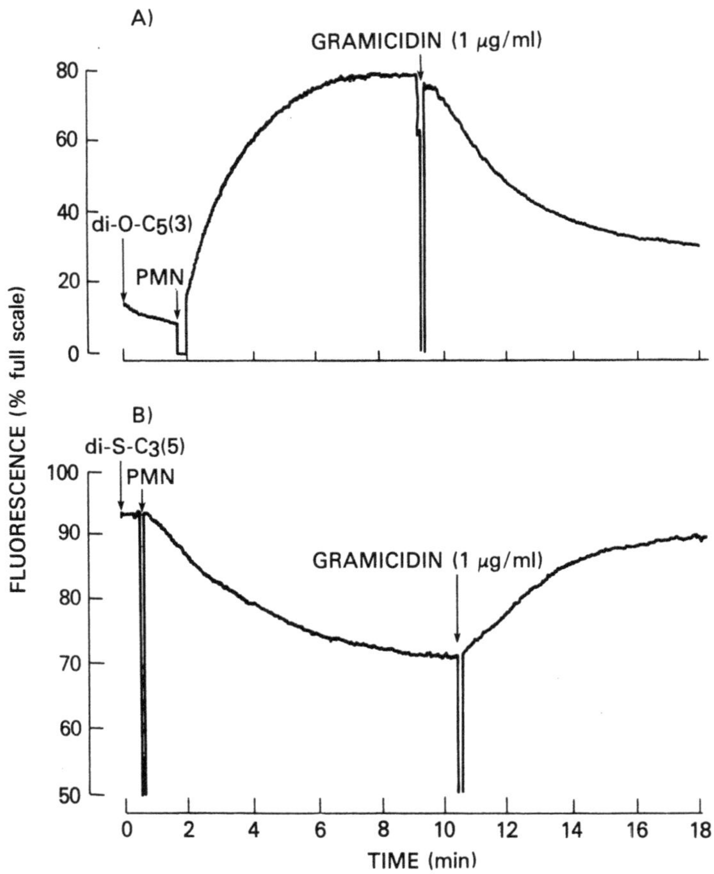

Fig. 2. Time course of fluorescence changes during equilibration of dyes with PMN and subsequent treatment with gramicidin D. As repeated in subsequent figures showing time courses of dye fluorescence, total fluorescence is plotted on the vertical axis as a percent of the full-scale fluorescence and time is recorded along the horizontal axis (min). Addition of dye, PMN, and gramicidin D is indicated. The spikes seen in the fluorescence tracings at the point of each addition are caused by the momentary closing of the photomultiplier tube. (A) Effect of PMN and gramicidin D on di-O-C_5(3) fluorescence. (B) Effect of PMN and gramicidin D on di-S-C_3(5) fluorescence.

probe inside is greater than the concentration outside and since solvent environment changes, the possibility exists that dye may form dimers when accumulated by cells. Fluorescence is also affected by the solvent environment, independent of dimerization. Nonpolar solvents enhance cyanine dye fluorescence so that it is possible for limited dimerization and fluorescence quenching to occur without a net decrease in total fluorescence. The absorbance spectra shown in Fig. 3A indicate that there is no dimerization of di-O-C_5(3). Since no dimerization occurs in the resting cell and since the fluorescence of cell-associated di-O-C_5(3) molecules is greater than that of molecules remaining in the supernate, the total fluorescence recorded results primarily from cell-associated dye fluorescence as previously reported.[5]

In contrast, the accumulation of di-S-C_3(5) (10^{-6} M) results in quenching of the total suspension fluorescence (Fig. 2B) and a large change in the absorbance spectra (Fig. 3B) compatible with the formation of nonfluorescent dimers of di-S-C_3(5). The fluorescence which is monitored is therefore that of di-S-C_3(5) remaining in the supernate and not that of probe molecules associated with the cells, the opposite of the di-O-C_5(3) mechanism of fluorescence. This is

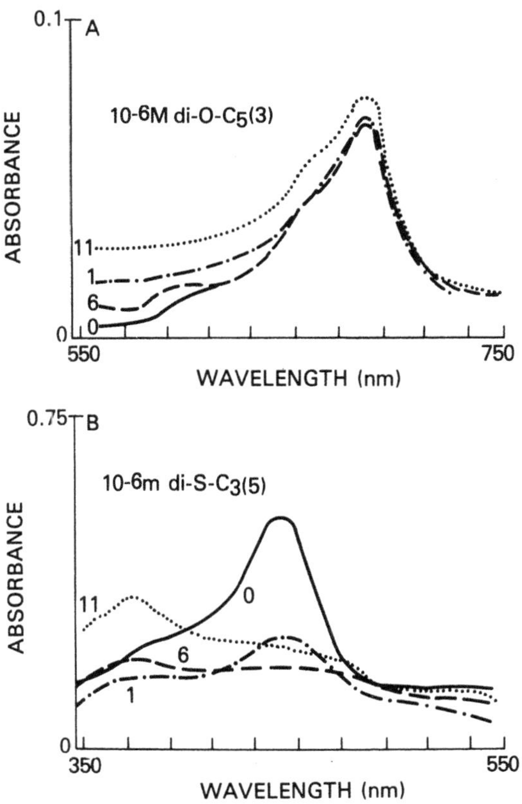

Fig. 3. Effect of equilibration of PMN on the absorbance spectra of di-O-C_5(3) and di-S-C_3(5). Successive absorbance spectra of both di-O-C_5(3) (A) and di-S-C_3(5) (B) are plotted. Spectra were obtained prior to addition of PMN (0 time) and subsequent to addition at 1, 6, and 11 min. Absorbance is plotted along the vertical axis and wavelength along the horizontal axis.

demonstrated in Fig. 2A and B by the effect the sodium/potassium ionophore gramicidin D (1 μg/ml) has on di-O-C_5(3) and di-S-C_3(5) fluorescence. This ionophore typically depolarizes cells, causing the cell membrane potential to become less negative, resulting in loss of positively charged probes. Gramicidin D decreased di-O-C_5(3) fluorescence and increased di-S-C_3(5) fluorescence consistent with the opposite fluorescence mechanisms of these two probes.

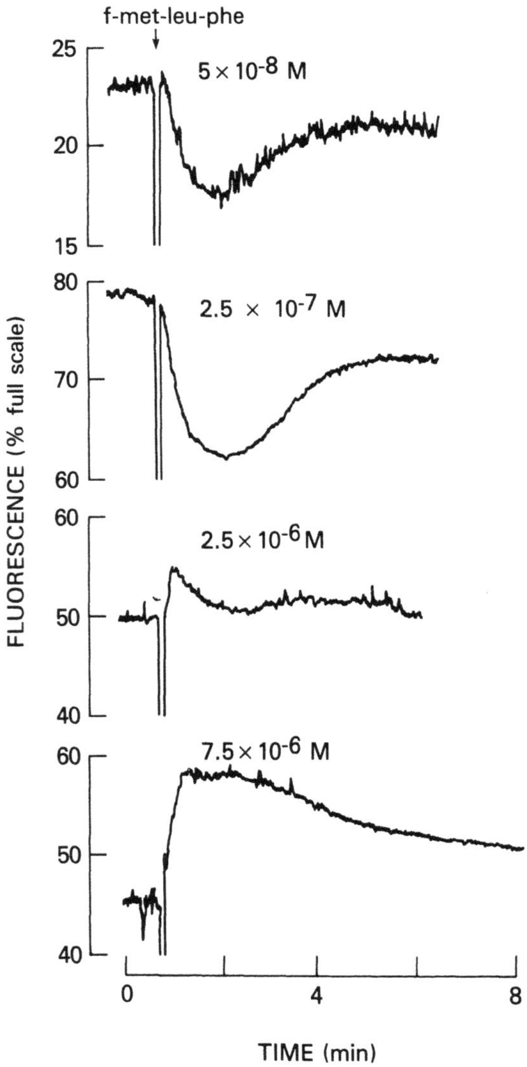

Fig. 4. Effect of di-O-C_5(3) concentration on the fluorescence response elicited by the chemoattractant f-met-leu-phe. Time courses of fluorescence are shown beginning 1 min before addition of stimulus (5×10^{-7} M f-met-leu-phe). di-O-C_5(3) concentration was varied from 5×10^{-8} M (top tracing) to 7.5×10^{-6} M (bottom tracing) as indicated. Photomultiplier sensitivity was varied to maintain the recordings on scale.

An additional demonstration of the behavior of di-O-C_5(3) is shown in Fig. 4. Dimerization occurs only when the dye concentration is sufficiently high. Therefore, by raising the concentration in a stepwise manner and stimulating with an agent that causes potential changes, a progressive change in the fluorescence mechanism toward dimerization should occur, resulting in an inversion of the response. The chemoattractant f-met-leu-phe causes a change in di-O-C_5(3) fluorescence which is reflected in loss of di-O-C_5(3) fluorescence followed by an increase as previously reported[5,6,18] and shown in Fig. 4 (top panel). This result has been interpreted as an apparent depolarization followed by a hyperpolarization. When the concentration of di-O-C_5(3) was raised successively up to 2.5×10^{-6} M, the fluorescence response became progressively more complex (mixed response) and then inverted to the mirror image (bottom panel). Thus, the fluorescence response behaved as predicted due to the change in dye fluorescence mechanism of cell-associated di-O-C_5(3) to a mechanism involving dimer formation. Records of absorbance spectra taken immediately before and at 2 min following activation confirmed that dimerization did occur at the higher but not lower di-O-C_5(3) concentrations (results not shown).

The two opposite fluorescence mechanisms of di-O-C_5(3) (low dye concentrations) and di-S-C_3(5) were used to advantage in subsequent experiments in order to demonstrate that the apparent changes in membrane potential stimulated by various activating substances were not artifacts due to cellular effects on dye fluorescence. As shown in Fig. 5 (control tracing), the di-S-C_3(5) fluorescence response to f-met-leu-phe is the exact opposite of the di-O-C_5(3) response (Fig. 4). Taking into account the different fluorescence mechanisms, these responses are both interpreted to reflect a transient loss of probe (apparent depolarization of membrane potential), followed by partial reaccumulation of probe (apparent repolarizing hyperpolarization of membrane potential).

Effect of Reactive Species on Cyanine Dye Fluorescence

Activation of PMN is associated with the production of reactive species such as $O_2^{\overline{\cdot}}$, H_2O_2, hydroxyl radical ($OH^{\cdot}$), and singlet oxygen ($^1O_2$) as well as secretion of granule contents under appropriate conditions. Therefore, control studies were conducted to establish whether neutrophil products affected dye fluorescence, independent of membrane potential. We have shown previously that there is no effect of reactive species on di-O-C_5(3) fluorescence,[6] and therefore the studies presented here deal mainly with di-S-C_3(5).

A cell-free system of xanthine plus xanthine oxidase, producing predominately $O_2^{\overline{\cdot}}$ but also H_2O_2, $OH^{\cdot}$, and $^1O_2$, was employed in addition to PMN. As shown in Fig. 6A, there is no effect of reactive species on di-O-C_5(3) fluorescence. However, in striking contrast,

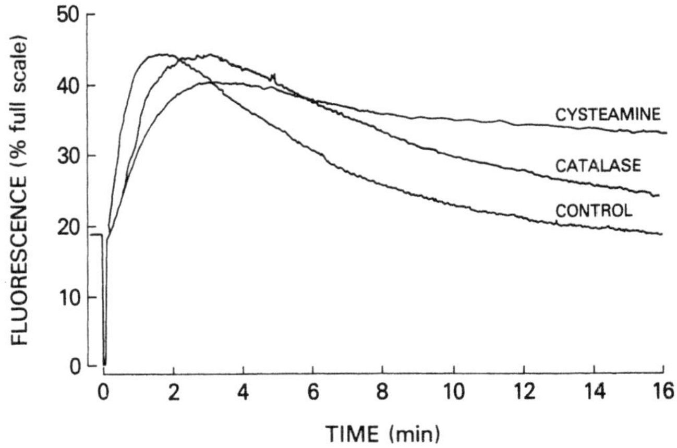

Fig. 5. Effect of the chemoattractant f-met-leu-phe on the time course of di-S-C_3(5) fluorescence. f-Met-leu-phe (5×10^{-7} M) was added at 0 time. The fluorescence responses obtained in the absence of scavengers (control) or presence of cysteamine (10 mM) or catalase (250 U/ml) are shown.

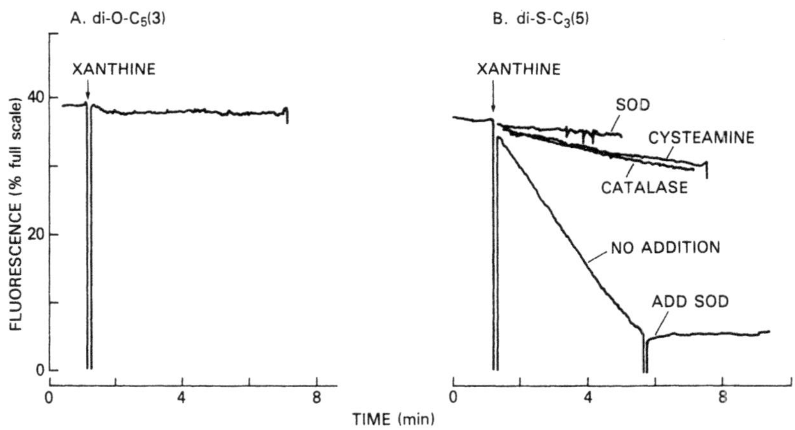

Fig. 6. Effect the cell-free generation of reactive species by xanthine plus xanthine oxidase has on dye fluorescence. (A) Fluorescence of di-O-C_5(3) (2.5×10^{-7} M) plus xanthine oxidase (5 U/ml) prior and subsequent to the addition of xanthine (10 mM). (B) Fluorescence of di-S-C_3(5) (10^{-6} M) plus xanthine oxidase prior and subsequent to the addition of xanthine either without scavengers present (no addition) or with SOD (300 U/ml), catalase (250 U/ml), or cysteamine (10 mM) present. In the case of no addition SOD was added subsequently at 6 min, demonstrating that the quenching of fluorescence could be stopped by this scavenger alone. di-S-C_3(5)

fluorescence was quenched completely by this cell-free $O_2^{\cdot}/H_2O_2/OH^{\cdot}$ generating system (Fig. 6B). Addition of SOD to the system reduced quenching, as did catalase and cysteamine. None of these agents actually inhibit the production of $O_2^{\cdot}$ or H_2O_2 but instead they utilize and thereby destroy $O_2^{\cdot}$ or H_2O_2 in competition with the agent [cytochrome c, di-S-C_3(5), or scopoletin plus horseradish peroxidase] used to detect their production. Therefore, the actual effectiveness of the scavengers on the cell-free generation of $O_2^{\cdot}/H_2O_2$ was measured (results not shown). SOD effectively scavenged all the detectable $O_2^{\cdot}$ produced and enhanced the amount of H_2O_2 generated (a product of the reaction of $O_2^{\cdot}$ with SOD). Catalase reduced the amount of $O_2^{\cdot}$ detected and scavenged all the detectable H_2O_2 produced. Cysteamine, reported to scavenge $O_2^{\cdot}$, reduced cytochrome c itself, and thus its effect on $O_2^{\cdot}$ could not be directly determined. However, cysteamine prevented any detectable formation of H_2O_2 and also inhibited chemiluminescence of PMN stimulated with PMA. These results suggest that quenching of di-S-C_3(5) fluorescence by the cell-free system is mediated primarily by $O_2^{\cdot}$ or possibly $OH^{\cdot}$; otherwise, SOD would not have been as effective as it was in preventing quenching of fluorescence. However, H_2O_2 also apparently quenches di-S-C_3(5) fluorescence since catalase also had an effect. Experiments with di-S-C_3(5) and PMN were therefore carried out with and without these scavengers present.

The di-S-C_3(5) response to f-met-leu-phe in the presence of cysteamine or catalase shown in Fig. 5 demonstrates that generation of reactive species affects the response, though not to a great extent. This is consistent with the observation that only a relatively small and transient amount of these reactive species is generated by PMN in response to f-met-leu-phe. In contrast, stimulation of PMN with PMA (20 ng/ml) results in the production of large amounts of $O_2^{\cdot}$ and H_2O_2 lasting for up to 2 hr. In these studies the amount of H_2O_2 formed in response to PMA was more than ten times that produced by the cell-free system and by 10 min the H_2O_2 formed was also ten times the total amount produced by PMN stimulated with f-met-leu-phe. Fig. 7A shows the di-S-C_3(5) fluorescence response of PMN stimulated with PMA. When no scavengers were present, the response consisted of an initial increase in fluorescence (apparent depolarization) followed by a complete quenching of fluorescence to zero. Catalase dramatically reduced the quenching of fluorescence while cysteamine prevented it entirely and SOD had minimal effect. The di-S-C_5(3) response in the presence of cysteamine is identical to that of di-O-C_5(3) (Fig. 7B), except inverted, as expected because of the different fluorescence mechanisms of the two probes. Thus, with appropriate scavengers present the interpretation of the di-S-C_3(5) responses to both stimuli are identical to the di-O-C_5(3) responses even though the actual fluorescence changes are opposite.

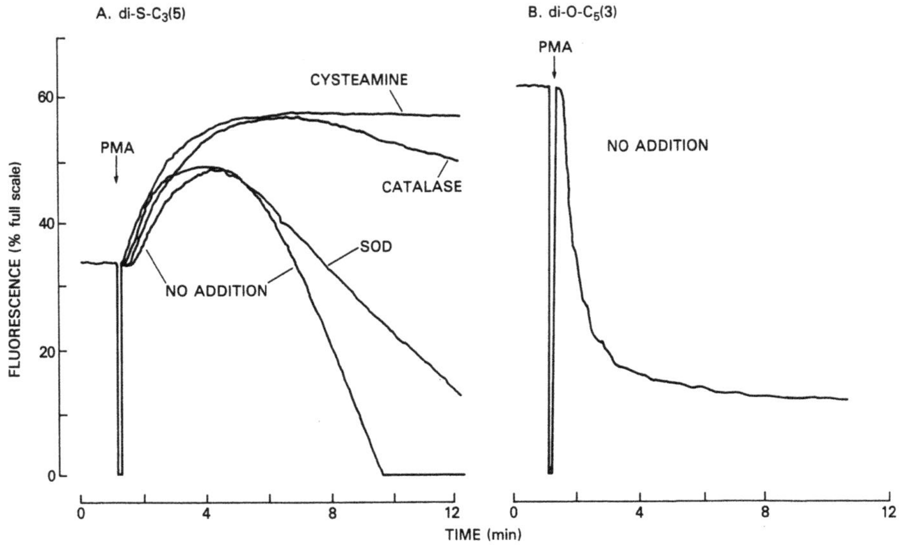

Fig. 7. Effect of the secretagogue PMA on the time course of dye fluorescence. Time courses of fluorescence are shown beginning approximately 1 min prior to the addition of PMA (20 ng/ml). (A) Effect on di-S-C_3(5) (10^{-6} M) fluorescence without any scavengers present and in the presence of cysteamine (10 mM), catalase (250 U/ml), or SOD (300 U/ml). Note that without any addition of scavengers di-S-C_3(5) fluorescence was completely quenched by 8 min after addition of PMA. (B) Effect on di-O-C_5(3) (2.5×10^{-7} M) fluorescence without any scavengers present.

DISCUSSION

The results shown in Figs. 2-4 demonstrate the differences between the di-O-C_5(3) and di-S-C_3(5) fluorescence mechanisms when used as we have described. di-O-C_5(3) fluorescence originates primarily from cell-associated dye molecules while di-S-C_3(5) fluorescence originates primarily from molecules remaining in the supernate. As shown in Fig. 4, this difference is caused by the assay conditions selected, and at high concentrations of di-O-C_5(3) the mechanism is the same as with di-S-C_3(5). Recognition of the factors influencing fluorescence is necessary for proper interpretation and understanding of the results. Accumulation of di-O-C_5(3) by cells is reflected by an increase in total fluorescence. In contrast, accumulation of di-S-C_3(5) is reflected by a decrease in total fluorescence as fluorescent dye is lost from the extracellular environment. Therefore, the interpretation of the "opposite" responses to f-met-leu-phe of di-O-C_5(3) and di-S-C_3(5) fluorescence shown in Figs. 4 and 5 is the same, reflecting loss of both dyes followed by their partial reaccumulation. Intracellular effects on dye fluorescence cannot explain these results since both probes behave identically but only di-O-C_5(3) fluorescence is subject to intracellular effects.

Extension of the dual probe approach to PMA demonstrated an important point bearing on interpretation of di-S-C_3(5) fluorescence results. $O_2^{\bar{}}$ or related species quenches di-S-C_3(5) but not di-O-C_5(3) fluorescence (Figs. 5-7), necessitating control experiments incorporating scavengers, particularly cysteamine. The response of di-S-C_3(5) to PMA in the absence of scavengers is misleading (Fig. 7A), but with scavengers present the interpretation is the same as di-O-C_5(3) (Fig. 7A and B).

It is obvious that the nature of the scavenger used is important. Neither catalase nor SOD is sufficiently effective. In contrast, cysteamine appears to be an effective scavenger of the reactive products produced by the PMN system which affect di-S-C_3(5) fluorescence. This may be due to the fact that cysteamine, unlike catalase and SOD, is freely permeable across the cell membrane and hence distributes itself in the same intracellular spaces as di-S-C_3(5). We conclude that use of di-S-C_3(5) as a probe of neutrophil membrane potential must be accompanied with control experiments utilizing scavengers, particularly cysteamine. Control experiments with catalase or SOD are not sufficient.

The fluorescence responses observed using the cyanine dyes may also be influenced by changes in the lipid content of the plasma membrane. Both stimuli used in this study alter PMN arachidonate metabolism,[19,20] and it is possible that this accounts for some of the changes in probe distribution. Alternatively, changes in membrane lipids may change ion permeability and membrane potential.

To conclude, we have shown that di-S-C_3(5) and di-O-C_5(3) respond by different fluorescence mechanisms and that reactive species produced by PMN affect di-S-C_3(5) but not di-O-C_5(3) fluorescence. However, by controlling this we have been able to demonstrate that both probes give the same results, which is strong evidence, particularly in view of previous studies, that f-met-leu-phe and PMA cause PMN to undergo changes in membrane potential.

REFERENCES

1. S. T. Hoffstein, H. M. Korchak, and J. E. Smolen, Rapid membrane and cytoskeletal changes in human neutrophils during stimulus induced secretion, in: "Membranes and Cytoskeleton: Role in Pathologic Processes," N. S. McNutt and S. T. Hoffstein, eds., Fed Proc. 40:209 (1981).

2. G. S. Jones, K. Van Dyke, and V. Castranova, Purification of human granulocytes by centrifugal elutriation and measurement of transmembrane potential, J. Cell Physiol. 104:425 (1980).
3. H. M. Korchak and G. Weissmann, Changes in membrane potential of human granulocytes antecede the metabolic responses to surface stimulation, Proc. Natl. Acad. Sci. U.S.A. 75:3818 (1978).
4. H. M. Korchak and G. Weissmann, Activation of polymorphonuclear leukocytes (PMN): role of membrane potential and extracellular Na^+, Fed. Proc. 38:1362 (1979).
5. B. Seligmann, E. K. Gallin, D. L. Martin, W. Shain, and J. I. Gallin, Interaction of chemotactic factors with human PMN: studies using a membrane potential sensitive cyanine dye, J. Membr. Biol. 52:257 (1980).
6. B. Seligmann and J. I. Gallin, Use of lipophilic probes of membrane potential to assess human neutrophil activation, J. Clin. Invest. 66:493 (1980).
7. B. Seligmann and J. I. Gallin, Use of indirect probes of membrane potential in human neutrophils: comparison of di-O-C_5(3) and triphenyl methyl phosphonium ion, in: "Membrane and Cytoskeleton: Role in Pathologic Processes", N. S. McNutt and S. T. Hoffstein, eds., Fed. Proc. 40:211 (1981).
8. L. Simchowitz, J. P. Atkinson, and I. Spilberg, Stimulus-specific deactivation of chemotactic factor-induced cyclic AMP response and superoxide generation by human neutrophils, J. Clin. Invest. 66:736 (1980).
9. P. Tatham, P. Delves, L. Shen, and I. Roitt, Chemotactic factor-induced membrane potential changes in rabbit neutrophils monitored by the fluorescent dye 3-3'-dipropylthiadicarbocyanine iodide, Biochim. Biophys. Acta 602:285 (1980).
10. K. Utsumi, K. Sugiyamm, M. Miyahara, M. Naito, M. Auai, and M. Inone, Effect of concanavalin A on membrane potential of PMN monitored by fluorescent dye, Cell Struct. Funct. 2:203 (1977).
11. J. L. Whitin, C. B. Chapman, B. A. Simons, M. Choaniec, and H. J. Cohen, Correlation between membrane potential changes and superoxide production in human granulocytes stimulated by phorbol myristate acetate, J. Biol. Chem. 255:1874 (1980).
12. E. K. Gallin and J. I. Gallin, Interaction of chemotactic factors with human macrophages: induction of transmembrane potential changes, J. Cell Biol. 75:277 (1977).
13. E. K. Gallin, B. Seligmann, and J. I. Gallin, Alteration of macrophage and monocyte membrane potential by chemotactic factor, in: "Mononuclear Phagocytes: Functional Aspects," R. Van Furth, ed., Martinus Nijhoff Publishers, The Hague, Netherlands, vol. I, p. 505 (1980).

14. E. K. Gallin and D. R. Livengood, Inward rectification in mouse macrophages; evidence for a negative resistance region, Am. J. Physiol.: Cell Physiol. In press (1981).
15. J. Kouri, M. Noa, B. Diaz, and E. Ninbo, Hyperpolarization of rat peritoneal macrophages phagocytasing latex particles, Nature 283:868 (1980).
16. E. K. Gallin, Calcium spikes in cultured human reticular cells from peritoneal exudates. J. Cell. Physiol. In press (1981).
17. A. Boyum, Isolation of mononulcear cells and granulocytes from human blood, Scand J. Clin. Lab. Invest. 97 (Suppl. 29):77 (1968).
18. B. Seligmann and J. I. Gallin, Secretagogue modulation of the response of human neutrophils to chemoattractants: studies with a membrane potential sensitive cyanine dye, Mol. Immunol. 17:191 (1980).
19. R. Snyderman and M. C. Pike, Role of transmethylation reactions in cellular motility and phagocytosis, Mol. Immunol. 17:209 (1980).
20. F. Hirata, B. A. Corcoran, R. Venkatasabramanium, E. Schiffmann, and J. Axelrod, Chemoattractants stimulate degradation of methylated phospholipids and release of arachidonic acid in rabbit leukocytes, Proc. Natl. Acad. Sci. U.S.A. 76:2640 (1979).

DISCUSSION
Edited by: R.E.Basford

Dr. De Chatelet:
Have you examined the alteration in membrane potential in CGD cells exposed to f-Met-Leu-Phe and PMA?

Dr. Seligmann:
Yes. The report appears in the Sept. issue of the J. Clin. Invest., 1980. We find that the potential response elicited by f-Met-Leu-Phe is severely reduced or entirely absent in CGD. Similarly, the PMA response is severely diminished. In contrast to these stimuli, the calcium ionophore A23187 produced a nearly normal response, as might be expected, since this agent depolarizes all cells independently of any active cellular response mechanisms.

Dr. Segal:
Could you please define the abnormalities you have detected in CGD.

Dr. Seligmann:
Refer to the reference in the answer to Dr. De Chatelet's question (above).

Dr. Elsbach:
Are your results with respect to kinetics and with respect to differences between normal and CGD neutrophils different in any way from those of the Boston group?

Dr. Seligmann:
Whitin _et al_ have also recently reported that the response of CGD PMN to stimulation with PMA is reduced, and in this respect their results are totally consistent with our findings.

The only difference lies in the PMA response stimulated from normal cells. They employed di-S-C_3(5) and found a biphasic response of apparent depolarization (increase in fluorescence) followed by hyperpolarization (quenching of fluorescence). They found that anaerobic conditions or catalase reduced but by no means eliminated the quenching of fluorescence, or apparent hyperpolarization. It is in this respect that our results differ. We can entirely eliminate this component with catalase or cysteamine, an intracellular scavenger of O_2^-, so that the PMA-di-S-C_3(5) response is reduced to a simple monophasic response of a permanent increase in fluoroescence or depolarization, identical to the di-O-C_5(3) response.

Dr. Valerius:
Did you study individuals heterozygotic with respect to CGD, and did you include in your series of CGD patients also patients with the non-X-linked type? What were the results?

Dr. Seligmann:

We have studied a total of 11 patients with CGD, three of whom are female. All have the same abnormality in elicited membrane potential responses.

We are presently studying the families of these patients to determine whether heterozygote conditions can be detected. In two instances we believe, based on O_2^- levels, that we have identified heterozygotes; in these instances the potential responses also appear to be intermediate between normal and CGD PMN's. Thus we feel the membrane potential response is reduced in heterozygotes. We have also studied other syndromes such as MPO deficient subjects, Chediak Higashi, and hyper IgE syndrome. In these patients the responses are normal, though in CHS the kinetics of the response are slowed down.

Dr. Roos:

A comment and a question. Dr. Wim Bleeker in our department has collected data that completely confirm your results on the degradation of the di-S compound and on the abnormal response by CGD cells. In contrast, he found a normal response by cells with the so-called "initiation defect" (Weening *et al*., J. Lab. Clin. Med., 1976) which again proves that these cells are different from CGD cells.

My question is: do you agree that, judging from the normal signal caused by A23187 in CGD cells, the change in membrane potential is a necessary but not sufficient signal for activation of neutrophils?

Dr. Seligmann:

Yes. However, we feel that the A23187 is slightly abnormal due to the absence in CGD of an activated cellular mechanism affecting membrane potential, which *is* activated by A23187 in normal cells. Further evidence that depolarization alone is not sufficient to stimulate O_2^- is that gramicidin D, which causes a large sodium-dependent depolarization, does *not* stimulate PMN's to produce O_2^-.

APPARENT Km OF LEUKOCYTE O_2^- AND H_2O_2 FORMING ENZYME FOR OXYGEN

Katsuko Kakinuma and Mizuho Kaneda

The Tokyo Metropolitan Institute of
Medical Science,
Honkomagome, Bunkyo-ku, Tokyo, 113

INTRODUCTION

Polymorphonuclear leukocytes (PMN) exhibit during phagocytosis a marked burst in oxidative metabolism, i.e., an increase in both oxygen uptake and in glucose oxidation through the hexose monophosphate shunt (HMP)[1], that involves the generation of H_2O_2 and O_2^-.[2,3] Studies of the enzyme mechanism underlying the activation of the oxidative metabolism in PMN have regarded NADH [4-6], or NADPH[7,8] as electron donor for the reduction of oxygen.

We have previously analyzed the kinetic properties of NAD(P)H oxidase activities by following the rates of H_2O_2 and O_2^- production. We observed a greater involvement of an NADPH- oxidase occuring in the granular (particle) fraction of phagocytosing PMN.[9] Both H_2O_2 and O_2^- generation by the particle fraction showed apparent Km for NADPH and NADH of approx. 26 and 124 μM, respectively.[10] A greater involvement of an NADPH-oxidase activity is supported by the five-fold lower Km for NADPH, in agreement with the data reported by Rossi and his coworkers.[7,8]

Here, we analyzed further the kinetic properties of the NADPH oxidase under various concentrations of oxygen by following the H_2O_2 and O_2^- production. The Km of the NADPH oxidase for oxygen was compared with Km of intact cells for oxygen.

MATERIALS AND METHODS

PMN were isolated from guinea pig peritoneal cavity.[1,11] The particle fraction (a 20,000 x g pellet) was isolated from resting and phagocytosing PMN.[7,12]

Detection of O_2^- and H_2O_2 generation by PMN, under various oxygen concentrations were carried out in a continuous gas flowing cuvette

(1.0 cm light path, 1.5 ml of reaction medium) as illustrated in Fig. 1. The cuvette holding a Clark-type oxygen electrode(k) was attached to a cell mixer (the top), in which a small wind mill(a) affixed to a glass rod(e & h) is driven by a continuous air pressure through a pipe(c) connected to an air pump.[13] Here(b) is a gas inlet connected to a gas proportioner and here(d) is a gas outlet. The reaction mixture(j) was bubbled with a gas mixture containing oxygen and nitrogen at a variable ratio through a gas proportioner until a desired oxygen concentration was reached. The O_2 concentration in the cuvette(g) was recorded by the oxygen electrode(k). Spectrophotometric assays were performed at 37°C in a Hitachi 556 double-beam spectrophotometer.

When measuring a very low oxygen concentration (0.1 to 10 μM), a method employing the bioluminescence of luminous bacteria was applied with high accuracy (see the text).

In attempt to measure H_2O_2 production by cellular and subcellular fractions of PMN, we employed the cytochrome c peroxidase (CCP) method.[9] Fluorometric methods have been used to detect H_2O_2 generation. Fluorometric assays involve the use of fluorescent dyes that act as hydrogen donor (AH_2) for the horseradish peroxidase (HRP)-reaction as shown in the following formulae:

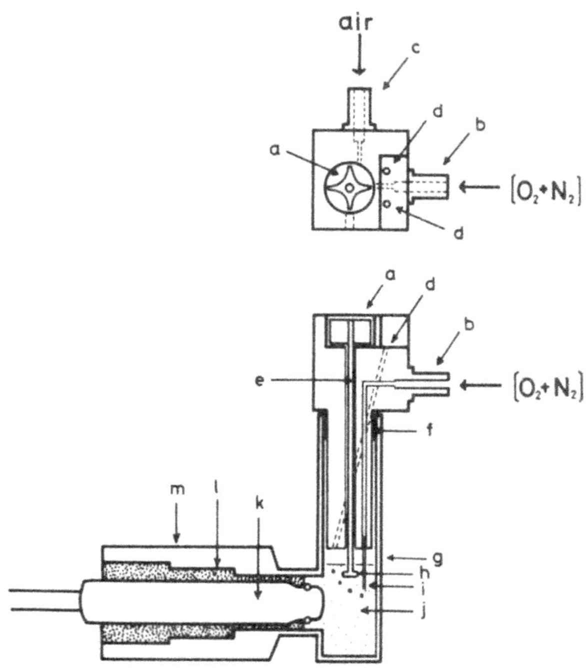

Fig. 1. A continuous gas flowing cuvette.

$$HRP + H_2O_2 \longrightarrow HRP\text{-}H_2O_2(I) \quad (1)$$

$$HRP\text{-}H_2O_2 + AH_2 \longrightarrow AH\cdot + HRP\text{-}H_2O_2(II) \quad (2)$$

$$HRP\text{-}H_2O_2 + AH\cdot \longrightarrow A + 2\,H_2O + HRP \quad (3)$$

where $HRP\text{-}H_2O_2$ (I) and (II) are compound I and II of the enzyme-substrate complexes.[14,15] In the HRP-reaction, the reduced form of the fluorescent dye (AH_2) reacts with compound I and II to form the oxidized dye which detect the presence of H_2O_2. However, fluorometric methods are not useful for the accurate measurement of H_2O_2 generation by subcellular fractions, since, in all probability, subcellular fractions contain endogenous hydrogen donors which react with compound I and II. In addition, the H_2O_2 generation by the subcellular fractions of PMN was initiated by the addition of NAD(P)H as hydrogen donor, which are also able to compete with the dye.

By contrast, the CCP method avoids interference from endogenous hydrogen donors, because the high specificity of the enzyme-H_2O_2 complex for ferrous cytochrome c as hydrogen donor gives great stability to the spectrophotometric indicators:

$$CCP + H_2O_2 \longrightarrow CCP\text{-}H_2O_2 \quad (4)$$

$$CCP\text{-}H_2O_2 + 2\text{ cyt }c^{2+} \longrightarrow CCP + 2\text{ cyt }c^{3+} + 2HO^- \quad (5)$$

This procedure measures the increase of the absorbance at 419 nm of the $CCP\text{-}H_2O_2$ complex with a reference wave-length at 407 nm, which is the absorbance maximum of free CCP.[16,17] CCP was purified from Baker's yeast according to the method of Yonetani.[16,17]

For the assay of O_2^- production by subcellular fractions, we prepared a modified cytochrome c with the amino groups 63% acetylated, and measured the reduction of the acetylated cytochrome c.[12] The specificity of ferri-cytochrome c (cyt c^{3+}) reduction is not high, since native cyt c can undergo several redox changes by many enzymatic processes other than O_2^- (eg. microsomal NADPH cyt c reductase & mitochondrial cyt c reductase) as shown in Fig. 2. By contrast, the

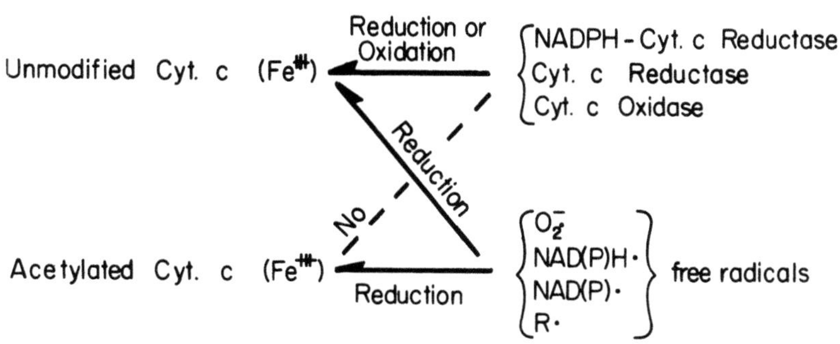

Fig. 2. Reactions of native and acetylated cytochrome c

enzymatically inactive acetylated cyt c reacts only with free radicals such as O_2^-, NAD(P)H· and semiquinone compounds. The use of the enzymatically inactive cyt c is necessary to evaluate a real steady state concentration of O_2^-.

RESULTS

Fig. 3-a shows the rate of H_2O_2 release from phagocytosing PMN, plotted against O_2 concentration in the cuvette. The H_2O_2 generation was progressively higher at increasing O_2 concentration until a maximum was reached at about 60 µM. Fig. 3-b shows a Lineweaver-Burk plot of the H_2O_2 generation as a function of O_2 concentration from the datum of Fig. 3-a. A value of 27 µM seems to be the apparent Km for oxygen.

In parallel experiments, the rate of O_2^- release by phagocytosing PMN was measured under various O_2 concentrations. Fig. 4-a shows the O_2^- generation and its release from intact cells: O_2^- production increases rapidly up to the O_2 concentration of about 10 µM, thereafter, it shows a slow increase, without reaching a real saturation up to 200 µM. The Km value obtained by a Lineweaver-Burk plot (Fig. 4-b) was about 27 µM, in good agreement with the Km of H_2O_2 production for oxygen.

In subsequent experiments, we assayed H_2O_2 and O_2^- production by subcellular fractions under various O_2 concentrations. We always observed a higher NADPH-induced production of H_2O_2 and O_2^- by the particle fraction isolated from phagocytosing PMN than by other subcellular fractions of PMN.

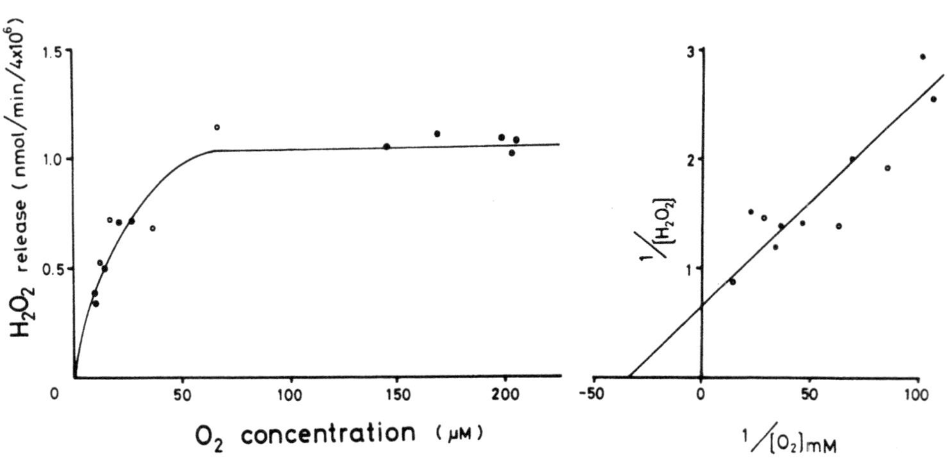

Fig. 3. (a) The rate of H_2O_2 release from phagocytosing PMN under various O_2 concentrations (left).
(b) A Lineweaver-Burk plot of the H_2O_2 release as a function of O_2 concentration (right). Heat killed E. coli were added as phagocytic particles.

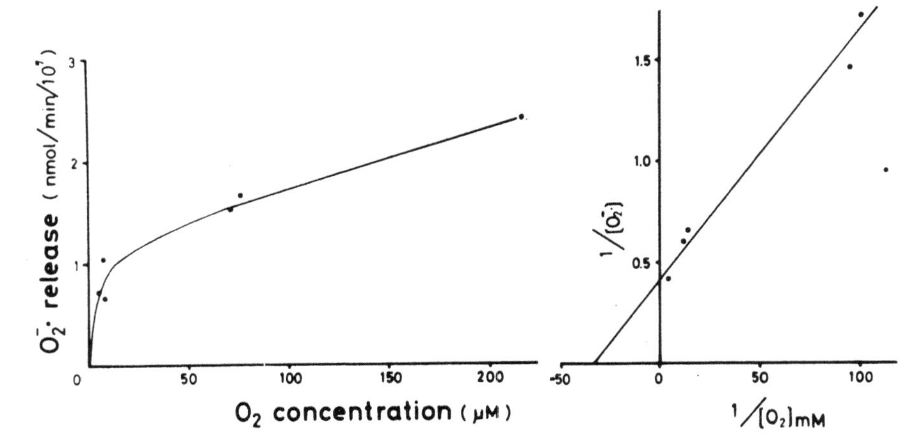

Fig. 4. (a) The rate of O_2^- release from phagocytosing PMN under various O_2 concentrations (left). (b) A Lineweaver-Burk plot of the datum of Fig. 4-a (right).

Fig. 5-a shows the rate of H_2O_2 production by the particle fraction isolated from phagocytosing PMN upon addition of 0.1 mM NADPH. The H_2O_2 generation by this fraction was progressively higher at increasing O_2 concentration until a maximum was reached at about 90 μM. The reciprocal plot (Fig. 5-b) gives Km of the NADPH oxidase for O_2 of approx. 25 μM. This effect is similar to that obtained with intact phagocytosing cells.

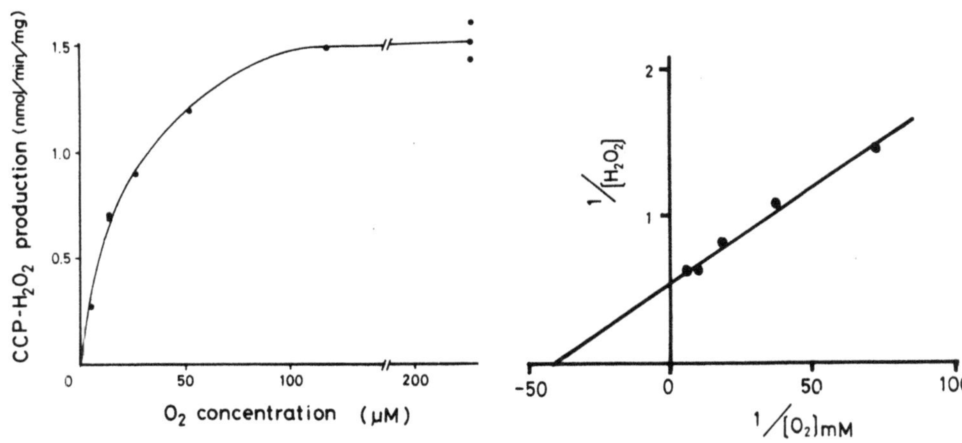

Fig. 5. (a) The rate of H_2O_2 generation by the particle fraction of phagocytosing PMN under various O_2 concentrations(left). (b) A Lineweaver-Burk plot of the datum of Fig. 5-a (right).

In parallel experiments, the rate of O_2^- production by the particle fraction was measured as evaluated by the acetylated cyt c reduction method. The reduction of acetylated cyt c by the subcellular fraction, however, showed little decline until very low oxygen concentration was reached (less than 10 μM). However, the polarographic method does not provide a precise value of oxygen concentration in the low oxygen concentration range of 1 to 10 μM.

Therefore, we utilized the luminescence of marine bacteria for the determination of O_2 concentration.[18] The intensity of luminescence of <u>photobacterium phosphoreum</u>, is known to be oxygen dependent. As shown in Fig. 6-a, the emission spectrum of bacterial luminescence has a peak at 472 nm. We used an airtight-quartz cuvette to measure

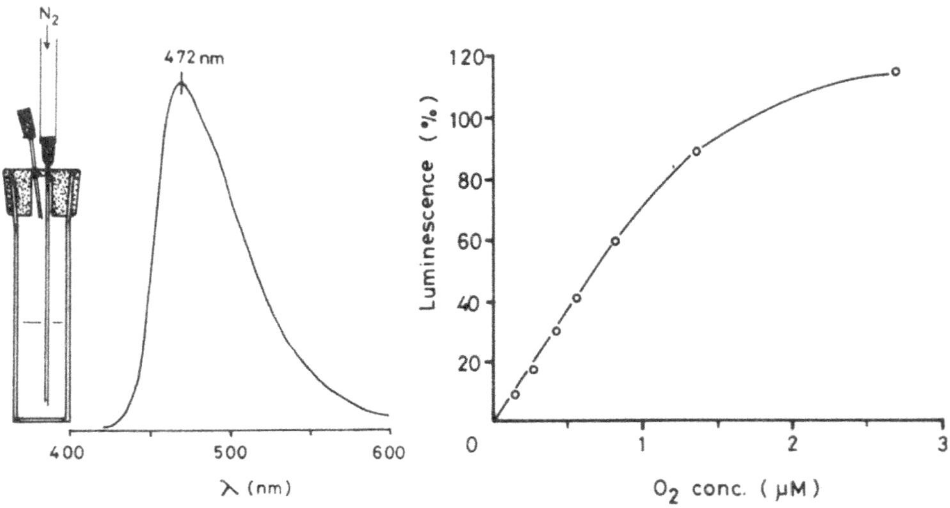

Fig. 6 (a) The emission spectrum of bacterial luminescence (middle); an airtight-cuvette for the measurement of the luminescence (left). (b) A calibration curve of the luminescence plotted against oxygen concentration(right).

luminescence of the bacteria (Fig. 6-a, left). Prior to the experiments, the reaction medium was bubbled with purified argon gas. To the anaerobic medium, the marine bacteria (ca. 50 μg protein) were added. The calibration curve was obtained by plotting L_i/L_{max} as a function of oxygen concentration after the addition of various amounts of air saturated buffer. The intensity of luminescence is maximal above 10^{-5} M oxygen and a nearly linear function of oxygen concentration below 10^{-6} M as shown in Fig. 6-b. The bacterial luminescent method can be applied with accuracy to the low oxygen concentration range of 0.1 to 5.0 μM.

With this bioluminescence method, the reduction of acetylated cyt c by the particle fractions isolated from phagocytosing (P) and resting (R) PMN was measured in the presence and absence of superoxide dismutase(SOD). The reduction of acetylated cyt c (ac cyt c) by the P-fraction was about 50% of the maximal value in the presence of about 5 µM oxygen as shown in Fig. 7-a. Fig. 7-b shows a Lineweaver-Burk plot of O_2^- production by the P-fraction as a function of oxygen concentration: a value of about 5 µM seems to be the apparent Km for oxygen.

DISCUSSION

In the present research, we have obtained evidence that the Km for O_2 of the O_2^- generation differs at cellular (ca27 µM) and subcellular fraction(ca. 5 µM) level, whereas the Km for O_2 of the H_2O_2

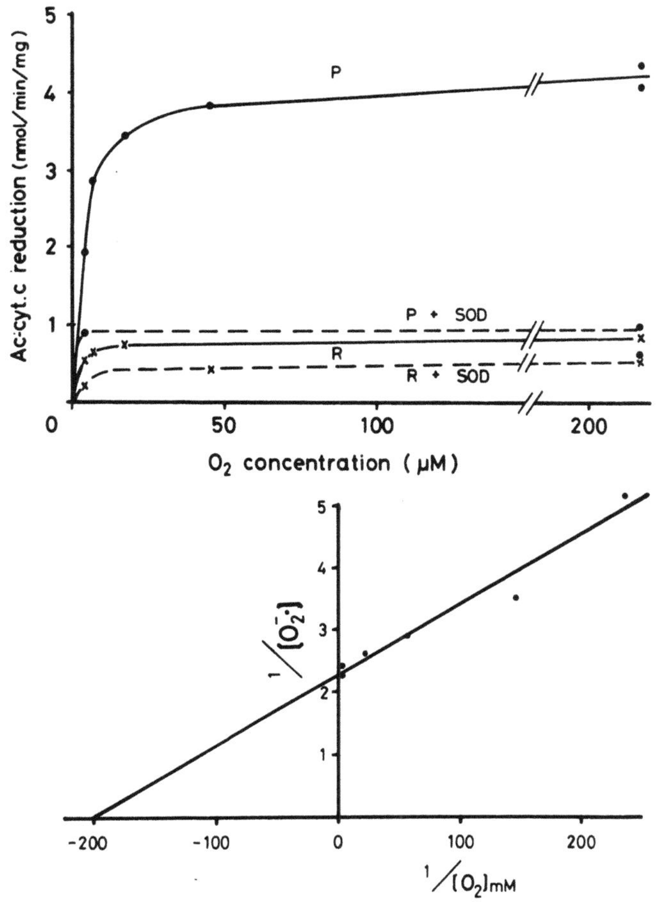

Fig. 7. (a) Reduction of acetylated cyt c by the particle fractions isolated from resting (R) and phagocytosing (P) PMN in the presence of 0.1 mM NADPH with and without SOD. (b) A lineweaver-Burk plot of the O_2^- production by the P-fraction as a function of oxygen concentration.

generation remains the same in both systems (25-27 μM). This discrepancy helps to interpret two current problems in the oxidative metabolism of PMN: [1] the possibility of semiquinone-intermediate, in addition of O_2^-, in the NADPH-oxidase reaction, and [2] the reaction site of NADPH-oxidase in cells.

[1] The lower Km for O_2 of the O_2^- production by NADPH-oxidase suggests that a portion of the cyt c reduction is almost O_2 independent. Thus, the presence of enzyme-NADPH semiquinone complex (E-NADPH) could account for this discrepancy in ac.cyt c-reduction under anaerobic conditions. Fig. 8 illustrates this difference in terms of aerobic and anaerobic reduction of ac cyt c. It seems that some NADPH -intermediate, with a higher redox potential than O_2^-, may directly reduce ac cyt c under anaerobic conditions:

$$\text{ac cyt c}(Fe^{3+}) + \text{E-NADPH}\cdot \xrightarrow{k_6} \text{ac cyt c}(Fe^{2+}) + \text{E-NADP} + H^+ \quad (6)$$

$$\text{ac cyt c}(Fe^{3+}) + O_2^- \xrightarrow{k_7} \text{ac cyt c}(Fe^{2+}) + O_2 \quad (7)$$

where the rate constants k_6 and k_7 seem to be about 10^6 and 10^8 M^{-1} sec^{-1}, respectively, if the data of reactivities of NADH· and O_2^- with ferri-cyt c apply to the reactions of E-NADPH· and O_2^- with ac cyt c as well.[19,20]

The E-NADPH· may react with O_2 to form O_2^- and NADP: the rate constant seems to be more than 2×10^9 M^{-1} sec^{-1}(react.(1), Fig. 8).[21]

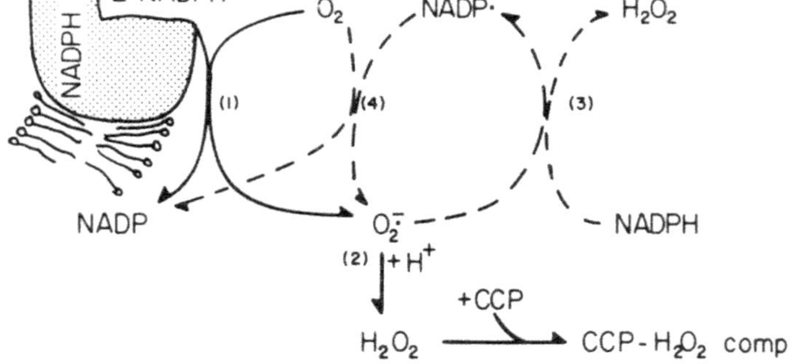

Fig. 8. The NADPH-oxidase reaction by the particle fraction.

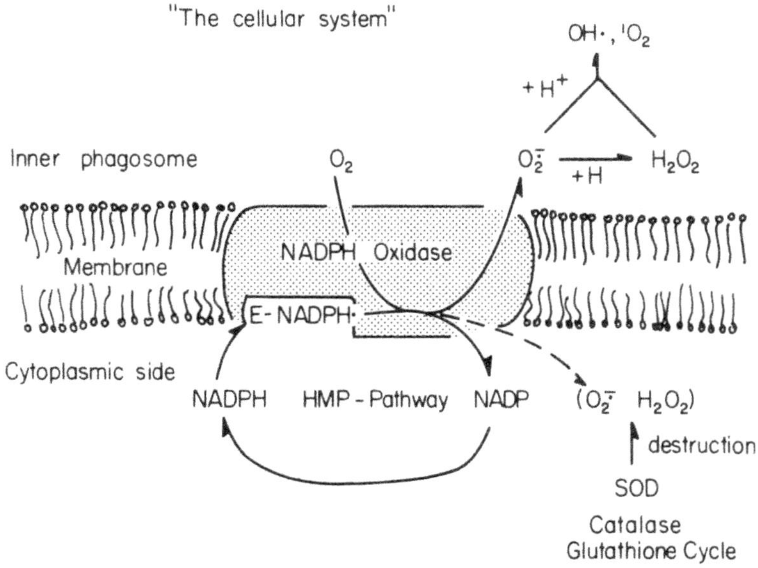

Fig. 9. The cellular system of O_2^- and H_2O_2 production.

The dismutation of O_2^- to H_2O_2 and O_2(react. (2), Fig. 8) shows a rate constant of about 10^5 /M. sec in neutral pH by the non-enzymatic reaction, and 2×10^9 $M^{-1}sec^{-1}$ in the presence of SOD.[22] On the other hand, the O_2^- yielded may react non-enzymatically with free NADPH in the reaction medium and form NADP· and H_2O_2 (react. (3), Fig. 8): the rate constant for this reaction, however, is less than 27 $M^{-1}sec^{-1}$ in neutral pH.[21] Therefore, reaction (3) may be negligible in the pH 7.0 buffer without addition of Mn^{2+}in our system ; consequently reaction (4) may also be negligible (dotted lines, Fig. 8)

[2] On the other hand, the present results seem to provide a clue for the reaction site of NADPH-oxidase with NADPH, presumably located at the cytoplasmic side of the cellular membrane, as shown in Fig. 9. The NADPH-oxidase reacts with NADPH produced through the HMP pathway, generating O_2^-, which may be released into the phagosome (opposite side of cytoplasm). Even if O_2^- were released into the cytoplasmic side of the membrane, the oxygen intermediates, O_2^- and H_2O_2, can be destroyed by cytoplasmic enzymes such as SOD, catalase, and glutathione peroxidase.

Since neither CCP nor ac.cyt c permeates into cells, the reaction of these detectors with either O_2^- or H_2O_2 should take place only in the outer medium, so that Km value of intact cells for O_2 may be identical either by following O_2^- or H_2O_2 generation.

REFERENCES

1. A. J. Sbarra and M. L. Karnovsky, J. Biol. Chem. 234:1355 (1959).
2. G. Y. N. Iyer, M. F. Islam and J. H. Quastel. Nature, 192:535 (1961).
3. B. M. Babior, R. S. Kipness and J. T. Curnutte, J. Clin. Invest. 52:741 (1973).
4. W. H. Evans and M. L. Karnovsky, Biochemistry, 1:159 (1962).
5. R. H. Cagan and M. L. Karnovsky, Nature, 204:255 (1964).
6. R. L. Baehner, N. Gilman and M. L. Karnovsky, J. Clin. Invest. 49:692 (1970).
7. P. Patriarca, R. Cramer, S. Moncalbo, F. Rossi and D. Romeo, Arch. Biochem. Biophys. 145:255 (1971).
8. F. Rossi, D. Romeo and P. Patriarca, J. Reticuloendothel. Soc. 12:127 (1972).
9. K. Kakinuma, A. Boveris and B. Chance, Febs Lett. 74:295 (1977).
10. K. Kakinuma and M. Kaneda, Febs. Lett. 111:90 (1980).
11. K. Kakinuma, J. Biochem. 68:177 (1970).
12. K. Kakinuma and S. Minakami, Biochim. Biophys. Acta, 538:50 (1978).
13. K. Kakinuma, T. Yamaguchi, M. Kaneda, K. Shimada, Y. Tomita and B. Chance, J. Biochem. 86:87 (1979).
14. B. Chance, Arch. Biochem. 22:224 (1949).
15. H. Theorell, Enzymologia,10:250 (1941).
16. T. Yonetani, in "Methods in Enzymology", R. W. Estabrook and M. E. Pullman, eds., Academic Press, New york, 10:336 (1967).
17. T. Yonetani, in "The Enzymes" 3rd ed. P. D. Boyer, ed., Academic Press, New York, 13:345 (1976).
18. R. Oshino, N. Oshino, M. Tamura, L. Kobilinsky and B. Chance, Biochim. Biophys. Acta, 273:5 (1972).
19. J. Butler, G. G. Jayson and A. J. Swallow, Biochim. Biophys. Acta, 408:215 (1975).
20. M. G. Simic, I. A. Taub, J. Tocci and P. A. Hurwitz, Biochem. Biophys. Res. Comm. 62:161 (1975).
21. E. J. Land and A. J. Swallow, Biochim. Biophys. Acta, 234:34 (1971).
22. H. Forman and I. Fridovich, Arch. Biochem. Biophys. 158:396 (1973)

FORMATION OF SUPEROXIDE ANIONS AND HYDROGEN PEROXIDE BY POLYMORPHONUCLEAR LEUKOCYTES STIMULATED WITH CYTOCHALASIN

Shigeki Minakami, Zeenat F. Nabi, Bernard Tatscheck and Koichiro Takeshige

Department of Biochemistry
Kyushu University School of Medicine
Fukuoka 812, Japan

Cyanide-insensitive respiration of polymorphonuclear leukocytes is induced by the D or E type of cytochalasins, a family of antibiotics which inhibits various cellular motile functions including membrane movement(Nakagawara et al.,1974). The respiratory stimulation is accompanied by the release of superoxide anions and the activated hexose monophosphate oxidative pathway but without any formation of phagosomes(Nakagawara et al.,1976). The cytochalasins D and E are ideal reagents for studying phagocytotic metabolic changes, because they inhibit endocytotic process at the same time. We are going to show the properties of the burst reactions triggered with cytochalasin D or E and compare the reactions with those occur during the ingestion of bacteria.

OXYGEN UPTAKE AND THE FORMATION OF ACTIVE OXYGEN

The uptake of oxygen and the release of superoxide anions by guinea pig exudate polymorphonuclear leukocytes stimulated with cytochalasin E or bacteria are shown in Fig. 1. The oxygen uptake traces were obtained by using a Clark-type oxygen electrode and the superoxide release is shown as the reduction of exogenously added cytochrome c which was measured at 550 nm relative to 540 nm with a dual-wavelength spectrophotometer equipped with an end-on photomultiplier tube. The oxygen uptake induced with the cytochalasin is weak with a long lag-time compared with the release of

*This work was supported in part by grants from Yamanouchi Foundation of Metabolism and Disease, the Ministry of Education, Science and Culture, and the Ministry of Welfare, Japan

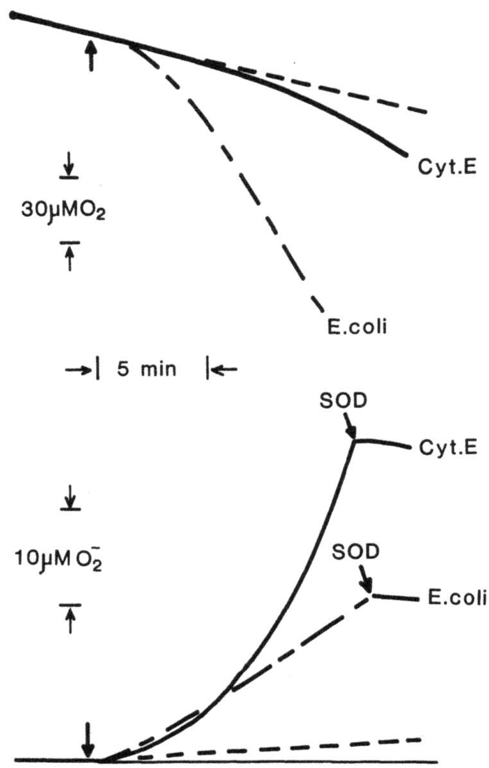

Fig. 1. Oxygen uptake and superoxide release of guinea pig exudate leukocytes(1.4 x 10^7 cells in 2.0 ml Krebs Ringer phosphate solution at 37°C). (from Nakagawara et al.,1976)

Table 1. Quantitative Comparison of the Oxidative Metabolism (from Nabi et al.,1979)

Stimulants	Oxygen Uptake	Superoxide Release	Hydrogen Peroxide Release
	nmol/min per 10^6 cells (mean ± S.E.M.)		
Cytochalasin D	4.5 ± 0.3	4.0 ± 0.3	1.9 ± 0.1
E. coli	8.0 ± 1.0	0.2 ± 0.04 (2.1 ± 0.3)	2.1 ± 0.1

superoxide anions which has also a lag but becomes very active afterward. The respiration stimulated by bacteria is very active while the superoxide release in this condition starts without an appreciable lag but tends to be attenuated to a negligible rate.

The quantitative comparison of cytochalasin D and bacteria as the stimulants of the oxygen uptake and the release of superoxide anions and hydrogen peroxide is given in Table 1. The appearance of hydrogen peroxide in the suspending medium was determined by measuring the oxidation of homovanillic acid in the presence of horse-radish peroxidase, with a fluorometer which detects fluorescence change on the surface of the cuvette(Takeshige et al.,1979). The steady-state rates in the presence of 2 mM-azide are given, except that the rate of the superoxide release in parenthesis is the initial transient rate during the ingestion of bacteria. If we compare the rates of the formation of superoxide and hydrogen peroxide, the ratio is 2 : 1 for the cytochalasin-treated cells, whereas the ratio is 1 : 10 for the cells ingesting bacteria. The ratio for the cytochalasin-treated cells indicates the dismutation of the released superoxide anions in the medium, while the ratio for the phagocytosing cells suggests that most of the hydrogen peroxide is not the one formed in the medium from the released superoxide but is the one directly released from the cells.

If we assume that the additional oxygen molecules utilized by the stimulated cells are quantitatively converted to superoxide anions and then to hydrogen peroxide, the ratio of the oxygen uptake and the hydrogen peroxide formation in the presence of azide should be 1 : 1. The ratios observed were 2 : 1 for the cytochalasin-treated cells and 4 : 1 for the phagocytosing cells. We, however, wish to propose that the additional oxygen molecules utilized are converted to superoxide anions, because if we measure the oxygen uptake and the hydrogen peroxide formation of the cytochalasin-treated cells in the presence of superoxide dismutase, the respiratory rate decreases(Minakami et al.,1979) and the rate of the hydrogen peroxide formation increases(Nabi et al.,1979), the ratio approaching to 1 : 1. Similar conclusion is obtained by Root and Metcalf(1977) on human leukocytes pretreated with cytochalasin B and stimulated with bacteria.

EFFECT OF AN IMPERMEANT SH-INHIBITOR

The above observation suggests that the cytochalasin-treated cells form superoxide anions by a superoxide-forming enzyme of the plasma membrane, whereas the enzyme is internalized by the invagination of the plasma membrane when the cells are ingesting bacteria.

This supposition can be ascertained by a comparison of the effect of an impermeant SH-inhibitor, p-chloromercuribenzenesulfonate, on the hydrogen peroxide formation by the cytochalasin-treated cells and the phagocytosing cells. When the inhibitor was added after the steady-state rate of the hydrogen peroxide formation had been attained, the formation induced by cytochalasin D was completely inhibited by the reagent at about 50 μM, whereas that induced by phagocytosis was not affected(Fig. 2). The inhibition can not be ascribed to the inhibition of the triggering mechanism, because the hydrogen peroxide formation has already been induced and the inhibitor is not effective for the phagocytosing cells. The most probable explanation is that the impermeant inhibitor attacks the superoxide-forming enzyme of the cytochalasin-treated cells which is localized on the plasma membrane, without attacking the enzyme internalized by the phagosome formation. A permeant SH-inhibitor, p-chloromercuribenzoate, inhibited the hydrogen peroxide formation of both cytochalasin-treated cells and phagocytosing cells.

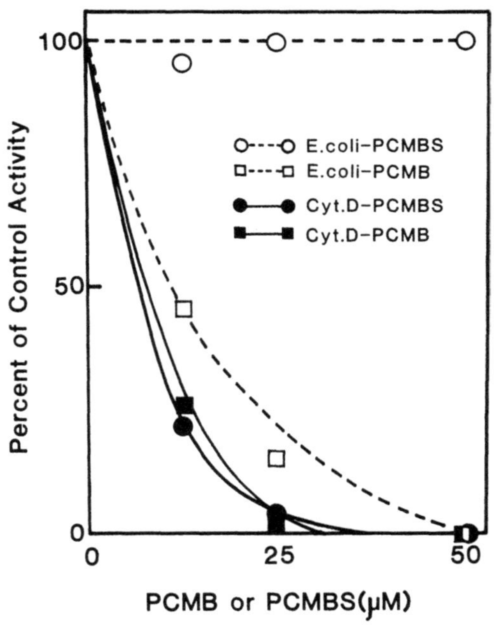

Fig. 2. Effect of SH-inhibitors on the hydrogen peroxide formation. p-Chloromercuribenzenesulfonate(PCMBS) or p-chloromercuribenzoate(PCMBS) was added 4 min after the triggering of the metabolism. (from Nabi et al.,1979)

HYDROGEN PEROXIDE FORMATION IN THE PRESENCE OF FERRICYTOCHROME C AND SUPEROXIDE DISMUTASE

Superoxide dismutase does not activate the steady-state rate of the hydrogen peroxide formation of the phagocytosing cells but does activate the formation of the cytochalasin-treated cells (Nabi et al.,1979). This is in agreement with the observation that the reduced oxygen is released to the medium as hydrogen peroxide when the cells are stimulated with bacteria and as superoxide anions when the cells are treated with cytochalasin D or E. The situation can be more clearly shown when the hydrogen peroxide formation is studied in the presence of ferricytochrome c, a scavenger of superoxide radicals.

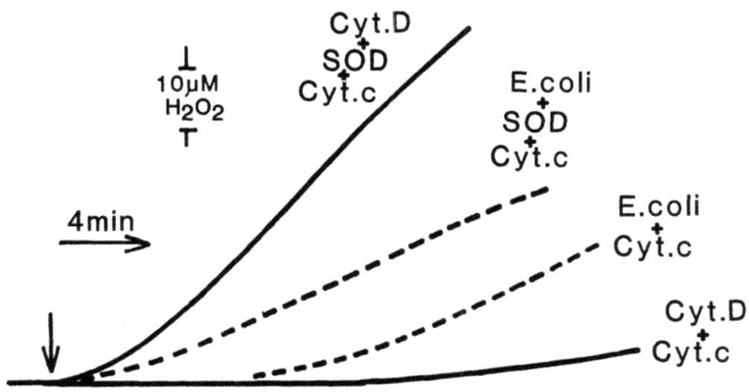

Fig. 3. Hydrogen peroxide formation in the presence of 100 µM-cytochrome c. (from Nabi et al.,1979)

As shown in Fig. 3, the hydrogen peroxide formation of the cytochalasin-treated cells in the presence of ferricytochrome c is essentially negligible until most of the pigment is reduced, because superoxide radicals are scavenged before being dismutated spontaneously. When superoxide dismutase is added to the medium, the hydrogen peroxide formation is activated very much because the enzyme is active enough to overcome the scavenging effect. In contrast, when the cells are ingesting bacteria, the steady-state rate was not affected by superoxide dismutase and only the disappearance of lag was observed. The disappearance can be attributed to the dismutation of superoxide radicals which are only transiently released until the phagosomes are sealed. The observation is explained by a scheme given in Fig. 4.

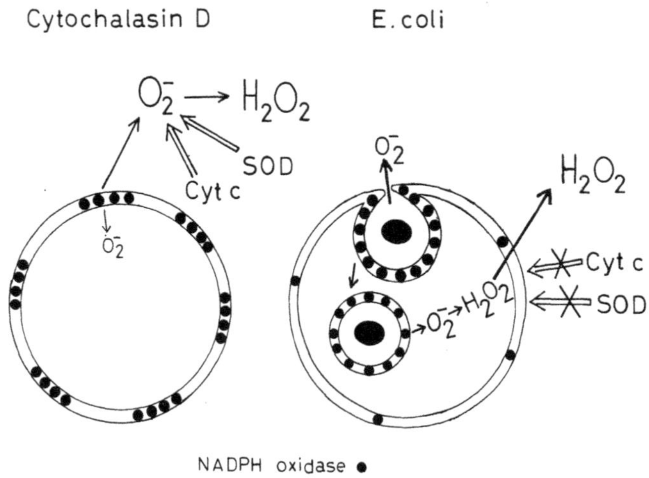

Fig. 4. A scheme showing the difference of hydrogen peroxide formation in the cytochalasin-treated cells and the phagocytosing cells.

ELECTRON MICROSCOPIC OBSERVATIONS

Electron microscopic observations kindly carried out by Drs. Hatae and Shibata of the Anatomy Department have revealed that the action of cytochalasin D and E is characterized by the disappearance of pseudopods and the formation of cytoplasmic protuberations and vacuole-like structures(Nakagawara et al.,1976; Nabi et al., 1979). We were interested to detect opening in the vacuole-like structures because of their similarity to phagosomes and openings were clearly shown in some of the vacuole-like structures, suggesting that they are formed by cutting the protuberations horizontally(Fig. 5). This was further ascertained by the addition of tannic acid in a glutaraldehyde solution for the fixation. Tannic acid which is impermeant of the plasma membrane was lining the inner surface of the structures as well as the cell surface, indicating that they are exposed to outside medium(Nabi et al.,1979). Superoxide anions can be directly released to outside even if the superoxide-forming enzyme happens to be localized on the surface of the vacuole-like structure.

CELLULAR LOCALIZATION OF NADPH-OXIDASE

Because the observations shown above indicate a possible localization of the superoxide-forming enzyme on the plasma membrane, we

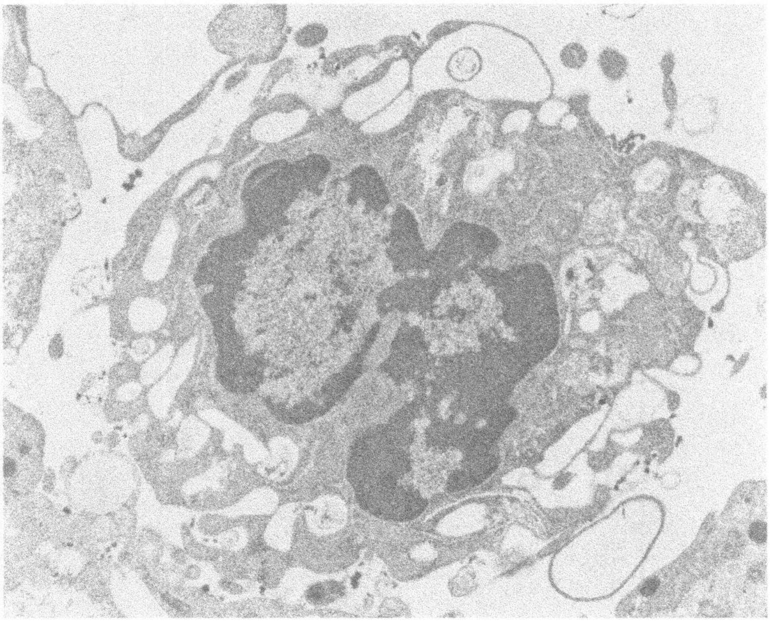

Fig. 5. Electron micrograph of guinea pig polymorphonuclear leukocytes after 10 min of incubation at 37°C with 10 μM-cytochalasin D. The cells were fixed with 2.5 % glutaraldehyde solution containing 0.1 M cacodylate, pH 7.4 and postfixed with 1 % osmium tetraoxide. (coutsey of Dr. Hatae)

tried to test the activity in the cellular organella. We measured the NADPH-oxidase activity, because we agree with the Trieste group in that an NADPH-oxidase is responsible for the respiratory burst(Rossi and Zatti,1964). We have shown that the release of superoxide anions and the NADPH-oxidase activity of the cellular particulate fraction behave in parallel: they are both activated by the treatment of the cells with cytochalasins D and E, but not with cytochalasins B and C, and are suppressed by the treatment of the cells with cyclic AMP and deoxyglucose(Nabi et al.,1979).

When we homogenize the cytochalasin-treated cells with a Polytron homogenizer and the post-nuclear supernatant was fractionated by a differential centrifugation in 0.34 M-sucrose solution, the NADPH-oxidase activity was found in an intermediate fraction between the fraction rich in 5'-AMPase, a marker enzyme of the plasma membrane and the fraction rich in myeloperoxidase, a marker enzyme of granules. We explain this observation by a hypothesis that by a stimulation of the cells, the NADPH-oxidase and 5'-AMPase move to different portions of the membranes: the stimulated part becomes rich in the NADPH-oxidase while another part becomes rich in 5'-AMPase.

In order to support the hypothesis and confirm the possible localization of the NADPH-oxidase on the plasma membrane, we labelled the plasma membrane before the incubation of the cells with cytochalasin D. The leukocytes were treated with radioactive iodide, lactoperoxidase and hydrogen peroxide, washed with saline and then incubated with or without the cytochalasin. The cells were homo-

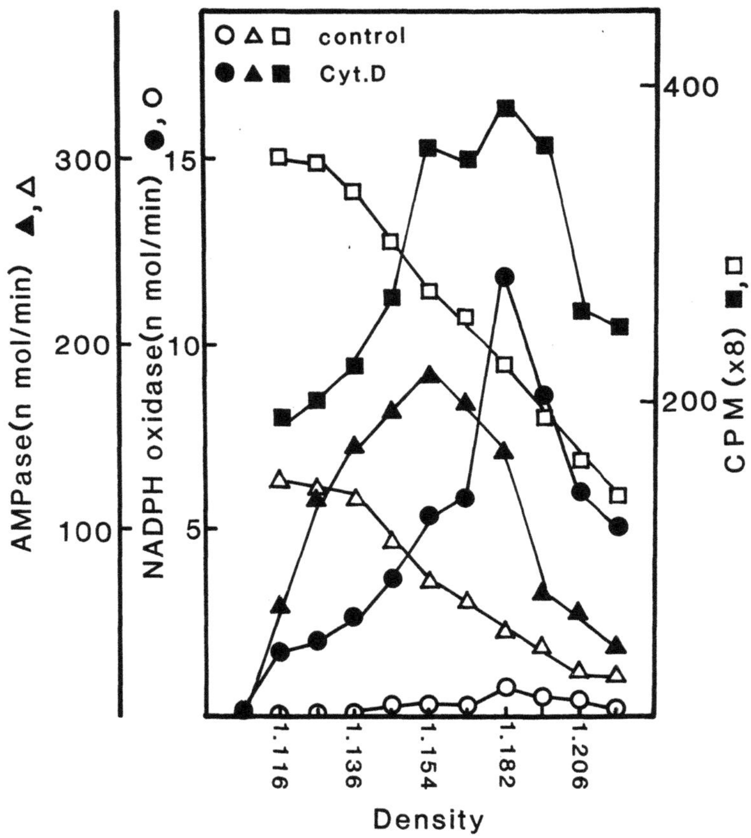

Fig. 6. Subcellular localization of the activated NADPH-oxidase. Guinea pig leukocytes were iodinated with radioactive iodide, lactoperoxidase and hydrogen peroxide. The cells were activated with cytochalasin D and homogenized. The post-nuclear supernatant was centrifuged on a linear sucrose-density gradient from 30 - 60 % at 77 000 g for 60 min.

genized and the fractions were obtained by an isopycnic centrifugtion. As shown in Fig. 6, the radioactivity of the control cells incubated without the cytochalasin is located in the fractions of low density, corresponding to the localization of 5'-AMPase. When the cells were stimulated with the cytochalasin, the radioactivity is found in heavier fractions forming 2 peaks, a lighter peak corresponding to 5'-AMPase and a heavier one corresponding to a fraction rich in the activated NADPH-oxidase. The result may support a hypothesis that the NADPH-oxidase and 5'-AMPase are uniformly distributed on the plasma membrane when the cells are in a resting state, whereas when the cells are stimulated, the stimulated portion of the membrane becomes rich in the NADPH-oxidase.

CONCLUDING REMARK

We have compared the oxidative metabolism of guinea pig exudate polymorphonuclear leukocytes treated with cytochalasin D(or E) and that of the cells ingesting bacteria. The results support hypotheses that an NADPH-oxidase on the plasma membrane is responsible for the respiratory burst and that the activated enzyme is internalized during invagination of the plasma membrane and the formation of phagosomes.

REFERENCES

Minakami,S.,Takeshige,K., and Nabi,Z.F.,1979, Oxidative metabolism of polymorphonuclear leukocytes induced by phagocytotic metabolic mimickers, in: "Phagocytosis, Its Physiology and Pathology",Y.Kokubun and N.Kobayashi,eds,University of Tokyo Press, Tokyo.

Nabi,Z.F.,Takeshige,K.,Hatae,T., and Minakami,S.,1979, Hydrogen peroxide formation of polymorphonuclear leukocytes stimulated with cytochalasin D, Exp.Cell Res.124:293

Nakagawara,A.Takeshige,K., and Minakami,S.,1974, Induction of a phagocytosis-like metabolic pattern in polymorphonuclear leukocytes by cytochalasin E, Exp. Cell Res.87:392

Nakagawara,A.,Shibata,Y.,Takeshige,K., and Minakami,S.,1976, Action of cytochalasin E on polymorphonuclear leukocytes of guinea pig peritoneal exudates, Exp.Cell Res.101:225

Root,R.K., and Metcalf,J.,1977, H_2O_2 release from human granulocytes during phagocytosis. Relationship to superoxide anion formation and cellular catabolism of H_2O_2:Studies with normal and cytochalasin B-treated cells, J.Clin.Invest.60:1266

Rossi,F., and Zatti,M.,1964, Biochemical aspects of phagocytosis in polymorphonuclear leukocytes. NADH and NADPH oxidation by granules of resting and phagocytizing cells,Experientia 20:21

Takeshige,K.,Matsumoto,T.,Shibata,R., and Minakami,S.,1979, Simple and rapid method for the diagnosis of chronic granulomatous disease, measuring hydrogen peroxide and superoxide anions released from leukocytes in whole blood, Clin.Chim.Acta 92:329

HYDROGEN PEROXIDE PRODUCTION IN A CELL-FREE SYSTEM: EVIDENCE FOR THE INVOLVEMENT OF A CHAIN REACTION

Lawrence R. DeChatelet, Charles E. McCall, and Pamela S. Shirley

Departments of Biochemistry and Medicine
Bowman Gray School of Medicine
Winston-Salem, N.C. 27103, U.S.A.

INTRODUCTION

During the process of phagocytosis, the polymorphonuclear leukocyte undergoes remarkable alterations in cyanide-insensitive oxidative metabolism, collectively referred to as the "respiratory burst."[1] The importance of these events to the bactericidal activity of the cell is underscored by the clinical syndrome, chronic granulomatous disease (CGD). Neutrophils from patients with CGD can ingest bacteria normally or near-normally, but are markedly deficient in their ability to kill most types of engulfed bacteria.[2] The biochemical lesion in this disease is characterized by a complete lack of respiratory burst activity.[3]

Although a number of candidates for the respiratory burst enzyme have been proposed, attention has focused on a reduced pyridine nucleotide oxidase which appears to preferentially utilize NADPH as originally proposed by Patriarca et al.[4]

Activity of the initiating oxidase has been measured in a variety of ways including monitoring oxygen consumption,[4] decrease in absorbance at 340 nm due to oxidation of NADPH,[5] and measurement of superoxide anion generation by means of the SOD-inhibitable reduction of cytochrome c.[6] Each of these assays has both strengths and limitations inherent in it. In our laboratory, we have generally measured oxidase activity by following the appearance of oxidized nucleotide, e.g. NADP, employing either isotopic[7] or fluorometric[8] assay procedures. These assays are more sensitive than others commonly employed, but questions have been raised concerning their specificity. It is well-known that the neutrophil possesses a number of diaphorase type enzymes capable of oxidizing either NADH or NADPH[9]; the existence of alternate routes of pyridine nucleotide oxidation could potentially obscure measurement of oxidase activity by these assays. Accordingly, we attempted to assay for oxidase activity by another means. We elected to measure hydrogen peroxide formation, since it is a final product of the oxidase and the cell would not be likely to possess multiple mechanisms for formation of this toxic compound. Further, a large body of evidence suggests the importance of H_2O_2 in the actual bactericidal event.[10]

ASSAY PROCEDURE

Initial attempts to measure NADPH-dependent H_2O_2 production by a cell-free system using the scopoletin assay of Root et al.[11] were unsuccessful. This lack of success appeared to be due to interference in the assay system by the pyridine nucleotides themselves.

Iyer et al.[12] first employed the catalatic oxidation of ^{14}C-formate to $^{14}CO_2$ to measure hydrogen peroxide production by phagocytizing cells. This assay is dependent upon the following reaction:

$$H^{14}COOH + H_2O_2 \xrightarrow{\text{catalase}} {}^{14}CO_2 + 2H_2O$$

The reaction, however, is complicated by the competing reaction of catalase which decomposes H_2O_2 in the following fashion:

$$2H_2O_2 \xrightarrow{\text{catalase}} 2H_2O + O_2$$

Thus under conditions where formate is in excess, the oxidation of formate to CO_2 is favored; however if H_2O_2 were present in excess, the decomposition reaction would be expected to occur. In initial experiments we observed no appreciable oxidation of formate when a bolus of reagent H_2O_2 (0-1500 nmol) was added to the complete assay system. Similar results have been reported by Klebanoff and Clark.[10] In contrast, Fig. 1 illustrates the oxidation of ^{14}C-formate under conditions where low levels of hydrogen peroxide are continuously generated. In this case where varying levels of H_2O_2 were achieved by utilizing different levels of glucose oxidase in the presence of glucose, the correspondence between formate oxidation and enzyme concentration was very linear. Similar results were obtained using xanthine-xanthine oxidase as the H_2O_2 generating system (data not shown). The generation of H_2O_2 by NADPH oxidase was expected to be analogous to that of these model systems, i.e. continuous generation of low levels of H_2O_2. Experiments were designed to test this hypothesis.

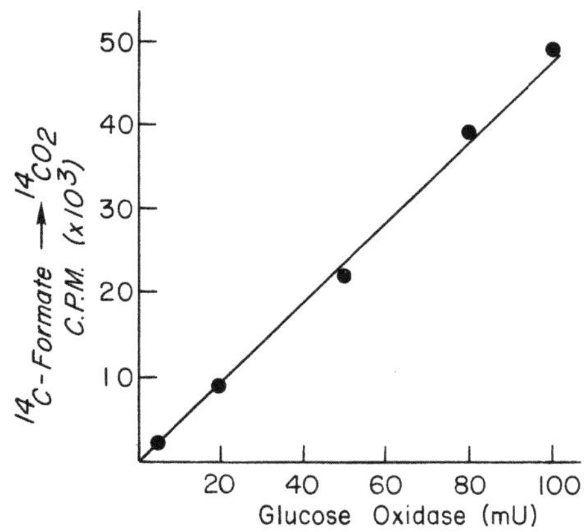

Fig. 1. Standard curve for hydrogen peroxide production by glucose oxidase. Each flask contained 0.10 M MES buffer, pH 6.0; 4200 units catalase; 0.5 μCi (0.05 mM) ^{14}C-formate; and 0.3 mM glucose in a total volume of 3.0 ml. Reaction was initiated by the addition of the indicated quantity of glucose oxidase and stopped after 30 min by the addition of hydrochloric acid. Each point represents the mean of triplicate determinations; the experiment illustrated is representative of three separate experiments.

The formate oxidation assay was conducted in a total volume of 3.0 ml. The standard assay system contained 0.1 M MES buffer, pH 6.0; 4200 units of catalase; 0.50 μCi (0.05 mM) ^{14}C-formate; and reduced pyridine nucleotide (usually 0.40 mM). Reaction was initiated by the addition of 0.20 mg of granule fraction from either resting or phagocytizing neutrophils[7] and was terminated after 30 to 60 min by the addition of 0.50 ml of 1N HCl. $^{14}CO_2$ released during the course of the reaction was trapped in a center well containing 0.50 ml of 2N KOH and radioactivity determined in a liquid scintillation spectrometer.

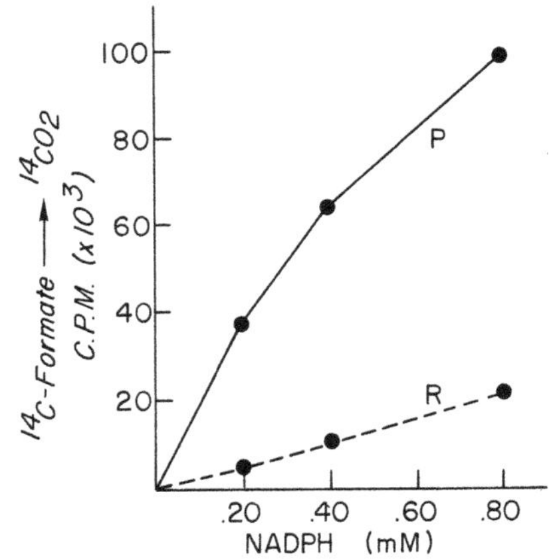

Fig. 2. Formate oxidation by a granule fraction derived from resting or phagocytizing cells. Each value represents the mean of triplicate determinations.

EXPERIMENTS WITH GRANULE FRACTIONS

Preliminary experiments employed a granule fraction from phagocytizing cells. Formate oxidation was absolutely dependent upon the presence of both NADPH and granule fraction and was enhanced at least threefold by the addition of catalase. The reaction was linear with protein concentration up to at least 0.40 mg and with incubation time for at least 30 min. The pH optimum for the reaction was between 5.5 and 6.0. Based on these observations, we selected a pH of 6.0, protein concentration of 0.20 mg, and incubation time of 30 min as standard conditions.

Figure 2 illustrates formate oxidation by granule fractions derived from resting versus phagocytizing cells. At all levels of NADPH examined, the fraction from phagocytizing cells showed much greater activity than a similar fraction from resting cells.

These results are further extended in Table 1. Again, the fraction from phagocytizing cells shows a greater activity toward formate oxidation than a similar fraction from resting cells. This increased activity was observed with either NADH or NADPH as substrate. However, at equal concentrations of substrate, NADPH always was preferred over NADH, in accord with previous results using a fluorometric assay for oxidase activity.[8] The data presented in Fig. 2 and Table 1 demonstrate that a neutrophilic enzyme dependent upon reduced pyridine nucleotides can generate H_2O_2 and appears to be activated by phagocytosis. The data are consistent with the hypothesis that a single oxidase is present which can act upon either NADPH or NADH, but is significantly more active toward the former.

Table 1. Formate Oxidation Dependency on Reduced Pyridine Nucleotides

Conditions	$H^{14}COOH \longrightarrow {}^{14}CO_2$ C.P.M.	
	NADH	NADPH
Resting Fraction	2,631	6,850
Phagocytizing Fraction	9,851	25,713

The experiment is representative of six separate experiments; each value represents the mean of triplicate determinations. The reaction was initiated by the addition of 0.20 mg of granule protein derived from either resting cells or cells which had phagocytized opsonized zymosan.

Recently, we were able to examine the generation of hydrogen peroxide by fractions derived from a patient with well-documented chronic granulomatous disease. Appreciable generation of H_2O_2 was detected in a resting fraction from normal cells which was markedly increased by phagocytosis; this was observed at both 0.20 and 0.40 mM NADPH. In contrast, the resting granule fraction from the patient failed to generate H_2O_2 at either substrate concentration; this was still markedly depressed when the patient cells had been exposed to a phagocytic stimulus. Thus the enzyme studied fulfills two important criteria as the initiating oxidase; it is activated by phagocytosis in normal cells, and it is defective in cells obtained from patients with chronic granulomatous disease.

Table 2. Effect of Inhibitors on Formate Oxidation

Description	$H^{14}COOH \longrightarrow {}^{14}CO_2$ C.P.M.
Complete System	45,980 ± 3,140 (3)
+ SOD	13,554 ± 1,731 (3)
+ Cytochrome c	7,250 ± 2,420 (2)
+ Mn^{2+}	212,434 ± 3,324 (3)

The complete system contained 0.40 mM NADPH and 0.20 mg protein of a granule fraction derived from cells which had phagocytized opsonized zymosan. Values represent the mean ± S.E. for the number of experiments shown in parentheses. Each value was determined in triplicate in each experiment.

We next examined the effect of several compounds on H_2O_2 production by this system. The superoxide scavengers SOD and cytochrome c markedly inhibited the generation of H_2O_2 as shown in Table 2. On the other hand, Mn^{2+} caused a marked stimulation of H_2O_2 production which is consistent with the observation that this metal ion can catalyze a superoxide-mediated chain reaction involving NADPH.[13] In contrast, inhibitors of hydroxyl radical such as mannitol and DABCO did not affect the reaction (data not shown). These results suggest that the mechanism of H_2O_2 formation may involve a chain reaction. The simplest sequence to account for the respiratory burst is as follows.

1) $2O_2 + NADPH \xrightarrow{oxidase} 2O_2^- + NADP^+ + H^+$

2) $O_2^- + O_2^- + 2H^+ \longrightarrow O_2 + H_2O_2$

If this were, in fact, the mechanism operating, it seems unlikely that the reaction would be stimulated by Mn^{2+}. Of more importance, however, it would be impossible for SOD to inhibit the reaction; if anything, the inclusion of superoxide dismutase would be expected to accelerate the generation of H_2O_2 by increasing the rate of reaction 2. The data suggest that a free radical chain reaction is involved in the actual formation of H_2O_2. Such a reaction might be of the following type:

1) $2O_2 + NADPH \xrightarrow{oxidase} 2O_2^- + NADP^+ + H^+$

2) $O_2^- + NADPH + H^+ \longrightarrow H_2O_2 + NADP\cdot$

3) $NADP\cdot + O_2 \longrightarrow NADP^+ + O_2^-$

4) $O_2^- + O_2^- + 2H^+ \xrightarrow{SOD} H_2O_2 + O_2$

According to this mechanism, the respiratory burst is initiated by NADPH oxidase which converts molecular oxygen to superoxide anion, as suggested by Babior.[14] The reaction is then amplified by two nonenzymatic chain reactions (reactions 2 and 3) which serve to increase the generation of both hydrogen peroxide and $NADP^+$. In this sequence, inhibition by SOD would be expected since it serves as a chain termination step, by removing superoxide from the overall sequence. This concept of an initial enzymatic step followed by a nonenzymatic chain reaction has been previously suggested by Patriarca et al.[15,16]

EXPERIMENTS WITH MODEL SYSTEMS

We next attempted to test this concept by using model systems to generate hydrogen peroxide and superoxide in the presence of NADPH. Glucose-glucose oxidase was used to generate H_2O_2 directly without generating detectable amounts of O_2^-. Significant quantities of H_2O_2 were detected in this system by means of the formate oxidation assay. The generation of H_2O_2, however, was not increased by the addition of NADPH and was not significantly inhibited by SOD in the presence or absence of the pyridine nucleotide. A markedly different picture emerged when xanthine-xanthine oxidase was employed in the assay; this enzyme generates both H_2O_2 and O_2^-. This enzyme generated significant H_2O_2 in the absence of NADPH; under these conditions, the inclusion of SOD failed to alter the production of hydrogen peroxide. In contrast to the results with glucose oxidase, the addition of NADPH to the xanthine oxidase system resulted in a twofold increase in H_2O_2 production which was completely ablated by the inclusion of SOD in the assay mixture. The addition of increasing amounts of NADPH to the xanthine oxidase system resulted in a progressive increase in detectable H_2O_2; the increased production was reduced to the baseline level in the presence of SOD, but the basal level, itself, was completely unaffected.

These results with model systems are in agreement with the results obtained with the crude granule fractions. When both hydrogen peroxide and superoxide anion are produced together (as in the neutrophil or in the xanthine oxidase model), the inclusion of NADPH results in a marked increase in H_2O_2 which can be completely inhibited by SOD. When H_2O_2 alone is produced (as in the glucose oxidase system), neither NADPH nor SOD, singly or in combination, has any effect.

Taken together, these studies suggest that hydrogen peroxide generation by a particulate fraction of human neutrophils is initiated by activation of a reduced pyridine nucleotide oxidase and amplified by a chain reaction involving NADPH and O_2^-.

ACKNOWLEDGMENTS

This research was supported by grant AI-10732 from the National Institute of Allergy and Infectious Diseases. The excellent editorial assistance of Ms. Tommye Campbell and Ms. Gwen Charles is gratefully acknowledged.

REFERENCES

1. L. R. DeChatelet, Initiation of the respiratory burst in human polymorphonuclear neutrophils: a critical review. J. Reticuloendothel. Soc. 24:73 (1978).
2. P. G. Quie, J. G. White, B. Holmes, and R. A. Good, In vitro bactericidal capacity of human polymorphonuclear leukocytes: diminished activity in chronic granulomatous disease of childhood. J. Clin. Invest. 46:668 (1967).
3. B. Holmes, A. R. Page, and R. A. Good, Studies of the metabolic activity of leukocytes from patients with a genetic abnormality of phagocytic function. J. Clin. Invest. 46:1422 (1967).
4. P. Patriarca, R. Cramer, S. Moncalvo, F. Rossi, and D. Romeo, Enzymatic basis of metabolic stimulation in leukocytes during phagocytosis: the role of activated NADPH oxidase. Arch. Biochem. Biophys. 145:255 (1971).
5. B. B. Paul, R. R. Strauss, A. A. Jacobs, and A. J. Sbarra, Direct involvement of NADPH oxidase with the stimulated respiratory and hexose monophosphate shunt activities in phagocytizing leukocytes. Exp. Cell Res. 73:456 (1972).
6. B. M. Babior, J. T. Curnutte, and R. S. Kipnes, Pyridine nucleotide-dependent superoxide production by a cell-free system from human granulocytes. J. Clin. Invest. 56:1035 (1975).
7. L. R. DeChatelet, L. C. McPhail, D. Mullikin, and C. E. McCall, An isotopic assay for NADPH oxidase activity and some characteristics of the enzyme from human polymorphonuclear leukocytes. J. Clin. Invest. 55:714 (1975).
8. D. Iverson, L. R. DeChatelet, J. Spitznagel, and P. Wang, Comparison of NADH and NADPH oxidase activities in granules isolated from human polymorphonuclear leukocytes with a fluorometric assay. J. Clin. Invest. 59:282 (1977).
9. B. Holmes, and R. A. Good, Laboratory models of chronic granulomatous disease. J. Reticuloendothel. Soc. 12:216 (1972).
10. S. J. Klebanoff, and R. A. Clark, "The Neutrophil: Function and Clinical Disorders," North Holland Publishing Co., Amsterdam (1978).

11. R. K. Root, J. Metcalf, N. Oshino, and B. Chance, H_2O_2 release from human granulocytes during phagocytosis. I. Documentation, quantitation, and some regulating factors. J. Clin. Invest. 55:945 (1975).
12. G. Y. N. Iyer, D. M. F. Islam, and J. H. Quastel, Biochemical aspects of phagocytosis. Nature 192:535 (1961).
13. J. T. Curnutte, M. L. Karnovsky, and B. M. Babior, Manganese-dependent NADPH oxidation by granulocyte particles. The role of superoxide and the nonphysiological nature of the manganese requirement. J. Clin. Invest. 57:1059 (1976).
14. B. M. Babior, J. T. Curnutte, and B. J. McMurrich, The particulate superoxide-forming system from human neutrophils. Properties of the system and further evidence supporting its participation in the respiratory burst. J. Clin. Invest. 58:989 (1976).
15. P. Patriarca, P. Dri, K. Kakinuma, F. Tedesco, and F. Rossi, Studies on the mechanism of metabolic stimulation in polymorphonuclear leukocytes during phagocytosis. I. Evidence for superoxide anion involvement in the oxidation of $NADPH_2$. Biochim. Biophys. Acta 285:380 (1975).
16. P. Bellavite, G. Berton, and P. Dri, Studies on the NADPH oxidation by subcellular particles from phagocytosing polymorphonuclear leukocytes: evidence for the involvement of three mechanisms. Biochim. Biophys. Acta 591:434 (1980).

DISCUSSION

Edited by: B.M. Babior

JOHNSTON: What pH did you use for the xanthine oxidase experiments?
DeCHATELET: pH 5.5.
JOHNSTON: Did you try the experiment at a higher pH?
De CHATELET: Not with the xanthine oxidase system. We did run a complete pH curve with the crude particulate system from activated cells and observed optimal activity in the range from 5.5 to 6.0, but we have not yet performed these experiments with the model systems.
BASFORD: In the past we have had problems with a high blank due to the non-enzymatic production of $^{14}CO_2$ from H^{14}-COOH. How do you get around this?
DeCHATELET: There are a number of trocks which have been circulated by word of mouth, but have never really been published. First, bacterial contamination must be avoided because of the rapid oxidation of formate by virtually all bacteria. The reaction relies on catalase, and on exogenous source of the enzyme should be provided. The ^{14}C-formate should be fresh and immediately before use should be acidified to pH ~4 and readjusted to neutrality to blow off any $^{14}CO_2$ present. Finally, it is essential to use KOH as a trap; other substances--e.g.,hydroxide of Hyamine--give very high blank values, probably because they cap trap the volatile formic acid itself.
BASFORD: If one assumes that Mn^{++} is a part of the NADPH oxidase, it isn't necessary to postulate an NADP free radical in a chain reaction leading to O_2^- or H_2O_2 production.
De CHATELET: I quite agree, but to the best of my knowledge no one has yet demonstrated a metal ion requirement for the oxidase. In fact, once the enzyme is activated by phagocytosis, we have observed that inclusion of 1 mM EDTA has no effect on oxidase activity, measured by a variety of techniques. Thus, I think the enzyme probably does not contain manganese or another metal ion activator.
PATRIARCA: How sensitive is your assay method for H_2O_2, and how does it compare with other methods used for assaying the oxidase activity (O_2 consumption, or the method described by Dr.Kakinuma for example)?
DeCHATELET: I do not have exact numbers available, but it seems to compare favorably with the scopoletin assay of Root et al. using the glucose oxidase system as a source for H_2O_2. In other words, it will pick up nanomolar amounts of hydrogen peroxide. We have not yet compared this assay in a quantitative fashion with other assays but intend to do so. Qualitatively, it compares very favorably in that activity is increased by prior phagocytosis in normal but not CGD cells, NADPH is a better substrate than NADH, etc.

PATRIARCA: Thank you for citing our previous work, which suggested that the oxidation of NADPH by the granules might occur via a chain reaction involving the participation of the NADPH free radical. Is there any technique for directly demonstrating the formation of such a radical?

DeCHATELET: I am not familiar with a method which would allow direct demonstration of a free radical of NADP in such a biological system. Theoretically, it should be possible to use electron spin resonance for such a direct demonstration, but I am unaware that anyone has attempted this.

THE RESPIRATORY BURST IN HUMAN POLYMORPHONUCLEAR LEUCOCYTES STIMULATED BY PARTICLES*

Michèle Markert and Jörg Frei

Laboratoire Central de Chimie Clinique
Centre Hospitalier Universitaire Vaudois (CHUV)
CH-1011 Lausanne, Switzerland

INTRODUCTION

A great deal of information has accumulated regarding the respiratory burst of stimulated polymorphonuclear leucocytes (PMN). Some general features can be outlined from the flood of results obtained by measurements of the different manifestations of the stimulated oxygen metabolism namely superoxide (O_2^-)[1], hydrogen peroxide (H_2O_2) and hydroxyl radical production, or chemiluminescence generation which is a consequence of the production of these activated species. Indeed these manifestations show some common events characterized by a similar time course of reaction, a lag period before the stimulation apparears, a short linear maximal rate and a late decline.

The relationship between the oxygen consumption and the metabolites produced has been worked out from stoichiometric studies which revealed that superoxide anion was an important product during the burst[2]. By treating the phagocytes with cytochalasin B (CB)[3], it was demonstrated that all the measurable oxygen consumed was recovered as O_2^-. The quantitative recovery of superoxide as compared to oxygen consumption has recently been questioned[4], since one has to take into account the stoichiometry of the reactions involved in O_2^- dismutation and H_2O_2 degradation.

As most of the assays for superoxide production by particle stimulated neutrophils are discontinuous, so that it is difficult to

*This research was supported by "Le Fonds National Suisse de la Recherche Scientifique", grant No. 3.271-0.74

make accurate observation on the time course of activation, we have recently investigated O_2 consumption and O_2^- production by zymosan stimulated human neutrophils during continuous and simultaneous recordings[5]. The present communication shows the effect of CB, a known inhibitor of ingestion of particles, but not of their attachment to the cell sulface[6], on the kinetics of O_2 uptake and O_2^- production.

RESULTS AND DISCUSSION

Fig. 1,a shows a typical original polarographic trace of O_2 uptake and the simultaneous optical recording of superoxide production as measured by cytochrome c reduction. A delay in response to the stimulation exists for both reactions, but cytochrome starts to be reduced by activated PMN shortly before the oxygen uptake is detected. This delay cannot be abolished even by raising zymosan concentration. After one minute, a change in the rate of cytochrome c reduction occurs, this corresponding to the beginning of the O_2 consumption. It was suggested[7], that cells deplete intracellular O_2 in the first stage. The change in the rate of O_2^- production could indicate the moment at which the cells take up their O_2 from the medium. This modification in the rate could also indicate the penetration of the particles. The binding of the particle, which is sufficient for oxidase activation[8,9], triggers the initial burst ; during the first minute a good recovery of O_2^- occurs before complete formation of the vacuole. At the end of the reaction, it can be seen that cytochrome c is slowly reoxidized. After 6 minutes, around 14 % of cytochrome has been reoxidized, so that the recovery of O_2^- is not maximal when the cells are incubated for a too long time. This reoxidation can be prevented by the addition of catalase or better catalase plus ethanol[5], which does not modify the O_2 uptake. The addition of superoxide dismutase (SOD) (EC 1.15.11) (fig. 1,b) returns O_2 uptake towards the control level and completely abolishes cytochrome c reduction.

The addition of CB to the medium greatly inhibits O_2 consumption (fig. 2,a). A reduced cell-bacteria association was demonstrated in the presence of CB which correlated with the decreased oxygen consumption[3,9]. The presence of cytochrome c in the medium further inhibits the oxygen uptake to a very low level (fig. 2,b). It can be observed that when cytochrome c has been fully reduced, the rate of O_2 uptake is restored to the level measured in the absence of cytochrome. In addition both lag periods are identical but slightly increased compared with normal cells and the kinetics of O_2^- production do not show the biphasic character. In that case, the cells only take up oxygen from the medium, and as CB prevents penetration of the particules, the extracellular recovery of O_2^- is not modified.

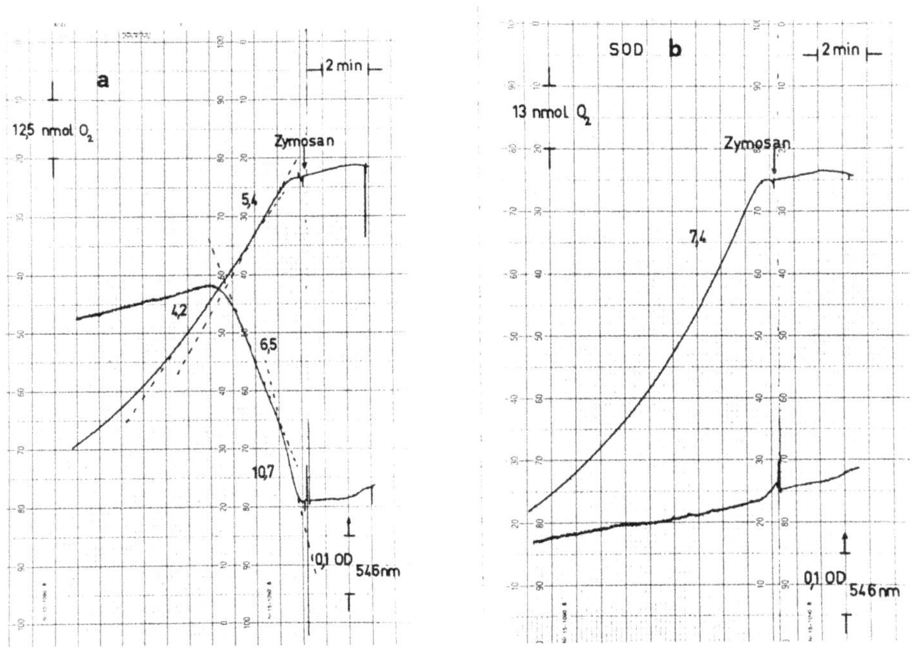

Fig. 1. (a,b) Continuous and simultaneous recording of O_2 consumption (upper trace) and O_2^- production (lower trace). The experiment was carried out in a stirred, thermostatically controlled (37 °C) glass cuvette of 0,42 ml and 0,6 cm light path length with reactants at the following final concentrations : PMN ($3x10^6$/ml) in Dulbecco's buffer pH 7,4 containing $CaCl_2$ (0,7 $mmol.l^{-1}$), $MgCl_2$ (0,9 $mmol.l^{-1}$) and glucose (12mmol.l^{-1}), dimethylsulfoxide (DMSO) (0,1 % v/v), cytochrome c (100 $\mu mol.l^{-1}$). Zymosan was opsonized by incubating in human autologous serum at 37 °C for 30 min and injected through a vent in the stopper of the cuvette (5 mg/ml).
(b) + SOD (68 µg/ml). The figures on the traces give the rates of the linear portions of the curves in nmol O_2 or $O_2^-.min^{-1}.10^{-6}$ PMN.

The quantitative results obtained from the analysis of the recordings are summarized on table 1. The maximal rates were determined from the initial linear portion of the curve. The addition of CB inhibits the rate of oxygen consumption by 75 % without modifying significantly the rate of production of superoxide. The inhibitory effect of adding cytochrome c can be seen in both normal and CB-

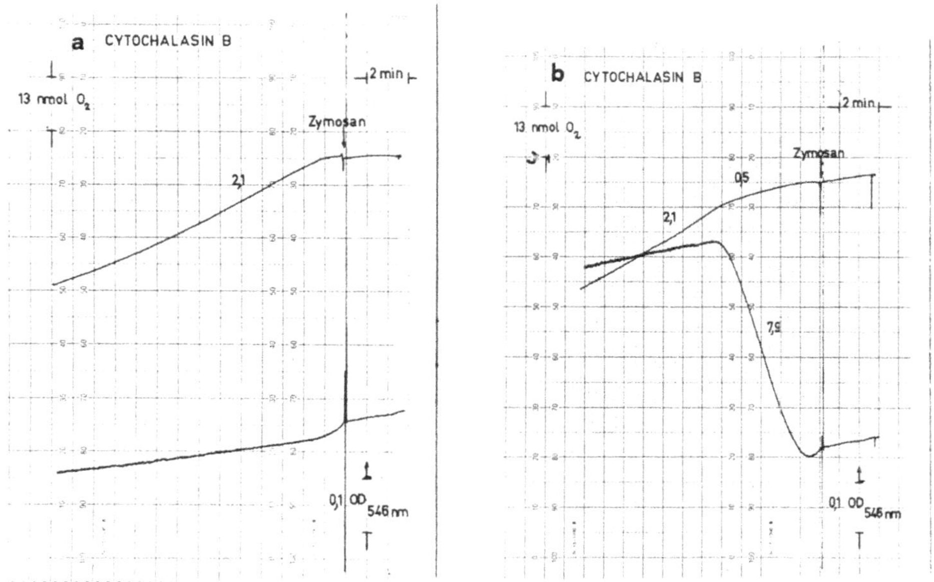

Fig. 2. (a,b) Kinetic of O_2 consumption (upper trace a,b) in the presence of cytochalasin B (5 µg/ml) and the simultaneous optical recording in the absence of cytochrome (lower trace a) or in the presence of cytochrome c (lower trace b). Experimental conditions as described in fig. 1.

treated cells, but to a greater extent in the latter, 27 and 80 % respectively. This more marked decrease in oxygen consumption in the presence of cytochrome c in the case of CB-treated cells is suggestive of a plasma membrane location of the O_2^--generating system. Indeed, as vacuolization is prevented, O_2^- is easily released and a higher recovery of O_2^- induces a more marked inhibition of O_2 uptake.

The stoichiometry between the extracellular O_2^- detected and the oxygen given back by the cytochrome c reduction is 3.7 O_2^- for 1 O_2 in normal cells and 4,1 in CB-treated cells. This experimental ratio measured in the absence of azide is identical to that found by Dri et al.[4], for guinea-pig PMN which means that the H_2O_2 arising from the extracellular O_2^- dismutation is degraded by catalase[4]. The addition of SOD prevents the reduction of cytochrome c and the decrease of O_2 consumption in both normal and CB-treated cells.

The extracellular recovery of O_2^- accounts for 147 % of the O_2 consumed during the initial burst, but 99 % if calculated with the subsequent rate of O_2^- production. With CB treatment 328 % of the

Table 1. Effect of Cytochalasin B on the relationship between the rate of O_2 consumption and O_2^- production by zymosan stimulated leucocytes[a].

CELLS	ADDITIONS	RATE OF OXYGEN CONSUMPTION mean	SD	RATE OF SUPEROXIDE PRODUCTION mean	SD
NORMAL (4)[b]	None	7,69	0,33		
	Cytochrome c	5,63	0,35	11,28[1] 7,63[2]	3,88
	Cytochrome c + SOD	6,97	1,51	0	
CYTOCHALASIN B (4)	None	1,89	0,67		
	Cytochrome c	0,38	0,37	6,20	2,58
	Cytochrome c + SOD	2,05	0,69	0	

[a]Conditions as described in fig. 1. Results in nmol O_2 or $O_2^-.min^{-1}.10^{-6}$PMN.
[b]Number of paired experiments in parenthesis.
[1]Rate from the initial linear portion of the curve (fig. 1,a).
[2]Rate from the subsequent linear portion of the curve.

O_2 consumed is recovered as O_2^-. According to Dri et al.[4], the amount of O_2^- released and the consumption of O_2 measured account for a small part of the total O_2 reduced. A recalculation based on a ratio between O_2 consumption and the total O_2^- generated in the absence of azide, namely 2,6[4], gives a recovery of 56 % and 38 % for normal cells whereas 126 % is found in CB-treated cells. These findings are consistent with the production of O_2^- at the surface of the cells and the results provide evidence that the O_2 consumed is all reduced to O_2^- and that the O_2^- is totally released and trapped extracellularly.

The total amount of O_2^- produced at the end of cytochrome c reduction as shown by the plateau on the recording and the total oxygen consumed after the same time were also measured (table 2). The mean time for the total reduction is 4 minutes in the normal cells and 10 minutes in those containing CB. The same observations can be drawn as for the rates, 40 % inhibition of O_2 uptake upon addition of CB, 20 % inhibition of O_2 consumption in the presence of cytochrome c in normal cells and 40 % in CB cells. The calculated ratios between the extracellular O_2^- and the cytochrome c-inhibitable O_2 consumption are found to be much higher than 4. As shown on fig. 3, the inhibitory action of added cytochrome c is only detected above a concentration of 50 μmol.l^{-1}. After 2 to 3 minutes of reaction, the concentration of cytochrome c is halved and its inhibitory action when reacting with O_2^- is abolished, so that the conditions are optimal at the early stage of the stimulation, but additional reactions which return O_2 to the medium cannot be excluded.

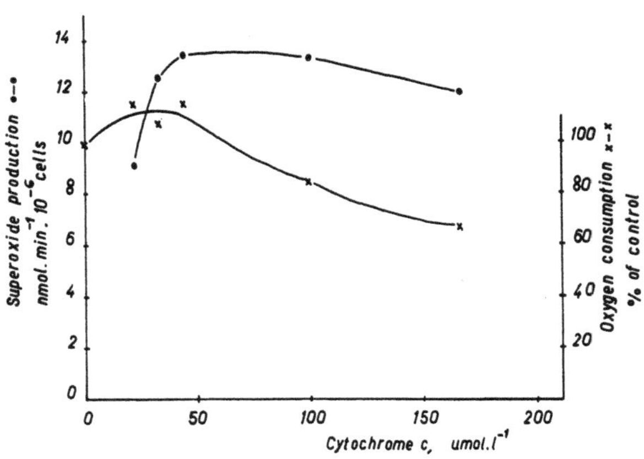

Fig. 3. The effect of varying cytochrome c concentration on the initial rate of O_2^- production (upper curve) and the rate of O_2 consumption in % of control (lower curve) by normal activated cells. For experimental conditions, see legend to fig. 1.

Table 2. Effect of cytochalasin B on the relationship between the total amount of superoxide produced and the total amount of oxygen consumed[a].

CELLS	ADDITIONS	OXYGEN CONSUMPTION mean	OXYGEN CONSUMPTION SD	SUPEROXIDE PRODUCTION mean	SUPEROXIDE PRODUCTION SD
NORMAL (4)[b]	None	24,17	4,48		
	Cytochrome c	19,53	2,27	25,74	2,03
	Cytochrome c + SOD	23,43	2,17	0	
CYTOCHALASIN B (4)	None	10,39	1,00		
	Cytochrome c	6,47	0,88	27,63	0,49
	Cytochrome c + SOD	11,39	1,10	0	

[a]Conditions as described in fig. 1.
[b]Number of paried experiments in parenthesis.
Results in nmol O_2 or $O_2^-.10^{-6}$ PMN measured at the end of cytochrome c reduction as shown by the plateau on the recording (fig. 1,a and 2,b). The total oxygen consumed was measured after the same time. Mean time for cytochrome c reduction was 4 minutes for normal cells and 10 minutes for CB-treated cells.

The recovery of O_2^- in the extracellular medium calculated with a ratio of 1 O_2 consumed for 2,6 O_2^- generated as discussed above is 41 % for normal cells and 100 % for CB-treated cells. Other authors[3,10] have also measured a much better recovery of O_2^- in the presence of CB, but their values are only less than half ours. These differences might be due to a better efficiency of our system. Indeed both parameters are measured under strictly identical conditions, continuous stirring and simultaneous recording. It was shown[11] that phagocytosis and therefor release of oxygen metabolites was markedly increased by continuous agitation. Treatment of cells with CB significantly inhibits O_2 uptake, and hence the total O_2^- generation of zymosan-stimulated PMN, so that the better recovery of O_2^- is due to an enhanced release of the oxygen products. When using the soluble stimulus Concanavalin A[12], the rise in O_2^- recovery in the presence of CB seems to be rather due to an increased generation (stimulation of O_2 uptake) rather than an enhanced release of the anion, whereas another soluble stimulus phytohaemaglutinin[12] enhances both generation and release of O_2^-.

REFERENCES

1. B. M. Babior, Oxygen-dependent microbial killing by phagocytes, N. Engl. J. Med. 298:659 (1978).
2. B. M. Babior, R. S. Kipnes, and J. T. Curnutte, The production by leukocytes of superoxide, a potential bactericidal agent, J. Clin. Invest. 52:741 (1973).
3. R. K. Root, and J. A. Metcalf, H_2O_2 release from human granulocytes during phagocytosis. Relationship to superoxide anion formation and cellular catabolism of H_2O_2: studies with normal and cytochalasin B-treated cells, J. Clin. Invest. 60:1266 (1977).
4. P. Dri, P. Bellavite, G. Berton, and F. Rossi, Interrelationship between oxygen consumption, superoxide anion and hydrogen peroxide formation in phagocytosis guinea pig polymorphonuclear leukocytes, Mol. Cell. Biochem. 23:109 (1979)
5. M. Markert, M.-J. Allaz, and J. Frei, Continuous monitoring of oxygen consumption and superoxide production by particle-stimulated human polymorphonuclear leukocytes, FEBS Lett. 113:225 (1980).
6. A. T. Davis, R. Estensen, and P. Quie, Cytochalasin B III. Inhibition of human polymorphonuclear leukocytes phagocytosis, Proc. Soc. Exp. Biol. Med. 137:161 (1971)
7. K. Takanaka, and P. J. O'Brien, Generation of activated oxygen species by polymorphonuclear leukocytes, FEBS Lett. 110:283 (1980).

8. I. M. Goldstein, D. Roos, H. Kaplan, and G. Weissmann, Complement and immunoglobulins stimulate superoxide production by human leukocytes independently of phagocytosis, J. Clin. Invest. 56:1155 (1975).
9. G. Zabucchi, M. R. Soranzo, G. Berton, D. Romeo, and F. Rossi, The stimulation of the oxidative metabolism of polymorphonuclear leukocytes: effect of colchicine and cytochalasin B, J. Reticuloendothel. Soc. 24:451 (1978)
10. D. Roos, J. W. T. Homan-Muller, and R. S. Weening, Effect of cytochalasin B on the oxidative metabolism of human peripheral blood granulocytes, Biochem. Biophys. Res. Commun. 68:43 (1976).
11. R. K. Root, J. Metcalf, N. Oshino, and B. Chance, H_2O_2 release from human granulocytes during phagocytosis I. Documentation, quantitation, and some regulating factors, J. Clin. Invest. 55:945 (1975).
12. M. S. Cohen, J. A. Metcalf, and R. K. Root, Regulation of oxygen metabolism in human granulocytes: relationship between stimulus binding and oxidative response using plant lectins as probes, Blood, 55:1003 (1980).

DISCUSSION
Edited by B.M. Babior

ROOS: You are not measuring total quantities of oxygen consumed or superoxide produced, but steady state rates. Oxygen is constantly being consumed, and at the same time is being regenerated by the reaction of superoxide with cytochrome c and the reaction of hydrogen peroxide with catalase. Cytochrome c is constantly being reduced by superoxide and at the same time is being oxidized by hydrogen peroxide. Therefore, I really don't see how can calculate stoichiometry from these numbers.

MARKERT: The presence of cytochrome c in the reaction medium prevents superoxide from dismuting to hydrogen peroxide. Cytochrome c reoxidation is thereby prevented. Hydrogen peroxide could diffuse from the inside of the cell, but it is known that very little peroxide is recovered unless NaN_3 is present, suggesting that it is mostly destroyed within the cell. In our experiments, reoxidation of cytochrome c is not detected until the reaction reaches completion, at which time the superoxide is no longer trapped by cytochrome c and can dismute to hydrogen peroxide.

Superoxide production as measured by cytochrome c reduction corresponds to the oxygen which is consumed but not detected because of the reaction of O_2^- with cytochrome c. This undetected oxygen plus the oxygen consumption measured in the presence of cytochrome c gives the total amount of oxygen consumed in the reaction. With regard to the oxygen uptake measured in the absence of cytochrome c, this represents only 25-50% of the total oxygen consumption as measured in the presence of cytochrome c. This discrepancy is due to the O_2^-- and H_2O_2-consuming reactions which occur during the burst in the absence of cytochrome c, as was shown by Rossi. Our data confirm his approach.

STIMULATION OF THE HEXOSE MONOPHOSPHATE SHUNT ACTIVITY IN HUMAN POLYMORPHONUCLEAR LEUKOCYTES

J. Glette and H.H. Bassøe

Medical Department B, Haukeland sykehus
University of Bergen
Bergen, Norway

SUMMARY

The glucose metabolism in human polymorphonuclear (PMN) leukocytes was stimulated by latex particles, zymosan particles, gamma-G globulin and formylmet-leu-phe (FMLP) peptide. Maximum stimulation occured when zymosan particles in amounts above 0.25 mg/ml per 4 x 10^6 leukocytes were used. Latex particles, gamma-G globulin and FMLP stimulated the glucose oxydation less than zymosan particles. The glucose oxydation increase in proportion to the amount of latex particles and gamma-G globulin in the measured range, while the dose response curve using FMLP shows maximum activity at a consentration of the peptide of 5 x 10^{-5}M per 4 x 10^6 leukocytes.

It is likely that FMLP and gamma-G globulin act on the glucose metabolism via cell surface receptors. The synthetic polystyrene particles (latex) do not occupy these receptors, but stimulate the hexose monophosphate (HMP) shunt activity of the PMN via intracellular mechanisms. Opsonized zymosan particles seem to act on the leukocyte glucose metabolism via cell surface receptors as well as via intracellular mechanisms.

INTRODUCTION

It has been shown that latex particles and zymosan particles are phagocytosed by polymorphonuclear leukocytes.

Phagocytosis of latex particles are followed by a four to five fold increase in the metabolic rate of the leukocytes measured as $^{14}CO_2$ produced from (1-^{14}C) glucose. This is a measure of the HMP shunt activity of the phagocytosing cells, which is the major energy yielding process of the granulocyte intracellular killing of bacteria[2].

Chemotaxis, attachment of particles the cell surface and the process of phagocytosis are influenced by serum factors such as complement and immunoglobulins. The chemotactic active peptide FMLP cause PMN leukocytes to emit light, with maximum activity at a consentration of 10^{-7}M[3]. Latex are polystyrene particles with negatively charged sulphate groups incorporated at the surfase. These sulphate groups are partly removed by dialysis. Zymosan is derived from yeast membranes. By phagocytosing opsonized zymosan particles, the PMN leukocytes emit light. Chemiluminescense seems to be a good measure of the ability of these cells to ingest and kill bacteria[4].

In the present investigation we have exermined the effect of latex particles, zymosan particles, FMLP peptide and gamma-G globulin on the glucose oxydation via the HMP shunt of human PMN leukocytes.

MATERIALS AND METHODS

Leukocytes. All experiments were carried out with leukocytes from normal persons without any tendency to infections and who were not taking drugs.

Leukocytes were isolated by dextran sedimentation of heparinized blood[5]. The cells were resuspended in Krebs-Ringer bicarbonate (KRB) buffer, pH 7.4, to give a final leukocyte consentration of 1 x 10^7 cells/ml.

Condition for incubation. The incubation mixture consisted of leukocytes in buffer (0.4 x 10^7 cells pr. ml), (1-^{14}C)glucose and unlabeled glucose. In addition latex particles, zymosan particles FMLP peptide or gamma-G globulin to a total volume of 1 ml.

Latex particles. (0.81 µm in diameter, Difco Lab., Ditroit Mich, USA) were dialyzed for two days with deionized water and diluted 1 : 9 in KRB buffer before used. For incubation 100 µl of this suspension were used.

Zymosan particles, (Sigma Chemical Company, St. Louis, Mo. USA) were opsonized in human serum and resuspended in KRB buffer to a final consentration of 10 mg/ml. Of this suspension 200 µl were used in the incubation mixtures. To determind the glucose oxydation of the zymosan particles, blind samples without leukocytes were used.

FMLP peptide was obtained from Sigma Chemical Company, St. Louis USA. A stock solution of the peptide (2 x 10^{-4}M) was made in destilled water and kept at - 70°C until used. The stock solution was dilluted in KRB buffer to get the desired final consentrations.

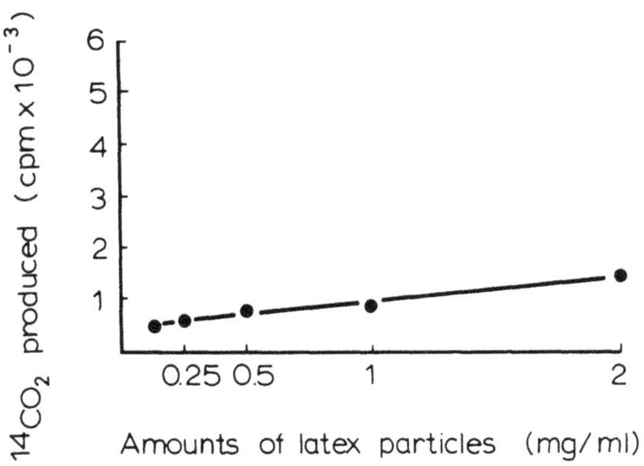

Fig. 1. Effect of increasing amounts of latex particles on $^{14}CO_2$ production from (1 -^{14}C) glucose by 4 x 10^6 leukocytes. Mean of three experiments in dublicate. All experiments are carried out with blood cells from the same donor.

All experiments were carried out with a glucose concentration of 2.78 mM and 0.1 µCi (1-^{14}C) glucose (specific activity 56.8 mCi/mmol Radiochemical Center, Amersham).

The incubation and the CO_2 measurements were done according to a method previously described.[2]

RESULTS

The ($^{14}CO_2$) production increases in proportion to the amount of latex particles in the measured range (Fig 1). Phagocytosis of zymosan particles induces CO_2 production to a greater extend than do phagocytosis of latex particles. (Fig 2). Zymosan itself shows no glucose oxydation activity. Maximal HMP shunt activity is obtained when zymosan consentrations above 0.5 mg/ml per 4 x 10^6 leukocytes are used. The dose response curve using gamma-G globulin as stimulatry agent shows a similar progress. The gamma-G globulin has a saturable stimulatory effect on the leukocyte hexose monophosphate shunt activity. (Fig. 3).

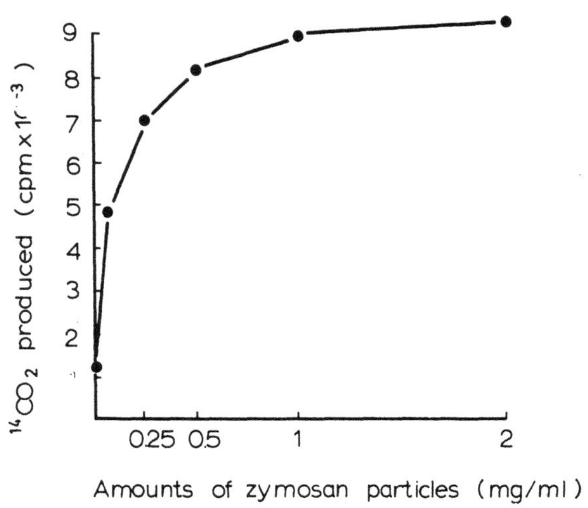

Fig. 2 Effect of increasing amounts of zymosan particles on $^{14}CO_2$ production from (1-^{14}C) glucose by 4 x 10^6 leukocytes. Mean of three experiments in dublicate. All experiments are carried out with blood cells from the same donor.

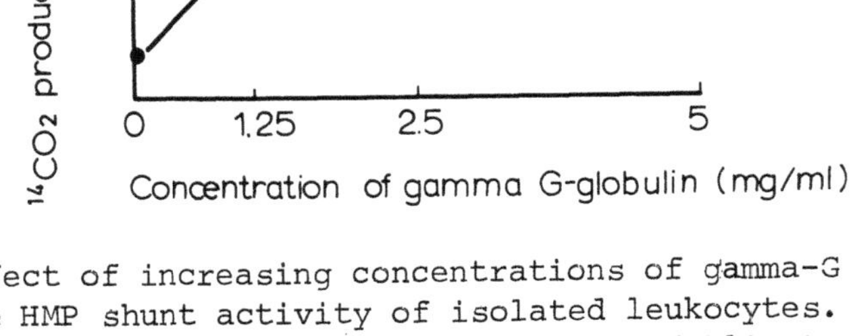

Fig.3. Effect of increasing concentrations of gamma-G globulin on the HMP shunt activity of isolated leukocytes. The results are the mean of three experiments in dublicate carried out with 4 x 10^6 cells/ml.

The FMLP peptide induces (CO_2) production from ($1-^{14}C$) glucose with a maximum activity at a consentration of 5 x 10^{-5}M per $4x10^6$ leukocytes. Higher consentrations of the peptide causes a decrease in the HMP shunt activity of the leukocytes (Fig.4). Cells incubated in the precens of both FMLP and latex particles show a greater glucose oxydation activity than cells incubated with any of these substances alone. The effect of latex and FMLP together seems to be the sum of the effect of the individual componds (Table 1). Opsonized zymosan particles induce a six to seven fold increase in the glucose oxydation by leukocytes compared to both FMLP and latex particles. Neither FMLP nor latex particles incubated together with zymosan cause any further increase in the leukocyte HMP shunt activity compared to samples incubated with opsonized zymosan alone. (Table 1).

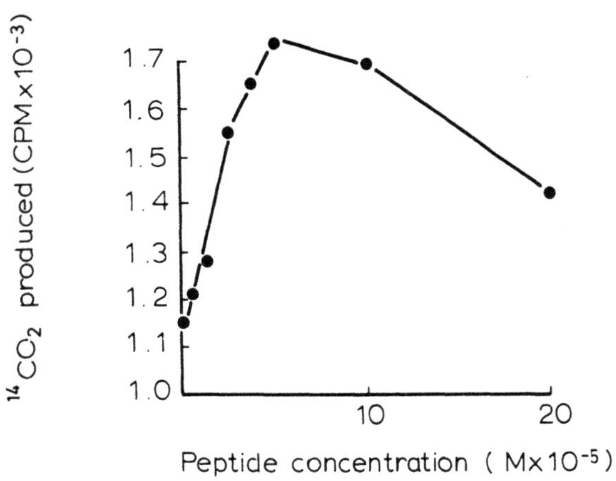

Fig. 4. Effect of the FMLP peptide on $^{14}CO_2$ production from ($1-^{14}C$) glucose by 4 x 10^6 leukocytes during 1 hour of incubation. All experiments are carried out with blood cells from the same donor. The results are the mean of three experiments in dublicate.

Table 1. CPM in the $^{14}CO_2$ fraction produced from (1-^{14}C) glucose by isolated leukocytes upon stimulation with various stimuli.

	CPM
RESTING CELLS	420
FMLP	1132
LATEX PARTICLES	1425
LATEX PARTICLES + FMLP	2547
ZYMOSAN PARTICLES	5508
ZYMOSAN PARTICLES + LATEX PARTICLES	5991
ZYMOSAN PARTICLES + FMLP	5238
GAMMA G-GLOBULIN (5 mg/ml)	1627

DISCUSSION

The CO_2 production from (1-^{14}C) glucose is proportional to the amount of latex particles used, while increasing amounts of zymosan particles have a saturable effect on the glucose oxydation by the leukocytes. Calculations indicate that the reason for this observation may be saturation of the leukocyte enzyme systems rather than glucose or oxygen deficiency of the incubation mixture.

The high stimulation of the metabolic rate of the leukocytes by zymosan particles compared to latex particles is not clearly understood. The incubation mixture contained about the same amount of particles whether zymosan or latex are used. This means the process of phagocytosis or the stimulation of the metabolic rate of the leukocytes depends on physical or chemical properties of the particles to be phagocytosed. Opsonization of the particles enhances the process of phagocytosis. More particles are phagocytosed per leukocyte. This results in an increase in the $^{14}CO_2$ produced from(1-^{14}C) glucose by the leukocytes. The effect of opsonization is greater when zymosan is used than when latex is used. Even nonopsonized zymosan particles stimulate the HMP shunt to a greater extent than do opsonized latex particles. The reason for this observation may be that the yeast derived zymosan particles but not the synthetic polystyrene particles (latex) are involved in the oxydation reduction reactions taking place in the leukocytes during phagocytosis and intracellular killing. This is in agreement with the fact that opsonized zymosan particles initiate chemiluminescense by phagocytosis while opsonized latex particles (0.81 µm) do not. Zymosan is shown to be a substrate in the light emmiting reactions of polymorphonuclear leukocytes[3].

The FMLP peptide stimulate the phagocyte to produce CO_2 via the HMP shunt with maximum effect at a consentration of 50 µM. This peptide has previously been shown to have stimulatory effect on leukocyte chemotaxis and leukocyte chemiluminescense[3].

The reason why the dose response curve reaches a maximum and then declines is not cleary understood. Glucose oxydation via the HMP shunt is limited by the rate at which $NADP^+$ becomes available through the oxydation of NADPH. At low consentration of the peptide the initial rate at which NADPH are produced are lower than the reactivation rate of $NADP^+$ by the cell. At higher consentrations of the peptide the initial NADPH production rate are greater than the reactivation rate of $NADP^+$. NADPH are than accumulated in the cell, and the HMP shunt activity are inhibited. This may be one explanation of the observed shape of the dose response curve. Another explanation may be that the peptide is toxic to the leukocytes at the highest consentrations used. It is likely that the leukocyte response to the peptide is caused by activation of receptors on the leukocyte plasma membrane. These reseptors are blocked by the opsonized zymosane particles, since no increase in HMP shunt activety is observed caused by the peptide when leukocytes are incubated with peptide together with zymosan. The effect of latex particles seems to be added to the effect of the peptide. This indicate that latex particles do not occupy the peptide reseptor site on the leukocyte plasma membrane.

Previous investigations have postulated two mechanisms of leukocyte chemiluminescense induction, one inhibited by superoxide dismutase (SOD), and the other inhibited by azide.[6] The present investigation indicate that there may be two mechanisms of HMP shunt activation in leukocytes, one via cell surfase reseptors and the other via intracellular activation mechanisms.

REFERENCES

1. A. W. Segal, J. Dorling and S. Coade, Kinetics of Fusion of the Cytoplasmic Granules with Phagocytic Vacuoles in Human Polymorphonuclear Leukocytes, J.Cell Biology, 85:42 (1980)
2. B. Kjøsen, H.H. Bassøe and O. Myking, The Glucose Oxydation in Isolated Leukocytes from Female Patients Suffering from Overweight or Anorexia Nervosa, Scand. J. clin.Lab. Invest. 35:447 (1975)
3. G. E. Hatch, D.E. Gardner and D.B. Menzel, Chemiluminescence of Phagocytic Cells Caused by N-formylmetionyl Peptides, J. exp. Med. 147:182 (1978)
4. R. C. Allen, R.L. Sternholm and R. H. Steele, Evidence of the Generation of Excitation State(s) in Human Polymorphonuclear Leukocytes and its Participation in Bactericidal Activity, Biochem. biophys. Res. Commun. 47: 679 (1972)
5. A. Bøyum, Separation of Leukocytes from Blood and Bone Marrow, Scand. J. Clin. and Lab. Invest, Suppl. 97, 21:1 (1968)
6. H. Rosen and S. J. Klebanoff, Chemiluminescence and Superoxide Production by Myeloperoxydase-deficient Leukocytes, J.Clin. Invest. 58:50 (1976)

THE NATURE AND FUNCTION OF THE MICROBICIDAL OXIDASE SYSTEM OF NEUTROPHILS

Anthony W. Segal, Angela Harper, Rodolpho Garcia
University College London
Department of Haematology, University College London
London W.C.1.

Owen T.G. Jones, Andrew R. Cross
University of Bristol
Department of Biochemistry, University of Bristol
University Walk, Bristol

Michael Geisow
National Institute for Medical Research
Department of Biophysics, National Institute for Medical Research, Mill Hill, London N.W.7

Stimulated neutrophils demonstrate a burst of non-mitochondrial respiration.[1,2] This appears to be important for the killing of certain bacteria which are killed less efficiently under anaerobic conditions,[3] or by cells from patients with Chronic Granulomatous Disease (CGD) which fail to demonstrate this respiratory burst.[4]

The nature of the oxidase system has been the subject of considerable interest and controversy. The leading contenders have been NADH[5] and NADPH[6] oxidases. There is no specific reason for implicating NADH. NADPH could be involved because of a dramatic increase in the activity of the hexose monophosphate shunt in stimulated cells[7] and a CGD like state in severe glucose 6 phosphate dehydrogenase deficiency.[8] However, there is no evidence that the actual substrate is NADPH - it could be another compound that receives its reducing equivalents from NADPH.

We have identified a cytochrome b in human leukocytes that we believe to be a major component of the oxidase system and probably the oxidase itself. The properties of this cytochrome are as follows:[9-13]

1. It is present in neutrophils, monocytes and eosinophils in concentrations of about 50-100 pmols/mg protein. It is absent from, or present in very low concentrations in basophils, fibroblasts, adipocyte membranes, erythrocyte ghosts, pancreatic membranes, hepatocyte membranes and lymphocytes, both B and T cells.

2. In neutrophils it is located predominantly in the plasma membrane,[9,10,11] with a smaller amount associated with the specific granules, and becomes incorporated into the phagocytic vacuoles.

3. A role as a component of the oxidase system comes from the observations that activation of the cells with phorbol myristate acetate (PMA) reduces the cytochrome[12] and that the cytochrome is either absent or functionally abnormal in patients with CGD.[13] PMA stimulation causes the rapid reduction of the cytochrome, an effect that is enhanced by anaerobiosis and reversed by the reintroduction of O_2, indicating reversible reduction and oxidation by a substance that stimulates the respiratory burst. We have studied eight patients in whom the disease is inherited in an X-linked manner and all are lacking spectral evidence of the cytochrome. Three patients with an autosomal recessive mode of inheritance have been studied, their cells appear to have a

normal cytochrome b, however, they fail to reduce it after stimulation with PMA. This indicates that they are probably lacking a proximal electron donor or are defective in some activation mechanism.

4. The cytochrome has a $Em_{7.0}$ = -245 mV. This is the lowest mid-point potential of any mammalian b cytochrome. It is pH dependant with characteristics of a 1 electron transfer. The same mid-point potential was observed in neutrophils, monocytes, macrophages and eosinophils. The latter also seemed to contain a b cytochrome with an $Em_{7.0}$ = -186 mV which we believe is probably a conversion product from b_{245}, because we have observed a mid-point potential of -186 mV in the b of neutrophils after solubilisation with detergents.

5. The neutrophil b_{245} has properties compatible with those of an oxidase. It binds CO to form a complex which is photodissociable and demonstrates a reassociation half-time of 6 m sec. After reduction with dithionite the reintroduction of oxygen causes rapid reoxidation of the cytochrome.

6. It contains a protohaem group.

7. The cytochrome is not reduced upon the addition of NADH or NADPH to membrane preparations under anaerobic conditions.

The function of this oxidase system has not been established. It is possible that this electron transport system does exert its effect through the production of oxidative or reductive reduced oxygen species. As an alternative or supplementary system, it could enhance bacterial killing by optimising conditions within the phagocytic vacuole for killing by another mechanism. Cationic proteins from the granules of human neutrophils are very toxic to bacteria in vitro, but only at pHs above 7.0.[14]

We exploited the indicator properties of fluorescein[14] to determine the intravacuolar pH and found that there is normally a rise in pH to levels of about 7.8 before a gradual fall to the previously described level of 6.0 - 6.5.[15] In cells from patients with CGD, or in the absence of oxygen, the initial rise in pH was absent and the fall was much faster and to lower levels than normal. In addition, digestion of killed Staphylococcus aureus and subsequent release of solubilised bacterial products was much lower in cells from the patients. Whereas phagocytic vacuoles normally enlarge considerably in the first few minutes after engulfment, those in the cells of patients with CGD remained small - an observation that has been previously reported. If bacteria are replaced by latex particles as the object of phagocytosis, swelling of the phagocytic vacuoles is not observed.

We have interpreted these results as indicating that:

1. The oxidase system elevates intravacuolar pH.
2. The rise in pH activates microbial killing by cationic proteins[17] and permits digestion by enzymes acting optimally at neutral pH.
3. The digestion products osmotically attract water into the vacuole, causing it to swell and eventually exocytose its contents to the exterior.
4. In the absence of oxidase activity, such as occurs in CGD and anaerobic cells, the rapid fall in pH inactivates killing and digestive systems. Killing, digestion and vacuolar size in CGD cells were all returned towards normal values in the presence of methylamine which also elevated the vacuolar pH.[16]

We believe that the oxidase system contains a very low potential cytochrome, $\underline{b}_{245}$ which probably functions as the terminal oxidase molecule itself. The oxidase may function to directly kill the ingested organism, but an alternative or supplementary role could be to elevate and regulate the intravacuolar pH so as to optimise conditions for killing by some other mechanism such as the granule proteins.

Key Words:
1. Oxidase
2. Cytochrome b
3. Vacuolar pH
4. Chronic Granulomatous Disease
5. Microbicidal mechanisms.

REFERENCES

1. Baldridge C.W. and Gerard R.W. (1933)
The extra respiration of phagocytosis.
Am. J. Physiol. 103, 235-236.

2. Sbarra A.J. and Karnovsky M.L. (1959)
The biochemical basis of phagocytosis. 1. Metabolic changes during the ingestion of particles by polymorphonuclear.
J. Biol. Chem. 234, 1355-1362.

3. Salvaraj R.J. and Sbarra A.J. (1966)
Relationship of glycolytic and oxidative metabolism to particle entry and destruction in phagocytosing cells.
Nature 211, 1272-1276.

4. Holmes B., Page A.R. and Good R.A. (1967)
Studies of the metabolic activity of leukocytes from patients with a genetic abnormality of phagocytic function.
J. Clin. Invest. 46, 1422-1432.

5. Cagan R.H. and Karnovsky M.L. (1964)
Enzymic basis of the respiratory stimulation during phagocytosis.
Nature 204, 255-257.

6. Iyer G.Y.N. and Quastel J.H. (1963)
NADPH and NADH oxidation by guinea pig polymorphonuclear leukocytes.
Canadian Journal of Biochemistry and Physiology, 41, 427-434.

7. Rossi F. and Zatti M. (1964)
Changes in the metabolic pattern of polymorphonuclear leukocytes during phagocytosis.
Brit. J. Exp. Path. 45, 548-559.

8. Gray G.R., Klebanoff S.J., Stamatoyannopoulos G., Austin T., Naiman S.C., Yoshida A., Kliman M.R. and Robinson G.C.F. (1973)
Neutrophil dysfunction, chronic granulomatous disease and non-spherocytic haemolytic anaemia caused by complete deficiency of glucose-6-phosphate dehydrogenase.
Lancet 2, 530-534.

9. Segal A.W. and Jones, O.T.G. (1978)
A novel cytochrome b system in phagocytic vacuoles from human granulocytes.
Nature 276, 515-517.

10. Segal A.W. and Jones O.T.G. (1979)
The subcellular distribution and some properties of the cytochrome b component of the microbicidal oxidase system of human neutrophils.
Biochem. J. 180, 33-44.

11. Segal A.W. and Jones O.T.G. (1980)
Rapid incorporation of the human neutrophil plasma membrane cytochrome b into phagocytic vacuoles.
Biochem. Biophys. Res. Commun. 92, 710-715.

12. Segal A.W. and Jones O.T.G. (1979)
Reduction and subsequent oxidation of a cytochrome b of human neutrophils after stimulation with phorbol myristate acetate.
Biochem. Biophys. Res. Commun. 88, 130-134.

13. Segal A.W. and Jones O.T.G. (1980)
Absence of cytochrome b reduction in stimulated neutrophils from both female and male patients with Chronic Granulomatous Disease.
FEBS Letters 110, 111-114.

14. Ohkuma S. and Poole B. (1978)
Fluorescence probe measurement of the intralysosomal pH in living cells and the perturbation of pH by various agents.
Proc. Natl. Acad. Sci. USA. 75, 3327-3331.

15. Mandell G.L. (1970)
Intraphagosomal pH of human polymorphonuclear neutrophils
Proc. of the Soc. for Exp. Biol. and Med. 134, 447-449.

16. Quie P.G., White J.G., Holmes B. and Good R.A. (1967)
In vitro bactericidal capacity of human polymorphonuclear leukocytes: Diminished activity in chronic granulomatous disease of childhood.
J. Clin. Invest. 46, 668-679.

18. Odeberg H. and Olsson I. (1975)
Anti-bacterial activity of cationic proteins from human granulocytes.
J. Clin. Invest. 56, 1118-1124.

Acknowledgements

We would like to thank the Wellcome Trust and Medical Research Council for Financial Support.

DISCUSSION

Edited by: B.M. Babior

ALLEN: What is the contribution of bacterial metabolism to vacuolar pH?

SEGAL: None. The bacteria are dead.

ALLEN: In your pH schema, electrical neutrality is not maintained. How can you explain this?

SEGAL: I believe that this system maximizes the function of a variety of enzymes and proteins with various pH optima. Initially, the pH is raised through the action of the oxidase. It then gradually falls as degranulation continues.

DeCHATELET: I am concerned about the observation that the CGD vacuoles are smaller than normal. You ascribe this to decreased degradation of vacuolar contents, yet most of the lysosomal hydrolases are most active at acid pH. This might be expected to result in increases osmotic pressure, leading to CGD vacuoles that are larger than normal rather than smaller.

SEGAL: We have no real knowledge of the pH vs. activity profile of lytic enzymes in human neutrophils.

HAMERS: I want to comment on the pH changes you showed us. In measurements made within 4 minutes, we obtained exactly the opposite results. In my opinion, our results make more sense, in that the rapid acidification of the phagosome/phagolysosome would help the acid hydrolases. I also want to ask you a question. How do you think that just a few granules--as you showed--fusing with a large bacterium-containing vacuole could bring about a pH drop of close to 1.5 units? In my opinion, the excess protons inside the granules would be diluted out and buffered immediately if no active pumping mechanism were present in the phagosome.

SEGAL: The observation was of a rise in the vacuolar pH. I believe that this results from the activity of the oxidase system. In the absence of this activity, I believe that the pH falls as a result of the discharge of the acidic granular contents. In the normal cells the fall in pH was very slow, and continued for a long time after the cessation of the respiratory burst.

ELSBACH: Any assignment of importance to oxygen-independent systems obviously pleases me. However, I wish to make a few comments:

Our own findings suggest that oxygen-dependent bactericidal systems are dominant in the killing of at least some Gram positive bacterial species.

With respect to Gram negative bacteria, I wish to remind you that the cationic proteins that we consider important are very toghtly associated with the granule membrane. Your measurement of pH in the fluid phase of a cell compartment (intravacuolar) may have little relevance to the proton concentration at an interphase, which is probably what counts.

CHEMILUMINESCENCE AND THE STUDY OF PHAGOCYTE REDOX METABOLISM

Robert C. Allen

U.S. Army Institute of Surgical Research
and Clinical Investigation Service
Brooke Army Medical Center
Fort Sam Houston, Texas 78234 U. S. A.

INTRODUCTION

The phenomenon of polymorphonuclear leukocyte (PMNL) chemiluminescence (CL) is a natural consequence of the redox metabolism that follows phagocytosis or chemical stimulation. These metabolic alterations are collectively referred to as the respiratory burst. As the term implies, there is a rapid increase in non-mitochondrial O_2 consumption, and a corresponding increase in glucose metabolism via the hexose monophosphate (HMP) shunt (Karnovsky, 1968; Rossi et al., 1971).

Of the mechanisms advanced to explain these metabolic changes, the proposal of Rossi et al., has best stood the test of time (Rossi and Zatti, 1964, 1966). Simply stated, phagocytic or chemical stimulation of the PMNL effects activation of NADPH:O_2 oxidoreductase (NADPH oxidase) as measured by decrease in Km. Activity of the HMP shunt dehydrogenases is controlled by the availability of oxidized cofactor, $NADP^+$. Therefore, increase in the $NADP^+$/NADPH ratio following oxidase activation results in increase in dehydrogenase activity (Patriarca et al., 1971). The consumption of O_2 results from its reduction by the oxidase. In this regard, Iyer et al., (1961) proposed that H_2O_2 is a product of the respiratory burst. The generation of H_2O_2 was later confirmed by the work of Paul and Sbarra (1968) and Zatti et al., (1968).

---

*The opinions or assertions contained herein are the private views of the author and are not to be construed as reflecting the views the Department of the Army or the Department of Defense.

The H_2O_2 could serve as substrate for myeloperoxidase (MPO), a PMNL lysosomal enzyme. Cell-free preparations of MPO are highly microbicidal when presented with H_2O_2 and an oxidizable halide co-factor in a mild acid environment. (Klebanoff 1968, 1971; McRipley and Sbarra, 1967).

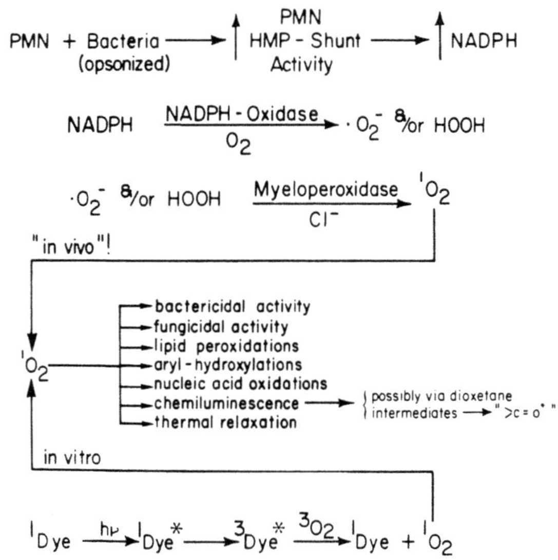

Fig. 1. Initially proposed mechanism for PMNL-CL (Allen et al., 1972).

These antecedent observations provided the basis for the following inferential construction. Phagocyte microbicidal activity is an energy requiring process. The primary event in this process is utilization of the potential of NADPH for reduction and electronic alteration of O_2. This enzymatic action could best be explained by a flavoprotein oxidase, and the evidence available at the time suggested that this was the case. Firstly, dehydrogenation of NAD(P)H is commonly effected by flavoproteins. The coupling of NAD(P)H oxidation to flavoprotein reduction is consistent with the reduction potentials of these redox reactants. Secondly, unlike most metalloprotein oxidases, flavoproteins are not sensitive to inhibition by cyanide. Lastly, flavoproteins are capable of directly catalyzing the univalent reduction of O_2 to yeild the superoxide anion, $\cdot O_2^-$ (Knowles et al., 1969).

The schema of figure 1 was proposed as a possible mechanistic explanation of the oxidative microbicidal action of the PMNL. Note that superoxide, $\cdot O_2^-$, was proposed as a product of NADPH oxidase activity (Allen et al., 1972, 1973), but for reasons that will be discussed, it was not considered as a direct microbicidal reactant. Instead it was assigned an intermediate function in microbicidal action. Acid disproportionation of $\cdot O_2^-$ had been reported to yield H_2O_2 and singlet molecular oxygen, $^1O_2$ (Khan, 1970; Sander and Stauff, 1971). The electrophilic reactivity of $^1O_2$ might be directed to microbicidal oxygenations. Additionally, H_2O_2, the other product of disproportionation, could serve as substrate for the MPO-catalyzed oxidation of Cl^- as originally suggested by Agner (1958). The products of halide oxidation, such as OCl^- or Cl_2, might further react with H_2O_2 to yield $^1O_2$ (Kasha and Khan, 1970). Thus, $^1O_2$ might be generated by two different PMNL mechanisms.

Photodynamic oxidations were known to effect potent microbicidal action since the turn of the century (Raab, 1900; Jodlbauer and von Tappeiner, 1905; Blum, 1941). In 1939, Kautsky proposed that photodynamic oxygenation was effected by the action of $^1O_2$. Chemical support for this $^1O_2$ mechanism was presented by Foote and Wexler (1964) and Corey and Taylor (1964). Thus if $^1O_2$ were generated by the PMNL, it could serve as a microbicidal agent.

$^1O_2$ is an electrophilic reactant capable of reacting with the substituted double bonds and electron rich regions of a substrate. In many cases, dioxygenation of substrate results in the generation of a dioxetane or dioxetanone product. These dioxygen-ring compounds break down to yield electronically excited carbonyl containing products in relatively high yield. The excited products can relax to ground state by photon emission or chemiluminescence (McCapra, 1968).

Therefore, if such electrophilic dioxygenation reactions are involved in phagocyte microbicidal action, there should be an associated light emission or chemiluminescence (CL). A CL response is observed from phagocytosing PMNL. This CL can be quantified by single photon counting techniques, and the integral CL response correlates with the extent of metabolic activation as measured by radiolabeled glucose studies (Allen et al., 1972).

The then proposed mechanism for PMNL-CL was summarized in the following statement taken from the original report. "The $^1O_2$ may react as an electrophile at certain sites of high electron density, π systems, within the cell and/or bacteria to form labile dioxetane structures. These structures may then cleave with the formation of electronically excited carbonyl groups which may then relax with light emission," (Allen et al., 1972).

Although additional data support the suggested role of $^1O_2$ as a microbicidal reactant (Krinsky, 1974; Rosen and Klebanoff, 1976), evidence is now emerging which suggests that the mechanism of dioxygenation may be more complex, and that the typical, gas-phase $^1\Delta gO_2$ may not be the reactant. For example, the preliminary results of Foote (1980) and the unpublished research of N. Krinsky suggest that one of the major products of PMNL oxygenation of cholesterol cannot be adequately explained by either a radical or a $^1O_2$ mechanism. Alternative mechanisms, both radical and non-radical, for effecting dioxygenations yielding light emission will be considered.

NADPH OXIDASE: SUPEROXIDE SYNTHETASE AND PROTON PUMP

Babior et al. (1973) demonstrated that activation of PMNL does result in the generation of $\cdot O_2^-$ as measured by superoxide dismutase-inhibitable cytochrome c reduction. Nitroblue tetrazolium reduction by $\cdot O_2^-$ has also been reported in the PMNL (Allen et al., 1974). Superoxide is generated as a product of NAD(P)H oxidase (Allen et al., 1974; Johnston et al., 1975; Babior et al., 1975, 1976). Evidence also suggests that the oxidase is membrane associated, and that the generation of $\cdot O_2^-$ is extracytoplasmic (Goldstein et al., 1975).

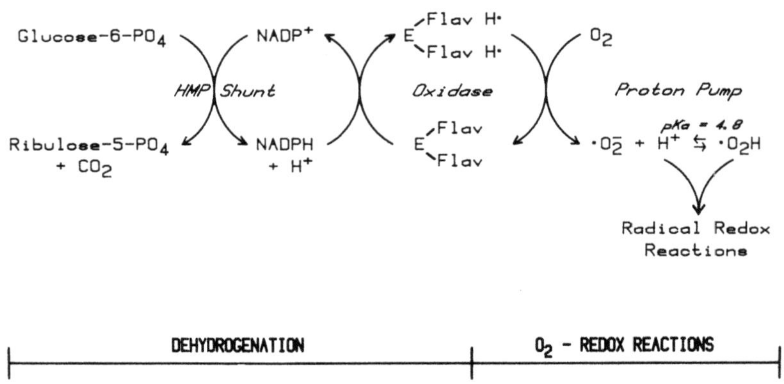

Fig. 2. Redox chain of the PMNL with the oxidase as a proton pump.

The schema presented in figure 2 depicts the central role of the oxidase in phagocyte metabolism. Note that O_2 reduction proceeds by one electron plus one proton; that is, the oxidase catalyzes the one equivalent reduction of O_2. As such, the product of reduction is hydrodioxylic acid, the conjugate acid of $\cdot O_2^-$. Therefore, in generating $\cdot O_2^-$, the oxidase can function as a proton pump (Allen, 1979).

This oxidase mechanism predicts that phagolysosomal acidification will be rapid and metabolically driven. In accordance with the pKa of $\cdot O_2H$, and with the pH dependence of $\cdot O_2^-$ disproportionation, the pH of the phagolysosomal space should approach a steady-state value of 4.8 (Allen, 1979). These theoretical predictions are in agreement with the experimental results reported by Ohkuma and Poole (1978).

RADICAL AND NON-RADICAL MECHANISMS FOR DIOXYGENATION

With respect to the chemistry of $\cdot O_2^-$, it is important to appreciate that $\cdot O_2^-$ is not a potent oxidant. In fact, it is a relatively good reductant, and is commonly quantified by its reducing activity (McCord and Fridovich, 1969). It is likewise important to appreciate that radicals are not necessarily reactive. For example, atmospheric O_2 is a triplet multiplicity, diradical molecule. It is the relative electronegativity of O_2, not its radical character, that allows it to serve as the terminal electron acceptor of redox metabolism. Radicals tend to react with radicals. The radical character of atmospheric O_2 imposes a mechanistic barrier to thermodynamically allowed reactions of O_2 with non-radical biological molecules. However, organic and inorganic radicals readily react with atmospheric O_2 (Allen, 1979).

The hydroxyl radical, $\cdot OH$ has been proposed as a PMNL-generated microbicidal agent by Johnston et al., 1975. These authors suggested that $\cdot OH$ could be generated by the reaction of $\cdot O_2^-$ and H_2O_2 as originally proposed by Haber and Weiss (1934). Although this reaction is improbable in "clean" chemical systems, it can proceed by metal catalysis (Halliwell, 1976; Koppenol and Butler, 1977). The hydroxyl radical is a very potent oxidizing agent. Its reactivity is a consequence of its great affinity for electrons. As such $\cdot OH$ can effect dehydrogenation of many biological substrates, and assuming that it is generated in close proximity to the target microbe, $\cdot OH$ could serve well as a microbicidal agent.

It seems probable that $\cdot OH$ does participate in PMNL microbicidal action. However, certain chemical realities must be kept in mind when designing experiments to "prove" the involvement of $\cdot OH$. One cannot realistically employ $\cdot OH$-scavengers in studies of cellular systems. These scavengers work well only in well defined systems, where the availability of susceptible substrates is limited. Perusual of the data collected by Dorfman and Adams (1973) will demonstrate that the common substrates of cellular systems are as susceptible or more susceptible to reaction with $\cdot OH$ than commonly employed scavengers. For example, the rate constant for the reaction of $\cdot OH$ with $\cdot O_2^-$ is $10^{10} M^{-1} s^{-1}$ (Sehested et al., 1968). Furthermore, it is difficult to determine whether inhibition of microbicidal action is due to scavenger action or derangement of physiology when scavengers are used at 10^{-1} to 10^{-2}M concentrations.

The reaction of $\cdot OH$ with $\cdot O_2^-$ has been proposed to yield OH^- plus the relatively low energy exicted state, $^1O_2$, but the reaction of $\cdot OH$ with organic substrate is not sufficiently exergonic to yield a high energy, electronically excited carbonyl product directly. However, $\cdot OH$, as a radical dehydrogenation agent, could participate with $\cdot O_2^-$ in substrate dioxygenation reactions yielding excited carbonyl functions by the mechanism proposed in figure 3.

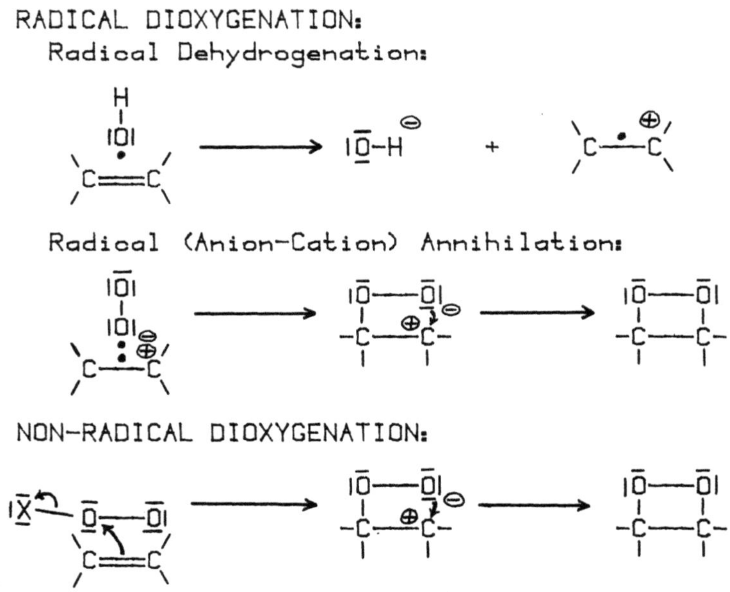

Fig. 3. Radical and non-radical mechanism of dioxygenation.

There are also non-radical, non- $^1O_2$ mechanisms for effecting substrate dioxygenation. It is likely that such a mechanism is operative in MPO-catalyzed dioxygenations. When presented with H_2O_2 and Cl^-, isolated MPO can effect a potent microbicidal action with an associated CL (Klebanoff, 1968; Allen, 1975a,b). MPO is a halide peroxidase, and the reaction of oxidized halogens with H_2O_2 is known to yield $^1O_2$ probably through the generation of $OOCl^-$ as an intermediate (Kasha and Khan, 1970). Although the liberation of $OOCl^-$ is not favored in acid conditions, electrophilic dioxygenation might proceed by a chloroperoxy mechanism as depicted in figure 3.

CHEMILUMIGENIC PROBING

The native CL of phagocytes provides a window for observing the energentics of microbicidal oxygenations. The pheonomenon of CL also demonstrates that human cells are capable of liberating energies greater than 50 kcal/mole in a single step (Allen, 1979). Chemilumigenic probing provides an expanded approach to the study of cellular O_2-redox reactions. In comparison to native CL studies, the use of chemilumigenic probes provides a greater than three order of magnitude increase in sensitivity of measurement, and in the same instance, allows a more specific measurement of O_2-redox activity (Allen and Loose, 1976; Allen, 1980a, 1980b; Wilson et al., 1978).

A chemilumigenic probe (CLP) is defined as an exogenous organic molecule with a high CL quantum yield that can serve as a substrate for CL measurement of certain types of O_2-redox activities. This bystander substrate must be non-toxic at the concentrations used for testing. Specific probes can be employed for assay of relatively specific oxygenation activities. For example, when luminol, 5-amino-2,3-dihydro-1,4-phthalazinedione, is employed as a CLP, CL is a result of oxidative dioxygenation; that is, luminol undergoes two equivalent dehydrogenation plus O_2 incorporation to yield electronically excited aminophthalate as product. However, when lucigenin, 10,10′-dimethyl-9,9′-biacridinium dinitrate (DBA^{++}), is employed as a CLP, CL is a result of reductive dioxygenation; that is, DBA^{++} undergoes a two equivalent reduction plus O_2 incorporation to yield electronically excited N-methyl acridone. Reaction can involve the radical or univalent reducted $\cdot DBA^+$ as intermediate. The

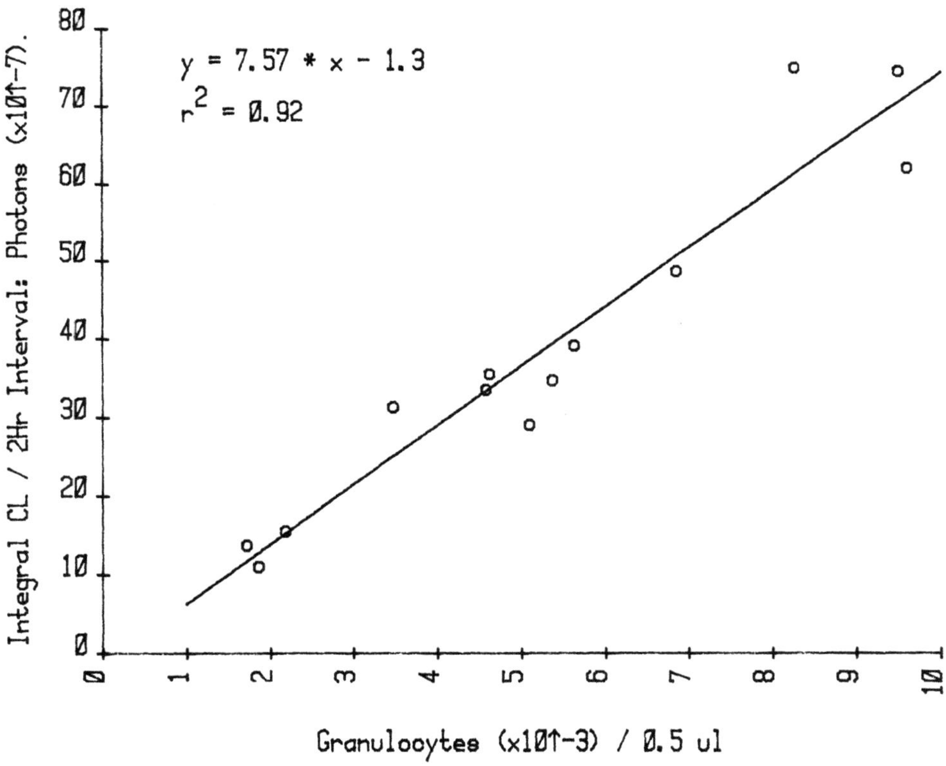

Fig. 4. Integral CL as a function of the number of granulocytes in 0.5μl of whole blood. Conditions: luminol 5.0μM; serum-opsonified zymosan, 50μg; albumin-veronal buffer complete with divalent cations and glucose to a final volume of 2.0ml. CL signal was detected by a single photon counting using an EMI 9829A bialkali photomultiplier tube.

radical-radical, anion-cation annihilation reaction of $\cdot DBA^+$ with $\cdot O_2^-$ can directly yield the dioxetane (Allen, 1980b). Unlike the luminol-CL response DBA^{++}-CL can be inhibited by unit quantities of superoxide dismutase (manuscript in preparation).

At this meeting, a great interest has been expressed regarding methods for quantifying phagocytic function. The data presented in figure 4 is offered as a preview of a CLP technique designed for routine quantification of phagocyte O_2-redox response to phagocytosis. The blood donors for this study were healthy laboratory personnel and healthy pregnant females in early labor. The latter group allowed investigation of a physiologic leukocytosis in the absence of infection. Microliter aliquots of EDTA-anticoagulated whole blood were removed from blood specimens collected from routine blood counts. The whole blood aliquots were then diluted one to one-hundred with normal saline, and the equivalent of one-half microliter of whole blood was added to counting vial containing the CLP luminol in albumin-veronal buffer complete with glucose and divalent cations. Stimulation was initiated by addition of serum-opsonified zymosan, and the CL response was automatically monitored every thirteen minutes over a two hour period.

The data shown was obtained in three days of routine testing. The post-venipuncture age of the blood specimens at the time of testing varied from two to eight hours, and the contribution of monocytes to CL was not taken into account. No CL response was obtained from any preparation following heat treatment at 56°C for 10 minutes. This and other microtechniques are presently being developed and tested as objective clinical laboratory procedures for quantifying the specific dioxygenation activity of phagocytes.

ACKNOWLEDGEMENTS

I wish to thank the U. S. Army Medical Research and Development Command for allowing my participation in this conference. The research data presented is a product of R&D project #3A161101A91C and D.C.I. project #C-5-79.

REFERENCES

Agner, K., 1958, Proc. Int. Congr. Biochem. 4th Vienna 15:64.
Allen, R. C., 1975a, Biochem. Biophys. Res. Commun. 63:675.
Allen, R. C., 1975b, Biochem. Biophys. Res. Commun. 63:684.
Allen, R. C., 1979, in: "Frontiers in Biology, Vol. 48", J.T. Dingle, P. J. Jacques, and I. H. Shaw, eds., p. 197, North Holland, Amsterdam.

Allen, R. C., 1980a, in: "Liquid Scintillation Counting: Recent Applications and Developments, Vol. 2," C. T. Peng, D. L. Horrocks, and E. L. Alpen, eds., p. 37, Academic Press, NY.
Allen, R. C., 1980b, "Bioluminescence and Chemiluminescence," M. DeLuca and W. D. McElroy, eds., (in press) Academic Press, New York.
Allen, R. C., and Loose, L. D., 1976, Biochem. Biophys. Res. Commun. 69:245.
Allen, R. C., Stjernholm, R. L., and Steele, R. H., 1972, Biochem. Biophys. Res. Commun. 47:679.
Allen, R. C., Stjernholm, R. L., Benerito, R. R., and Steele, R. H., 1973, in: "Chemiluminescence and Bioluminescence," M. J. Cormier, D. M. Hercules, and J. Lee, eds., p. 498, Plenum Press, New York.
Arneson, R. M., 1970, Arch. Biochem. Biophys. 136:352.
Babior, B. M., Kipnes, R. S., and Curnutte, J. T., 1973, J. Clin. Invest. 52:741.
Babior, B. M., Curnutte, J. T., and Kipnes, R. S., 1975, J. Clin. Invest. 56:1035.
Babior, B. M., Curnutte, J. T., and McMurrich, B. J., 1976, J. Clin. Invest. 58:989.
Blum, H. F., 1941, "Photodynamic Action and Diseases Caused by Light," Reinhold, New York.
Corey, E. J., and Taylor, W. C., 1964, J. Amer. Chem. Soc. 86:3881.
Dorfman, L. M., and Adams, G. L., 1973, "Reactivity of the Hydroxyl Radical in Aqueous Solution," Natl. Bureau Stds., NSRDS-46.
Foote, C. S., and Wexler, S., 1964, J. Amer. Chem. Soc. 86:3879.
Foote, C. S., Abakerli, R. B., Clough, R. L., and Shook, F. C., 1980, in: "Biological and Clinical Aspects of Superoxide and Superoxide dismutase," W. H. Bannister and J. V. Bannister, eds., p. 222, Elsevier/North Holland, New York.
Goldstein, I. M., Roos, D., Kaplan, H. B., and Weissmann, G., 1975, J. Clin. Invest. 56:1155.
Haber, F., and Weiss, J., 1934, Proc. Roy. Soc., Ser A 147:332.
Halliwell, B., 1976, FEBS Lett. 72:8.
Iyer, G. Y. N., Islam, M. F., and Quastel, J. H., 1961, Nature 192:535.
Jodlbauer, A., and von Tappeiner, H., 1905, Deut. Arch. Klin. Med. 82:520.
Johnston, R. B., Keele, B. B., Misra, H. P., Lehmeyer, J. E., Webb, L. S., Bachner, R. L., and Rajagopalan, K. V., 1975, J. Clin. Invest. 55:1357.
Karnovsky, M. L., 1968, Semin. Hematol. 5:156.
Kasha, M., and Khan, A. U., 1970, Ann. N.Y. Acad. Sci. 171:5.
Kautsky, H., 1939, Trans. Faraday Soc. 35:216.
Khan, A. U., 1970, Science 168:476.
Klebanoff, S. J., 1968, J. Bacteriol. 95:2131.
Klebanoff, S. J., 1971, Annu. Rev. Med. 22:39.
Knowles, P. J., Gibson, J. F., Plick, F. M., and Bray, R. C., 1969, Biochem. J. 111:53.
Koppenol, W. H., and Butler, J., 1977, FEBS Lett. 83:1.

Krinsky, N. I., 1974, Science 186:363.
McCapra, F., 1968, J. C. S. Chem. Commun. 155.
McCord, J. M., and Fridovich, I., 1969, J. Biol. Chem. 244:6049.
McRipley, R. J., and Sbarra, A. J., 1967 J. Bacteriol. 94:1425.
Ohkuma, S., and Poole, B., 1978, Proc. Natl. Acad. Sci. USA. 75:3327.
Patriarca, P., Cramer, R., Moncalvo, S., Rossi, F., and Romeo, D., 1971, Arch. Biochem. Biophys. 145:255.
Paul, B., and Sbarra, A. J., 1968, Biochem. Biophys. Acta. 156:168.
Raab, O., 1900, Z. Biol. 39:524.
Rosen, H., and Klebanoff, S. J., 1977, J. Biol. Chem. 252:4803.
Rossi, F., and Zatti, M., 1964, Brit. J. Exp. Pathol. 45:548.
Rossi, F., and Zatti, M., 1966, Biochem. Biophys. Acta. 121:110.
Rossi, F., Patriarca, P., and Romeo, D., 1971, in: "The Reticulo-endothelial System and Immune Phenomena," N. R. DiLuzio, ed., 191, Plenum Press, New York.
Sander, U., and Stauff, J., 1971, Anales Asoc. Quim. Argent. 59:149.
Sehested, J., Rasmussen, O. L., and Fricke, H., 1968, J. Phys. Chem. 72:626.
Wilson, M. E., Trush, M. A., Van Dyke, K., Kyle, J. M., Mullett, J. Immunol. Meth. 23:315.

DISCUSSION

Edited by: R.C. Allen

ROMEO: I have a very short technical question. How long does it take to run one of the CL assays you have just described?

ALLEN: Using previously prepared vials, the assay requires approximately thirty minutes to set up fifty samples. The reaction is allowed to continue automatically for two hours; however, good correlations between CL and the number of functional granulocytes are obtained in one hour. Using one technician with two counters, we perform more than two hundred assays per day.

SEGAL: You showed that azide and cyanide inhibited luminol luminescence and enhanced that of lucigenin. You also showed that SOD inhibited lucigenin luminescence. This suggests that myeloperoxidase might be the natural substance that promotes the conversion of $O_2^{\cdot -}$ to H_2O.

ALLEN: Ohkuma and Poole (see references) and others have estimated the intralysosomal pH as approximately 4.8. At this pH, the rate of proton-disproportionation of $O_2^{\cdot -}$ yielding H_2O_2 plus O_2 is greater than $10^7 M^{-1}s^{-1}$. Therefore, $O_2^{\cdot -}$ can be non-enzymatically converted to H_2O_2, and then MPO can effect reduction of H_2O_2 to H_2O.

MPO is in many ways analogous to cytochrome oxidase of the mitochondria. Both enzymes contain two a-type cytochromes, each with a different reactivity, and both enzymes yield H_2O as a product.

WEVER: In one of your slides you showed that NADPH oxidase contains two flavins. Do you have any empirical evidence for that proposal?

ALLEN: There is no direct empirical evidence for this proposal. it is very difficult to obtain the enzyme in quantities that would allow analysis.

However, a two flavin system is consistent with a semiquinone mechanism of catalysis. Similar flavoprotein systems have this type of two flavin structure (Massey and Gibson, 1964, Fed. Proc. 23: 18; Masters et al., 1965, J. Biol. Chem. 240: 291). A two flavin structure might also obscure the detection of the semiquinone signal by electron spin resonance.

AN ALTERNATIVE MECHANISM FOR THE PRODUCTION OF HYDROXYL RADICALS BY STIMULATED NEUTROPHILS

M.J. Okolow-Zubkowska and H.A.O. Hill

Inorganic Chemistry Laboratory
South Parks Road
Oxford OX1 3QR

The importance of oxygen in the microbiocidal activity of polymorphic leukocytes has been established[1] for many years. Twenty years ago it was shown[2] that the two-electron reduction product of oxygen, hydrogen peroxide, is present in the medium surrounding stimulated cells. Later, the superoxide ion, the one-electron reuction product of oxygen was also found[3] to be produced by phagocytosing neutrophils. The three-electron reduction product, the hydroxyl radical, has since been found[4] to be produced by these activated cells. A flurry of mechanisms, seeking to account for the genesis of, and relationship between, these potentially reactive species has resulted. It is now generally accepted that the superoxide ion is produced[5] by a membrane-bound enzyme that requires NAD(P)H. It is not yet known whether the hydrogen peroxide that is found extracellularly arises by the direct two-electron reduction of oxygen or through the spontaneous disproportionation of superoxide:

$$O_2 + 2e + 2H^+ \longrightarrow H_2O_2 \qquad [1]$$

$$2O_2^- + 2H^+ \longrightarrow H_2O_2 + O_2 \qquad [2]$$

The origin of the hydroxyl radical is even more obscure. For a time it was believed[6] to be formed by the Haber-Weiss reaction [3]

$$H_2O_2 + O_2^- \longrightarrow OH^\cdot + OH^- + O_2 \qquad [3]$$

However, careful work by Weinstein and Bielski has shown[7] that this reaction is very slow indeed in the absence of trace amounts of metal ions. Hence no obvious explanation seems to exist for the formation of the hydroxyl radical. It had originally been detected using methional, which gives rise to ethylene on reaction with the hydroxyl radical, but this method has since been shown[8] to be non-specific.

We have used[9] the technique of spin trapping[10], whereby short-lived radicals form adducts with nitrones or nitroso compounds to yield relatively stable radicals, to investigate the formation of oxygen-derived radicals during phagocytosis. The 'trap' used was 5,5'-dimethylpyrroline-N-oxide (DMPO) whose adducts with the superoxide ion and the hydroxyl radical are well-characterised. All our assays contained DTPA, diethylenetriaminepentaacetic acid, as a metal ion chelating agent to avoid ambiguity in the interpretation of the results since the reaction of adventitious metal ions with H_2O_2 can give rise to the catalysed Haber-Weiss reaction [4] [5]

$$M^{n+} + H_2O_2 \longrightarrow OH^{\cdot} + OH^- + M^{(n+1)+} \qquad [4]$$

$$M^{(n+1)+} + O_2^- \longrightarrow O_2 + M^{n+} \qquad [5]$$

The spin trap was used at concentrations, 100mM (see (9) for full experimental details) that do not interfere with the respiratory burst. In our hands pure DMPO did not act as a stimulant to these cells. When the cells are stimulated with latex particles, opsonized with IgG, the hydroxyl radical adduct of DMPO is detected. However, if the cells are stimulated with latex IgG in the presence of superoxide dismutase, no radical adducts are observed. This observation is consistent with that of Tauber and Babior who found that superoxide dismutase inhibited the formation of ethylene from methional. In marked contrast, catalase has very little effect on either the spin trapping assay or on the formation of ethylene from methional. These results seem to indicate that the superoxide ion is the precursor of the hydroxyl radical. Yet the superoxide anion radical adduct was not detected[9] in the spin trapping assay using latex particles as stimulants. It has however been detected from neutrophils, in the presence of a particulate stimulus, using[11] the cytochrome c assay of McCord and Fridovich. Presumably this is because the rate of reaction of O_2^- with cytochrome c is considerably faster ($1.5 \times 10^6 M^{-1} s^{-1}$) than the rate of its reaction with DMPO ($10\ M^{-1} s^{-1}$) whilst that of $OH^{\cdot}$ with DMPO is[12] several orders of magnitude ($10^9\ M^{-1} s^{-1}$) faster. Thus it is not surprising that O_2^- is difficult to detect by spin-trapping. Circumstances do exist, however, where it can be observed: when a soluble stimulus is used; in the presence of azide; at reduced temperatures.

Further evidence for hydroxyl radicals is based on studies using scavengers which significantly decrease the amount of the hydroxyl-DMPO adduct detected. The same results can be obtained using the methional[4] method suggesting that the radicals are causally related to the respiratory burst and not simply formed as artifacts associated with the methods of detection.

Thus both the superoxide anion and the hydroxyl radical are detected extracellularly with the former being the precursor of the latter. The uncatalysed Haber-Weiss reaction is too slow to account for the formation of the hydroxyl radicals. It is possible that myeloperoxidase may be involved since extensive studies by Klebanoff and Rosen have shown [13] that, not only do MPO-deficient patients have a higher susceptibility to infection but that their neutrophils have a greatly impaired ability to form ethylene from methional.

The observation[14] by Long and Bielski that superoxide and hypochlorous acid can react together to give the hydroxyl radical [6].

$$HOCl + O_2^- \longrightarrow OH^{\cdot} + O_2 + Cl^- \qquad [6]$$

has led us to propose the following mechansim for the formation of the hydroxyl radical:

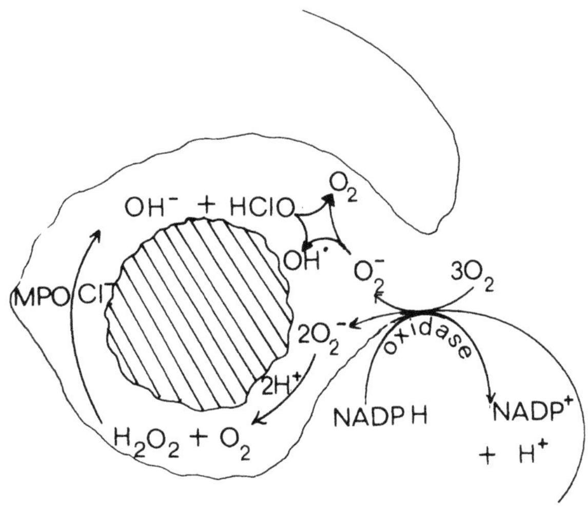

The figure illustrates a possible mechanism for the generation of hydroxyl radicals by phagocytosing neutrophils during the respiratory burst.

Oxygen is reduced by a membrane-bound enzyme to form the superoxide ion which then dismutes to form hydrogen peroxide and oxygen. In a reaction catalysed by myeloperoxidase, chloride ions and hydrogen peroxide react to give hypochlorous acid. The latter reacts with superoxide to form the hydroxyl radical. The detection of this product and the intermediates will depend critically on the concentrations and rates of reaction of these with each other and the compounds used to detect them. Superoxide dismutase would be expected to be a particularly effective inhibitor due to the high rate of the catalysed reaction. Catalase, on the other hand, may not be particularly effective due to its low turnover rate at low H_2O_2 concentrations. Azide, which inhibits myeloperoxidase, will prevent the formation of HOCl and consequently the hydroxyl radical. The soluble stimulus, phorbol myristate acetate, causes the release of the specific granules, not the azurophilic granules which contain myeloperoxidase. The reduction of the temperature also prevents de-granulation. The hydroxyl radical is so reactive that it is likely that scavengers, having to compete with a wide range of reactions, cannot be completely effective at concentrations where the neutrophils are undammaged. A noteworthy observation[13] is that neutrophils of MPO-deficient patients are unable to produce significant amounts of hydroxyl radicals. It is possible, therefore, that the hydroxyl radical represents the final, designed product of the respiratory burst. If biocidal activity is related to reactivity, then it is understandable that it is used as the killing agent. Whether these observations are relevant to all oxygen-dependent phagocytic processes remains to be proven.

ACKNOWLEDGEMENTS

We are grateful to the Science Research Council for a studentship to M.J. O-Z. We wish to thank our colleagues, Professor W.H. Bannister, and Drs. M.P. Esnouf, J.V. Bannister and A.E.G. Cass for help.

REFERENCES

1. C.W. Baldridge and R.W. Gerard, The extra respiration of phagocytosis, Am. J. Physiol., 103:235 (1933).
2. G..Y.N. Iyer, M.F. Islam and J.H. Quastel, Biochemical aspects of phagocytosis, Nature, 192:535 (1961).
3. B.M. Babior, R.S. Kipnes and J.T. Curnutte, The production by leukocytes of superoxide, a potential bactericidal agent. J. Clin. Invest., 52:741 (1973)
4. A.I. Tauber and B.M. Babior, Evidence for hydroxyl radical production by human neutrophils, J. Clin. Invest., 60:374 (1977).

5. T.G. Gabig and B.M. Babior, The superoxide forming oxidase responsible for the respiratory burst in human neutrophils, J. Biol. Chem., 254:9070 (1979)
6. F. Haber and J. Weiss, The catalytic decomposition of hydrogen peroxide by iron salts, Proc. R. Soc. Lond., 147:332 (1934).
7. J. Weinstein and B.H.J. Bielski, Kinetics of the interaction of HO_2 and O_2^- radicals with H_2O_2, the Haber-Weiss reaction, J. Amer. Chem. Soc., 101:58 (1979).
8. W.A. Pryor and R.H. Tang, Ethylene formation from methional, Biochem. Biophys. Research Commun., 81:498 (1978).
9. M.R. Green, H.A.O. Hill, M.J. Okolow-Zubkowska and A.W. Segal, The production of hydroxyl and superoxide radicals by stimulated human neutrophils - measurements by e.p.r. spectroscopy, FEBS Letts., 100:23 (1979).
10. C. Lagercrantz, Spin trapping of some short lived radicals by the nitroxide method, J. Phys. Chem., 75: 3466 (1971).
11. J.M. McCord and I. Fridovich, Superoxide dimsutase. An enzyme function for erythrocuprein (hemocuprein), J. Biol. Chem., 244:6049 (1969).
12. E. Finkelstein, G.M. Rosen, and E.J. Raukman, Spin trapping of superoxide and hydroxyl radical practical aspects, Arch. Biochem. Biophys., 200:1 (1980).
13. S.J. Klebanoff and H. Rosen, Ethylene formation by polymorphonuclear leukocytes: Role of myeloperoxidase, J. Exp. Med., 148:490 (1978).
14. C.A Long and B.H.J. Bielski, Rate of reaction of superoxide radical with chlroide containing species, J. Phys. Chem., 84:555 (1980).

SUPEROXIDE DISMUTASES AND THE OXIDATIVE BURST IN HUMAN BLOOD POLYMORPHONUCLEAR LEUKOCYTES

Martine Torres and Jacques Hakim

Laboratoire d'Immunologie et d'Hématologie -
CHU Xavier Bichat - Université Paris VII
170, bld Ney - 75877 PARIS CEDEX 18 - FRANCE

Both cyanide (CN)-sensitive and CN-resistant superoxide dismutases (SOD) have been identified in human polymorphonuclear leukocytes (PMN), and located in the cytosol and granules of these cells[1-4]. These enzymes catalyse the dismutation of superoxide anion ($O_2^{\bar{\cdot}}$) to hydrogen peroxide (H_2O_2), and are thus thought to be protective in PMN against the highly reactive $O_2^{\bar{\cdot}}$, as they are in other tissues and cells of aerobic organisms[2,4-6]. However, the functional activity of PMN, i.e. their microbial killing activity, depends largely on the ability of PMN to generate $O_2^{\bar{\cdot}}$ during their stimulated oxidative burst[7] (OB). Moreover, the OB of the PMN not only kills the ingested bacteria, but is also a suicidal event for the PMN, suggesting that the SOD might lose their activity during the OB. On the other hand it is well known that SOD activity inhibits O_2-uptake, NAD(P)H oxidation, and $O_2^{\bar{\cdot}}$ formation by isolated subcellular fractions from PMN[8-10].

The purpose of this communication, mainly based on previous work in our laboratory[3,10-12], is to show that in resting PMN, SOD inhibits the occurrence of the OB, whereas in stimulated PMN, it is the abolition of this inhibition which leads to the OB and to the associated generation of $O_2^{\bar{\cdot}}$ and H_2O_2 by releasing a cyclic chain reaction. The experimental results which led to this scheme are reported below in two parts : first, we show that the granule-rich fraction (GRF) isolated from PMN is able to carry from O_2 and NAD(P)H, a large CN-resistant production of $O_2^{\bar{\cdot}}$ and H_2O_2. This production occurs through a cyclic chain reaction initiated by a small amount of $O_2^{\bar{\cdot}}$. This reaction is inhibited by SOD or $O_2^{\bar{\cdot}}$ scavengers, as shown by O_2-uptake, $O_2^{\bar{\cdot}}$ and H_2O_2 formation, and NAD(P)H oxidation . The second point of the present report shows that both granular and cytosolic CN-resistant SOD are inhibited by the cytosol of stimulated PMN.

I - CYCLIC CHAIN REACTION IN GRF FROM PMN -

Several pertinent studies[7] reported that once generated, $O_2^{\bar{\cdot}}$ gives rise to H_2O_2, in accordance with one of the following two reactions :

$$O_2^{\bar{\cdot}} + O_2^{\bar{\cdot}} + 2\,H^+ \longrightarrow O_2 + H_2O_2 \qquad \text{(Reaction 1)}$$

$$O_2^{\bar{\cdot}} + 1\,e^- + 2\,H^+ \longrightarrow H_2O_2 \qquad \text{(Reaction 2)}$$

If this was the case for PMN granules, H_2O_2 production should be half or at most equal to that of $O_2^{\bar{\cdot}}$. This, however, is not the case, and the amount of H_2O_2 produced is apparently much larger than that of $O_2^{\bar{\cdot}}$. This apparent discrepancy suggests that either H_2O_2 is produced directly from O_2 without $O_2^{\bar{\cdot}}$ as an intermediate, or that a cyclic chain reaction is initiated by $O_2^{\bar{\cdot}}$ and is terminated by the scavenging of $O_2^{\bar{\cdot}}$ during its measurement. In other words, the measurement of $O_2^{\bar{\cdot}}$ formation by a scavenger dye must lead to a lower estimate of its rate of formation if a cyclic chain reaction is involved, because the scavenger used for the measurement terminates the cyclic chain reaction. Both these possible processes of H_2O_2 production were tested with NADH and NADPH as electron donors.

1. Cyclic chain reaction with NADH as electron donor -

15,000 g sedimentable fraction hereafter referred to as GRF was prepared from resting PMN as previously described[13]. The isolated GRF was strained through a column filled with glass wool, and dialyzed against a Ca^{2++}-free 0.1 M phosphate buffer, pH 5.5, containing 10^{-4} M Na_2 EDTA. Measurements were made in 0.1 M citrate-phosphate buffer, pH 5.0, containing 2 mM KCN, 0.1 % bovine serum albumin, 2.5 mM NADH, and intact or boiled GRF at a final concentration of 100 µg protein per ml. Variations in this reaction mixture are stated in the results. Table I shows that under our experimental conditions, H_2O_2 was formed by GRF at a rate of 69 ± 6 nmol per min and per mg of GRF protein (Mean ± 1 SD of 20 experiments), and that the stoichiometry between NADH oxidation, O_2-uptake and the H_2O_2 generated was close to 1. Although SOD is inhibited by CN, its addition to the preparation reduced O_2-uptake, NADH oxidation and H_2O_2 formation by 72, 83 and 77 % respectively. Ferricytochrome c (0.1 M) or NBT (0.1 M) inhibited H_2O_2 formation, NADH oxidation and O_2-uptake stoichiometrically by more than half. These results show that with NADH as electron donor, at least 75 % of H_2O_2 formation by GRF occurs via $O_2^{\bar{\cdot}}$ as an intermediate.

Parallel measurements of O_2-uptake and $O_2^{\bar{\cdot}}$ formation with the

Table 1. GRF[a] mediated NADH oxidation, O_2-uptake, and H_2O_2 formation -

Additions to the incubation medium[b]	NAD^+ produced[c]	O_2-uptake[c]	H_2O_2 formed[c]
	in nmol per min per ml		
Intact GRF (100 µg/ml)	7.23	6.61	6.88
+ SOD (62.5 µg/ml)	2.02	1.19	1.51
+ boiled SOD (62.5 µg/ml)	7.09	6.54	6.74
+ Fe^{3+} cytochrome c (0.1 mM)	3.48	3.04	2.96
+ NBT (0.1 mM)	2.68	2.51	2.48
+ KCl (0.1 mM)	7.21	6.64	6.91
Boiled GRF (100 µg/ml)	0.11	0.10	0.09
Phosphate buffer	0.10	0.08	0.11

[a]GRF are dialyzed granule-rich fraction from resting PMN, prepared as in ref.13.

[b]Incubation medium was 0.1 M citrate phosphate buffer, pH 5.0, containing 2 mM KCN, 0.1 % bovine serum albumin, 2.5 mM NADH and 0.1 ml (80 µg protein) of the GRF (or boiled GRF) or phosphate buffer.
In the results given for intact GRF the blanks (results obtained with boiled GRF) have been substracted. Results are the mean of at least 3 different experiments.

[c]Techniques are described in ref.13.

addition of increasing amounts of ferricytochrome c (from 0.1 to 0.4 mM) showed that for each mol of $O_2^{\cdot}$ scavenged and revealed by the reduction of ferricytochrome c, about 8 mol of O_2 were not taken up (Tab. 2). In these experiments the stoichiometry of O_2-uptake, NADH oxidation and H_2O_2 formation remained close to 1. These results suggested that, under our experimental conditions a cyclic chain reaction occurs. The following two processes (reactions 3 and 4) known to form cyclic reaction[14] were postulated to be essential for the production of H_2O_2 by GRF.

$$HO_2^{\cdot}\ (\text{protonated } O_2^{\cdot}) + NADH + H^+ \rightarrow H_2O_2 + NAD\cdot \quad \text{(Reaction 3)}$$

$$NAD\cdot + O_2 + H^+ \longrightarrow HO_2^{\cdot} + NAD^+ \quad \text{(Reaction 4)}$$

The reality of this cyclic chain reaction was verified by the following experiments (Tab. 3) : first, in the absence of NADH and GRF (or in the presence of boiled GRF) we generated $O_2^{\cdot}$ in the

Table 2. Effect of ferricytochrome c on O_2-uptake, NADH oxidation and H_2O_2 formation by GRF[a] -

	Ferricytochrome added to incubation medium[b]				
	None	0.1 mM	0.2 mM	0.3mM	0.4 mM
O_2-uptake[c] nmol/min/mg protein	66.1	33.4	26.6	20.0	15.5
Ferricytochrome c reduced[c] nmol/min/mg protein	0.0	4.3	5.0	5.7	6.1
O_2-uptake inhibition / Ferricytochrome reduced	-	7.6	7.9	8.1	8.3

For[a and b] see Table 1. [c]Techniques used were as in ref.12. Blanks with boiled GRF have been substracted and were less than 5 % of the reported activities. Stoichiometry of NADH oxidation and H_2O_2 formation with O_2-uptake was close to 1.

Table 3 . Effect of O_2^- generated by water photolysis[a] on O_2-uptake and H_2O_2 formation by NADH and GRF -

Additions[b] to	O_2-uptake	H_2O_2-production
	in nmol per min per ml	
I- NADH- and GRF-free incubation medium :		
+ phosphate buffer	0.90	0.92
+ boiled GRF	0.90	0.93
+ SOD + boiled GRF	0.45	0.44
II- GRF-free incubation medium containing 2.5 mM NADH		
+ phosphate buffer	1.88	1.80
+ boiled GRF	6.10	6.21
+ intact GRF	13.32	14.0

[a]O_2^- was generated by water photolysis as in ref.12. Reduction of ferricytochrome c was 0.9 nmol/min/ml, and was completely inhibited by SOD.

[b]Additions were 100 µg GRF, 20 µg SOD, or phosphate buffer 0.1 ml i.e. the volume in which the other additions were dissolved.

incubation medium by water photolysis. The O_2^- generated, measured by the reduction of ferricytochrome c, was 0.9 nmol per min and per ml. The stoichiometry of O_2-uptake, H_2O_2 formation and the O_2^--produced in this system was 1, suggesting that H_2O_2 was generated via reduction of O_2^- by reaction 2 and not by dismutation of two O_2^- (reaction 1). Addition of SOD halved O_2-uptake and H_2O_2 formation, indicating that the enzyme was fully active under our experimental conditions, and that in its presence, H_2O_2 formation resulted from O_2^- dismutation (reaction 1). Secondly, in the incubation medium containing 2.5 mM of NADH but no GRF, O_2-uptake doubled compared to that in the incubation medium which was NADH-free. Additions of boiled or intact GRF (100 μg protein per ml) increased O_2-uptake by factors of about 6.5 and 14.5 respectively. If we take into account the O_2^- produced by intact GRF, the cyclic chain reaction induced by GRF is the same, whether this fraction is boiled or not. These results indicate that GRF contains an undializable, heat stable factor which promotes a cyclic chain reaction of about 7 to 8 cycles under our experimental conditions.

2. Cyclic chain reaction with NADPH as electron donor -

GRF from resting and zymosan-stimulated PMN (1 min) were prepared as previously described[10,11]. Measurements were made in 0.1 M phosphate buffer, pH 5.5, containing 0.5 mM MnCl2, 2 mM KCN, 1.25 mM NADPH, and intact or boiled GRF at a final concentration of 100 μg per ml. Variations in the composition of the reaction medium are stated in the results. Mn^{++} was used in all experiments because in our hands its omission caused no NADPH oxidation, O_2-uptake or H_2O_2 formation. It was also shown that 0.5 mM Mn^{++} alone, i.e. without GRF, induced both O_2-uptake (= 0.8 nmol per min per ml) and H_2O_2 generation (= 0.8 nmol per min per ml) which were abolished by ferricytochrome c and halved by SOD. This suggests that, in the absence of SOD, H_2O_2 formation occurred via reaction 2, and in its presence, via reaction 1. In the presence of intact GRF from resting PMN, O_2-uptake rose to 5.0 nmol per min and per mg protein ; the stoichiometry between O_2-uptake and NADPH oxidation was approximately 1. SOD inhibited O_2-uptake by 2/3. This suggested that the O_2^- produced in the presence of intact granules underwent a cyclic chain reaction with NADPH which, unlike that with NADH, was not observed with boiled GRF. Consequently, when NADPH was electron donor, the cyclic chain reaction observed with GRF was dependent on a heat-labile factor, under our experimental conditions. This factor has been reported to be an enzymatic protein whose substrates are $HO_2^{\cdot}$ and NADPH[10].

For further investigation of Mn^{++}-dependent NADPH oxidation, O_2-uptake and H_2O_2 formation by the GRF, we studied the interaction of subcellular fractions isolated from resting and stimulated PMN. Results (Tab.4) showed that : 1. the activity in the 350 g super-

Table 4. CN-resistant NADPH-oxidation activity in subcellular fractions from resting and stimulated PMN -

	Resting PMN	Zymosan-stimulated PMN
350 g supernatant	1.56	3.96
15,000 g pellet	2.78	2.96
15,000 g supernatant	0.36	0.99
Cytosol : 100,000 g supernatant	0.00	0.00
15,000 g pellet :		
+ cytosol from resting PMN	1.56	1.98
+ cytosol from stimulated PMN	2.80	3.12

Experimental conditions are as described in ref.11. Activities are expressed in nmol of $NADP^+$ produced per min per 10^6 PMN. CN concentration was 2 mM and Mn^{++} 0.5 mM.

natant fraction was higher in stimulated than in resting cells, whereas in the 15,000 g pellet this was not the case. 2. The activity in the 350 g supernatant was lower than that of the 15,000 g pellet in resting cells, whereas the reverse was observed in stimulated cells. 3. Addition of the cytosolic fraction from resting cells to a 15,000 g pellet of stimulated cells reduced the activity of the pellet and 4. addition of the cytosolic fraction from stimulated cells to a 15,000 g pellet from resting cells raised the activity of the pellet.

These results suggest that in resting cells, an inhibitor of NADPH oxidation was present in the cytosol and that phagocytosis induced the formation of a factor which abolished the activity of the inhibitor. As suggested above, we hypothesized that the inhibitor of NADPH oxidation might be the CN-resistant SOD.

3. NADH and NADPH-dependent cyclic chain reaction -

Differences and similarities are observed in the CN-resistant cyclic chain reactions with GRF, depending on whether NADH or NADPH is added. Both require a small amount of O_2^- to initiate a chain reaction of about 6 to 8 cycles which is inhibited by SOD or O_2^- scavengers. The GRF-mediated cyclic chain reaction with NADH is dependent on an undialyzable heat stable factor, whereas the reaction with NADPH is dependent on a heat labile factor which has the properties of an enzymatic protein. In the case of both the

NADH and NADPH-dependent reactions, the cytosol and the GRF, in resting PMN, contain a cyclic chain reaction inhibitor. In stimulated PMN, a cytosolic factor is generated which abolishes the activity of this inhibitor. As regards the inhibitor's identity, we postulated that it was CN-resistant SOD, and that phagocytosis induced a factor which cancelled out the CN-resistant SOD activity. This is why we measured CN-resistant SOD in the GRF and cytosol of resting and stimulated PMN.

II - CN-RESISTANT SOD ACTIVITIES IN RESTING AND STIMULATED PMN -

CN-resistant SOD activities were measured in resting PMN, and after 1 min of stimulation with opsonized zymosan as previously described[3]. In resting PMN, CN-resistant SOD activity was found both in the 15,000 g supernatant and in the 15,000 g pellet of the post-350 g supernatant (Tab.5). The activity of the 15,000 g supernatant was approximately 27 % of that of the 15,000 g pellet. The sum of the activities of these two fractions was slightly higher than that found in the 350 g supernatant which includes these two fractions. In stimulated cells, the values indicated in Table 5 show that the stimulation reduced by about 70 % the CN-resistant SOD activity of the 350 supernatant. The effect was similar in the 15,000 g supernatant, whereas SOD activity in the 15,000 g pellet was unaffected. The presence of the 15,000 g supernatant of

Table 5. CN-resistant SOD in subcellular fractions from resting and stimulated PMN -

	Resting PMN	Zymosan-stimulated PMN
350 g supernatant	0.59	0.18
15,000 g supernatant	0.15	0.04
15,000 g pellet	0.55	0.50
+ 15,000 g supernatant from stimulated cells	0.26	0.22

Experimental conditions were as in ref. 3. Results of SOD are expressed in Sigma units per 10^6 PMN. 50 % inhibition of NBT reduction were obtained with 0.48 Sigma units, under our experimental conditions.

stimulated cells is thus required for the observation of a decrease in CN-resistant SOD activity. This was confirmed by cross reactions with 15,000 g pellets from stimulated and resting cells to which a 15,000 g supernantant from stimulated cells was added (Tab.5). These results clearly show that a factor which apparently cancels CN-resistant SOD activity, is generated in the cytosol of stimulated PMN, and that this factor does not bind firmly to the GRF fraction. Recently, however, Hassan et al[15] reported that apparent inhibition of SOD might be caused by molecules acting as electron carriers from $O_2^{\bar{\cdot}}$ to cytochrome c or to NBT. The effect of such carriers was not visible in the absence of SOD, because they do not affect the amount of NBT reduced but only increase the speed of electron transfer from $O_2^{\bar{\cdot}}$ to NBT. In contrast, in the presence of SOD, the electron carriers compete with SOD for $O_2^{\bar{\cdot}}$ and the result is an apparent inhibition of SOD activity. While bearing in mind the above possibility, we think that the inhibition of SOD activity might be important in explaining the OB associated with phagocytosis.

The following evidence supports this conclusion : 1. GRF contains a biochemical apparatus able to generate large amounts of $O_2^{\bar{\cdot}}$ and H_2O_2 from a small amount of $O_2^{\bar{\cdot}}$, through a cyclic chain reaction ; 2. SOD inhibits the aerobic oxidation of NAD(P)H and the production of $O_2^{\bar{\cdot}}$ and H_2O_2 by the GRF ; 3. The GRF and the cytosol contain CN-resistant SOD which loses its activity after stimulation of the PMN and 4. The reduction in SOD activity is caused by a factor found in the cytosol of the stimulated-PMN. Our proposal that the OB of PMN is partly due to induced inhibition of SOD leading to the release of cyclic chain reactions, seems to us of great physiological importance. Consequently, further studies are intended to establish that SOD is really inhibited by stimulation of normal PMN, and that the inhibition of SOD does not occur in PMN from patients with chronic granulomatous disease. Should the cytosolic factor to which we assign a SOD inhibitory effect prove to be a potent electron carrier competing with SOD, our scheme would of course have to be revised, but the physiological role of this still unidentified molecule would also merit investigation.

Acknowledgements : The authors wish to thank Miss M. Dumont for her skilled secretarial help. This study was supported by INSERM grant N° ATP 61.78.93.

REFERENCES

1. G. Beckman, E. Lundgren, and A. Tarnvik, Superoxide dismutase isozymes in different human tissues, their genetic control and intracellular localization, Hum. Hered. 23:338 (1973).
2. L.R. DeChatelet, C.E. McCall, L.C. McPhail, and R.B. Johnston, Jr, Superoxide dismutase activity in leukocytes, J. Clin. Invest. 53:1197 (1974).

3. C. Auclair, J. Hakim, and P. Boivin, Subcellular superoxide dismutase activity in phagocytozing human blood polymorphonuclear leukocytes, Febs Letts. 79:390 (1977).
4. M.L. Salin, and J.M. McCord, Superoxide dismutases in polymorphonuclear leukocytes, J.Clin. Invest. 54:1005 (1974).
5. J.M. McCord, B.B. Keele, Jr, and I. Fridovich, An enzyme-based theory of obligate anaerobiosis : the physiological function of superoxide dismutase, Proc. Natl. Acad. Sci. U.S.A 68: 1024 (1971).
6. I. Fridovich, Superoxide radical and the bactericidal action of phagocytes, N. Engl. J. Med. 290:624 (1974).
7. B.M. Babior, Oxygen-dependent microbial killing by phagocytes, N. Engl. J. Med. 298:659 (1978).
8. T. Odojima,Myeloperoxidase of the leukocyte of normal blood. II. The oxidation reduction reaction mechanism of the myeloperoxidase, Biochim. Biophys. Acta 235:52 (1971).
9. J.T. Curnutte, M.L. Karnovsky, and B.M. Babior, Manganese-dependent NADPH oxidation by granulocyte particles. The role of superoxide and the non physiological nature of the manganese requirement, J. Clin. Invest. 57:1059 (1976).
10.C. Auclair, E. Cramer, J. Hakim, and P. Boivin, Studies on the mechanism of NADPH oxidation by the granule fraction isolated from human resting polymorphonuclear blood cells, Biochimie 58:1359 (1976).
11.C. Auclair, M. Torres, J. Hakim, and H. Troube, NADPH-oxidation activities in subcellular fractions isolated from resting or phagocytozing human polymorphonuclears, Amer. J. Hemat. 4:113 (1978).
12.M. Torres, C. Auclair, and J. Hakim, NADH oxidation and oxygen reductions by the granule-rich fraction isolated from human polymorphonuclear blood cells. Involvement of a cyclic chain reaction, Febs Letts. 97:50 (1979).
13.C. Auclair, M. Torres, E. Cramer, and J. Hakim, Cyanide-insensitive NADH oxidation by subcellular fractions isolated from human polymorphonuclear blood cells, Enzyme 23:225 (1978).
14.P.C. Chan, and B.H.J. Bielski, Enzyme-catalyzed free radical reactions with nicotinamide adenine nucleotides. II. Lactate dehydrogenase catalyzed oxidation of reduced nicotinamide adenine dinucleotide by superoxide radicals generated by xanthine-oxidase, J. Biol. Chem. 249:1317 (1974).
15.H.M. Hassan, H. Dougherty, and I. Fridovich, Inhibitors of superoxide dismutases : A cautionary tale, Arch. Biochem. Biophys. 199:349 (1980).

DISCUSSION
Edited by R.E.Basford

Dr. Roos:
Could it be that the SOD activity in stimulated neutrophils was low because the superoxide production of the cells obscured the SOD assay?
Dr. Hakim:
I do not think that your suggestion is possible for the following reasons: 1. SOD is in a solubilized form after extraction with chloroform-ethanol from the 15,000 g pellet of the post 350 g supernatant. As you know, solubilized superoxide synthetase activity cannot be obtained so easily from neutrophils. 2. There is no electron donor (NADH or NADPH) in the assay medium used for measurement of SOD activity. 3. Internal standards with Cu-Zn SOD and in the absence of CN were run. The results obtained excluded the possibility that an extra production of O_2^- occurred.

Dr. De Chatelet:
I am puzzled by one aspect of your data. When you measured O_2 consumption with NADPH there was no apparent difference between the resting and phagocytozing granule fraction. Isn't this contrary to the usual reports in the literature?
Dr. Hakim:
Yes, I am aware that our results are contrary to the usual reports since with our experimental procedure the increase in "NADPH oxidase" was less than 20%, in the 15,000 g pellet. This discrepancy in my opinion is of great interest because it might be due to differences in the experimental procedures used. First our pellet is washed before use and second we use a 15,000 g pellet and not a 27,000 g pellet. Washing could possibly remove some of the SOD adherent to the GRF of the resting cells and also remove a cytosolic inhibitor which could have adhered to the 15,000 g pellet in stimulated cells. In studies with subcellular fractions isolated by differential centrifugation (Auclair et al., Amer. J. Hematol., 4:113-120, 1978) we observed that the 100,000 g pellet of the post 15,000 g supernatant was twice as active in stimulated cells as in resting cells. Interaction between these fractions was not studied but might be important since the usual 27,000 g pellet contains some fractions that we recover in 100,000 g pellet.
Dr. Rossi:
1) Dr. Hakim I believe that the reactions you show occur in your cuvettes but I do not believe that they occur in the cell during the respiratory burst. The conditions that you use are far from those within the cells. (pH 5.5, NADH at high concentration, manganese, etc.).
2) Secondly, we have measured SOD after stimulation, and in preliminary experiments we have found that the activity does not

change significantly -- we must continue this investigation.
3) As a third point I ask whether the leukocytes of the subject with high content of SOD (Down's syndrome) have a respiratory burst less marked than that of normal leukocytes.

Dr. Hakim:
To answer your first question, I agree that the reaction I describe occurs in our cuvettes but I personally do not understand how you can know the conditions which are at the cellular site where the reaction occurs. If it is in the phagosome or in the membrane of the phagosome the pH is probably closer to pH 5.5 than pH 7.0 and if it occurs somewhere else the reverse might be true. I agree with you that NADH concentration expressed per whole cell is less than 2.5 mM, however, I do not know the concentration of this nicotinamide coenzyme at the site of the reaction.

With regard to your second question, we observed a decrease of SOD activity mainly in the CN-resistant form of the enzyme. I hope that your further investigations on SOD will confirm our results.

As for your third question we have not studied neutrophils from patients with Down's syndrome. I, however, would not be puzzled if the oxidative burst of their neutrophils was not less marked than that or normal neutrophils because our suggestion is that during this event an inhibition of SOD activity occurs and I wonder if a larger amount of SOD would not be inhibited as well. If we have the occasion to study neutrophils of patients with Down's syndrome we shall try to obtain an answer to your question.

Dr. Seligmann:
We have studied cells from patients with Down's syndrome. In the patients studied there was an abnormality in SOD activity found in all cells examined. It is unfortunate that in these particular patients their PMN's were not studied. However, based on the assumption that SOD activity should also be abnormal in PMNs as in all their other cells, we compared O_2^- production stimulated with f-Met-Leu-Phe and PMA to normal cells to determine if SOD has a modulatory role. We did <u>not</u> find any difference in either rate or extent of O_2^- production by these particular stimuli. This finding argues against a major or any modulatory role of SOD on stimulated PMN O_2^- production.

Dr. Hakim:
This is, I think, a statement which answers Dr. Rossi's question, but as I said it does not still puzzle me too much. It suggests however that we should study CN-resistant SOD in resting and stimulated neutrophils from these patients.

Dr. Hamers:

Could it be that the CN-sensitive SOD you found in the granular fraction originates from contaminating mitochondria which contain the CN-sensitive manganese SOD or is it really localized in the granules. I ask this because it is crucial for the reactions in your scheme that SOD must be CN-insensitive.

Dr. Hakim:

The chloroform-ethanol extract from the 15,000 g pellet contains probably the Mn (CN-resistant)-SOD of the mitochondria. Mitochondria, however, are few in the neutrophils and the amount of SOD extracted is too large to be that of the mitochondria only. On the other hand, others (Rister R. and Baehner R.L., J. Clin. Invest., 58, 1174-1184, 1976) have shown after selective lysis of the mitochondria that a large fraction of CN-resistant SOD is located in the granules of the neutrophils. The location of SOD is however less crucial for my scheme than the observation that the cytosol, after neutrophil stimulation, contains a factor which apparently inhibits SOD activity, because cytosolic SOD might also act as an inhibitor of the oxidative burst.

ROLE OF SERINE PROTEASES IN SUPEROXIDE PRODUCTION

BY HUMAN NEUTROPHILS, MONOCYTES AND BASOPHILS

Seiichi Kitagawa and Fumimaro Takaku

Department of Hematology, Jichi Medical School
Minamikawachi-machi, Kawachi-gun
Tochigi-ken, 329-04, Japan

Superoxide anion (O_2^-) is generated by neutrophils not only during phagocytosis but also on exposure of cells to various surface active agents, suggesting that perturbation of the plasma membrane is an adequate stimulus to activate NAD(P)H oxidase. Recent reports indicate that the O_2^- producing system, including NAD(P)H oxidase, is located at the plasma membrane of the neutrophils[1,2]. However, very little information is available on the mechanisms which couple membrane perturbation to NAD(P)H oxidase activation. The possibility that membrane-bound serine proteases might be involved in the O_2^- production by human neutrophils, monocytes and basophils is the subject of this presentation.

Effect of Serine Protease Inhibitors and Synthetic Substrates for Serine Proteases on O_2^- Production by Human Neutrophils and Monocytes Stimulated by Concanavalin A and Cytochalasin E

Concanavalin A (Con A) and cytochalasin E (Cyt E) were able to induce O_2^- production not only by human neutrophils but also by human monocytes, and the O_2^- production was markedly enhanced by the simultaneous addition of Con A and Cyt E[3-5] (Fig. 1). It appears that marked enhancement of O_2^- production induced by the simultaneous addition of Con A and Cyt E may result from the direct or indirect interaction between Con A-receptor complexes and Cyt E-receptor complexes on the cell surface membrane[6].

We first explored the effect of serine protease inhibitors and synthetic substrates for serine proteases on O_2^- production by human neutrophils and monocytes induced by the simultaneous addition of Con A and Cyt E. O_2^- was assayed by superoxide dismutase inhibitable cytochrome C reduction spectrophotometrically, and the continuous

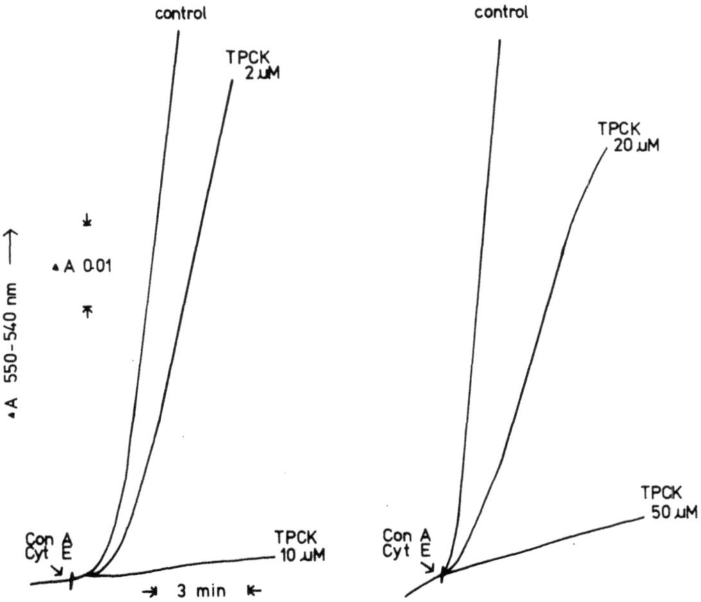

Fig. 1. The inhibitory effect of TPCK on the O_2^- production by human neutrophils (Left) and monocytes (Right) stimulated by the simultaneous addition of Con A and Cyt E. Cell suspensions containing $2x10^5$ neutrophils/ml or $2x10^6$ mononuclear cells /ml (monocytes 15%) were preincubated with TPCK for 3 min at 37°C before Con A (50 µg/ml) and Cyt E (5 µg/ml) were added. The O_2^- production was calculated from an apparently linear portion.

assay was performed in a Hitachi 557 spectrophotometer (a double wave-length spectrophotometer), equipped with thermostatted cuvette holder as described[5]. As shown in Figs. 1-3, O_2^- production by human neutrophils and monocytes was inhibited in a dose dependent fashion by various serine protease inhibitors and synthetic substrates for serine proteases, including the active-site serine sulfonylating agent phenylmethylsulfonylfluoride (PMSF); active-site histidine alkylating agents L-1-tosylamido-2-phenylethyl-chloromethyl ketone (TPCK) and N-α-p-tosyl-L-lysine-chloromethyl ketone (TLCK); naturally occurring macromolecular inhibitors aprotinin and soybean trypsin inhibitor (SBTI); a substrate for chymotrypsin type protease N-benzoyl-L-tyrosine ethyl ester (BTEE) and a substrate for trypsin type protease p-tosyl-L-arginine methyl ester (TAME). The O_2^- production by human neutrophils and monocytes was inhibited by the inhibitor and substrate for chymotrypsin type protease (TPCK and BTEE) as well as those for trypsin type protease (TLCK and TAME), and the inhibitory effect of the former was much greater than that of the latter, suggesting that chymotrypsin-like serine proteases are

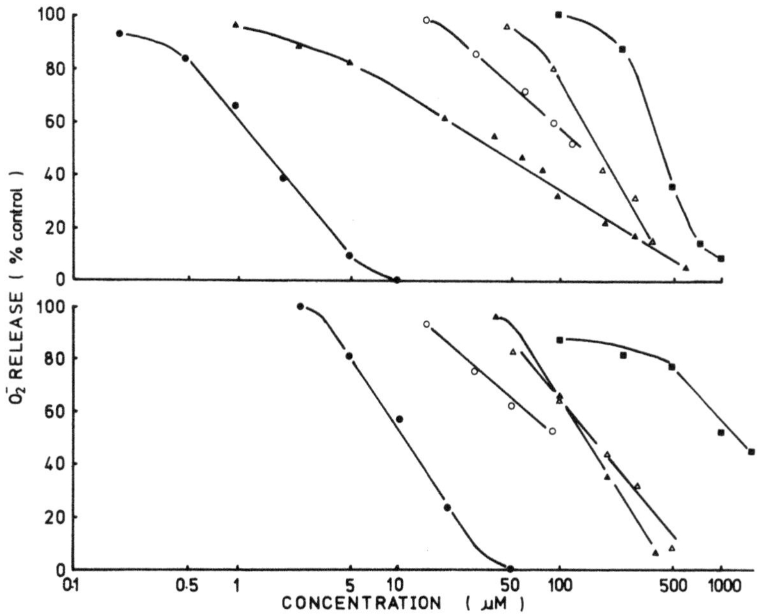

Fig. 2. The inhibitory effect of various serine protease inhibitors on the O_2^- production by human neutrophils (upper panel) and monocytes (lower panel) stimulated by the simultaneous addition of Con A and Cyt E. Cell suspensions were pre-incubated with protease inhibitors for 3 min at 37°C before Con A (50 µg/ml) and Cyt E (5 µg/ml) were added. ●, TPCK; ○, aprotinin; ▲, TLCK; △, SBTI; ■, PMSF.

involved in O_2^- production by human neutrophils and monocytes. Chloromethyl ketone derivatives of amino acids such as TPCK and TLCK also possess a high reactivity with sulfhydryl groups and can inhibit thiol-dependent enzymes such as ficin, papain or cathepsin B_1. Therefore, the present results may indicate that chymotrypsin-like proteases and/or thiol-dependent enzymes are required for neutrophil and monocyte O_2^- production. However, the fact that the O_2^- production was also inhibited by an amino acid ester (BTEE and TAME) and other serine protease inhibitors (PMSF, aprotinin and SBTI) may strengthen the hypothesis that serine proteases are involved in O_2^- production by human neutrophils and monocytes. O_2^- production was sufficiently inhibited by ester substrates (BTEE and TAME), suggesting that the esterase activity may contribute to the O_2^- production, and that these synthetic substrates may inhibit O_2^- production by competing with natural substrates for the proteases that are involved in O_2^- production by human neutrophils and monocytes. O_2^- production was also inhibited by macromolecular inhibitors (aprotinin and SBTI). These findings suggest that proteases involved in O_2^- production may be located on the outer surface of plasma membrane, which is accessible to these macromolecular inhibitors.

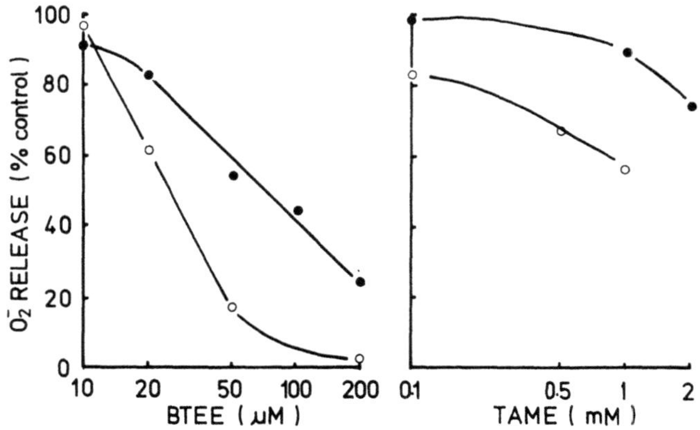

Fig. 3. The inhibitory effect of synthetic substrates for serine proteases BTEE (Left) and TAME (Right) on the O_2^- production by human neutrophils and monocytes stimulated by the simultaneous addition of Con A and Cyt E. Cell suspensions were preincubated with BTEE or TAME for 10 min at 37°C before Con A (50 μg/ml) and Cyt E (5 μg/ml) were added. ○, neutrophils; ●, monocytes.

Compared to the inhibitory effect on neutrophil O_2^- production, several times higher concentrations of protease inhibitors and synthetic substrates were required to obtain the same inhibition of monocyte O_2^- production (Figs. 2 and 3). The relative resistance of monocytes to protease inhibitors and synthetic substrates may imply that proteases involved in the O_2^- production by monocytes may not be identical to those of neutrophils. Another possibility is that the difference of the inhibition profiles of the O_2^- production in human neutrophils and monocytes may reflect the difference of the O_2^- producing system between human neutrophils and monocytes[6]. PMSF-, TPCK- or TLCK-mediated inhibition was irreversible, an observation consistent with their proposed mechanism that they ultimately form covalent bonds at the active sites of the enzymes. These above findings suggest that chymotrypsin-like serine proteases, which exist in an enzymatically active form on the cell surface membrane, are involved in the O_2^- production by human neutrophils and monocytes stimulated by Con A and Cyt E.

Effect of Serine Protease Inhibitors and Synthetic Substrates for Serine Proteases on O_2^- Production by Human Neutrophils and Monocytes Stimulated by Various Surface Active Agents

In addition to Con A and Cyt E, phorbol myristate acetate (PMA, an active principle of croton oil), phospholipase C, N-formyl-methionyl-phenylalanine (FMP, a chemotactic peptide), calcium iono-

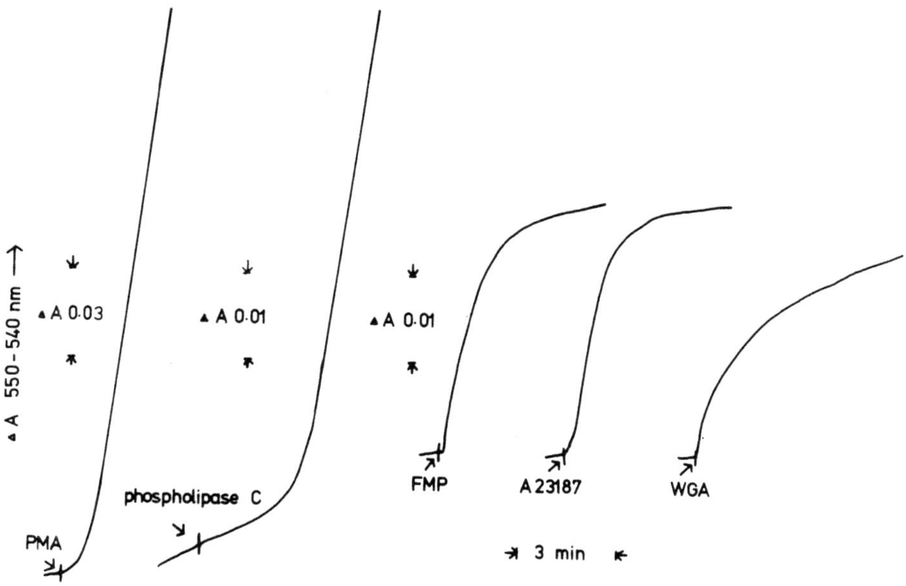

Fig. 4. O_2^- production by human neutrophils stimulated by various surface active agents. Cell concentrations used were $8x10^5$/ml for PMA, $1x10^7$/ml for phospholipase C and $2x10^6$/ml for FMP, A23187 and WGA. The O_2^- production was calculated from an apparently linear portion in PMA and phospholipase C induced O_2^- production, and was calculated from cytochrome C reduced for 2 min after the addition of FMP or A23187 and for 5 min after the addition of WGA.

phore A23187 and wheat germ agglutinin (WGA, a lectin) were also able to stimulate human neutrophils and monocytes to release O_2^-[7]. However, A23187 and phospholipase C were toxic for mononuclear cells at the concentrations used in the present experiments. Therefore, A23187 and phospholipase C were excluded from the monocyte O_2^- production studies. The representative time courses of neutrophil O_2^- production are shown in Fig. 4. The almost similar patterns were observed in monocyte O_2^- production. Con A binds to α-D-mannose and α-D-glucose and WGA binds to $(\beta(1\rightarrow4)$-D-N-acetylglucosamine$)_2$ on the cell surface membrane. Cytochalasins, FMP and PMA are also suggested to bind to the cell surface membrane[8], although it is uncertain what molecules are the binding sites. It is probable that membrane perturbation provoked by the binding of these active agents on the cell surface membrane is responsible for the O_2^- production. On the other hand, phospholipase C acts on the membrane lipid and hydrolyzes phosphatidylcholine[9], which is a major component of the outer layer of the cell membrane. It is unknown how phospholipase C is able to provoke the oxidative metabolism, while it is possible that the hydrolysis of phosphatidylcholine results in the conformational changes of the

Table 1. Inhibitory Effect of Serine Protease Inhibitors and Synthetic Substrates for Serine Proteases on Neutrophil and Monocyte O_2^- Production Stimulated by Various Surface Active Agents

	Neutrophil O_2^- Production (% Control)[a]				
	WGA (100 µg/ml)	FMP (20 µM)	PMA (20 ng/ml)	A23187 (10 µM)	Phospho-lipase C (0.25 U/ml)
PMSF (1 mM)	34.5 ± 7.3	23.2 ± 5.5	82.6 ± 2.1	14.5 ± 2.8	n.d.
TPCK (10 µM)	27.0 ± 1.4	31.2 ± 6.3	50.9 ± 1.4	22.5 ± 8.7	11.1 ± 3.2
TLCK (100 µM)	49.1 ± 5.8	46.7 ± 1.6	74.4 ± 1.4	64.4 ± 4.0	20.4 ± 2.6
BTEE (50 µM)	n.d.	27.0 ± 4.2	77.8 ± 4.5	n.d.	n.d.
TAME (1 mM)	n.d.	79.8 ± 9.0	82.7 ± 2.8	n.d.	n.d.

	Monocyte O_2^- Production (% Control)[a]		
	WGA (100 µg/ml)	FMP (20 µM)	PMA (20 ng/ml)
PMSF (1 mM)	50.8 ± 3.9	54.7 ± 6.4	68.5 ± 5.4
TPCK (10 µM)	27.3 ± 2.9	67.6 ± 1.5	52.4 ± 8.8
TLCK (100 µM)	60.4 ± 3.6	63.9 ± 6.6	65.1 ± 3.3
BTEE (50 µM)	n.d.	23.2 ± 5.6	69.3 ± 7.4
TAME (1 mM)	n.d.	69.1 ± 3.3	88.6 ± 7.9

a Cell suspensions were preincubated with protease inhibitors for 3 min or with synthetic substrates for 10 min at 37°C before various stimulating agents were added.

membrane and allows the redistribution of the macromolecules in the surface membrane, which in turn activates the NAD(P)H oxidase. A23187 acts as a freely mobile carrier to equilibrate divalent cation concentrations across membrane via an electroneutral process, and is suggested to initiate the oxidative metabolism by promoting Ca^{++} influx from the milieu[10]. These active agents appear to initiate the oxidative metabolism of phagocytic cells through different mechanisms.

As summarized in Table 1, irreversible serine protease inhibitors (PMSF, TPCK and TLCK) and synthetic substrates for serine proteases (BTEE and TAME) inhibited the O_2^- production by human neutrophils and monocytes induced by various surface active agents, including WGA, FMP, PMA, A23187 and phospholipase C. These findings indicate that the serine proteases play an essential role for the O_2^- production by human neutrophils and monocytes. Extracellular Ca^{++} plays an important role and enhances the oxidative metabolism induced by N-formylmethionyl peptides, phospholipase C and A23187 [9-11], whereas PMA-induced O_2^- production is never affected by extracellular Ca^{++} [11]. WGA-induced O_2^- production by human neutrophils was also enhanced by extracellular Ca^{++}, and it was inhibited by TPCK whether or not Ca^{++} was present in the milieu. These findings suggest that the serine proteases may not be implicated in Ca^{++} influx[12].

Recently, Washida et al[13] have shown that O_2^- production by guinea pig peritoneal neutrophils stimulated by antigen-antibody complex or Con A is inhibited by diisopropylfluorophosphate (DFP), an active-site serine phosphorylating agent, whereas sodium dodecyl sulfate-induced O_2^- production is not affected by DFP. In addition, TPCK and tetrahydrofuran can stimulate guinea pig peritoneal neutrophils to release O_2^- at high concentrations, and TPCK- or tetrahydrofuran-induced O_2^- production is not inhibited by DFP, whereas they inhibit antigen-antibody complex induced O_2^- production at low concentrations. These findings suggest that NAD(P)H oxidase could be activated by some unphysiological agents through other mechanisms than the serine protease-mediated process. However, in contrast to guinea pig peritoneal neutrophils, TPCK could not induce O_2^- production by human neutrophils and high concentrations of TPCK (such as 145 µM or greater) were even toxic for human neutrophils. These differences may reflect the difference of species.

Potent inactivators of serine proteases (esterases) can inhibit many functions of phagocytic cells, including chemotaxis, phagocytosis and degranulation as well as O_2^- production[14-18]. The available evidence suggests that active and stimulus-activated esterases are required for chemotaxis and phagocytosis[14-16]. The inhibition profiles of chemotaxis of rabbit neutrophils given by the several phosphonate esters indicate that the esterases are similar to chymotrypsin rather than trypsin, although not completely similar to chymotrypsin[14]. The chymotrypsin-type proteases are also required for the phagocytosis of antigen-antibody complexes by guinea pig peritoneal macrophages[17]. Tsung et al[19] have demonstrated that a chymotrypsin-like esterase is present in the membrane fractions isolated from rabbit neutrophils and that the molecular weight of the esterase is about 200 000. It remains to be determined whether serine proteases (esterases) involved in many functions of phagocytic cells are identical one another or not.

It is unknown how the serine proteases might be involved in the activation of NAD(P)H oxidase. If the precursor of NAD(P)H oxidase would be a natural substrate for serine proteases, the precursor might gain access to the proteases by conformational changes of the surface membrane induced by stimuli and might be activated. Another possibility is that the serine proteases might be associated with the movement of the macromolecules in the surface membrane, as it has been suggested that the triggering of the metabolic activation

Table 2. Inhibitory Effect of Serine Protease Inhibitors and Synthetic Substrates for Serine Proteases on Basophil O_2^- Production Stimulated by Anti-IgE

	O_2^- Production (% Control)[a]	
	Patient 1 (F.A.)	Patient 2 (K.T.)
PMSF (1 mM)	41.6 ± 1.3	20.8 ± 3.1
TPCK (5 µM)	n.d.	47.2 ± 4.1
(10 µM)	30.5 ± 1.8	33.0 ± 0.1
(50 µM)	0.8 ± 0.1	n.d.
TLCK (100 µM)	77.6 ± 2.1	73.0 ± 4.7
(500 µM)	17.2 ± 3.6	28.1 ± 0.6
SBTI (400 µM)	65.7 ± 5.9	n.d.
BTEE (50 µM)	52.0 ± 1.6	38.1 ± 2.1
TAME (1 mM)	96.5 ± 2.3	108.4 ± 8.1

a Human basophils were obtained from two leukemic patients (F.A. and K.T.) with basophilia. Cell suspensions were preincubated with protease inhibitors for 3 min or with synthetic substrates for 10 min at 37°C before anti-IgE (10 µl/ml) was added. The O_2^- production was calculated from cytochrome C reduced for 5 min after the addition of anti-IgE.

of phagocytic cells is provided by a surface redistribution of the ligand-receptor complexes, and Smith et al[20] have recently shown that FMP-induced redistribution of binding sites for albumin-coated latex beads on human neutrophil surface membrane is blocked by 100-200 µM TLCK; such concentrations of TLCK also inhibit O_2^- production by human neutrophils and monocytes (Figs. 2 and Table 1). If this is the case, it may be possible that impaired movement of the macromolecules in the surface membrane may also contribute to the inhibition of chemotaxis and phagocytosis in addition to O_2^- production by serine protease inhibitors, inasmuch as chemotaxis and phagocytosis also accompany the movement of the macromolecules in the surface membrane[21,22].

Effect of Serine Protease Inhibitors and Synthetic Substrates for Serine Proteases on O_2^- Production by Human Basophils Stimulated by Anti-IgE

We further investigated whether the similar mechanisms are also involved in the O_2^- production by human basophils, since human basophils release not only histamine but also O_2^- in response to rabbit anti-human IgE[23,24]. As summarized in Table 2, anti-IgE-induced O_2^- production by human basophils was inhibited by PMSF, TPCK, TLCK, SBTI and BTEE. No significant inhibition was observed by 1 mM TAME. These findings suggest that chymotrypsin-like serine proteases are also involved in basophil O_2^- production as has been inferred for O_2^- production by human neutrophils and monocytes. A requirement of intact serine protease (esterase) activity has been suggested for antigen-induced histamine release from human basophils, since histamine release is inhibited by DFP[25]. The present results show a similar requirement in anti-IgE-induced O_2^- production by human basophils.

REFERENCES

1. I.M.Goldstein, M.Cerqueira, S.Lind, and H.B.Kaplan. Evidence that the superoxide-generating system of human leukocytes is associated with the cell surface. J. Clin. Invest. 59:249 (1977).
2. B.Dewald, M.Baggiolini, J.T.Curnutte, and B.M.Babior. Subcellular localization of the superoxide-forming enzyme in human neutrophils. J. Clin. Invest. 63:21 (1979).
3. D.Romeo, G.Zabucchi, and F.Rossi. Reversible metabolic stimulation of polymorphonuclear leukocytes and macrophages by concanavalin A. Nature (New Biol.) 243:111 (1973).
4. A.Nakagawara, B.Z.F.Nabi, and S.Minakami. An improved procedure for the diagnosis of chronic granulomatous disease, using concanavalin A and cytochalasin E. Clin. Chim. Acta 74:173 (1977).
5. S.Kitagawa, F.Takaku, and S.Sakamoto. Evidence that proteases are involved in superoxide production by human polymorphonuclear leukocytes and monocytes. J. Clin. Invest. 65:74 (1980).
6. S.Kitagawa, F.Takaku, and S.Sakamoto. A comparison of the superoxide releasing response in human polymorphonuclear leukocytes and monocytes. J. Immunol. 125:359 (1980).
7. S.Kitagawa, F.Takaku, and S.Sakamoto. Serine protease inhibitors inhibit superoxide production by human polymorphonuclear leukocytes and monocytes stimulated by various surface active agents. FEBS Lett. 107:331 (1979).
8. L.T.Williams, R.Snyderman, M.C.Pike, and R.J.Lefkowitz. Specific receptor sites for chemotactic peptides on human polymorphonuclear leukocytes. Proc. Natl. Acad. Sci. U.S.A. 74:1204 (1977).
9. P.Patriarca, M.Zatti, R.Cramer, and F.Rossi. Stimulation of the respiration of polymorphonuclear leukocytes by phospholipase C. Life Sci. 9:841 (1970).
10. D.Romeo, G.Zabucchi, N.Miani, and F.Rossi. Ion movement across leukocyte plasma membrane and excitation of their metabolism. Nature 253:542 (1975).
11. J.E.Lehmeyer, R.Snyderman, and R.B.Johnston,Jr. Stimulation of neutrophil oxidative metabolism by chemotactic peptides: influence of calcium ion concentration and cytochalasin B and comparison with stimulation by phorbol myristate acetate. Blood 54:

35 (1979).
12. L.Simchowitz, J.Mehta, and I.Spilberg. Chemotactic factor-induced superoxide radical generation by human neutrophils. Requirement for protease (esterase) activity. J. Lab. Clin. Med. 94:403 (1979).
13. N.Washida, A.Sagawa, K.Tamoto, and J.Koyama. Comparative studies on superoxide anion production by polymorphonuclear leukocytes stimulated with various agents. Biochim. Biophys. Acta 631:371 (1980).
14. P.A.Ward, and E.L.Becker. Mechanisms of the inhibition of chemotaxis by phosphonate esters. J. Exp. Med. 125:1001 (1967).
15. S.Aswanikumar, E.Schiffmann, B.A.Corcoran, and S.M.Wahl. Role of a peptidase in phagocyte chemotaxis. Proc. Natl. Acad. Sci. U.S.A. 73:2439 (1976).
16. R.A.Musson, and E.L.Becker. The role of an activatable esterase in immune-dependent phagocytosis by human neutrophils. J. Immunol. 118:1354 (1977).
17. K.Nagai, T.Nakamura, and J.Koyama. Characterization of macrophage proteases involved in the ingestion of antigen-antibody complexes by the use of protease inhibitors. FEBS Lett. 92:299 (1978).
18. E.L.Becker, and H.J.Showell. The ability of chemotactic factors to induce lysosomal enzyme release. II. The mechanism of release. J. Immunol. 112:2055 (1974).
19. P.-K.Tsung, S.W.Kegeles, and E.L.Becker. The evidence for the existence of chymotrypsin-like esterase activity in the plasma membranes of rabbit neutrophils and the specific chemotactic peptide binding activity of the subcellular fractions. Biochim. Biophys. Acta 541:150 (1978).
20. C.W.Smith, and J.C.Hollers. Motility and adhesiveness in human neutrophils. Redistribution of chemotactic factor-induced adhesion sites. J. Clin. Invest. 65:804 (1980).
21. G.B.Ryan, J.Z.Borysenko, and M.J.Karnovsky. Factors affecting the redistribution of surface-bound concanavalin A on human polymorphonuclear leukocytes. J. Cell Biol. 62:351 (1974).
22. R.D.Berlin, and J.M.Oliver. Analogous ultrastructure and surface properties during capping and phagocytosis in leukocytes. J. Cell Biol. 77:789 (1978).
23. W.R.Henderson, and M.Kaliner. Immunologic and nonimmunologic generation of superoxide from mast cells and basophils. J. Clin. Invest. 61:187 (1978).
24. S.Kitagawa, F.Takaku, and S.Sakamoto. Serine protease inhibitors inhibit superoxide production by human basophils stimulated by anti-IgE. Biochem. Biophys. Res. Commun. 95:801 (1980).
25. J.J.Pruzansky, and R.Patterson. The diisopropylfluorophosphate inhibitable step in antigen-induced histamine release from human leukocytes. J. Immunol. 114:939 (1975).

DISCUSSION

Edited by: B.M. Babior

HAMERS: Did you consider the possibility that the protease inhibitors you used in your studies might be interacting with sulfhydryl residues or other essential groups in the dehydrogenase portion of the redox system responsible for the respiratory burst? It is known that mercurials such as N-ethylmaleimide are inhibitory in this system. Since these protease inhibitors can react with sulfhydryl groups, it is possible that their effect on the burst might have nothing to do with their antiprotease activity. I think controls to rule out this possibility are extremely important.

KITAGAWA: The chloromethyl ketone antiproteases TPCK and TLCK react rapidly with sulfhydryl groups and in this way inhibit thiol dependent enzymes such as ficin, papain and cathepsin B. Results with these compounds suggest that either serine proteases, thiol-dependent enzymes or both are required fo O_2^- production. However, the fact that O_2^- production is also inhibited by more specific serine protease inhibitors (PMSF, aprotinin and SBTI) as well as by synthetic substrates for serine proteases (BTEE and TAME) strengthens the hypothesis that serine proteases are involved in O_2^- production. It is of interest that whereas TPCK at low concentrations may inhibit O_2^- production by guinea pig peritoneal exudate PMNs, TPCK at high concentrations induces O_2^- production by these same cells, though not by human PMNs. TPCK-induced O_2^- production is not prevented by serine protease inhibitors (Dri and Patriarca at this meeting; Washida et al in recent issue of Biochim. Biophys. Acta). These findings may indicate that the protease inhibitors used in our experiments inhibit by their antiprotease activity, not by interaction with sulfhydryl or other groups in the dehydrogenase portion of the redox system responsible for the respiratory burst.

RELEASE OF THE MEMBRANE-CALCIUM AND ITS RELATION TO THE SUPEROXIDE FORMATION BY POLYMORPHONUCLEAR LEUKOCYTES

Koichiro Takeshige, Tadashi Matsumoto, Zeenat F. Nabi, and Shigeki Minakami

Department of Biochemistry
Kyushu University School of Medicine
Fukuoka 812, Japan

Polymorphonuclear leukocytes produce active oxygens such as superoxide anions and hydrogen peroxide during phagocytosis or by treatment with reagents for example cytochalasin D(Nakagawara et al., 1976), digitonin(Cohen and Chovaniec, 1978) and a calcium ionophore A23187(Romeo et al., 1975). The role of calcium in the activation of the oxidative metabolism has been discussed, based on the observations that the enhancement of the oxygen consumption was induced by the calcium ionophore(Romeo et al. 1975) and the superoxide release by digitonin was inhibited by EGTA(Cohen and Chovaniec, 1978). Supposing that cytoplasmic calcium ions function as a second messenger to regulate the metabolism, we should consider the possibility of an intracellular translocation of the ions between an intracellular storage pool and cytosol as well as the flux of calcium ions through plasma membrane. We show in this report the intracellular mobilization of calcium ions from storage pool and its relation to the oxidative metabolism of leukocytes by using an intracellular calcium antagonist, 8-(N,N-diethylamino)-octyl-3,4,5-trimethoxybenzoate(TMB-8) and a fluorescent probe for divalent cations, chlortetracycline(CTC).

EFFECT OF TMB-8 ON THE SUPEROXIDE RELEASE

A new intracellular calcium antagonist, TMB-8 has been used in the studies on muscles(Malagodi et al., 1974a, 1974b) and platelets (Charo et al., 1976). We used this reagent in our study on leukocytes.

The addition of a calcium ionophore A23187 to a guinea pig exudate polymorphonuclear leukocyte suspension induced the release

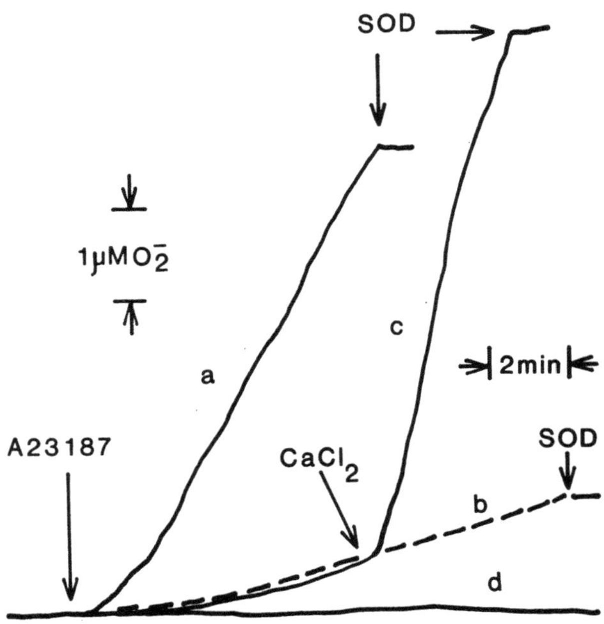

Fig.1. The superoxide release induced with ionophore A23187 of guinea pig exudate leukocytes($2x10^6$ cells in 1 ml calcium free Tris-Ringer solution at 37°C). a; without TMB-8, b; with 100μM TMB-8, c; with 100μM TMB-8 and 1mM $CaCl_2$, d; resting.(from Matsumoto et al., 1979)

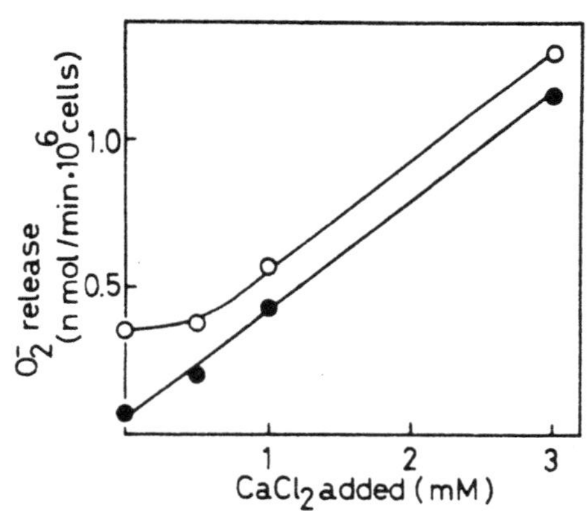

Fig.2. The effect of added calcium concentration on the superoxide release induced by A23187 in the presence and absence of TMB-8. (●) with 100μM TMB-8, (O) without TMB-8. (from Matsumoto et al., 1979)

of superoxide even without the addition of calcium to the medium(Fig. 1, curve a). The superoxide release is shown as the reduction of exogenously added cytochrome c which was measured at 550-540nm with a dual-wavelength spectrophotometer equipped with an end-on photomultiplier tube. The reduction could be stopped by the addition of superoxide dismutase. The release was inhibited by TMB-8(100μM), as shown by the curve b and was restored by the subsequent addition of calcium ions, as shown by the curve c.

As shown in Fig.2, the effect of extracellular calcium concentration of the superoxide release induced by A23187 was different in the presence and absence of TMB-8. The activity of the cells to release the superoxide was essentially proportional to extracellular calcium concentrations when the cells were treated with TMB-8, whereas the activity of the cells not treated with TMB-8 was apparently independent of extracellular calcium when the calcium concentration was low but it increased linearly with the calcium concentration at higher calcium concentrations. The release in the absence of TMB-8 and at low concentration of extracellular calcium may be ascribed to the effect of A23187 to accelerate the release of calcium from an intracellular storage site to the cytosol, whereas the superoxide release at higher concentration of extracellular calcium may essentially be due to the transport function of the ionophore through plasma membrane. Thus, the inhibition was strong in calcium-free medium and it decreased gradually with the increase of extracellular calcium concentration. This indicates that TMB-8 inhibits the superoxide release by inhibiting the release of calcium from storage pool.

Table 1. The effect of TMB-8 on the superoxide release activity induced by E.coli, cytochalasin D or A23187. (from Matsumoto et al., 1979)

	Activated with		
	E.coli (0.9 mg)	Cytochalasin D (10 μg)	A23187 (5 nmol)
Control	0.84	1.20	2.49
TMB-8 50 μM	0.49	0.40	2.49
100 μM	0.42	0.13	1.06
200 μM	0.14	0	0.51
300 μM	0.04	0	0.07

The inhibition of the superoxide release by TMB-8 was not limited to the activity induced by the calcium ionophore. As shown in Table 1, the superoxide release induced by the ingestion of heat-killed E.coli or by the interaction of cytochalasin D was similarly inhibited by TMB-8. This observation suggests the general role of intracellular calcium in the regulation of the superoxide release reaction. Inhibitors such as verapamil which have been used to inhibit the influx of extracellular calcium showed only weak inhibition of the superoxide release induced by cytochalasin D.

FLUORESCENCE CHANGE OF CTC-LOADED LEUKOCYTES BY STIMULATION

If calcium ions are stored in a membrane compartment and released by the stimulation, we may be able to monitor the mobilization by using a fluorescent probe, chlortetracycline(CTC). Fluorescence of CTC is extremely sensitive to the concentration of divalent cations within the cellular hydrophobic environment(membrane) into which it is preferentially partitioned. The calcium chelate gives stronger fluorescence than the magnesium chelate and the spectra of both chelates are different(Caswell and Hutchinson, 1971). CTC has been used as a probe to investigate intracellular events associated with the divalent cations and their interactions with membranes such as

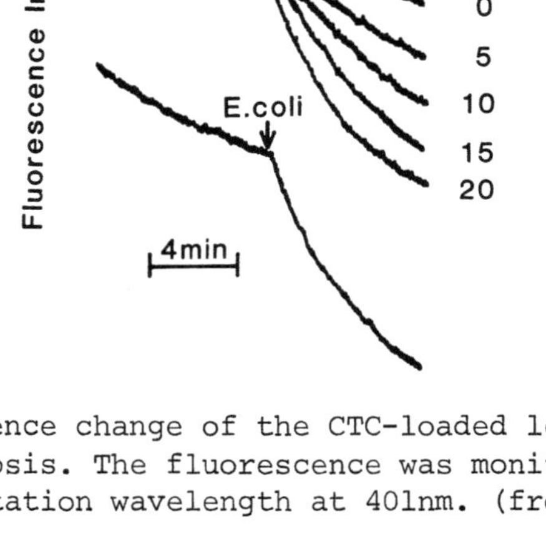

Fig.3. Fluorescence change of the CTC-loaded leukocytes during phagocytosis. The fluorescence was monitored at 514nm with the excitation wavelength at 401nm. (from Takeshige et al., 1980)

mitochondria and sacoplasmic reticulum(Caswell, 1972; Caswell and Pressman, 1972; Chandler and Williams, 1978).

Leukocytes were preincubated for 30min at 30°C in the modified Krebs-Ringer phosphate solution(calcium: 0.6mM, pH7.4) with 100μM CTC and 2mM glucose, washed once , resuspended in the medium containing no calcium, and transferred to a spectrofluorometer with a constant temperature cuvette holder(37°C). As shown in Fig.3, the CTC-loaded leukocytes exhibited a fluorescence which decreased slowly with time. A rapid decrease of fluorescence was observed when cytochalasin D or heat-killed E.coli was added to a suspension of CTC-loaded cells. Because calcium ions were omitted from the suspending solution, the rapid change in fluorescence is reflecting the intracellular events. The cells loaded with oxytetracycline, a fluorescent reagent structually similar to CTC but insensitive to divalent cations, did not display such fluorescence change during phagocytosis. Because dose-responses of the fluorescence change and the superoxide release of the CTC loaded cells were parallel as shown in Fig.4, both reactions are closely related.

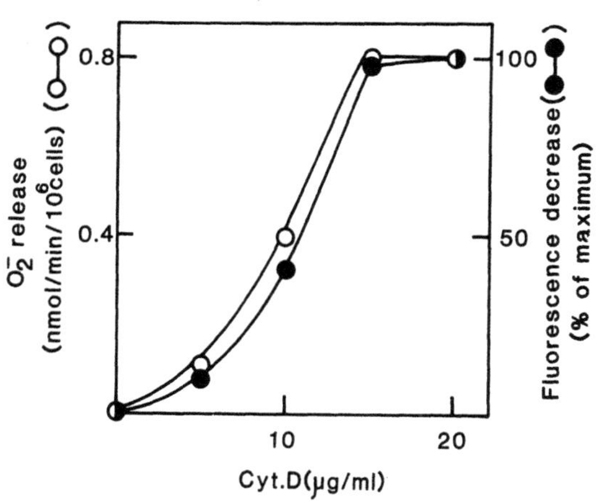

Fig.4. Dose dependence of the fluorescence change(●) and the superoxide release(O) induced by cytochalasin D. The rates of the fluorescence change(arbitrary unit) and the superoxide release(0.8nmol/min per 10^6 cells) at 20μg/ml cytochalasin D are set as 100%. (from Takeshige et al., 1980)

CHANGES OF FLUORESCENCE SPECTRA

The spectra differences between the calcium and magnesium ions were utilized to distinguish which ions are chelated to CTC in the cells. The excitation and emission maxima of the magnesium chelate

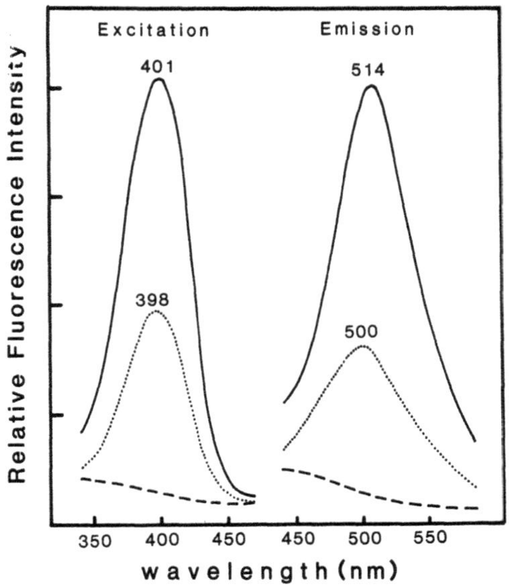

Fig.5. Excitation and emission spectra of the CTC-loaded leukocytes. The spectra were recorded at 5min(solid lines) and at 15min after(dotted lines) the addition of cytochalasin D. The spectra of the cells without loading CTC are shown by dashed lines. The excitation and emission wavelengths were set at 401 and 514nm for the excitation and emission spectra, respectively.(from Takeshige et al., 1980)

are about 10nm shorter in wavelength than those of the calcium chelate (Chandler and Williams, 1978). Fig . 5. shows the excitation and emission spectra of the CTC-loaded cells before and after the stimulation with cytochalasin D. Spectra taken 5min before the stimulation had excitation and emission maxima at 401nm and 514nm, respectively, and the fluorescence intensity of the cells decreased greatly after the stimulation, with excitation and emission maxima shifted to 398nm and 500nm, respectively. Similar fluorescence changes were observed when the cells were stimulated with E.coli, whereas the resting cells displayed only a moderate change in fluorescence intensity without the shift of the maxima even after incubation at 37°C for 20min. These observations indicate that a large part of the calcium ions in hydrophobic environment(membrane) is displaced by magnesium ions when the cells are stimulated. Essentially similar results and interpretations have been reported for peritoneal neutrophiles stimulated with chemotactic factors(Nacchache et al., 1979).

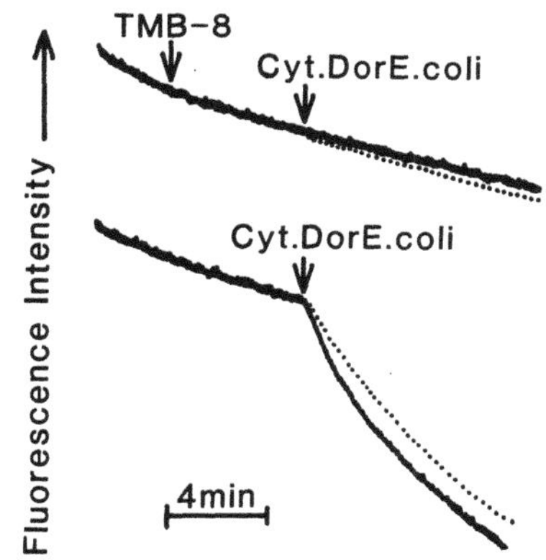

Fig.6. Effect of TMB-8 on the fluorescence of the CTC-loaded cells. (from Takeshige et al., 1980)

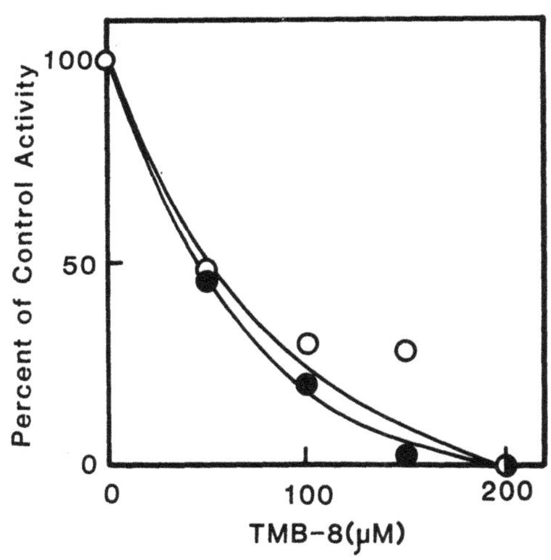

Fig.7. Effects of TMB-8 on the fluorescence change(●) and the superoxide release(O) of the CTC-loaded cells induced by cytochalasin D. (from Takeshige et al., 1980)

EFFECT OF TMB-8 ON THE FLUORESCENCE CHANGE

The relation between the release of the membrane-calcium and the phagocytotic metabolic change of releasing superoxide was further studied by using TMB-8. As shown in Fig.6, when the CTC-loaded cells were treated with TMB-8(100μM) for 5min, the fluorescence decrease and the shift of the excitation and emission maxima due to the stimulation disappeared. The titration of the cells with TMB-8 gave essentially the same curve both for the fluorescence change and the superoxide release as shown in Fig.7. This finding suggests that the release of calcium ions from the intracellular hydrophobic pool to cytosol is an essential process in the oxidative metabolic change of the stimulated cells.

SUMMARY

The relationship between the intracellular translocation of calcium from the storage pool and the oxidative metabolism was studied. An intracellular calcium-antagonist, TMB-8, inhibited the release of the superoxide induced by a calcium ionophore A23187 and the inhibition was relieved by the addition of calcium ions. The release induced by cytochalasin D or by the ingestion of bacteria was similarly inhibited by TMB-8. The mobilization of intracellular divalent cations of leukocytes was monitored by a fluorescent probe, CTC. When the CTC-loaded cells were stimulated with cytochalsin D or E.coli, a fluorescence change ascribable to the release of calcium from the intracellular hydrophobic environment was observed. The dose-response curve of the fluorescence change and that of the superoxide release of the cells were very similar. TMB-8 inhibited both metabolic and fluorescence changes in parallel. The results support the hypothesis that an intracellular translocation of calcium is stimulating the oxidative metabolism of leukocytes.

REFERENCES

Caswell,A.H., and Hutchinson,J.D., 1971, Visualization of membrane bound cations by a fluorescent technique. Biochem. Biophys. Res. Commun. 42:43.

Caswell,A.H., 1972, The migration of divalent cations in mitochondria visualized by a fluorescent chelate probe. J. Memb. Biol., 7:345.

Caswell,A.H., and Pressman,B.C., 1972, Kinetics of transport of divalent cations across sarcoplasmic reticulum vesicles induced by ionophores. Biochem. Biophys. Res. Commun., 49:292.

Charo,I.F., Feiman,R.D., and Detwiler,T.C., 1976, Inhibition of platelet secretion by an antagonist of intracellular calcium. Biochem. Biophys. Res. Commun., 72:1462.

Chandler,D.E., and Williams,J.A., 1978, Intracellular divalent cation release in pancreatic acinar cells during stimulus-secretion coupling, I. Use of chlorotetracycline as fluorescent probe. J. Cell Biol., 76:371.

Cohen,H.J., and Chovaniec,M.E., 1978, Superoxide production by digitonin-stimulated guinea pig granulocytes. The effects of N-ethylmaleimide, divalent cations, and glycolytic and mitochondrial inhibitors on the activation of the superoxide generating system. J. Clin. Invest., 61:1088.

Malagodi,M.H., and Chiou,C.Y., 1974a, Pharmacological evaluation of a new Ca^{2+} antagonist, 8-(N,N-diethylamino)-octyl-3,4,5-trimethoxybenzoate hydrochloride(TMB-8): Studies in smooth muscles. Eur. J.Pharmacol., 27:25.

Malagodi,M.H., and Chiou,C.Y., 1974b, Pharmacological evaluation of a new Ca^{2+} antagonist, 8-(N,N-diethylamino)-octyl-3,4,5-trimethoxybenzoate hydrochloride(TMB-8): Studies in skeletal muscles. Pharmacol. 12:20.

Matsumoto,T., Takeshige,K., and Minakami,S., 1979, Inhibition of phagocytotic metabolic changes of leukocytes by an intracellular calcium-antagonist 8-(N,N-diethylamino)-octyl 3,4,5-trimethoxybenzoate. Biochem. Biophs. Res. Commun., 88:974.

Naccache,P.H., Showell,H.J., Becker,E.L., and Sha'afi, R.I., 1979, Involvement of membrane calcium in the response of rabbit neutrophils to chemotactic factors as evidenced by the fluorescence of chlorotetracycline. J. Cell Biol., 83:179.

Nakagawara,A., Shibata,K., Takeshige,K., and Minakami,S., 1976, Action of cytochalasin E on polymorphonuclear leukocytes of guinea pig peritonear exudates. Exp. Cell Res., 101:225.

Romeo,D., Zabucchi,G., Miani,N., and Rossi,F., 1975, Ion movement across leukocyte plasma membrane and excitation of their metabolism. Nature, 253:542.

Takeshige,K., Nabi,Z.F., Tatschek,B., and Minakami,S., 1980, Release of calcium from membranes and its relation to phagocytotic metabolic changes: A fluorescence study on leukocytes loaded with chlortetracycline. Biochem. Biophys. Res. Commun., 95:410.

ATP-DRIVEN Ca^{2+} PUMP ACTIVITY OF MACROPHAGE AND NEUTROPHIL PLASMA MEMBRANE

Claudio Schneider, Cristina Mottola, Lucilla Dolzani and Domenico Romeo

Istituto di Chimica Biologica, Università di Trieste
Italy

Introduction

Elevation of cytosolic Ca^{2+} concentration elicits a variety of responses from macrophages and neutrophils. These include activation of cell locomotion (1-4), secretion of granule components such as hydrolytic enzymes (3,5-7), and increased reduction of oxygen to cytocidal superoxide anion and hydrogen peroxide (7-10). The activity of cytosolic Ca^{2+} in these processes has been in part explained at the molecular level. For example, Ca^{2+} controls gelsolin-dependent gel-sol transformation of actin, which modulates macrophage locomotion, secretion and endocytosis (11). Further, a Ca^{2+}-requiring plasma membrane phospholipase A_2 (12) can release arachidonic acid, whose metabolites are likely involved in the modulation of neutrophil chemotaxis and secretion (13,14).

Ionized calcium appearing in the cytosol is very likely mobilized from membrane-bound forms as a result of cell interaction with surface-interacting stimulants, such as phagocytosable particles, formyl-peptides and the complement fragment C5a (7,10,15). Restoration and/or maintenance of steady-state levels of cytosolic Ca^{2+}, characteristic for resting conditions of macrophages and neutrophils, thus requires the activity of pumps, which accumulate the ion within intracellular organels or extrude it across the plasma membrane (16-22).

Changes in the steady-state concentration of Ca^{2+}, critical for the control of such properties of leukocytes as assembly and directionality of the contractile apparatus or activation of the plasma membrane O_2^--generating enzyme (8), very likely occurs in the cortical cytoplasm. Thus, among the various intracellular Ca^{2+} buffering systems, a pump localized at the cell surface and capable of extruding calcium against a large electrochemical gradient, should exert a primary role in modulating macrophage and neutrophil response to environmental stimuli.

The presence and activity of a plasma membrane-localized Ca^{2+} transport system can be studied by different experimental approaches. First, the permeability of cell membranes can be increased by exposing intact cells to a specific ionophore (1,5,6,8,9,16-18,20). The cytosolic concentration of free calcium ions thus raises and this activates the Ca^{2+} pumps, including that present at the cell periphery. An extrusion of Ca^{2+} from the cells can therefore be detected, the magnitude of this process depending on the relative sizes of the pump and the ionophore-mediated passive fluxes.

Alternatively, the properties of the transport system can be investigated by using inside-out plasma membrane-derived vesicles,which expose to the medium the inner face of the cell surface membrane. If a Ca^{2+} pump is present in the membrane, the vesicles accumulate Ca^{2+}, when supplied with a suitable energy-generating system (21,23,24).

Finally, if the transport process is driven by ATP, it will manifest itself in isolated plasma membrane fractions as a Ca^{2+}-activated ATPase (21,25-27). This activity can be utilized as convenient marker during extraction of the transport system from the membrane, its purification and reconstitution into phospholipid vesicles (28).

All the three approaches have been either partially or completely applied to the study of the peripheral Ca^{2+} pump system of macrophages and neutrophils. This paper will review the most significant aspects of these studies.

Ca^{2+} Transport across the Plasma Membrane of Intact Macrophages and Neutrophils.

Exposure of macrophages or neutrophils to A23187, an ionophore for bivalent cations, causes a mobilization of calcium ions from cellular stores, as reported for other cell types (17,18,20). This can be shown by following the movement of $^{45}Ca^{2+}$ between cells and the incubation medium. For example, addition of the ionophore A23187 to neutrophils in the presence of 0.1-1 mM Ca^{2+} elicits a prompt increase in the amount of cell-associated $^{45}Ca^{2+}$, which peaks at about 60 sec. (29). This rise in neutrophil Ca^{2+} is not due to a net influx of the ion, as indicated by measurements of Ca^{2+} movements carried out with a highly sensitive, Ca^{2+}-selective electrode (30). This suggests, therefore, that immediately after neutrophil exposure to the ionophore there is an exchange of calcium between readily accesible pools, possibly localized near the cell surface, and the $^{45}Ca^{2+}$-containing extracellular milieu.

The A23187-induced mobilization of Ca^{2+} from the stores is followed by a rapid cell extrusion of the bivalent cation, which can be easily detected as net cation outflux by a Ca^{2+}-selective electrode (29). The initial rate at which the cation is extruded from macrophages and neutrophils is about 0.1-0.2 μg-ions/min/ml of cell water, and the fraction of total cell calcium released under these conditions is lower than 15%.

The extracellular release of Ca^{2+} is driven by ATP, as suggested by experiments carried out with metabolic inhibitors. For example, oligomycin, which causes a greater than 90% ATP depletion in lung macrophages (6), does not produce per se any detectable perturbation of Ca^{2+} distribution between macrophages and the medium. However,when the ionophore A23187 is added to the macrophages suspension 2-3 min after oligomycin, a 75-95% inhibition of the rate and the extent of Ca^{2+} extrusion is seen (29).

In some cell types, beside the ATP-dependent Ca^{2+} pump, there is a mechanism for Ca^{2+} extrusion dependent on the opposite Na^+ electrochemical gradient (24,31,32). This mechanism does not appear to be present on the macrophage and neutrophil peripheral membrane. In fact, elimination of the inwardly directed Na^+ gradient, obtained by replacing extracellular Na^+ with choline, does not affect the A23187-induced cell Ca^{2+} extrusion.

Calcium Transport by Inside-out Plasma Membrane Vesicles.

Utilization of intact cells for the study of the Ca^{2+} pump does not permit the evaluation of some biochemical parameters of the pump, such as the affinity for Ca^{2+} and ATP. This goal can be achieved by measuring Ca^{2+} accumulation by inside-out plasma membrane vesicles, whose matrix is not accessible to freely diffusable ions.

To this aim, Lew and Stossel (23) have used phagocytic vesicles, containing oil droplets and thus easily purified in an intact state by floatation. Internalization of these droplets by macrophages turns the plasma membrane inside out, with exposure of the inner membrane surface to the bathing medium. By this approach a high affinity ATP-dependent Ca^{2+} pump, with a Km for free calcium of about 0.5 μM, has been characterized. The presence of both Mg^{2+} and ATP is essential for calcium uptake by the vesicles, and the optimal pH for uptake lies between 6.6 and 7.2.

The activity of the pump is stimulated by calmodulin, the ubiquitous Ca^{2+}-binding protein that confers calcium sensitivity to several enzymes and stimulates active calcium transport into inside-out vesicles of red blood cell membranes (27).

Presence of a High-affinity Ca^{2+}-Dependent ATPase Activity in the Plasma Membrane of Macrophages and Neutrophils.

The Ca^{2+} pump activity of erythrocyte ghosts, cardiac sarcolemma and sarcoplasmic reticulum, and brain microsomes manifests itself as a Ca^{2+}-dependent ATPase, which catalyzes a reaction sequence including phosphorylation of the enzyme by ATP (21,25,26,33). We have shown that also plasma membrane fractions purified from macrophages and neutrophils catalyze an ATPase reaction, specifically stimulated by Ca^{2+} and characterized by the formation of a phosphorylated intermediate (34-37).

Highly purified plasma membrane fractions can be isolated from lysates of rabbit alveolar macrophages or from homogenates of human or bovine neutrophils.

In the former case, the macrophages are disrupted by lysis in 1 mM $NaHCO_3$, and the particulate material separated from the cell lysate by centrifugation at 1000-1500g is subfractionated by centrifugation through a discontinuous sucrose gradient. The membrane fraction, collected at the 30/40% sucrose interface, shows an about 20-fold enrichment in specific activity of the plasma membrane marker enzyme alkaline phosphodiesterase I over the original cell lysate, with a significant reduction in the specific activities of enzymes associated with intracellular organels.

With neutrophils, post-nuclear supernatants derived from homogenized cells are subfractionated by centrifugation through either continuous (36) or discontinuous (37) sucrose density gradients.

In the plasma membrane fractions of human and bovine neutrophils, the specific activity of the plasma-membrane associated enzyme alkaline phosphatase in enriched 10- and 35-fold, respectively (36,37), with respect to the original cell homogenate.

An ATPase activity, which is specifically stimulated by low concentrations of Ca^{2+}, but not of Mg^{2+}, and is distinct from myosin ATPase, copurifies with the plasma membrane marker enzymes (34-37). At pH 7, half-maximal velocity of the ATPase reaction is reached at 0.7 μM Ca^{2+} (macrophages), 0.08 μM Ca^{2+} (human neutrophils) and 0.22 μM Ca^{2+} (bovine neutrophils). These concentrations of ionized calcium fall in the range of concentrations of the bivalent ion, which are thought to be characteristic for the cytosol of animal cells (22,31).

Unlike other Ca^{2+}-stimulated ATPases (21,25,26), the activity of the enzyme of macrophages and neutrophils does not depend on the addition of Mg^{2+} to the assay mixture. However, since the preparation of the plasma membrane of both macrophages and neutrophils is carried out in the presence of Mg^{2+}, it is likely that the membrane-derived material binds saturating amounts of this ion. Grounds for this conclusion are also offered by the observation that the fluorescence excitation spectrum of chlorotetracycline-binding plasma membranes shows a peak at 375 nm, characteristic for Mg^{2+}-chlorotetracycline complexes (38).

Splitting of ATP by the macrophage and neutrophil plasma membrane Ca^{2+}-dependent enzyme occurs though a very rapid phosphorylation of a membrane polypeptide component, followed by decay of the complex and generation of inorganic phosphate. This mechanism of reaction is suggested by the following observations(35,36).

When incubated in the presence of γ [^{32}P] ATP, the plasma membrane of macrophages and neutrophils shows a Ca^{2+}-stimulated incorporation of radioactive phosphate into a protein subunit with an apparent mol.wt. of 135000, as determined by SDS/polyacrylamide gel electrophoresis. The half-maximal activating affect of Ca^{2+} is found at a concentration of 0.1-0.3 μM. Addition of Mg^{2+} causes a slight enhancement in the steady-state value of Ca^{2+}-dependent ^{32}P labelling, but no significant increase in the basal phosphorylation of membranes.

Ca^{2+}-Dependent ^{32}P incorporation into the membrane is very rapid at 0°C, attaining the steady-state value by less than 30 sec. Addition of excess ADP after 30 sec results in a rapid loss of radioactivity, the amount of membrane-bound ^{32}P falling to less than 20% within 10 sec. The rate of decay of the ^{32}P-membrane complex is under the influence of Ca^{2+} concentration, in a fashion similar to that observed for the phosphorylation step of the membrane.

Radioactivity of labelled plasma membrane is readily removed by exposure to hydroxylamine, suggesting that the terminal ^{32}P phosphate group of ATP is incorporated into an acyl-phosphate bond.

Finally, when the Ca^{2+}-dependent labelling of macrophage plasma membrane is carried out in the presence of 20-100 mM Na^+, a slight increase in the extent of phosphorylation occurs. In contrast, 5-100 mM K^+ markedly decreases the steady-state value of membrane phosphorylation, this effect being primarily caused by acceleration of decay of the ^{32}P-membrane complex.

In conclusion, there is circumstantial evidence that the Ca^{2+}-ATPase reaction catalyzed by the enzyme associated with the plasma membrane of macrophages and neutrophils is the enzyme manifestation of the Ca^{2+} pump activity shown by phagocytic vesicles. Both the ATPase and the pump have a very similar, high affinity for Ca^{2+}, and exert their maximal activity around pH 7 (23,24). Furthermore, the Ca^{2+}-ATPase reaction occurs through the formation of a phosphorylated intermediate, characteristic for transport ATPases (21,25,26,33):

1. $ATP + Enz \rightleftharpoons Enz.P + ADP$ 2. $Enz.P \rightarrow Enz + P$

Reaction 1., both in the onward and backward directions, is stimulated by concentrations of Ca^{2+} in the range of 10^{-8}-10^{-6}M. Reaction 2., at least in macrophages, is strongly accelerated by physiological concentrations of K^{+}, which should thus stimulate the turnover of the intermediate and possibly the efficiency of the transport system, associated with the intermediate formation.

Definite experimental proof for the hypothesis that the Ca^{2+}-ATPase is the expression of a Ca^{2+} pump should derive from purification of the enzyme and its reconstitution into lipid vesicles capable of accumulating Ca^{2+}.

Experiments along this line are in their beginning. The macrophage plasma membrane has been extracted at 0°C with a variety of detergents, such as Na-deoxycholate or taurodeoxycholate, Triton X-100 and Lubrol PX or WX, at a concentration of 5-10 mg/ml. This treatment causes the solubilization. of 20-40% protein and permits the recovery of 20-50% of the Ca^{2+}-dependent ATPase activity in soluble form. A partial purification of the enzyme can be achieved by affinity chromatography through a column of Lentil lectin-Sepharose 4B.Ten-twenty% of protein and 70-80% of the Ca^{2+}-stimulated ATPase activity are retained by the immobilized lectin and can subsequently be eluted by 1M α-methyl mannopyranoside.

This behaviour of the enzyme indicates that the Ca^{2+}-ATPase is a glycoprotein. It has been suggested (39) that transmembrane oligomeric glycoproteins are required for transport across the membranes of eukariotic cells, the presence of carbohydrate chains helping fix the transmembrane arrangement of the protein. The glycoprotein nature of the macrophage Ca^{2+}-ATPase would thus be consistent with the hypothesis that the activity of this enzyme is linked with an active transport of Ca^{2+}.

Further purification of the Ca^{2+}-ATPase has been hampered by its extreme instability in the soluble form, even in the presence of detergents. New procedures are thus under study to overcome this problem.

Conclusions

The results of the investigations here described permit us to conclude that macrophages and neutrophils have a peripheral, outwardly directed Ca^{2+} extrusion system, which is very similar to the well known Ca^{2+} pump of the red cell, with regard to capacity and mechanism (16,21,29). In fact, all the three cell types have similar maximum pumping rates (about 0.1-0.2 μg-ions Ca^{2+}/min/ml cells) and use ATP for extruding Ca^{2+}. Furthermore, the plasma membrane of all the three cell types catalyzes a Ca^{2+}-dependent ATPase reaction,which is very likely the enzyme manifestation of the Ca^{2+} pump activity.

Further investigation is needed to establish whether the peripheral Ca^{2+} pump system of macrophages and neutrophils is utilized to restore steady-state levels of cytosolic Ca^{2+} upon cell stimulation, or is somehow involved in the triggering of cell response to various stimuli. In fact impairment of the pump activity by a cell stimulant would unbalance Ca^{2+} passive leaks and active cation extrusion, thereby leading to higher steady-state levels of Ca^{2+} in the cytosol and to stimulation of Ca^{2+}-dependent functions (1-12).

Acknowledgements

This work was supported by the Italian Research Council and by the University of Trieste. C.M. is a recipient of a fellowship from the A.Villa Rusconi Foundation.

REFERENCES

1. Wilkinson, P.C. 1975. Leucocyte locomotion and chemotaxis. The influence of divalent cations and cation ionophores. Exp.Cell. Res. 93: 420.
2. Bouceck, M.M. and R.Snyderman. 1976. Calcium influx requirement for human neutrophil chemotaxis. Inhibition by lanthanum chloride. Science 193: 905.
3. Petroski, R.J., P.H.Naccache, E.L.Becker and R.I.Sha'afi. 1979. Effect of chemotactic factors on the calcium levels of rabbit neutrophils. Am.J.Physiol. 237: C43.
4. Cramer, E.B. and J.I.Gallin. 1979. Localization of submembraneous cations to the leading end of human neutrophils during chemotaxis. J.Cell Biol. 82: 369.
5. Zabucchi, G., M.R.Soranzo, F.Rossi and D.Romeo. 1975. Exocytosis in human polymorphonuclear leukocytes induced by A23187 and calcium. FEBS Lettr. 54: 44.
6. Schneider, C., R.Gennaro, G.de Nicola and D.Romeo. 1978. Secretion of granule enzymes from alveolar macrophages. Regulation by intracellular Ca^{2+}-buffering capacity. Exp.Cell Res. 112: 249.
7. Roos, D., M.de Boer and R.S.Weening. 1977. The role of calcium and magnesium in some functions of human neutrophils. In Movement, Metabolism and Bactericidal Mechanisms of Phagocytes, F.Rossi, P.Patriarca and D.Romeo, eds., p.223, Piccin Medical Books,Padova.
8. Romeo, D., G.Zabucchi, N.Miani and F.Rossi. 1975. Ion movement across leucocyte plasma membrane and excitation of their metabolism. Nature 253: 542.
9. Root, R.K. and J.A.Metcalf. 1977. Superoxide and hydrogen peroxide formation by human granulocytes: inter-relationships and activation mechanisms. In Movement, Metabolism and Bactericidal Mechanisms of Phagocytes, F.Rossi, P.Patriarca and D.Romeo, eds.,

p.185, Piccin Medical Books, Padova.
10. Matsumoto, T., K.Takeshige and S.Minakami. 1979. Inhibition of phagocytic metabolic changes of leukocytes by an intracellular calcium antagonist 8-(N,N-diethylamino)-octyl-3,4,5-trimethoxybenzoate. Biochem.Biophys.Res.Commun. 88: 974.
11. Yin, H.L. and T.P.Stossel. 1979. Control of cytoplasmic actin gel-sol transformation by gelsolin, a calcium-dependent regulatory protein. Nature 281: 583.
12. Franson, R., J.Weiss, L.Martin, J.Spitznagel and P.Elsbach. 1977. Phospholipase A activity associated with the membranes of human polymorphonuclear leucocytes. Biochem.J. 167: 839.
13. Turner, S.R., J.A.Tayner and W.S.Lynn. 1975. Biogenesis of chemotactic molecules by the arachidonate lipoxygenase system of platelets. Nature 257: 680.
14. Naccache, P.H., H.J.Showell, E.L.Becker and R.I.Sha'afi. 1979. Arachidonic acid induced degranulation of rabbit peritoneal neutrophils. Biochem.Biophys.Res.Commun. 87: 292.
15. Naccache, P.H., M.Volpi, H.J.Showell, E.L.Becker and R.I.Sha'afi. 1979. Chemotactic factor induced release of membrane calcium in rabbit neutrophils. Science 203: 461.
16. Ferreira, H.G. and V.L.Lew. 1976. Use of ionophore A23187 to measure cytoplasmic Ca buffering and activation of the Ca pump by internal Ca. Nature 259: 47.
17. Babcock, D.F., N.L.First and H.A.Lardy. 1976. Action of ionophore A23187 at the cellular level. Separation of effects at the plasma and mitochondrial membranes. J.Biol.Chem. 251: 3881.
18. Desmedt, J.E. and K.Hainaut. 1976. The effect of A23187 ionophore on calcium movements and contraction processes in single barnacle muscle fibers. J.Physiol., Lond. 257: 87.
19. Cittadini, A., D.Bossi, G.Rosi, F.Wolf and T.Terranova. 1977. Calcium metabolism in Ehrlich ascites tumour cells. Biochim.Biophys.Acta 469: 345.
20. Chen, J.J., D.F.Babcock and H.A.Lardy. 1978. Norepinephrine, vasopressin, glucagon, and A23187 induce efflux of calcium from an exchangeable pool in isolated rat hepatocytes. Proc.Natl.Acad.Sci. USA 75: 2234.
21. Schatzmann, H.J. and H.Bürgin. 1978. Calcium in human blood red cells. Ann.N.Y.Acad.Sci. 307: 125.
22. Rasmussen, H., C.Clayberger and M.C.Gustin. 1979. The messenger function of calcium in cell activation. In Symposia of the Society of Experimental Biology, vol.33, Secretory Mechanisms, p.161, Cambridge University Press, Cambridge.
23. Lew, P.D. and T.P.Stossel. 1980. Calcium transport by macrophage plasma membranes. J.Biol.Chem. 255: 5841.
24. Caroni, P. and E.Carafoli. 1980. An ATP-dependent Ca^{2+}-pumping system in dog heart sarcolemma. Nature 283: 765.
25. Robinson, J.D. 1976. Calcium-stimulated phosphorylation of a brain (Ca + Mg)-ATPase preparation. FEBS Lettr. 87: 261.
26. St.Louis, P.J. and P.V.Sulakhe. 1978. Protein analysis of cardiac

sarcolemma: Effects of membrane-perturbing agents on membrane proteins and calcium transport. Biochemistry, 17: 4540.
27. Larsen, F.L. and F.F.Vincenzi. 1979. Calcium transport across the plasma membrane: Stimulation by calmodulin. Science 204: 306.
28. Haaker, H. and E.Racker. 1979. Purification and reconstitution of the Ca^{2+}-ATPase from plasma membranes of pig erythrocytes. J.Biol. Chem. 254: 6598.
29. Mottola, C., L.Dolzani and D.Romeo. 1980. The peripheral Ca^{2+} pump activity of macrophages and neutrophils. Cell Calcium 1:371.
30. Flora, U., R.Gennaro and D.Romeo. 1980. Improved technique for the measurement of the kinetics of Ca^{2+} uptake by cells: The coupling of an amplifier with voltage regulator to a Ca^{2+} selective electrode. Anal.Biochem. 102: 77.
31. Baker, P.F. 1976. Regulation of intracellular Ca and Mg in squid axons. Fed.Proc. 35: 2589.
32. Lee, C.O., D.Y. Uhm, and K.Dresdner. 1980. Sodium-calcium exchange in rabbit heart muscle cells: Direct measurement of sarcoplasmic Ca^{2+} activity. Science 209: 699.
33. Martonosi, A. 1969. Sarcoplasmic reticulum:VII. Properties of a phosphoprotein intermediate implicated in calcium transport. J. Biol.Chem. 244: 613.
34. Gennaro, R., C.Mottola, C.Schneider and D.Romeo. 1979. Ca^{2+}-Dependent ATPase activity of alveolar macrophage plasma membrane. Biochim.Biophys.Acta 567: 238.
35. Schneider, C., C.Mottola and D.Romeo. 1979. Phosphoprotein intermediate in the Ca^{2+}-dependent ATPase reaction of macrophage plasma membrane. J.Supramol.Struct. 10: 433.
36. Schneider, C., C.Mottola and D.Romeo. 1979. Calcium ion-dependent adenosine triphosphatase activity and plasma-membrane phosphorylation in the human neutrophil. Biochem.J. 182: 655.
37. Mottola, C.,R.Gennaro, A.Marzullo and D.Romeo. 1980. Isolation and partial characterization of the plasma membrane of purified bovine neutrophils. Eur.J.Biochem. 111: 341.
38. Caswell, A.H. and J.D.Hutchinson. 1971. Visualization of membrane bound cations by a fluorescent technique. Biochem.Biophys.Res. Commun. 42: 43.
39. Bretscher, M.S. 1973. Membrane structure: Some general principles. Science 181: 622.

DISCUSSION

Edited by: B.M. BABIOR

SCHELL-FREDERICK: Would you comment on the ATP dependence of the ionophore effect on calcium release?

SCHNEIDER: At the low concentration (0.5 µM) and short time (<1 min) used in this study, A23187 did not alter the ATP content of the cell (see also C.Schneider et al., Exp. Cell Res. (1978) 112, 249-256). Its main effect was to mobilize calcium from internal stores, most probably mitochondria in the case of alveolar macrophages. Only if the ATP is lowered by mitochondrial inhibitors such as oligomycin, which is less than 2 minutes reduces the ATP content of alveclar macrophages by 90-100%, will the calcium mobilized by A23187 fail to be pumped out by the ATP-dependent calcium pump of the plasma membrane.

Ca^{2+} TRANSPORT AND SURFACE MEMBRANE ATPase IN MACROPHAGES

R.Pantaleo°, A.Fusaro°, D.Giordano°, S.Papa•

Institute of General Pathology° and Institute
of Biological Chemistry•, Faculty of Medicine
University of Bari, Bari 70124, Italy

Macrophages display a number of defense processes like endocytosis, chemiotactic motility and secretion of hydrolytic enzymes[1,2], which are functional expressions of the plasma membrane and contiguous structures like the surface coat and cytoskeleton components [3,4]. Elucidation of the mechanisms by which these diverse functions are controlled, so to respond in an orderly fashion to external and internal stimuli, is of primary importance for understanding of the role of macrophages in inflammation, immunological processes and neoplastic growth and citotoxicity[1,2,5].

A common denominator to processes like endo-/exocytosis and motility, which are essentially based on motion of molecules within and among cellular membranes, surface coat and the cytoskeleton, is thought to be represented by the intra- and extra-cellular level of Ca^{2+} and its interaction with the plasma membrane and contiguous structures[6-9]. It has been reported, for example, that Ca^{2+} influx in macrophages and neutrophils, promotes secretion of hydrolytic enzymes[7] and motility[6-10]. In addition the phagocytic activity depends on energy supply as indicated by the stimulatory action that this process exerts on respiration[11].

We have investigated Ca^{2+} transport in peritoneal macrophages and its relationship to plasma membrane ATPase. It is shown that these cells rapidly bind Ca^{2+} with relatively high affinity. Exogenous ATP causes a Ca^{2+} release from macrophages, which probably involves ATPase activity at the surface of the cell.

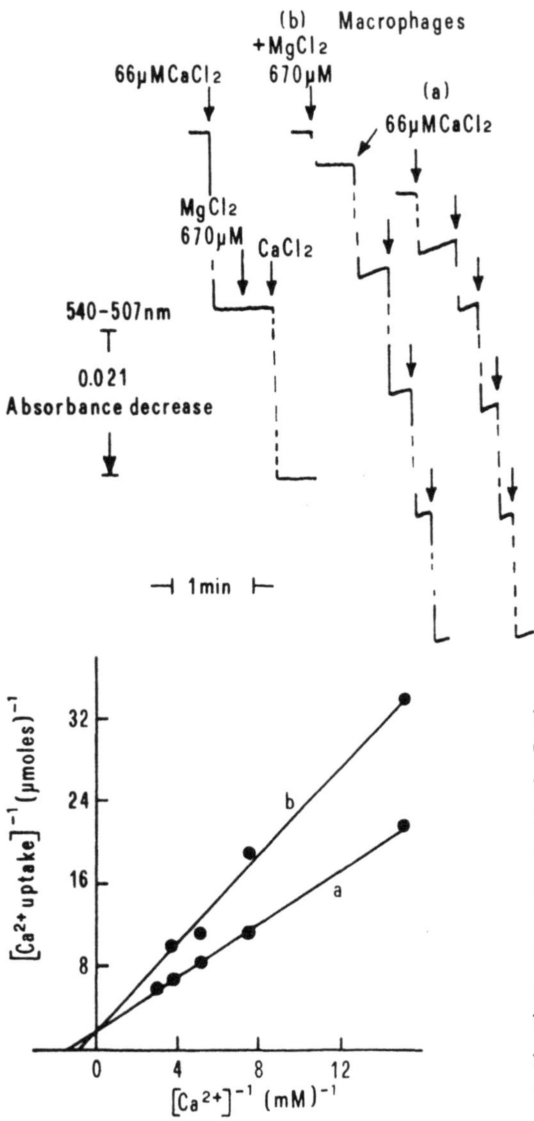

Fig.1. Ca^{2+} uptake by macrophages. Macrophages were obtained collecting the peritoneal exudate 3 days after intraperitoneal injection of 10% peptone in 0.9% NaCl to male Albino Wistar rats. The cells were washed three times with 0.9% NaCl and suspended in 0.25 M sucrose at 4°C. Microscopic controls showed that the cells of the exudate were 80% macrophages and 20% lymphocytes. Cells viability was checked with neutral red vital staining. Macrophages (8 mg prot/ml) were suspended in: 0.25 M sucrose, 3 mM Hepes, pH 7.4, 200 μM murexide. Temp. 23°C. Absorbance changes of the Ca^{2+}-murexide complex were followed with a dual wavelength spectrophotometer at 540-507 nm. Ca^{2+} uptake was calculated from the difference between the absorbance decrement caused by the addition of $CaCl_2$ to the reaction medium devoid of macrophages and the immediate absorbance decrease caused by addition of the same amount of $CaCl_2$ to the macrophages suspension.

Ca^{2+} translocation was monitored with the metallochromic indicator murexide. Subsequent additions of 66 μM $CaCl_2$ to a reaction medium containing murexide produced, in the absence of macrophages, identical absorbance decrements, which are due to the formation of the Ca^{2+}-murexide complex[12,13]. Addition of the same amount of $CaCl_2$ to a suspension of macrophages caused a smaller absorbance decrement than in the reaction mixture alone. Evidently part of added Ca^{2+} is taken up by the cells, this resulting in a lower free Ca^{2+} concentration. That this is,indeed,the case is shown by two facts: (i) the immediate absorbance decrement caused by $CaCl_2$ addition to the macrophage suspension was followed for 20-30 sec by absorbance increase, indicative of the occurrence of a slow phase of Ca^{2+} uptake; (ii) subsequent equal additions of $CaCl_2$ caused progressively larger absorbance decrements which tended to approach the constant decrement caused by the addition of the same amount of $CaCl_2$ to the reaction medium without macrophages. This shows saturability of the rapid phase of Ca^{2+} uptake by macrophages. A Lineaweaver Burk plot of rapid Ca^{2+} binding to macrophages vs the concentration of added Ca^{2+} produced a straight line, from which an apparent affinity constant of 720 μM Ca^{2+} and a maximum binding capacity of 60 nmoles $Ca^{2+} \cdot mg\ prot^{-1}$ could be calculated. Mg^{2+} caused a competitive inhibition of Ca^{2+} binding to the macrophage surface.

Fig.2 - Trace A shows that the addition of 330 μM ATP caused a significant release of Ca^{2+} from macrophages which had been preexposed to 330 μM $CaCl_2$. The immediate absorbance increase, which preceed upon ATP addition the absorbance decrease indicative of Ca^{2+} release from macrophages, was observed also in the absence of cells and was due to lowering of free Ca^{2+} in the medium by formation of Ca^{2+} ATP complex.

Trace B shows that the addition of the Ca^{2+}-ionophore A 23187 caused, when added to macrophages exposed to 330 μM Ca^{2+}, a rapid influx of this cation which in 2 min reached an extent of 7 ngions $Ca^{2+} \cdot mg\ prot^{-1}$ (cf. Schneider et al.[7]). Addition of ATP to Ca^{2+}-loaded macrophages resulted in a much faster and larger release of Ca^{2+}. The initial rate of Ca^{2+} release amounted to 6 natoms$\cdot min^{-1}$/mg protein. This stimulation of the ATP-induced Ca^{2+} release after stimulation of Ca^{2+} uptake by A 23187, which is known to equilibrate Ca^{2+} across biological membranes[13,14], shows that ATP does, in fact, promote extrusion of Ca^{2+} from the inner to the outer space. The ATP-linked Ca^{2+} extrusion, promoted by enhacement of intracellular Ca^{2+} concentration by A 23187, evidently predominated over Ca^{2+} re-entry mediated by the ionophore.

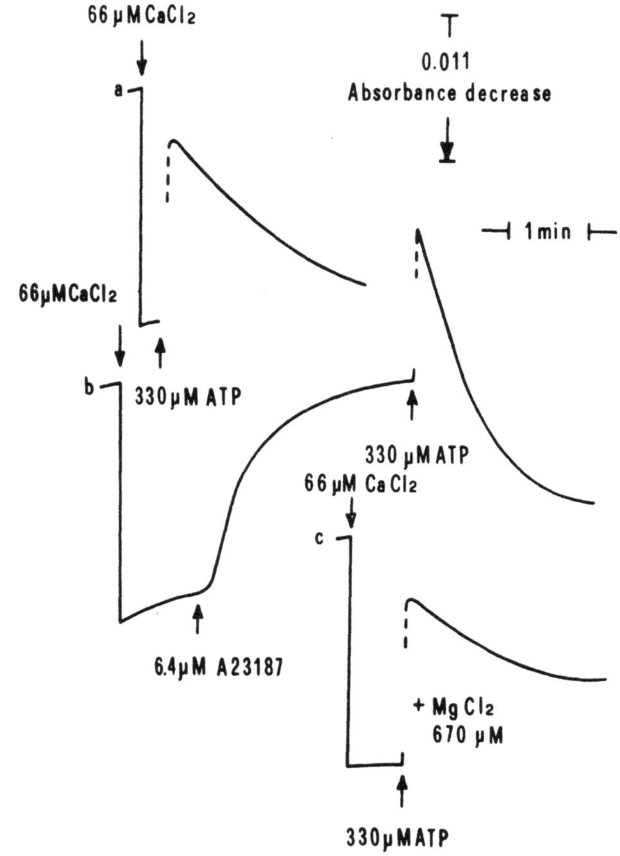

Fig.2. Ca^{2+} uptake by peritoneal macrophages induced by A 23187 and Ca^{2+} release caused by external added ATP. Macrophages (6.7 mg prot/ml) were suspended in the reaction mixture described in the legend to Fig.1. $CaCl_2$ was added as 66 μM aliquots to give a final concentration of 330 μM. A 23187 was added as ethanolic solution (2 μl of 4.8 mM solution). ATP was added together with an equimolecular amount of $MgCl_2$.

Trace C shows that the ATP-linked Ca^{2+} extrusion was inhibited by the addition of $MgCl_2$. The experiment of Fig.3 shows that the addition of 85 µM EGTA, producing by chelation the same decrease of free concentration of extracellular Ca^{2+}, did not result, at difference with what observed with ATP, in any subsequent release of Ca^{2+} from the cells.

Enzymatic assay showed that intact macrophages exhibited, upon addition of ATP in the reaction medium, an ATPase activity which was unaffected by ouabain[15,16]. The ATPase activity was determined on macrophages suspended in the reaction mixture described before. The activity was measured by following spectrophotometrically NADH oxidation in the presence of excess phosphoenol-pyruvate and added lactate dehydrogenase and pyruvatekinase[17]. Upon disruption of the cells by exposure to ultrasonic energy there occurred only a small increment of the ATPase activity. The ATPase activity exhibited by intact cells, which amounted to 10-20 nmoles ATP hydrolized$\cdot$min$^{-1}\cdot$ mg prot^{-1} at 1 mM ATP and 1 mM $MgCl_2$, thus behaves as an ectoenzyme, that is i.e. the active site for ATP faces the extracellular space, like the Mg^{2+}-ATPase of the plasma membrane of granulocytes [15,16].

The relatively high affinity Ca^{2+} binding capacity of macrophages, ges, disclosed in the present investigation, can have obvious influence on the state of aggregation and the functional activity of surface components of these cells. It might involve components of the surface coat or of the plasma membrane itself. Selective digestion by proteases and glycosidases and treatment with chemical non-permeant reagents could help to identify the surface components involved in Ca^{2+} binding.

In addition to rapid Ca^{2+} binding to the surface of the cells, a slow penetration of Ca^{2+} into macrophages takes place, which can be enhanced by the addition of the Ca^{2+}-ionophore A 23187. It has been proposed that penetration of Ca^{2+} into the cytosol of macrophages ges promotes secretion of hydrolytic enzymes[7,9].

The Ca^{2+} release from macrophages induced by exogenous ATP is a quite peculiar and interesting facet of these cells. This process could in principle represent either Ca^{2+} release from complexes with components of the cell surface or Ca^{2+} extrusion from the cytoplasm or both events. The observation that Ca^{2+} release can be greatly enhanced when, before adding ATP, macrophages are induced to take up Ca^{2+} by exposure to the Ca^{2+}-ionophore A 23187 in the presence of extracellular Ca^{2+}, provides evidence that the process represents, at least in part, net Ca^{2+} extrusion from the cytoplasm.

It is tempting to propose that the ATP-linked release of Ca^{2+}

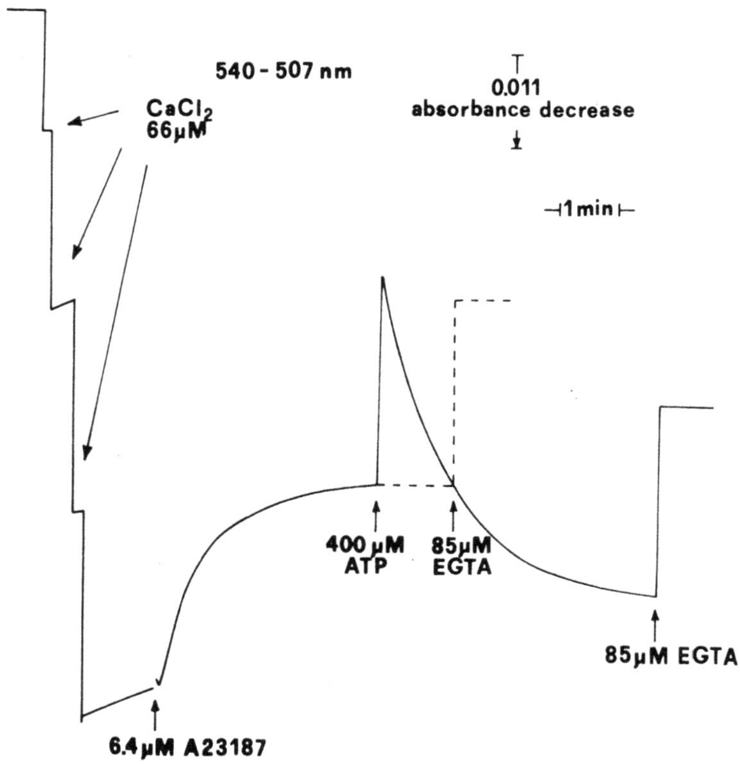

Fig.3. Effect of ATP and EGTA on Ca^{2+} translocation in macrophages. For experimental conditions see legend to Fig.s 1 and 2.

involves the ecto ATPase activity exhibited by the cells. One might conceive that this ATPase would act as a Ca^{2+} extrusion system of the plasma membrane, transporting Ca^{2+} from the cytoplasm to the outer space while reacting with ATP on the outer side of the membrane. ne. It is also possible that part of the ecto-ATPase is an activity of myosin of the plasma membrane of macrophages[18] (this ATPase is not activated by Mg^{2+}) and that molecular rearrangement of the actin and actin-binding protein of the peripheral motility system induced by myosin ATPase[18] results in Ca^{2+} release.

It can be mentioned that an ATPase activity has been detected in isolated plasma membrane from alveolar macrophages[19] which is specifically activated by Ca^{2+} at low concentration (10^{-7} - 10^{-5} M) and by Mg^{2+} at high concentration (10^{-4} - 10^{-3} M) and which has been proposed to act as a Ca^{2+} extrusion system [20]. However further informations will be necessary to clarify the relationship existing between the Ca^{2+} and Mg^{2+} activated ATPase activities and their possible action in the Ca^{2+} binding and transport in macrophages described in the present paper.

ACKNOWLEDGEMENTS

This work was supported by Grant N° 78.02854.96 PFCN to S.Papa, Consiglio Nazionale delle Ricerche, Italy.

REFERENCES

1. G.M.Green, Pulmonary clearance of infections afents, Annu. Rev. Med. 19: 315 (1968).
2. R.Snydermann and S.E.Mergenhagen, Chemotaxis of macrophages, in: "Immunology of the Macrophages" D.S.Nelson ed., Academic Press, New York (1976).
3. G.L.Nicolson, Transmembrane control of the receptors on normal and tumor cells. I - Cytoplasmic influence over cell surface components, Biochim.Biophys.Acta 457 : 57 (1976).
4. G.L.Nicolson, Transmembrane control of the receptors on normal and tumor cells. II - Surface changes associated with transformation and malignancy, Biochim.Biophys.Acta 458: 1 (1976).
5. P.J.Edelson and E.A.Cohn, Peroxidase mediated mammalian cell cytotoxicity, J.Exp.Med. 138: 318 (1973).
6. B.D.Gomperts, Calcium and cell activation, in: "The Plasma Membrane, Models for Structure and Function",P.Cuatrecasas and M.F.Greaves, eds., Academic Press, London (1977).
7. C.Schneider, R.Gennaro, G.De Nicola and D.Romeo, Secretion of granule enzymes from alveolar macrophages, Exp.Cell.Res.112: 249 (1978).
8. A.C.Allison, P.Davies and S.D.Petris, Role of contractile microfilaments in macrophage movement and endocytosis, Natura New Biol. 232: 153 (1971).
9. D.Romeo, C.Schneider, R.Gennaro and C.Mottola, The regulation of macrophages activities. Role of the energy-dependence in-

tracellular Ca^{2+}-buffering systems, in: Macrophages and Limphocytes, Nature, Functions and Interactions", M.E.Escober and M.Friedman, eds., Plenum Press, New York, in press.

10. M.M.Boucek and R.Snydermann, Calcium influx requirement for human neutrophil chemotaxis: inhibition of Lanthanum chloride, Science 193: 905 (1976).
11. G.Zabucchi and D.Romeo, The dissociation of exocytosis and respiratory stimulation in leucocytes by ionophores, Biochemical J. 156: 209 (1976).
12. T.Ohnishi and S.Ebashi, The velocity of calcium binding of isolated sarcoplasmic reticulum, J.Biochem. (Tokio) 55: 599 (1964).
13. A.Scarpa, J.Baldassarre and G.Inesi, The effect of calcium-ionophores on fragmented sarcoplasmic reticulum, J.Gen.Physiol. 60: 735 (1972).
14. P.W.Reed and M.A.Lardy, A 23187 a divalent cation ionophore, J.Biol.Chem. 247: 6970 (1972).
15. M.F. Mustafa, C.E.Gross and J.A.Hardic, Localization of Na^{+}-K^{+}, Mg^{++}-adenosinetriphosphatase activity in pulmonary alveolar macrophages subcellular fraction, Life Sci. 9, part.I: 947 (1970).
16. J.E.Smolen and G.Weissmann, Mg^{2+}-ATPase as a membrane ectoenzyme of human granulocytes. Inhibitors, activators and response to phagocytosis, Biochim.Biophys.Acta 512: 525 (1978).
17. A.Pansini, F.Guerrieri and S.Papa, Control of proton conduction by the H^{+}-ATPase in the inner mitochondrial membrane. Eur.J. Biochem. 92: 545 (1978).
18. J.H.Hartwig and T.P.Stossel, Isolation and properties of actin, myosine and a new actin-binding protein in rabbit alveolar macrophages, J.Biol.Chem. 250: 5696 (1975).
19. R.Gennaro, C.Mottola, C.Schneider and D.Romeo, Ca^{2+}-dependent ATPase activity of alveolar macrophage plasma membrane, Biochim.Biophys.Acta 567: 238 (1979).
20. C.Schneider, C.Mottola and D.Romeo, Phosphoprotein intermediate in the calcium dependent ATPase reaction of macrophage plasma membrane, J.Supram.Struct. 10: 433 (1979).

EFFECT OF IONOPHORES ON LYMPHOCYTE CELLULAR METABOLISM

Paola Arslan, Alda Miolo, Cesare Montecucco and Tullio Pozzan
C.N.R. Unit for the Study of Physiology of Mitochondria
Institute of General Pathology, University of Padova
Via Loredan, 16, 35100 PADOVA, ITALY

SUMMARY

The effect of valinomycin, nigericin, gramicidin S and D, A23187 and X537A on respiration and cellular ATP content of rat spleen lymphocytes is presented. It has been found that while valinomycin and nigericin interfere with mitochondrial functions, gramicidin D does not show an appreciable effect. These results are explained in terms of different ability of ionophores to redistribute among intracellular membranes. A23187 and X537A, added with Ca^{2+}, strongly enhanced O_2 consumption and reduced cellular ATP content.

INTRODUCTION

Monovalent and divalent cation ionophores have been largely used to study the modifications of extracellular and cytoplasmic cation concentrations and/or to measure membrane potential on several biological systems[1]. In the present paper, the effects on rat spleen lymphocyte metabolism of some among the most widely used cation ionophores are presented as a model of ionophores action on cells which actively synthetize ATP from mitochondria. Valinomycin, for neutral ionophores, nigericin, for carboxylic ionophores and gramicidin, for so-called "channel formers" have been used. A23187, a unique ionophore predominantly selective for divalent over monovalent cations, and X537A, which complexes virtually every known cation, although with different affinity, have been used. Ionophores have been used to probe the role of ion fluxes and membrane potential on a variety of biological phenomena, valinomycin and nigericin have been shown to inhibit lymphocyte mitogenesis[2,3] valinomycin and A23187 inhibit the capping of surface immunoglobulins in lymphocytes[4]. On the other hand, ionophores seldom perturb the equilibrium concentration of the cation across the biological membranes without altering pH as well as other ion gradients, resulting in secondary and non considered cation or anion fluxes. Other metabolic effects may result on cells such as ATP depletion, interference with prostaglandin biosynthesis[5] which might render difficult to attribute an ionophore effect only to transport of a specific cation. In this study we show that several ionophores influence cell energy metabolism and present two easily measurable parameters to assess toxic effects.

MATERIALS AND METHODS

Rat spleen lymphocytes were prepared from spleens of male albino Wistar rats in Hank's minimal essential medium buffered with 20 mM Hepes, pH 7.3 and they were purified on Ficoll gradient. After two washings, the cells were counted, checked with the viability test and used only if the viability was higher than 95%. The cells were resuspended at the concentration of $2\text{-}3.10^7$ cells/ml. ATP was measured with the luciferase method, with a Dupont luminescence biometer[6]. O_2 consumption was measured either with a vibrating platinum electrode (American Instrument Co., U.S.A.) or with a Clark-type oxygen electrode. All the experiments were carried out at 37°C. The electrode signal was amplified with a linear amplifier (Johnson Foundation Instrum., U.S.A.). Oligomycin, FCCP, rotenone, valinomycin, nigericin and gramicidin S were purchased from Sigma (St. Louis, Ma., U.S.A.) and gramicidin D from Koch-Light Laboratories (Colnbrook, England), and added from freshly distilled DMSO stock solutions.

RESULTS AND DISCUSSION

In figure 1, the O_2 consumption of rat spleen lymphocytes in presence of oligomycin, FCCP and rotenone is presented. The O_2 consumption rate of rat spleen lymphocytes, without additions, varies from 80 to 110 nmoles/10^6 of cells/10^6/min, and the addition of oxidative phosphorylation uncouplers as FCCP induces only a small increase of the basal respiration rate. The addition of oligomycin, a specific inhibitor of mitochondrial ATPase lowers the rate of O_2 consumption to approximately 50% of the initial value and the addition of FCCP, after oligomycin, restores or slightly increases the initial O_2 consumption. The observation that 50% of the O_2 consumption is oligomycin sensitive and that the uncoupler FCCP is unable of increasing O_2 consumption indicate that O_2 consumption of resting lymphocyte mitochondria is nearly maximum and at least in resuspended lymphocytes, does not show a controlled or resting respiratory rate. Rotenone inhibits, after few minutes, the respiration, leaving a residual O_2 consumption of no more than 10%, indicating that mitochondria are respon-

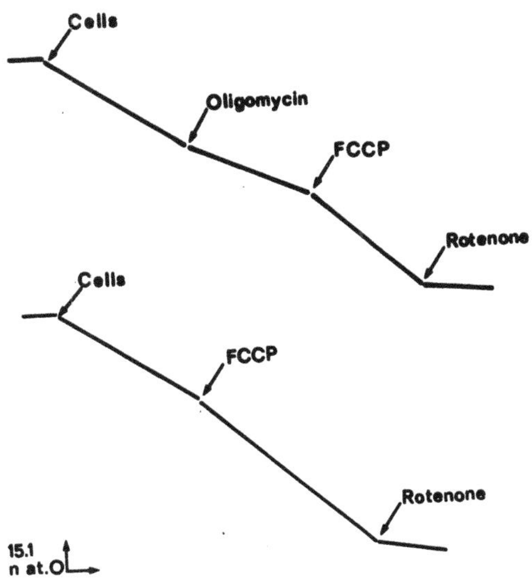

Fig. 1 - Effect of oligomycin, FCCP and rotenone on the O_2 consumption rate.

Medium: Hank's Hepes pH 7.4; temperature: 37°C.
Cells: 2 x 10^7/ml.
Final volume: 3 ml.
Oligomycin: 1 µg/ml, FCCP 1 µM.
Rotenone 5 µM.

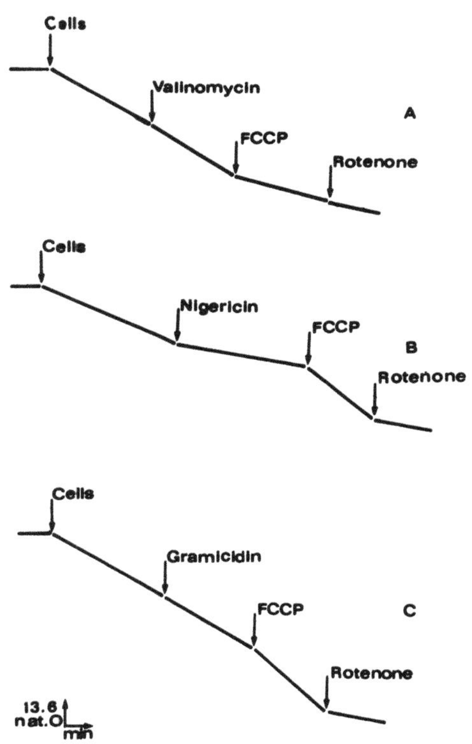

Fig. 2 - Effect of valinomycin, nigericin and gramicidin D on O_2 consumption rate.

Experimental conditions as in fig. 1. Cell concentration, oligomycin, FCCP and rotenone concentrations as in fig. 1. Valinomycin, nigericin and gramicidin D were respectively: 1 µM, 0.1 µM and 0.1 µM.

sible for most of the O_2 consumption and that substrates of lymphocytes are NAD-linked.

In figure 2, the effect of valinomycin (panel A), nigericin (panel B) and gramicidin D (panel C) are presented. 10^{-7}M valinomycin, a good uncoupler of isolated mitochondria respiration in high K^+ medium, induces in rat spleen lymphocytes only a small increase of basal O_2 consumption. The uncoupler FCCP, added after valinomycin, causes the release from cellular mitochondria of the accumulated K^+ in exchange for H^+, equilibrating H^+ across the mitochondrial membrane, the cellular O_2 consumption, consequently, is reduced or unaltered. 10^{-7}M nigericin, which exchanges K^+ for H^+, has an inhibitory effect on the basal O_2 consumption as it does, at this concentration, on isolated mitochondria in low K^+ medium, oxidizing NAD-linked substrates. Gramicidin, always at 10^{-7}M concentration, a potent uncoupler in isolated mitochondria, which rapidly and aspecificly permeabilizes membranes to monovalent cations, does not show any appreciable effect on basal O_2 consumption. These results indicate that the O_2 consumption measurements cannot distinguish between a normal high O_2 consumption leading to ATP synthesis (basal rate), from an increased O_2 consumption due to an uncoupling effect or to ion cycling. In order to appreciate the true effect of these ionophores on cellular O_2 consumption these drugs have been tested when mitochondrial ATP synthesis was inhibited by addition of oligomycin.

The figure 3 shows the effect of valinomycin (panel A) nigericin (panel B) and gramicidin D (panel C) on the O_2 consumption of rat spleen lymphocytes after addition of oligomycin. Valinomycin, in presence of oligomycin, is able to restore the initial respiration rate, in this respect the effect is similar to that observed when the uncoupler FCCP is added after oligomycin. Also, in isolated mitochondria, valinomycin acts as a respiration uncoupler in high K^+ medium, and K^+ cytoplasmic concentration is above 100 mM. Nigericin decrease the oligomycin-inhibited respiration. In isolated mitochondria, nigericin, at this concentration, inhibits the oxidation of NAD-linked substrates in low K^+ medium and the inhibition is relieved either by increasing nigericin concentration or by increasing the external K^+ concentration. Since cytoplasm is considered an high K^+ medium, nigericin effect, before and after oligomycin, was unexpected. A possible explanation is that nigericin depletes cytoplasmic K^+ by exchanging it with the high extracellular Na^+. If cytoplasmic K^+ concentration is decreased, nigericin shows an inhibitory effect on cellular mitochondria. In parallel with experiments performed in isolated mitochondria, also in isolated rat spleen lymphocytes, increasing the concentration of K^+ in the external medium or increasing the concentration of nigericin, the nigericin inhibitory effect is relieved. Gramicidin D does not affect cellular respiration and its lack of effect is not due to absence of interaction with membranes of rat spleen lymphocytes since, at the concentration used, it collapses

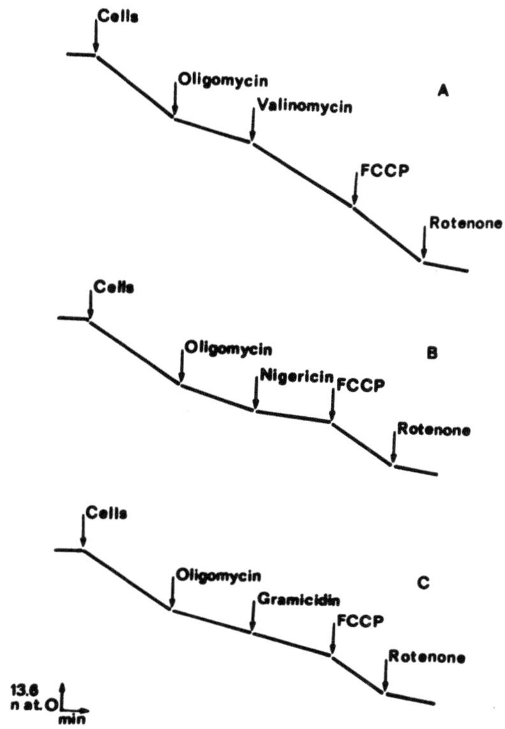

Fig. 3 - Effect of valinomycin, nigericin and gramicidin D on O_2 consumption rate after addition of oligomycin.

Experimental conditions as in Fig. 1.

membrane potential measured with fluorescent probes. Gramicidin S shows, on the contrary, an uncoupling effect on cellular O_2 consumption and lowers ATP cellular level.

The effect of monovalent cation ionophores and oxidative phosphorylation inhibitors on cellular ATP in rat spleen lymphocytes is shown in Table I. Cellular ATP has been measured under conditions similar to those used for measuring O_2 consumption. Cellular ATP is lowered to less than 30% of the control either by oxidative phosphorylation inhibitors (FCCP, rotenone) or by ATP synthesis inhibitor (oligomycin) or ionophores as valinomycin, nigericin and gramicidin S. Again, only gramicidin D does not affect cellular ATP (93% of the control). Our conclusion about cation ionophores are that valinomycin, nigericin and gramicidin S redistribute among intracellular membranes while gramicidin D confines itself, in short-terms incubation, prevalently to the plasma membrane.

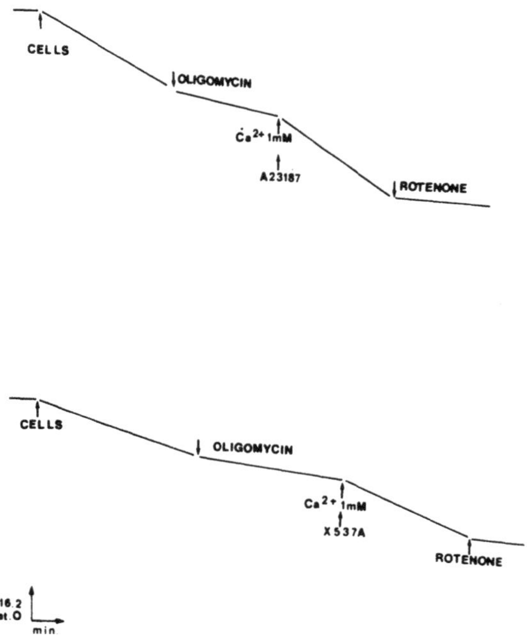

Fig. 4 - Effect of divalent cation ionophores, A23187 and X537A, on O_2 consumption rate in presence of oligomycin.

Medium: Hank's Hepes pH 7.4: temperature: 37°C. Final volume:3 ml, cells: 2 x 10^7/ml. Oligomycin: 1 µg/ml, rotenone 5 µM. Ca^{2+} 1.8 mM and 50 µM. A23187 10^{-6} M and 30 µM. X537A 10^{-6} M.

Table I. Effect of Monovalent Cation Ionophores and Mitachondial Poisons on Cellular ATP Content

		% ATP		S.D.
VALINOMYCIN	10^{-7} M	38.2	±	10.3
NIGERICIN	10^{-7} M	29.8	±	5.5
	10^{-5} M	20.9	±	5.8
GRAMICIDIN D	10^{-7} M	93.3	±	10.3
GRAMICIDIN S	10^{-7} M	30.1	±	7.9
FCCP	10^{-6} M	24.3	±	5.6
OLIGOMYCIN	5 x10^{-6} M	31.5	±	3.7
ROTENONE	5 x10^{-6} M	31.2	±	3.1

LYMPHOCYTES WERE INCUBATED 5 MINUTES WITH THE DRUGS AT 37°C BEFORE BLOCKING THE REACTION WITH PERCLORIC ACID.
ATP WAS ASSAYED WITH THE LUCIFERASE METHOD (STANLEY,P.C.AND WILLIAMS, S.G.(1969) ANAL.BIOCHEM. 29,381)
THE RESULTS ARE THE MEAN OF 6 INDEPENDENT EXPERIMENTS (± : S.D.).

Table II. Effect of divalent cation ionophores on cellular ATP content.

		% ATP		S.D.
A 23187	10^{-6} M	85	±	10.1
A 23187	10^{-6} M + Ca^{2+} 1.8×10^{-3} M	22	±	5.1
A 23187	30×10^{-5} M + Ca^{2+} 50×10^{-5} M	20	±	6.2
X 537 A	10^{-6} M	83	±	10.3
X 537 A	10^{-6} M + Ca^{2+} 10^{-3}M	24	±	5.4

Lymphocytes were incubated 5 minutes with the drugs at 37°C before blocking the reaction with percloric acid. ATP was assayed with the luciferase method (Stanley, P.E. and Williams, S.G. (1969) Anal. Bioch. 29, 381). The results are the mean of 4 independent experiments (± S.D.).

Another set of experiments was performed for divalent cation ionophores using A23187 and X537A. In figure 4, the effects of divalent cation ionophores on O_2 consumption after addition of oligomycin, are presented. In Ca^{2+} free medium, addition of A23187, a carboxylic antibiotic which transfers Ca^{2+} and Mg^{2+} across membranes, and X537A, another carboxylic antibiotic which binds and equilibrates monovalent and divalent cations, do not affect basal O_2 consumption. On the contrary, if ionophores A23187 and X537A are added at the same time as Ca^{2+}, a large increase of O_2 consumption is observed. This effect is reduced, although still present, if Ca^{2+} is added long before the ionophores.

In Table II, the effect of divalent cation ionophores on cellular ATP content is presented. Again, A23187 and X537A, if added without Ca^{2+} do not affect cellular ATP content in rat spleen lymphocytes, but if added with Ca^{2+}, they dramatically decrease the ATP cellular level. For A23187 the effect is the same either adding 1 μM A23187 and 1,8 mM Ca^{2+} or low concentration(50 μM)of Ca^{2+} and relatively high concentration (30 μM) of A23187. Similar effects with A23187 and X537A have been also obtained in PMN leucocytes by Romeo et al.[7]. This decrease in cellular ATP content may result from penetration of these ionophores into the cell reaching mitochondria and causing a rapid cycling of Ca^{2+}, with a consequent reduction in ATP synthesis as shown by Bygrave in isolated mitochondria[8].

REFERENCES

1. A. Gomez-Pujou and C. Gomez-Lojero, Curr. Topics Bioenerg. 16: 221 (1977)
2. R.P. Daniele and S.K. Holian, Proc. Natl. Acad. Sci. USA 73: 3599 (1976)
3. R.P. Daniele, S.K. Holian and P.C. Nowell, J. Exp. Med. 147: 571 (1978)
4. R.P. Daniele and S.K. Holian, Cell. Immunol. 26: 21 (1976)
5. P.W. Reed, Methods in Enzymol. 55: 435 (1979)
6. P.E. Stanley and S.G. Williams, Anal. Biochem. 29: 381 (1969)
7. D. Romeo, Z. Abucchi, N. Miani and F. Rossi, Nature 253: 542 (1975)
8. F.L. Bygrave, Biol. Rev. Cambridge Philos. Soc. 43: 43 (1978)

DISCUSSION
Edited by R.E. Basford

Dr. Seligmann:
We have found that 10^{-4}-10^{-6} M Gramicidin S makes cells including PMN, macrophages, lymphocytes, and glial cells permeable to Trypan blue within a few seconds of treatment, while leaving morphology intact and, in the case of adherent cells, leaving the cells attached. Even the highest concentrations of Gramicidin D did not have this affect, suggesting there are unique differences between the two Gramicidins at the level of their action on the plasma membrane.

Dr. Arslan:
Although the concentration of Gramicidins in our experiments is 10^{-7} M, I think that your observations are in agreement with the data we have presented.

Dr. Fonnesu:
Have you done any experiments using other types of isolated cells and, in particular, ascites tumor cells?

Dr. Arslan:
We have not tried tumor cells, we have tried human mobile spermatozoa, which are cells actively synthesizing ATP and these cells show a great similarity with isolated rat spleen lymphocytes in regard to the response of oxidative phosphorylation inhibitors, ATP synthesis inhibitors and ionophores.

Dr. Markert:
Do you have more direct evidence that mitochondrial inhibitors and uncouplers penetrate the outer membrane of whole cells, as their effect is very low compared to their effect in mitochondria.

Dr. Arslan:
I agree that before adding oligomycin, the effect on whole cells of mitochondrial inhibitors may appear low, but the effects appear clear and convincing when we consider oligomycin inhibited respiration. These cells are actively synthesizing ATP and therefore the uncoupler FCCP does not show a marked increase of respiration, because, as I said, the mitochondria of these cells show an almost maximum O_2 consumption.

Dr. Rossi:
I advise you to study those phenomena in lung macrophages which contain many mitochondria. The function of mitochondria in these cells is also that of control of exocytosis through a control of calcium disposal as Dr. Romeo has shown some years ago.

PEROXIDATIC ACTIVITY DISTINCT FROM MYELOPEROXIDASE IN HUMAN MONOCYTES CULTURED IN VITRO AND IN ALVEOLAR MACROPHAGES

Janine Breton-Gorius+, Jean Louis Vildé°, Josette Guichard+, William Vainchenker°, Françoise Basset++
+INSERM U.91. Hp Henri Mondor. 94010 Créteil, France
°Hp Claude Bernard. 75019 Paris, France
++INSERM U.82 CHU Bichat. 75877 Paris Cedex 18, France

INTRODUCTION

Peroxidase cytochemistry and electron microscopy have revealed that resident macrophages from serous cavities and tissues of several rodent species possess a peroxidatic activity (PA) which is exclusively localized in the rough endoplasmic reticulum (RER) including the nuclear envelope (NE). In contrast, exudate macrophages elicited into the peritoneal cavity with stimulants and which are considered as newly mobilized blood monocytes, only contain PA in granules [1-4]. On this basis, Daems and Brederoo[4] have suggested the existence of two separate lines of macrophages although Van Furth[5] had presented findings showing that resident macrophages also originate from monocytes. More recently, it was reported[6-8] that rodent and human monocytes may rapidly develop in vitro after adherence, a PA in RER which resembles that seen in resident macrophages. Moreover, intermediate forms between exudate and resident macrophages have been detected in vivo by several investigators[9, 10]. All these findings were more consistent with the theory that resident macrophages derive from monocytes. However, macrophages with PA pattern of resident macrophages have not yet been described in man while human monocytes may acquire a PA in RER after adherence. This result may be due to the inability to reveal an enzyme with low activity or easily inhibited by aldehyde prefixation. We have previously shown that the absence of fixation before the cytochemical method improves the detection of peroxidase-like enzymes in leukemic Hairy cells[11], megakaryoblasts[12] and adherent monocytes[8]. Using this sensitive method, we have investigated whether PA in RER may be triggered by other culture procedure than adherence and whether alveolar macrophages obtained by broncho-pulmonary lavage may present the PA pattern of resident macrophages.

RESULTS AND DISCUSSION

When circulating human monocytes are incubated in the Graham Karnovsky[13] medium without prefixation[14] a PA is only detected in granules ; RER is always devoid of reaction. Mitochondrial envelope and cristae are reactive as in other cell types treated by this method. Within 3 hours of adherence, a faint PA appears in RER of a portion of monocytes. The percentage of reactive cells, the intensity of PA increase with the time of culture. Thus, at 24 hours of culture, 90% of monocytes possess a PA in RER. Other investigators[6,7] have found a smaller proportion of reactive cells. Thus, the incubation of unfixed cells in diaminobenzidine medium appears to improve the detection of this PA which disappears progressively after one day of culture. The majority of cells lacks all peroxidase-positive granules between day 3 and day 5 and new peroxidase-negative granules are produced. After one week of culture, large macrophages derived from adult monocytes are in large majority, peroxidase-negative. In contrast, those derived from neonatal monocytes, often retain a faint PA in RER after one week of culture. PA is also detected in RER of adult monocytes cultured for two days in plasma clot technique[15]. This transient enzyme disappears later. In contrast, monocytes of newborn, cultured for one week under the same conditions, consistently exhibit PA in RER. Further explorations are required in order to assess the significance of these differences between adult and newborn cultured monocytes.

The true nature of the enzyme responsible for PA in RER and its function remain unknown in human as well as in rodent spread monocytes. However since monocytes from patients with congenital myeloperoxidase deficiency develop the same PA in RER, after adherence, this means that the enzyme differs from myeloperoxidase.It also differs from catalase which is not detected by the used method[14] and which is localized in small granules[16].

We have then investigated whether human alveolar macrophages with a PA pattern of resident macrophages could be identified. These cells were easily obtained by bronchopulmonary lavages in normal volunteers and in a serie of patients with various diseases.

In normal subjects, two categories of macrophages were observed : a majority had no PA while a small variable proportion had a PA within cytoplasmic granules. However, in pathological conditions, macrophages with the PA pattern of resident macrophages and intermediate forms between exudate and resident macrophages could be identified. The table below indicates the percentage of macrophages with PA.

Table 1. Localisation of peroxidatic activities in human alveolar macrophages

Patients	Number	Granules + (%)	RER + (%)	Granules and RER + (%)
Normal steady state	5	0.1-2.6	0	0
Histiocytosis X	4	1-16	3-8	0.5-6
Sarcoidosis	6	1.2-3.6	0-0.5	0-0.1
Dust exposed patients	3	0-8	0-1.5	0

It should be pointed out that in 4 patients with histiocytosis X, all cells with the typical X body which is identical to Langerhans' cell granules are strongly reactive (Fig. 1). Few of them possess the pattern of exudate-resident macrophages.

It was recently shown[17] that alveolar X cells were able to synthesize prostaglandin PGD_2 while alveolar macrophages of a control did not produce PGD_2. It is well known that a peroxidase activity essential for the prostaglandin synthesis is contained in endoplasmic reticulum of several cell types[18]. Thus, PA in RER would reflect the synthesis of prostaglandins.

In Guinea pig, Soranzo et al.[19] demonstrated that alveolar macrophages with PA in RER identical to resident macrophages of peritoneal cavity predominated. The situation is different in man since in normal subjects, all macrophages have peroxidase-negative RER. Since, we used a very sensitive cytochemical method to detect PA, one can exclude negative results by inhibition of the enzyme. Similar results were obtained in human pleural effusions under chronic inflammation[7]. Since human monocytes and macrophages may acquire a PA in RER after adherence it is possible that some substances produced in vivo may inhibit the appearance of PA[20]. However, under pathological conditions, a small proportion of alveolar macrophages may acquire the characteristics of resident macrophages and intermediate

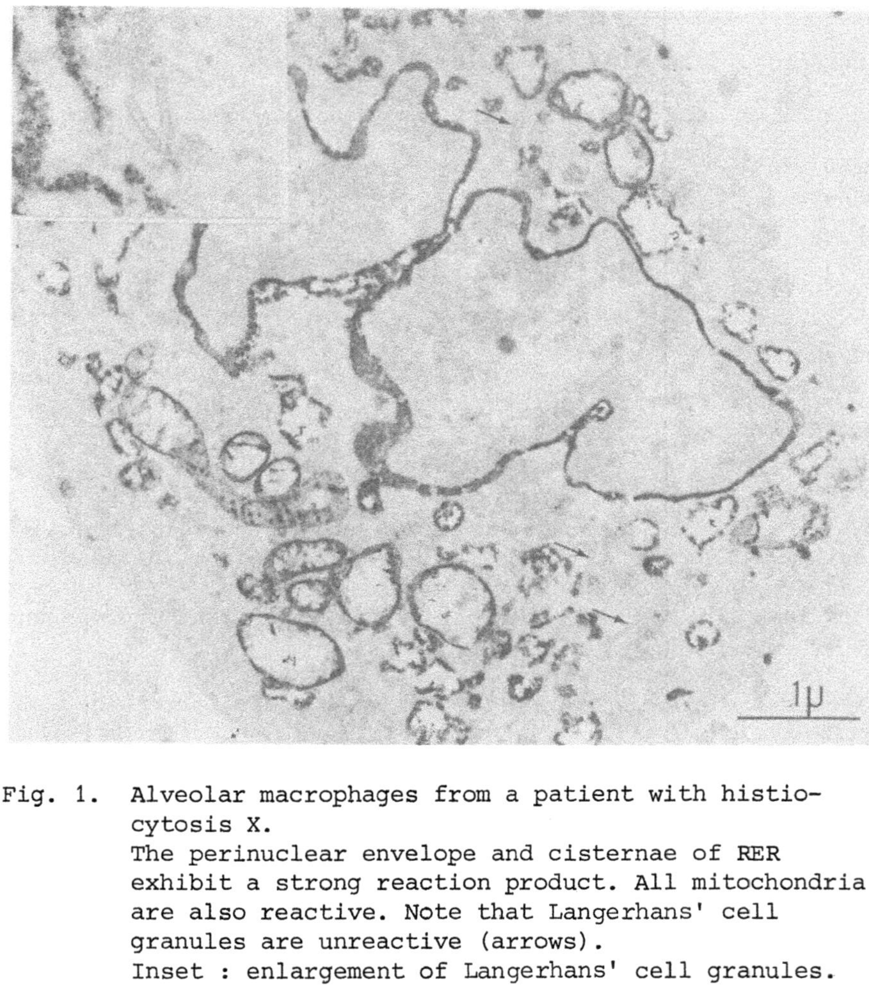

Fig. 1. Alveolar macrophages from a patient with histiocytosis X.
The perinuclear envelope and cisternae of RER exhibit a strong reaction product. All mitochondria are also reactive. Note that Langerhans' cell granules are unreactive (arrows).
Inset : enlargement of Langerhans' cell granules. Since the section are unstained, the structure inside the granules is not apparent.

forms between exudate and resident macrophages. This type of macrophages could be linked to a different functional state[20].

SUMMARY

Human monocytes develop a peroxidatic activity (PA) in rough endoplasmic reticulum (RER) after adherence or after culture in semi-solid medium. This enzyme activity disappears after three days of culture in the majority of macrophages derived from adult monocytes but persists for one week in macrophages derived from neonatal monocytes. The PA is due to an enzyme distinct from myeloperoxidase (MPO), since monocytes from a patient with MPO deficiency develop the same PA as that of normal monocytes after adherence. By its localization and other characteristics, PA of adherent monocytes resembles that of rodent macrophages. We therefore investigated whether human alveolar macrophages exhibit PA, using a sensitive cytochemical method which prevents inhibition by aldehyde in adherent monocytes. In various pathological cases, four types of macrophages could be identified : the majority were peroxidase-negative, a small percentage was of exudate type exhibiting a PA in granules as blood monocytes, while few macrophages were intermediate, possessing only PA in RER i.e. of type resident and a smaller proportion had PA in RER and in granules i.e. exudate-resident macrophages. These findings demonstrate that human macrophages and adherent monocytes may exhibit PA in RER as has been reported for rodent macrophages. The true nature and function of the enzyme responsible for this PA, which is distinct from MPO, remains unknown, but some arguments seem to suggest its role in prostaglandin synthesis.

REFERENCES

1. R.S. Cotran and M. Litt, Ultrastructural localization of horse-radish peroxidase and endogenous peroxidase activity in Guinea-pig peritoneal macrophages. J. Immunol. 105 : 1536 (1970).
2. H.D. Fahimi, The fine structural localization of endogenous and exogenous peroxidase activity in Kupffer cells of rat liver. J. Cell Biol. 47 : 247 (1970).
3. D. Robbins, H.D. Fahimi and R.S. Cotran, Fine structural cytochemical localization of peroxidase activity in rat peritoneal cell mononuclear cells, eosinophils and mast cells. J. Histochem. Cytochem. 19 : 751 (1971).
4. W.Th. Daems and P. Brederoo, Electron microscopical studies on the structure, phagocytic properties and peroxidatic activity of resident and exudate peritoneal macrophages in the Guinea pig. Z. Zellforsch, 144 : 247 (1973).

5. R. Van Furth, Origin and kinetics of monocytes and macrophages. Sem. Hemat. 7 : 125 (1970).
6. P.T. Bodel, B.A. Nichols and D.F. Bainton, Appearance of peroxidase reactivity within the rough endoplasmic reticulum of blood monocytes after surface adherence. J. Exp. Med. 145 : 264 (1977).
7. R.H.J. Beelen, M.B. Van't Veer, D.M. Fluitsma, and E.C.M. Hoefsmit, Identification of different peroxidatic activity patterns in human macrophages in vivo and in vitro. J. Reticuloendoth. Soc. 24 : 351 (1978).
8. J. Breton-Gorius, J. Guichard, W. Vainchenker and J.L. Vildé, Ultrastructural and cytochemical changes induced by short and prolonged culture of human monocytes. J. Reticuloendoth. Soc. 27 : 289 (1980).
9. R.H.J. Beelen, D.M. Broekhuis-Fluitsma, C. Korn and E.C.M. Hoefsmit, Identification of exudate-resident macrophages on the basis of peroxidatic activity. J. Reticuloendoth. Soc. 23 : 203 (1978).
10. W. Deimann and H.D. Fahimi, Appearance of transition forms between monocytes and Kupffer cells in the liver of rats treated with glucan. J. Exp. Med. 149 : 883 (1979).
11. F. Reyes, M.F. Gourdin, J.P. Farcet, B. Dreyfus, and J. Breton-Gorius, Synthesis of a peroxidase activity by cells of Hairy cell leukemia : a study by ultrastructural cytochemistry. Blood, 52 : 537 (1978).
12. J.P. Marie, J.P. Vernant, B. Dreyfus, and J. Breton-Gorius, Ultrastructural localization of peroxidases in "undifferentiated" blasts during the blast crisis of chronic granulocytic leukaemia. Brit. J. Haemat. 43 : 549 (1979).
13. R.C. Graham and M.J. Karnovsky, The early stages of absorption of injected horseradish peroxidase in the proximal tubules of mouse kidney : ultrastructural cytochemistry by a new technique. J. Histochem. Cytochem. 14 : 291 (1966).
14. F. Roels, E. Wisse, B. De Prest and J. Van der Meulen, Cytochemical discrimination between catalases and peroxidases using diaminobenzidine. Histochem. 41 : 281 (1975).
15. W. Vainchenker, J. Guichard and J. Breton-Gorius, Growth of human megakaryocyte colonies in culture from fetal, neonatal and adult peripheral blood cells : ultrastructural analysis. Blood Cells, 5 : 25 (1979).
16. J. Breton-Gorius, Y. Coquin and J. Guichard, Cytochemical distinction between azurophils and catalase-containing granules in leukocytes. I. Studies in developing neutrophils and monocytes from patients with myeloperoxidase deficiency : comparison with peroxidase deficient chicken heterophils. Lab. Invest. 38 : 21 (1978).
17. F. Gonzalez-Crussi and W. Hsueh, Prostanglandin production by histiocytosis X cells. Fed. Proc. 1038 : abstract 4055 (1980)

18. J. Gerrard, J.G. White, G.H.R. Rao and D. Towsend, Localization of platelet prostanglandin production in the platelet dense tubular system. Am. J. Path. 83 : 283 (1976).
19. M.R. Soranzo, M.K. Koerten and W.T.H. Daems, Peroxidatic activity and morphometric analysis of alveolar macrophages in the Guinea pig. J. Reticuloendoth. Soc. 23 : 343 (1978).
20. R.H.J. Beelen, D.M. Fluitsma, J.W.M. Van der Meer and E.C.M. Hoefsmit, Development of exudate-resident macrophages, on the basis of the pattern of peroxidatic activity in vivo and in vitro. In : "Mononuclear Phagocytes Functional aspects". Part I. R. Van Furth ed., Martinus Nijhoff publishers. The Hague, Boston, London (1980).

ACKNOWLEDGEMENTS

This work was supported by INSERM : ATP/CCP 71 78 103

DISCUSSION
Edited by Charles E. McCall, M.D.

"Peroxidatic Activity Distinct from Myeloperoxidase in Human Monocytes Cultured in vitro and in Alveolar Macrophages"

Presented by Dr. J. Breton-Gorius

Question 1 - Dr. Johnston

What are your ideas concerning the persistence of peroxidase in the endoplasmic reticulum of neonatal monocytes? Could this be related to a difference in prostaglandin synthesis?

Answer - Dr. Breton-Gorius

This difference between adult and neonatal monocytes remains unexplained. However, we cannot exclude that the persistence of peroxidase activity in the ER in adherent neonatal monocytes could be due to the presence of promonocytes in there peripheral cord blood which mature into monocytes later.

Question 2 - Dr. Harkness

Can you exclude the possibility that some of your changes are due to cytochromes; the distribution suggests that this is possible.

Answer - Dr. Breton-Gorius

The enzyme activity is not inhibited by cyanate. It, however, still remains a possibility and we will try to look for this with biochemical analysis.

Question 3 - Dr. Rossi

What is the evidence that the histochemical reaction in ER is enzymatic? If the peroxidatic activity of ER is related to the synthesis of prostaglandins, I wonder why the activity is not present in other cell types.

Answer - Dr. Breton-Gorius

The peroxidatic activity in ER is enzymatic since this activity is supressed if H_2O_2 is omitted from the incubation medium or if the cells are prefixed by aldehydes. The enzyme is also inactivated by the addition of aminotriazole or azide.

Mature granulocytes do have peroxidatic activity in ER. But, alveolar macrophages which synthesize PGE do not have peroxidatic activity in ER while alveolar macrophages from patients with histiocytos X, which possess Birbeck granules and peroxidatic activity in ER, synthesize PGD_2 and thromboxane as shown recently by Gonzalez and Crussi.

Dr. Beelen's comment that indomethacin does not inhibit peroxidatic activity is not a sufficient arguement against the role of this enzyme in prostaglandin synthesis. Dr. Gerrard from Minneapolis has demonstrated that prostaglandin synthesis is inhibited in platelets if peroxidase is blocked by aminotriazole. Indomethacin does not inhibit this reaction. A similar situation could occur in resident macrophages.

CHARACTERIZATION OF THE PEROXIDASE IN HUMAN EOSINOPHILS

R. Wever, M.N. Hamers*, C.J. de Graaf, R.S. Weening* and D. Roos*

Laboratory of Biochemistry, B.C.P. Jansen Institute, University of Amsterdam, Plantage Muidergracht 12, 1018 TV Amsterdam and *Central Laboratory of the Netherlands Red Cross Blood Transfusion Service and Laboratory of Experimental and Clinical Immunology of the University of Amsterdam, Plesmanlaan 125, 1066 CX Amsterdam (The Netherlands)

SUMMARY

Human eosinophil peroxidase is a cationic protein with a high content of arginine, the enzyme being poorly soluble in water. The purified enzyme is able to carry out the peroxidative chlorination of monochlorodimedon. Like myeloperoxidase the position of the pH optimum of this reaction depends on the ratio of the concentrations of chloride and H_2O_2. Compared to myeloperoxidase the pH optimum is shifted by 0.8 pH unit to more acid pH values. The physiological consequences of the properties of the eosinophil peroxidase are discussed.

INTRODUCTION

Recent studies [1-3] have shown that eosinophils are able to damage parasites such as Schistosoma mansoni and it is well documented (for reviews see[4,5]) that during parasitic infection the number of eosinophils in the blood circulation and in tissue is increased. This suggests that eosinophils may have a specialized role in host defence against parasites. It was shown by electron microscopy[2,3] that the eosinophil granule content is discharged onto the surface of the parasite and that the cells adhere to this layer of discharged granule material. This material, which is of lysosomal origin, contains also the eosinophil peroxidase[6].

In neutrophils the myeloperoxidase is involved in the killing of micro-organisms and the bactericidal activity of the myeloperoxidase system is due to the ability of the enzyme to form HOCl in the presence of Cl^- and H_2O_2 [7-9]. As shown by Sips and Hamers,[10] the mechanism of the bactericidal action of myeloperoxidase is due to permeabilization of the bacterial cell envelope by HOCl after which the bacteria die. The damage to the parasite surface by eosinophils may also be due to the enzymic activity of the eosinophil peroxidase. However, it is not known whether, like myeloperoxidase, eosinophil peroxidase is able to form the reactive and bactericidal hypochlorous acid. Knowledge of the human enzyme is still limited, mainly because eosinophils can only be obtained in significant amounts from blood of patients with eosinophilia and in order to be able to purify the peroxidase from these cells, sufficient patient blood must be available. This is generally not the case and therefore a method has been developed[11] which enables us to purify the eosinophil peroxidase from human leucocytes of normal healthy donors. In the present paper some of the properties of the purified eosinophil peroxidase are given and the enzymic activity of the enzyme is compared with that of myeloperoxidase from neutrophils.

RESULTS AND DISCUSSION

As we have shown before[12] by EPR and optical spectroscopy methods the concentration of the peroxidase in eosinophils is about 4 times higher than that in neutrophils. Since normally speaking about 4% eosinophils are present in the blood circulation, about 16% of the total amount of peroxidase present in leucocytes may be due to eosinophil peroxidase. Since we are able to purify myeloperoxidase in large amounts from a cetyltrimethylammonium bromide extract[13] of leucocytes, it is likely that also eosinophil peroxidase can be purified from this source. Table I shows a general outline of the procedure used by us; full details will be published elsewhere[11]. Unlike myeloperoxidase the purified eosinophil peroxidase is not well soluble in water and therefore a detergent must be present to keep the enzyme in solution. In one of the purification steps a sulphoprophyl-Sephadex column is used (Table I) and the eosinophil peroxidase, like myeloperoxidase, is strongly bound to this ion-exchange material. This demonstrates that at neutral pH the enzyme is a basic protein. Indeed the amino acid analysis shows that the eosinophil peroxidase contains approximately 12% arginyl residues. This was also found for other eosinophil cationic proteins[14].

From a physiological point of view the most interesting question is, of course: is the enzyme able to oxidize chloride in the presence of H_2O_2 and to carry out halogenation reactions? This can

Table 1. Isolation procedure of eosinophil peroxidase

Procedure:

1. Homogenise leucocytes in 0.34 M sucrose; centrifuge at 20.000 g (20').
2. Homogenise pellet in 0.5% cetab, buffer[a]); centrifuge at 20.000 g (20').
3. Homogenise pellet in 1.0% cetab, buffer[a]); centrifuge at 20.000 g (20').
4. Combine green-yellow supernatants.
5. Ammoniumsulphate precipitation 40-60%; suspend yellow-green pellet in 0.1 M potassium phosphate (pH 7.4) and 0.5% Tween 80.
6. Sulphoprophyl-Sephadex column; wash with 0.25 M potassium phosphate, 0.5% Tween; elute eosinophil peroxidase with 0.5 M potassium phosphate.

[a]) 0.1 M Na_2SO_4, 0.1 M KPi (pH 7.0)

easily be tested in the assay system with monochlorodimedon[15] which is chlorinated by hypochlorous acid leading to formation of dichlorodimedon. Since this compound, unlike the monochlorodimedon, has little or no absorbance at 290 nm the chlorination activity of the enzyme can be followed spectrophotometrically. When eosinophil peroxidase is incubated with H_2O_2, Cl^- and monochlorodimedon at slightly acid pH values, a rapid decrease in absorbance at 290 nm is found. The rate of the absorbance change increases linearly with the concentration of enzyme and is independent of the monochlorodimedon concentration from 5 to 100 µM. This demonstrates that eosinophil peroxidase indeed catalyses the peroxidation of Cl^-.

The rate of this chlorination reaction shows a strong pH dependency[11] with sharp optima, the position of which depends on the concentration of H_2O_2 and Cl^-. Myeloperoxidase of neutrophils also shows this enzymic behaviour and it was shown by Zgliczyński et al.[16] and Bakkenist et al.[17] that the position of the pH optima is a function of the $[Cl^-]/[H_2O_2]$ ratio. This pH dependency for myeloperoxidase and eosinophil peroxidase is illustrated in Fig. 1 where the position of the pH optima is plotted as a function of the ratio $[Cl^-]/[H_2O_2]$ on a logarithmic scale. At the same concentration ratio the pH optimum of eosinophil peroxidase is shifted by 0.8 pH unit to more acid pH values as compared to myeloperoxidase. Or, alternatively, at constant pH and chloride concentration such as found for example in

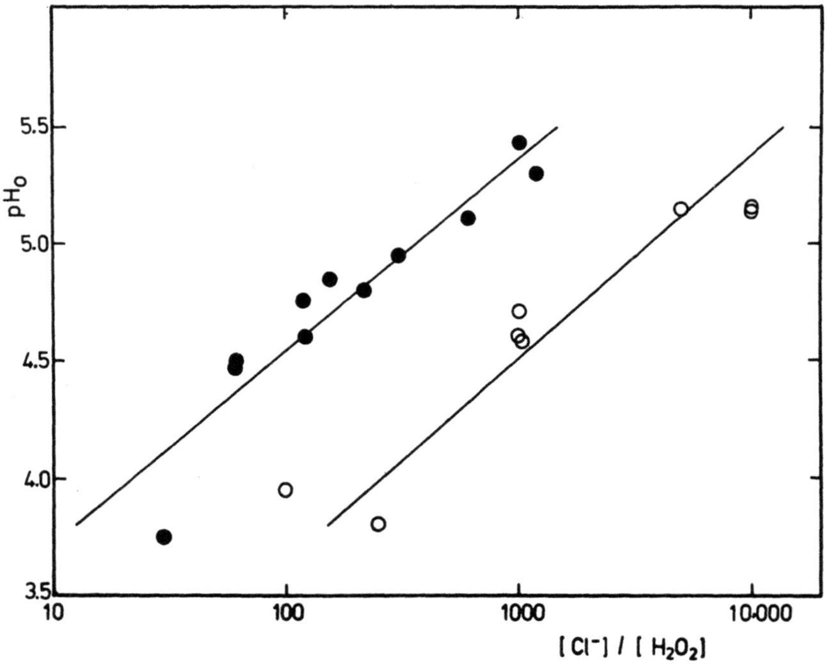

Fig. 1. Position of the pH optima in the chlorination reaction catalyzed by eosinophil peroxidase (O-O) and myeloperoxidase (●-●) as a function of the chloride / H_2O_2 ratio. The chlorination reaction of monochlorodimedon was measured in 0.2 M sodium sulphate, 0.1 M sodium acetate, 16 μM monochlorodimedon and 10 nM peroxidase. The concentration of potassium chloride varied from 5 to 500 mM and the concentration of H_2O_2 from 0.05 to 2.0 mM.

serum, eosinophil peroxidase should be supplied with a lower concentration of H_2O_2 than myeloperoxidase in order to be optimally active.

The kinetics of the chlorination reaction are complex and resemble those observed for myeloperoxidase[16,17]. It has been found that both chloride and H_2O_2 not only act as substrates for the enzyme, but also as inhibitors. The inhibition by chloride is more pronounced at low pH and the Lineweaver-Burk plot (Fig. 2A) of the chlorination rate versus the H_2O_2 concentration shows that at pH 4.4 chloride acts as a competitive inhibitor with respect to H_2O_2. The data from Fig. 2A replotted in a Dixon plot (Fig. 2B) show a common intercept from which an inhibition constant K_i of 70 mM can be obtained. The inhibition by chloride is not due to an effect of ionic strenght on the chlorination

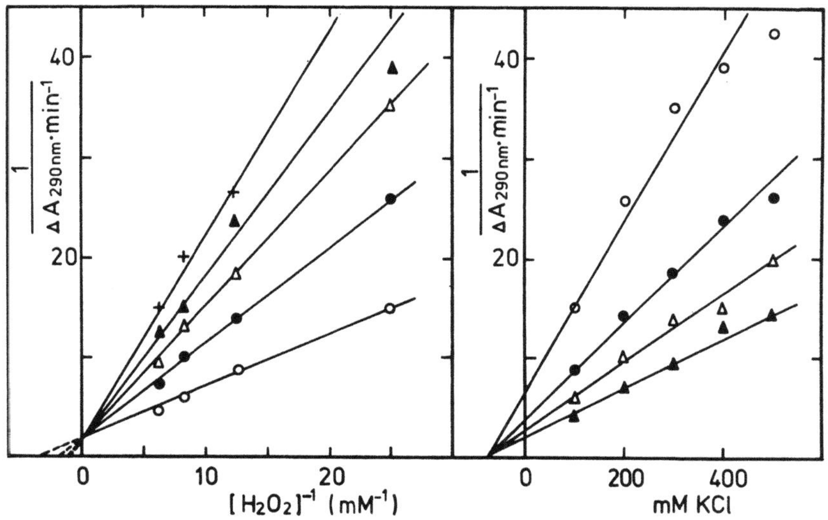

Fig. 2A. Lineweaver-Burk plot of the chlorination of monochlorodimedon at various chloride concentrations. The enzymic reaction was started by the addition of eosinophil peroxidase to a final concentration of 10 nM. 100 mM sodium acetate (pH 4.4), 16 μM monochlorodimedon; temp. 23°C. O-O, 100 mM KCl; ●-●, 200 mM KCl; Δ-Δ, 300 mM KCl; ▲-▲, 400 mM KCl; +-+, 500 mM KCl.

B. Dixon plot of the inhibition of eosinophil peroxidase by chloride. O-O, 40 μM H_2O_2; ●-●, 80 μM H_2O_2; Δ-Δ, 120 μM H_2O_2; ▲-▲, 160 μM H_2O_2. The total ionic strenght was kept at I = 1.0 by addition of sodium sulphate.

reaction, since the ionic strenght was kept constant at I = 1.0 M by addition of sodium sulphate. The maximal velocity calculated from the intercept on the abscissa of Fig. 2A (ΔA = 0.5/min) corresponds to formation of 0.1 μmol dichlorodimedon per minute. Since the concentration of eosinophil peroxidase is 25 pmol, the turnover is 67 s^{-1}. Nearly the same value can be calculated for myeloperoxidase at pH 4.5[17] and in this respect the enzymic activities of the two peroxidases are similar. However, the inhibition constant for chloride is clearly different. A value of 14 mM was found for myeloperoxidase at pH 4.5[17], whereas, as shown here, the inhibition constant is 70 mM for eosinophil peroxidase. This difference may be related to the observation that the position of the pH optima of both enzymes differs. As shown previously[12], the peroxidase content of an eosinophil is 4 times higher than that of a neutrophil and thus it may be concluded that under appropriate conditions both cell types can produce hypochlorous acid at a similar rate.

Our results show that the eosinophil peroxidase is positively charged and not very well soluble in water in the absence of a detergent. These properties may be of prime importance with regard to the manner in which parasites are inactivated by eosinophils. Once the peroxidase is extruded onto the surface of the parasite, it will be strongly bound. Indeed, when eosinophils were incubated with parasites[6], staining for peroxidase revealed that the peroxidase is present between the eosinophil and worm surface. As such this could provide a site where H_2O_2 is generated by the eosinophil and this substrate in turn could be used by the peroxidase for peroxidation of Cl^- to HOCl. In this way the reactive HOCl can be directed immediately to the membranes of the parasite.

REFERENCES

1. S.L. James and D.G. Colley, Eosinophil-mediated destruction of Schistosoma mansoni eggs, J. Reticuloendothel. Soc. 359 (1976).
2. J.P. Caulfield, G. Korman, A.E. Butterworth, M. Hogan and J.R. David, The adherence of human neutrophils and eosinophils to schistosomula: Evidence for membrane fusion between cells and parasites, J. Cell Biol. 86:46 (1980).
3. J.P. Caulfield, G. Korman, A.E. Butterworth, M. Hogan and J.R. David, Partial and complete detachment of neutrophils and eosinophils from schistosomula: Evidence for the establishment of continuity between a fused and normal parasite membrane, J. Cell Biol. 86:64 (1980).
4. P.B. Beeson and D.A. Bass, "The Eosinophil", vol. 14 in the series Major Problems in Internal Medicine, L.H. Smith, Jr., ed., W.B. Saunders Comp., Philadelphia (1977).
5. P.F. Weller and E.J. Goetzl, The human eosinophil, roles in host defense and tissue injury, Am.J. Pathol. 100:795 (1980).
6. D.J. McLaren, C.D. Mackenzie, F.J. Ramalho-Pinto, Ultrastructural observations on the in vitro interaction between rat eosinophils and some parasitic helminths (Schistosoma mansoni, Trichinella spiralis and Nippostrongylus braziliensis, Clin. Exp. Immunol. 30:105 (1977).
7. K. Agner, Biological effects of hypochlorous acid formed by "MPO"-peroxidation in the presence of chloride ions, in: "Structure and function of Oxidation-Reduction Enzymes", Å. Åkesson and A. Ehrenberg, eds., p. 329, Pergamon Press, Oxford (1970).
8. T. Stelmaszyńska and J.M. Zgliczyński, Myeloperoxidase of human neutrophilic granulocytes as chlorinating enzyme, Eur. J. Biochem. 45:305 (1974).
9. J.E. Harrison and J. Schultz, Studies on the chlorinating activity of myeloperoxidase, J. Biol. Chem. 251:1371 (1976).

10. H.J. Sips and M.N. Hamers, Mechanism of the bactericidal action of myeloperoxidase: permeabilization of Escherichia coli, Infect. Immun., in press.
11. R. Wever, H. Plat and M.N. Hamers, Human eosinophil peroxidase: a novel isolation procedure, spectral properties and chlorinating activity. FEBS Letters, submitted for publication.
12. R. Wever, M.N. Hamers, R.S.Weening and D. Roos, Characterization of the peroxidase in human eosinophils, Eur. J. Biochem. 108:491 (1980).
13. A.R.J. Bakkenist, R. Wever, T. Vulsma, H. Plat and B.F. van Gelder, Isolation procedure and some properties of myeloperoxidase from human leukocytes. Biochim. Biophys. Acta 524:45 (1978).
14. I. Olsson, P. Venge, J.K. Spitznagel and R.I. Lehrer, Arginine-rich cationic proteins of human eosinophil granules, Lab. Invest. 36:493 (1977).
15. L.P. Hager, D.R. Morris, F.E. Brown and H. Eberwein, Chloroperoxidase II. utilization of halogen anions, J. Biol. Chem. 241:1769 (1966).
16. J.M. Zgliczyński, R.J. Selvaraj, B.B. Paul, T. Stelmaszyńska, P.K.F. Poskitt and A.J. Sbarra, Chlorination by the myeloperoxidase-H_2O_2-Cl^- antimicrobial system at acid and neutral pH, Proc. Soc. Exp. Biol. Med. 154:418 (1977).
17. A.R.J. Bakkenist, J.E.G. de Boer, H. Plat and R. Wever, The halide complexes of myeloperoxidase and the mechanism of the halogenation reactions, Biochim. Biophys. Acta 613:337 (1980).

DISCUSSION

Edited by R. B. Johnston, Jr.

Segal: Can you be sure that the decrease in absorbance of monochlorodimedone is specific for chlorination and does not simply reflect oxidative damage to the indicator?

Wever: You may actually isolate the product of the reaction and it can be identified as the dichloro compound by I R spectroscopy and by its melting point. (For further details see Hager et al. *J. Biol. Chem.* 241:1769 (1966)). The reaction is specific since if any of the components is omitted, no reaction is observed. Furthermore, incubation of EPO with H_2O_2, Cl^- and alanine results in the formation of a chloroamine.

STIMULATION OR ACTIVATION OF EOSINOPHILS IN VIVO DURING EOSINOPHILIA: POSSIBLE ROLE OF ARACHIDONIC ACID METABOLISM

David A. Bass, William H. Grover, Jon C. Lewis,
Lawrence R. DeChatelet and Charles E. McCall

Bowman Gray School of Medicine
Winston-Salem, North Carolina 27103

INTRODUCTION

Eosinophil leukocytes obtained from the blood of patients with eosinophilia (EE), are abnormal which compared with eosinophils obtained from normal individuals (NE). Such EE have repeatedly been observed to contain variable numbers of vacuoles, reduced numbers of granules (suggesting some degree of degranulation in vivo) and increased segmentation of nuclear lobes (1). These morphologic abnormalities have suggested that eosinophils might be stimulated by a response to the pathologic process causing the eosinophilia. Such stimulation could lead to augmented (activated) or suppressed (deactivated or exhausted) responses of EE to subsequent stimuli. However, there have been few studies of biochemical, physiologic or functional comparisons between such EE and NE. Indeed, nearly all studies of eosinophil physiology have relied upon EE, as such cells have proven easier to purify in vitro and have provided the quantity of eosinophils necessary for most techniques available for examination of leukocyte physiology and function in vitro.

We have purified eosinophils from the blood of normal individuals or patients with eosinophilia of diverse etiologies, and recently examined cell surface charge, membrane transport of deoxyglucose, activation of acid phosphatase of specific granules and stimulation of the hexose monophosphate shunt (HMPS) to provide a profile which will allow comparison of NE with EE (2). The data indicated that circulating EE are stimulated, and in some respects activated (as opposed to deactivated), during pathologic processes which cause eosinophilia. Further studies employing the hexose transport assay suggest that this activation might involve an increased metabolism of arachidonic acid in EE.

Cell Surface Charge

Eosinophils or neutrophils were purified (>95%) by density gradient centrifugation. Cell surface charge was determined by electrophoretic mobility in μm/s/V/cm (2). Neutrophils had a surface charge of 1.98 $\pm$ .01 μm/s/V/cm (n=11), which was reduced to 1.69 $\pm$ .02 following stimulation by ZAS, in agreement with previous studies (3). Resting normal eosinophils were found to have significantly greater charge (2.46 $\pm$ .03 μm/s/V/cm, n=11) than that of neutrophils which was reduced by exposure to ZAS to 1.82 $\pm$ .02 μm/s/V/cm (n=3). Moreover, the surface charge of "resting" EE (1.89 $\pm$ .05 μm/s/V/cm, n=7) was not significantly different from that of stimulated NE and was not significantly altered (to 1.91 $\pm$ .02, n=7) by exposure to ZAS. This is compatible with the hypothesis that such EE had been stimulated in vivo.

A decrease in cell charge is not necessary for normal chemotactic responses of granulocytes. Although EE did not alter surface charge when exposed to ZAS, the chemotactic response of EE was equal to that of neutrophils (2).

Gallin (3) suggested that the reduced charge of stimulated neutrophils might involve the release of lysozyme by the stimulated cells. However, eosinophils do not contain lysozyme (1), which indicates that this substance is not responsible for the altered cell charge. It remains possible that the altered charge might be related to degranulation of another component of the granules of these cells.

Cytochemical Acid Phosphatase Activity of Specific Granules

Lysosomal acid phosphatase was determined by an electron microscopic cytochemical technique (2). In normal eosinophils and neutrophils, lysosomal acid phosphatase in mature cells is held in a latent form (4). Resting NE demonstrated weakly positive acid phosphatase activity in 7.8 $\pm$ 1.2% of the specific granules. NE, stimulated by opsonized staphylococci or the calcium ionophore A23187, developed rapid activation of acid phosphatase in approximately 80% of the granules throughout the cells (5). This activation occurred within the time required for fixation by glutaraldehyde (<30 sec) and studies demonstrating exclusion of the membrane marker ruthenium red indicated that the activation of acid phosphatase preceded degranulation of the cells. We therefore examined the acid phosphatase activity of "resting" eosinophils obtained from patients with eosinophilia. EE fixed immediately after isolation, without stimulation in vitro, revealed acid phosphatase activity varying from 2 to 95% (mean = 49%) of the granules of the EE preparations. 11 of 15 preparations of EE demonstrated significant acid phosphatase activation (2).

Hexose Monophosphate Shunt Activity

Studies of oxidative metabolism of human eosinophils have relied entirely on eosinophils obtained from patients with eosinophilia. These studies have repeatedly found that such EE are capable of an oxidative metabolic burst which appears qualitatively similar but in several aspects is more active in eosinophils than in neutrophils. Most of these studies demonstrated both a high resting activity of eosinophils with further stimulation by phagocytosis or membrane active agents such as phorbol myristate acetate. For example the HMPS of resting EE may approach that of normal neutrophils, yet the EE retain the ability for a further 7-20 fold increase in HMPS activity following stimulation in vitro (6). However, when normal eosinophils were isolated from 4 donors by centrifugation over a linear density gradient of sodium metrizamide, the HMPS of the NE was not significantly different from that of normal neutrophils, both in resting cells and cells stimulated by opsonized zymosan (2).

Membrane Hexose Transport

Granulocytes require energy derived from glucose for virtually all functional responses. Although cytoplasmic stores of glycogen provide one possible source for such glucose, it is of interest that rat neutrophils, chemotactically stimulated to migrate into a peritoneal exudate, were found to contain increased, rather than decreased, glycogen stores (7). We recently found that neutrophils (8-10), or eosinophils (11) respond to stimulation by chemotactic factors with an increase in the carrier-mediated, facilitated transport of the hexose deoxyglucose. Interestingly, insulin did not stimulate this response in granulocytes. Eosinophils were isolated (>95% purity) by centrifugation over linear density gradients of sodium metrizamide (2). "Resting" NE and EE demonstrated deoxyglucose uptake (in c.p.m. x $10^3/2 \times 10^5$ eosinophils), of 2.40 $\pm$ .40 (n=8) and 7.55 $\pm$ .58 (n=15), respectively. Both NE and EE demonstrated a similar degree of further stimulation after exposure to 1.6% ZAS, to uptakes of 5.44 $\pm$ .39 (n=6) and 15.3 $\pm$ .06 (n=15), respectively (1). Thus the deoxyglucose uptake of both "resting" and stimulated EE was roughly 3 times that of NE.

ROLE OF ARACHIDONIC ACID IN STIMULATION, IN VIVO AND IN VITRO OF HEXOSE TRANSPORT OF NE AND EE

In previous studies, we suggested that stimulation of deoxyglucose uptake by neutrophils might involve release of membrane-bound arachidonic acid and its subsequent metabolism by the lipoxygenase pathway (9-11). Stimulation of eosinophils in vitro by chemotactic factors is associated with increased formation and more rapid endogenous turnover of lipoxygenase-mediated arachidonate derivatives (12). This prompted the speculation that such events might also take place during the stimulation of deoxyglucose uptake by EE in vivo and/or in vitro. The initial studies examined the stimulation of eosinophils in vitro. Arachidonic acid stimulated deoxyglucose uptake in a concentration related manner (Figure 1). Structural analogs of arachidonic acid were not stimulatory (Table 1). Indeed, loss of the double bond at the 5 position or reduction of the terminal carboxyl group to an alcohol were sufficient to ablate stimulatory activity of the arachidonic acid molecule. Two inhibitors of lipoxygenase-mediated arachidonate metabolism, 5,8,11,14-eicosatraynoic acid [ETYA] and nordihydroguaiaretic acid [NDGA] inhibited the deoxyglucose uptake of resting eosinophils; moreover, in the presence of these agents, partially purified C5a caused no increase in deoxyglucose uptake (Table 2). C5a caused a similar deoxyglucose uptake by cells incubated in PBS or in 10 μM indomethacin, which inhibits cyclooxygenase-mediated arachidonate metabolism in human granulocytes (13). Furthermore, two lipoxygenase-mediated derivatives of arachidonic acid, 5-HETE and 11-HETE, cause a concentration-related stimulation of deoxyglucose uptake by eosinophils (11). The combined data suggest that arachidonic acid metabolism, apparently by the lipoxygenase

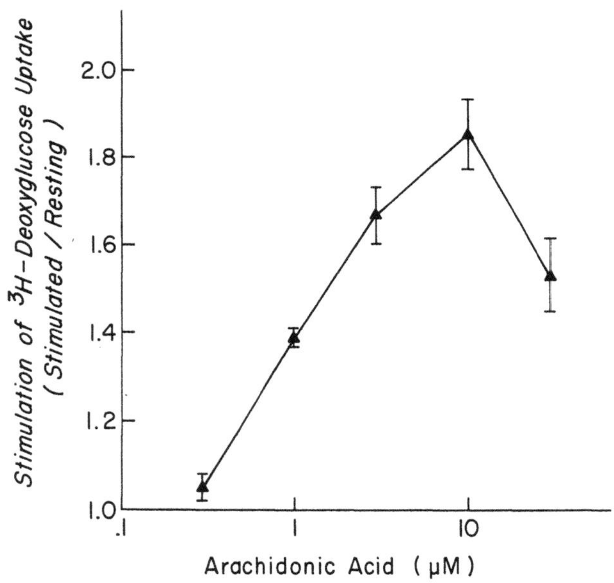

Figure 1. Stimulation of deoxyglucose uptake of eosinophils by varying concentrations of arachidonic acid. Points are means ± SE obtained for 4 separate eosinophil preparations.

Table 1. Stimulation of Deoxyglucose Uptake by ZAS, Arachidonic Acid and Arachidonic Acid Analogues

Stimulant	^{3}H-Deoxyglucose Uptake (% of Resting Control[a])
ZAS, 1.6%	202 ± 9 (30)
Arachidonic Acid (5,8,11,14-eicosatetraenoic acid), 10 μM	199 ± 5 (14)
8,11,14-eicosatrienoic acid, 10 μM	81 ± 6 (9)
11,14,17-eicosatrienoic acid, 10 μM	97 ± 6 (3)
Eicosanoic acid, 10 μM	100 ± 11 (3)
Arachidonyl alcohol, 10 μM	101 ± 8 (6)

a) Data are means ± SE of (n) determinations on purified eosinophils. Because of the variability of uptake of different preparations of "resting" eosinophils, data are expressed as per cent of the uptake of cells incubated in phosphate-buffered saline. Assays employed 2 x 10^5 purified (to >95%) eosinophils in a volume of 0.3 ml glucose-free phosphate buffered saline containing 0.9 mM Ca^{++}, 0.5 mM Mg^{++}. Eosinophils were incubated at 37°C with the indicated stimuli for 15 min prior to initiation of deoxyglucose uptake by the addition of 0.5 μCi ^{3}H-DOG. After 60 min, uptake was stopped by addition of iced PBS, the cells were washed and pelleted by centrifugation and radioactivity in the cell pellets determined (7).

Table 2. Effects of Inhibitors of Arachidonic Acid Metabolism on Deoxyglucose Uptake by Eosinophils

	DOG Uptake, % of Resting Control[a]	
Inhibitor	Resting	C5a
None	100	171 ± 6.2 (4)
ETYA, 30 μM	88 ± 6.0 (8)	91 ± 9.0 (4)
NDGA, 10 μM	61 ± 7.0 (7)	66 ± 7.0 (3)
Indo, 10 μM	144 ± 9.3 (7)	173 ± 17 (4)

a) Data are means ± SEM of (n) separate eosinophil preparations. Triplicate determinations were obtained with each preparation. Deoxyglucose uptake was measured as described in Table 1.

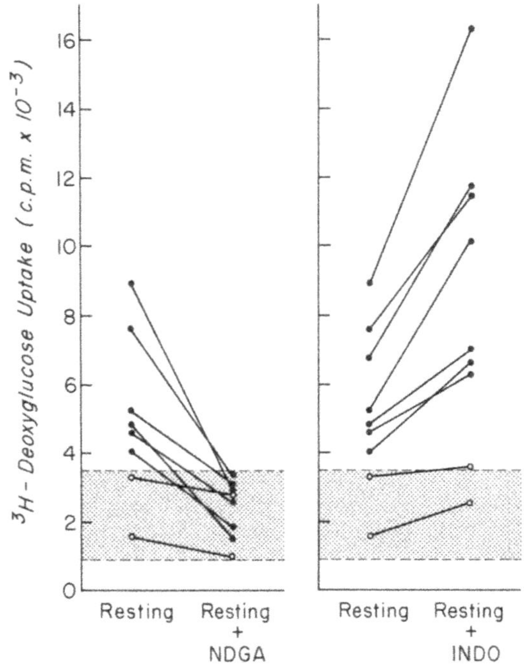

Figure 2. Effect of 15 min preincubation in 10 μM nordihydroguaiaretic acid (NDGA) or of 10 μM indomethacin (INDO) on deoxyglucose uptake of "unstimulated" eosinophils from normals (o—o) or from patients with eosinophilia (●—●). Points are means of closely-agreeing triplicate determinations on separate eosinophil preparations. The shaded area represents the range of deoxyglucose uptake of resting normal eosinophils, as determined on 8 separate normal eosinophil preparations.

pathway, may be linked with in vitro stimulation of deoxyglucose uptake in eosinophils.

The observations that these inhibitors of arachidonate metabolism altered deoxyglucose uptake by the "resting" eosinophils were particularly interesting, since they might suggest a linkage between endogenous arachidonate metabolism and the increased deoxyglucose uptake of "resting" EE. The effects of NDGA and indomethacin on the deoxyglucose uptake of unstimulated cells are further examined in Figure 2. It is apparent that the effects of these agents (inhibition by NDGA and stimulation by indomethacin) are related to the degree of activity of the "resting" cells. That is, cells with a higher "resting" deoxyglucose uptake were more markedly inhibited (back into the range) by NDGA and, conversely, were also more markedly stimulated by indomethacin. Indeed, the "resting" deoxyglucose uptake correlated with the net suppression caused by NDGA (R=.94, P=.0002, N=8) and, conversely, correlated with the percent enhancement of uptake produced by indomethacin (R=.84, P<.001, N=8). Thus, the increased activity of the resting cells is related to a NDGA-inhibitable process. Although the possibility of nonspecific effects of NDGA cannot be excluded, the inhibitory effect of 10 μM NDGA coincides with its ability to inhibit formation of lipoxygenase-mediated arachidonic acid derivatives by human eosinophils (12). Conversely, indomethacin causes an enhancement of deoxyglucose uptake which is also related to the uptake of the "resting" eosinophils. If indomethacin were acting by inhibition of cyclooxygenase-mediated arachidonic acid metabolism (14), it could simultaneously divert metabolism of arachidonic acid to the lipoxygenase pathway (15,16) with the resultant stimulation of deoxyglucose uptake.

Thus, the studies of cell charge and acid phosphatase activity indicate that EE have been stimulated in vivo. The studies of HMPS and deoxyglucose uptake reveal that this results in functional activation, i.e., EE not only demonstrate increased "resting" activities, but also generate augmented responses to subsequent stimulation. The possiblity exists that this in vivo activation might involve an increased turnover of arachidonic acid in the cells. This hypothesis is worthy of further study to clarify the activated state of eosinophils of patients with eosinophilia.

ACKNOWLEDGEMENTS

We thank Ms. Pamela Szejda, Susan Cousart, Pam Shirley and Linda Cowley for excellent technical assistance and Ms. Vickie Cox for typing the manuscript. This work was supported by National Institutes of Health grants AI 14929, AI 09169, HL 16769, AI 10732, and a grant from the National Foundation March of Dimes.

REFERENCES

1. Beeson, P.B. and Bass, D.A. The Eosinophil, W.B. Saunders Co., Philadelphia, 1977.
2. Bass, D.A., Grover, W.H., Lewis, J., Szejda, P., DeChatelet, L.R., and McCall, C.E. Comparison of human eosinophils from normals and patients with eosinophilia. J. Clin. Invest., in press, 1980.
3. Gallin, J.I. Degranulating stimuli decrease the negative surface charge and increase the adhesiveness of human neutrophils. J. Clin. Invest. 65:298-306, 1980.
4. Bainton, D.F. and Farquhar, M.G. Segregation and packaging of granule enzymes in eosinophilic leukocytes. J. Cell

Biol. 45:54-73, 1970.
5. Bass, D.A., Lewis, J.C. and Szejda, P. Immediate activation of eosinophil lysosomal acid phosphatase by phagocytic or chemotactic membrane stimuli. Clin. Res. 27:287A, 1979.
6. DeChatelet, L.R., Shirley, P.S., McPhail, L.C., Huntley, C.C., Muss, H.B., and Bass, D.A. Oxidative metabolism of the human eosinophil. Blood 50:525-535, 1977.
7. Scott, R.B. and Cooper, L.W. Leucocyte glycogen response in inflammatory exudates. Br. J. Haematol. 26:485-496, 1974.
8. McCall, C.E., Bass, D.A., Cousart, M.S., and DeChatelet, L.R. Enhancement of hexose uptake in human polymorphonuclear leukocytes by C5a. Proc. Nat. Acad. Sci. 76:5896-5900, 1979.
9. Bass, D.A., O'Flaherty, J.T., Szejda, P., DeChatelet, L.R., and McCall, C.E. The role of arachidonic acid in stimulation of hexose transport by human polymorphonuclear neutrophilic leukocytes. Proc. Nat. Acad. Sci. 77:5125-5129, 1980.
10. Bass, D.A., DeChatelet, L.R., Goetzl, E.J., Thomas, M., and McCall, C.E. Lipoxygenase-derived products of arachidonic acid mediate hexose uptake in human polymorphonuclear leukocytes. Submitted for publication.
11. Bass, D.A., Szejda, P., Goetzl, E.J., Love, S., O'Flaherty, J.T., and McCall, C.E. Stimulation of hexose uptake of human eosinophils by chemotactic factors. Submitted for publication.
12. Goetzl, D.J., Weller, P.F., and Sun, F.F. The regulation of human eosinophil function by endogenous mono-hydroxy-eicosatetraenoic acids (HETEs). J. Immunol. 124:926-933, 1980.
13. Spagnuolo, P.J., Ellner, J.J., Hassid, A., and Dunn, M.J. Thromboxane A_2 mediates augmented polymorphonuclear leukocyte adhesiveness. J. Clin. Invest. 66:406-414, 1980.
14. Flower, R.J. Drugs which inhibit prostaglandin biosynthesis. Pharmacol. Rev. 26:33-67, 1974.
15. Hamberg, M. and Samuelsson, B. Prostaglandin endoperoxides. Novel transformations of arachidonic acid in human platelets. Proc. Nat. Acad. Sci. 71:3400-3404, 1974.
16. Hamberg, M. On the formation of thromboxane B_2 and 12L-hydroxy-5,8,10,14-eicosatetraenoic acid (12ho-20:4) in tissues from the guinea pig. Biochim Biophys Acta 431: 651-654, 1976.

DISCUSSION
Edited by Charles E. McCall, M.D.

"Stimulation or Activation of Eosinophils in vivo During Eosinophilia: Possible Role of Arachidonic Acid Metabolism"

Presented by Dr. David A. Bass

Question 1 - Dr. Roos

I can add one parameter to the list you mentioned. Dr. Bos in our laboratory has compared the oxidation of iodide by H_2O_2, catalized by eosinophil peroxidase. The activity and the kinetic properties of this enzyme were similar in eosinophils from normal blood and from blood of patients with high counts of eosinophils. This reaction could be measured in mixed neutrophil-eosinophil suspensions because the peroxidase from eosinophils have different pH optima for the iodide oxidation.

Answer - Dr. Bass

That is an interesting observation.

Question 2 - Dr. Esmann

In PMNs the oxidation of 1-^{14}C-glucose may be safely taken as an indicator of the extent of the hexose monophosphate shunt. However, as far as I remember from some experiments done on eosinophils 20 years ago, the oxidation of 6-^{14}C-glucose in eosinophils is not negligible and would therefore influence the evaluation of shunt activity.

Answer - Dr. Bass

In earlier studies at our institution we examined this question. Although eosinophils oxidized glucose-6-^{14}C at a somewhat greater rate than observed with neutrophils, this pathway was only about 1/10 as active as glucose-1-^{14}C. Thus, the oxidation of glucose-1-^{14}C remains a reasonable criterion for hexose monophosphate shunt activity in the eosinophil.

Question 3 - Dr. Hammers

The acid d-glycerol phosphate-phosphatase that you measure is sensitive to sodium-potassium tartrate. In other words, does this acid phosphatase meet the classical criteria set for lysosomal acid phosphatases?

Answer - Dr. Bass

We demonstrated that the reaction was completely blocked by fluoride--a "classical" inhibitor of acid d-glycerol phosphatase.

There was also no depositon of reaction product if the b-glycerolphosphate substrate was omitted.

Question 4 - Dr. Weening

Is there a major difference in isolation procedure for eosinophils between patients with eosinophila and normal donors? Since it is conceivable that differences in isolation procedures may account for observed differences, this should be taken into account.

Answer - Dr. Bass

In each study (cell charge, acid phosphatase activity, deoxyglucose uptake, and hexose monophosphate shunt activity), normal and eosinophilic eosinophils were isolated by identical methods.

PLATELET LEUKOCYTE INTERACTIONS II
IN-VIVO CORRECTION OF CHEDIAK-HIGASHI LEUKOCYTE FUNCTION WITH SEROTONIN OR NORMAL PLATELETS

S.S. Kaplan, R.E. Basford, S.S. Boggs, and U.E. Zdziarski

Departments of Pathology, Biochemistry, Radiation Biology
University of Pittsburgh School of Medicine
Pittsburgh, PA 15261

INTRODUCTION

The formed elements of the peripheral blood often are characterized according to their separate primary functions in homeostasis. While these elements perform independently of each other, there is increasing evidence documenting an essential interaction between these formed elements. Some of these interactions, for example, have been shown to involve lymphocyte subtypes or lymphocytes and macrophages in cellular and humoral immunity[1], and more recently, the relationship of neutrophilic metabolic activation to the platelet release reaction has been demonstrated[2-7]. We recently have presented evidence for the reversed interaction between neutrophils and platelets where a component of platelet dense bodies affects neutrophil function[8]. In these studies, platelet-leukocyte interactions were evaluated in-vitro, using the beige mouse, an animal model of the Chediak-Higashi (CH) syndrome, where both platelet and leukocyte function are defective. These studies demonstrated that normal platelets, or serotonin, a constituent of normal platelet-dense bodies which is absent in CH platelets, corrects the abnormal microbicidal activity of CH peripheral blood leukocytes. The present study further tests the hypothesis that platelet serotonin acts as an effector of PMN function be evaluating the effect of intravenous administration of normal platelets or of serotonin on CH leukocyte microbicidal activity, relating our findings to resulting platelet and plasma serotonin levels. The fate of the injected serotonin also was determined, and it was demonstrated that large amounts of the metabolite 5 hydroxyindole acetic acid (5-HIAA) are formed. This substance, however, has no effect on leukocyte function. These studies confirmed our initial in-vitro observations and showed that CH leukocyte function may be corrected in-vivo by normal platelets or serotonin but not by serotonin breakdown products.

Methods

The mouse animal model of the Chediak-Higashi syndrome C57 Bl/6 (bg/bg) and coisogenic controls C57 Bl/6J (+/+) were obtained from Jackson Laboratories, Bar Harbor, ME. Peripheral blood from these animals was obtained as previously described[8] and its microbicidal capacity determined using Staphylococcus aureus[9]. Studies in the native state and at intervals following the injection either of serotonin or of concentrated peripheral blood platelets obtained from normal mice were done. Blood samples also were obtained for determination of plasma and platelet serotonin and 5HIAA concentrations[10] and for evaluation of intermediary metabolites[11]. In vitro studies to determine the effect of two serotonin breakdown products 5HIAA and 5 hydroxy-tryptophol also were done.

Results

Bacterial killing by leukocytes in diluted peripheral blood from bg/bg and +/+ control animals is shown in Table 1. At the end of 90 minutes of incubation, bg/bg leukocytes killed 66 $\pm$ 1.3 percent of the bacterial inoculum while +/+ leukocytes killed 86 $\pm$ 1.3 percent (P= <0.01). Intravenous injection of +/+ platelets into the bg/bg mouse resulted in a significant (P= <0.01) improvement of bg/bg leukocyte microbicidal activity 2 hours following the infusion, when 81 $\pm$ 1% of the inoculum was killed after 90 minutes of incubation. This enhanced bacterial killing is not seen after either 6 or 24 hours post platelet infusion, Table 2. Similar results were obtained after injection of serotonin, Table 2. Infusion either of 1 µM or 0.1 µM of serotonin resulted in improved

Table 1. Bacterial Killing of BG/BG and +/+ Leukocytes in Whole Blood*

	bg/bg	+/+
5-9 Staphylococci per PMN + 90 min incubation	66 $\pm$ 1.3	86 $\pm$ 1.3
n	35	21
	P= < 0.01	

* Percent of inoculum killed $\pm$ SEM

Table 2. Effect of Intravenous Serotonin or +/+ Platelets on BG/BG Leukocyte Bacterial Killing+

Interval	Platelets 1.5 x 10^8	Serotonin 1.0 μ Mole	Serotonin 0.1 μ Mole
2 hrs	81 ± 1.0* (4)++	82 ± 0.8* (5)	82 ± 3.6* (4)
6 hrs	66 ± 4.8x (3)	70 ± 3.5x (3)	
24 hrs	71 ± 3.2x (4)	69 ± 7.0x (6)	

+ Percent of inoculum killed after 90 minutes of incubation ± SEM
* P= <0.01
x Not significant
++ Number of experiments in parenthesis

bacterial killing 2 hours post injection. After the higher dose of serotonin 82 ± 0.8% (P= <0.01) of the bacterial inoculum were killed while 82 ± 3.6% (P= <0.01) were killed after the lower dose. At 6 and 24 hours post serotonin infusion, no consistent improvement of Staphylococcal killing could be seen.

Table 3. Effect of Serotonin Metabolites on BG/BG and +/+ Leukocyte Bacterial Killing*

	(n)	bg/bg	(n)	+/+
5 HIAA+ (mM)				
1	8	68 ± 2.5	4	85 ± 1.8
10	3	64 ± 6.5		
100	3	61 ± 3.1		
5 H-OL++ (mM)				
1	9	65 ± 3.3	9	80 ± 1.9
10	4	62 ± 5.1		

* In percent of inoculum killed ± SEM
+ 5 hydroxyindole acetic acid
++ 5 hydroxytryptophol
P= > 0.5 in all cases

TLC of extracts of incubation mixture containing neutrophils, bacteria, and serotonin demonstrated oxidation of all serotonin to 5-HIAA in the isopropyl ethyl acetate solvent. Using the ethyl acetate solvent, only serotonin could be seen after chromatography in one direction. Rechromatography at 90° to the first in the second solvent (isopropyl alcohol: ethyl acetate: ammonia) revealed a second spot which may be 5 hydroxy-tryptaldehyde. The effect of the end product of serotonin catabolism, 5-HIAA, and an intermediary product, 5 hydroxy-tryptophol, on peripheral blood microbicidal activity was determined. These results are summarized in Table 3 and demonstrated that neither 5-HIAA, nor 5 hydroxy-tryptophol have any significant effect on bacterial killing.

Serotonin concentrations in bg/bg and +/+ mouse plasma and in platelets were measured in uninjected mice and in mice receiving infusions of serotonin or of platelets. Normal mouse platelets were found to contain 3.4 ± 0.4 nanomoles of serotonin per 10^8 platelets while plasma contained 3.9 ± 0.2 nanomoles per ml. Platelets from uninjected bg/bg mice contained only $.13 \pm .02$ nanomoles of serotonin/10^8 platelets. Bg/bg plasma contained 3.0 ± 0.3 nanomoles/ml. These results and the plasma and platelet serotonin concentrations 2 hours post infusion when improved bacterial killing was seen, and 6 hours post injection when no improvement in bacterial killing was seen are summarized in Table 4. Significant increases both of platelet and plasma serotonin were seen both in bg/bg and +/+ specimens 2 hours after serotonin infusion but not at 6 hours post infusion. After injecting +/+ platelets into the bg/bg mice, the platelet associated serotonin from these bg/bg mice was found to be increased at 2 hours ($P= <0.01$) but not at 6. Plasma serotonin, however, was not significantly altered at any of the times studied up to 4 hours post injection. Platelets and plasma serotonin levels were not determined in +/+ animals given either +/+ or bg/bg platelets.

The kinetics of platelet and plasma serotonin changes following an intravenous injection of 1 μ mole of serotonin are shown in Fig. 1 and 2. Significant and similar rises in bg/bg and +/+ plasma serotonin concentrations were seen within the first 5 minutes of the injection. This increase appeared to plateau until about 60 minutes following injection and then was shown to fall, such that baseline levels were found at 4 hours following the injection. In platelets, a significant increase in the quantity of measurable serotonin also was seen within 5 minutes of the injection. In +/+ platelets, a small increment occurred which persisted for the 6 hours of observation. In bg/bg platelets, a larger increment was seen but the increase persisted for only 2 hours with a return to baseline values 4 hours post injection. 5-HIAA concentrations also were measured in platelets and plasma after infusion of serotonin. This substance also was found to increase and disappear in a manner parallel to that of serotonin. Peak 5-HIAA levels were seen within 60 minutes after the injection returning to baseline after 4 hours, Fig. 3. No 5-HIAA

Table 4. Quantity of serotonin in platelets[v] and plasma[vv] initial 2 and 6 hours post injection of serotonin or normal platelets

		Concentration of Platelet Serotonin			Concentration of Plasma Serotonin		
		Time			Time		
Species	Substance given	0	2 hrs	6 hrs	0	2 hrs	6 hrs
+/+	Serotonin	3.4 ± 0.4 (4)**	5.1 ± 0.6+ (3)	5.2 (1)	3.9 ± 0.2	7 ± 2.6+ (3)	3.6 (1)
	+/+ Platelets		N.D.[x]	N.D.		N.D.	N.D.
bg/bg	Serotonin	0.13 ± .02 (12)	0.61 ± .09∇ (4)	0.1 ± .08* (4)	3.0 ± 0.3 (9)	37 ± 19++ (3)	3 ± 1.2 (1)
	+/+ Platelets		0.28 ± .05++ (3)	0.13 ± .03* (3)		2.78 ± .15* (3)	N.D.

v nanomoles/10^8 platelets
vv nanomoles/ml plasma
\+ P= <0.05
++ P= <0.01
∇ P= <0.001
** number in parenthesis = number of experiments
* not significant
x not determined

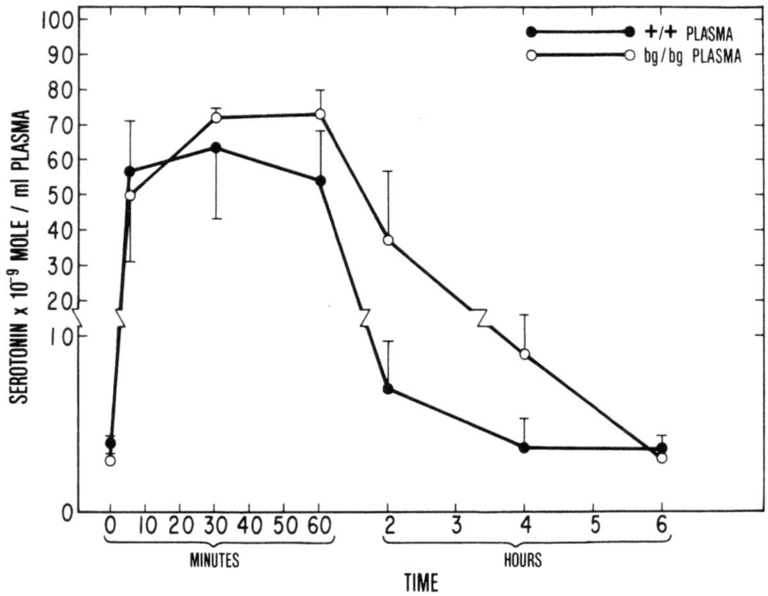

Fig. 1. Plasma serotonin concentrations before and after intravenous injection of 1 μ mole of serotonin are shown. Vertical bars show 1 SEM.

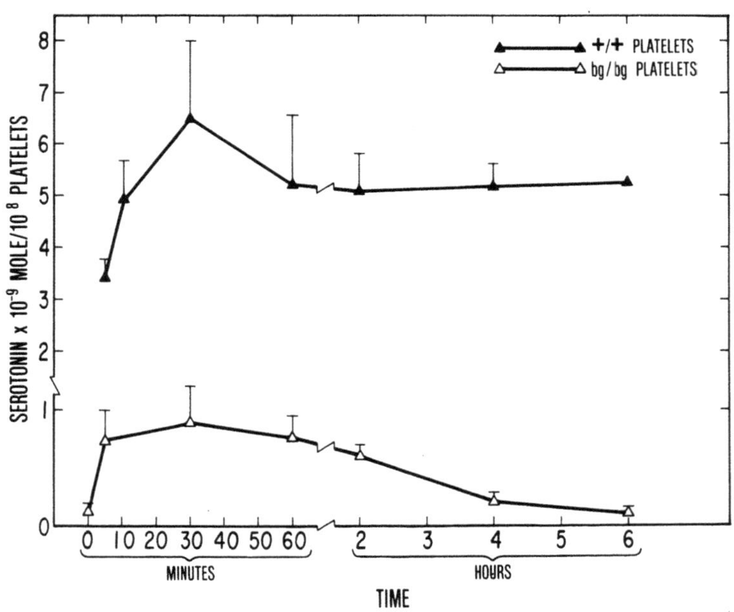

Fig. 2. Platelet serotonin concentrations before and after intravenous injection of 1 μ mole of serotonin are shown. Vertical bars show 1 SEM.

was found in platelet extracts. The appearance and disappearance of platelet and plasma serotonin from bg/bg mice after an infusion of +/+ platelets is shown in Fig. 4. Platelet-associated serotonin was increased until 4 hours post injection (P= <0.01). The concentration of plasma serotonin was unchanged.

Discussion

The beige mouse is an ideal animal model for the study of platelet leukocyte interactions since the beige platelet lacks serotonin[12,13] providing a readily substituted effector; and since neutrophil microbial killing is defective, providing a readily observable target[14,15]. Under normal conditions, it is not readily apparent that such an interaction occurs since deficiency of one cell type is not usually seen without concomitant severe deficiency of the other. Even in idiopathic thrombocytopenic purpura, sufficient platelets are likely to be present for an effective platelet leukocyte interaction since we have shown that as few as 2 platelets/PMN are sufficient[8]. Other recent work has documented the absence of life-threatening infection in patients with severe idiopathic neutropenia who have normal platelet counts[16]. This is in contrast to the well-known host defense defect when both neutropenia and thrombocytopenia are present. This can be partially explained by

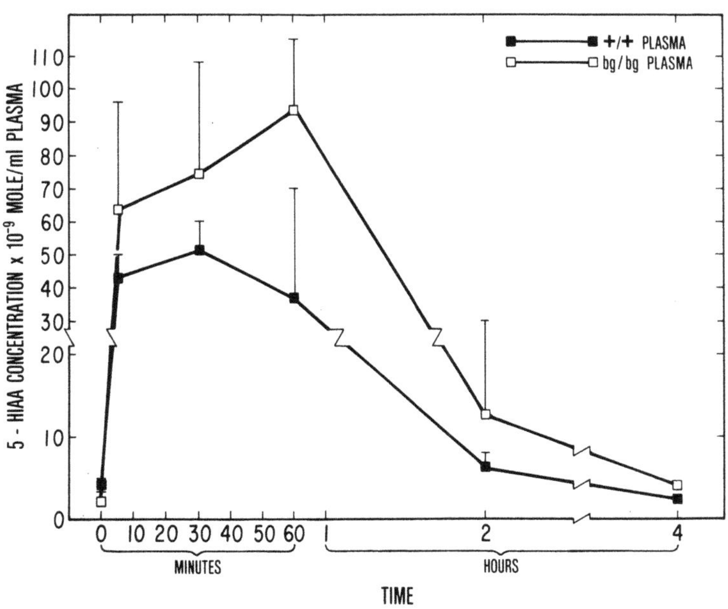

Fig. 3. Plasma 5HIAA concentrations before and after intravenous injection of 1 μ mole of serotonin are shown. Vertical bars show 1 SEM.

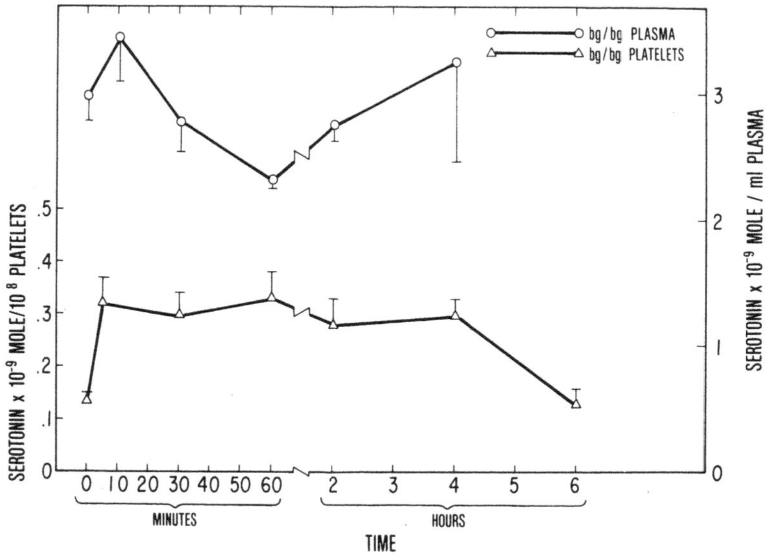

Fig. 4. Plasma and platelet serotonin concentrations before and after the concentration of serotonin in bg/bg plasma 0 —— 0 and platelets Δ —— Δ prior intravenous infusion of 1.5 x 10^8 +/+ platelets are shown. Vertical bars show 1 SEM.

monocyte-mediated host defense but also may be due to a greater degree of effectiveness of the remaining neutrophils.

The results of this study support our hypothesis that platelets and leukocytes may interact in a manner to enhance leukocyte bacterial killing and confirm our earlier study suggesting that it is the serotonin from normal platelets which acts as the mediator of neutrophil function since both normal platelets or serotonin correct the bg/bg microbial killing defect in-vivo and in-vitro. Serotonin concentrations rise following an intravenous infusion and remain elevated for over 2 hours. It is only during this period of serotonin increase that bg/bg leukocyte function is enhanced, with no improvement in leukocyte function observed 6 hours post injection when serotonin levels have fallen. This suggests that serotonin must be present at the time of challenge and that function cannot be "primed" for a later event. Serotonin released into the circulation is rapidly broken down by monoamine oxidase forming 5 hydroxy-tryptaldehyde. This is either reduced to 5 hydroxy-tryptophol or oxidized to 5 hydroxyindole acetic acid which is excreted in the urine[17]. With the exception of central nervous system serotonin, virtually all of it is stored in the dense bodies of platelets[18] where it is protected from breakdown by monoamine oxidase, and from which it may be delivered to polymorphonuclear leukocytes. Although serotonin is rapidly broken down once it leaves the platelet, it is unlikely that any of the breakdown products affect leukocyte function. Two breakdown products which we have evaluated fail to enhance microbial killing in diluted whole blood.

In addition to defective microbial killing, bg/bg leukocytes also show defective chemotactic activity[19,20] and there is strong evidence for defective microtubular polymerization[21]. It also has been shown that the use of substances which enhance microtubule polymerization, such as cyclic guanosine monophosphate (cGMP)[22] or ascorbic acid[23] which tends to decrease cyclic adenosine monophosphate (cAMP), enhances the host defense capability of CH leukocytes both in the beige mouse and in Chediak-Higashi disease of humans. Serotonin also tends to raise cellular cGMP[24,25] and, therefore, may mediate its effect by this mechanism. While other factors also may play a role in bg/bg neutrophil dysfunction, our data and that of others, supports the idea that the platelet defect and the leukocyte defect are related and further suggests that normal platelets are mediators of normal neutrophil function.

REFERENCES

1. J. F. A. P. Miller, Cellular interactions in immune responses, Seminars in Hematology 16:283 (1979).
2. R. A. Clark, and S. J. Klebanoff, Neutrophil platelet interaction mediated by myeloperoxidase and hydrogen peroxide, J. Immunol. 124:399 (1980).
3. R. A. Clark, and S. J. Klebanoff, Myeloperoxidase mediated platelet release reaction, J. Clin. Invest. 63:177 (1979).
4. P. H. Levine, R. S. Weinger, J. Simon, K. L. Scoon, and N. I. Krinski, Leukocyte platelet interactions - release of hydrogen peroxide by granulocyte as a modulator of platelet reactions, J. Clin. Invest. 57:955 (1976).
5. R. Rodoren, J. N. Lindon, and P. H. Levine, Physiology and ultrastrucute of the blood platelet following exposure to hydrogen peroxide, Br. J. Hematol. 33:19 (1976).
6. H. Holmsen, and L. Robkin, Hydrogen peroxide lowers ATP levels in platelets without altering adequate energy change and platelet function, J. Biol. Chem. 252:1752 (1977).
7. R. I. Handin, R. Karabin, and G. J. Boxer, Enhancement of platelet function by superoxide anion, J. Clin. Invest. 59:959 (1977).
8. S. S. Kaplan, S. S. Boggs, M. A. Nardi, R. E. Basford, and J. M. Holland, Leukocyte platelet interactions in a murine model of the Chediak-Higashi syndrome, Blood 52:719 (1978).
9. S. S. Kaplan, and R. E. Basford, Effect of vitamin B_{12} and folic acid deficiencies on neutrophil function, Blood 47:801, 1976.
10. G. Curzon, and A. R. Green, Rapid method for the determination of 5 hydroxy-tryptamine and 5 hydroxy-indole acetic acid in small regions of rat brain. Br. J. Pharmacol. 39:653 (1970).

11. G. Bartholini, A. Pletscher, and H. Broderer, Formation of hydroxy-tryptophol from endogenous 5 hydroxy-tryptamine by isolated blood platelets, Nature 203:1201 (1964).
12. J. M. Holland, Serotonin deficiency and prolonged bleeding in beige mice, Proc. Soc. Exp. Biol. Med. 151:32 (1976).
13. T. G. Bell, K. A. Myers, D. J. Prieur, A. S. Fauce, S. M. Wolff, and G. A. Padgett, Decreased nucleotide and serotonin storage associated with defective function in Chediak-Higashi syndrome, cattle and human platelets, Blood 48:175 (1976).
14. S. N. Wolff, D. C. Dale, R. A. Clark, R. K. Root, and H. R. Kimball, The Chediak-Higashi syndrome: studies of host defenses, Ann. Int. Med. 76:293 (1972).
15. R. K. Root, A. S. Rosenthal, and D. J. Balestra, Abnormal bactericidal metabolic and lysosomal function of Chediak-Higashi syndrome leukocytes, J. Clin. Invest. 51:649 (1972).
16. R. A. Kyle, Natural history of chronic idiopathic neutropenia, N. E. J. M. 302:908 (1980).
17. S. Udenfriend, E. Titus, H. Weissback, and R. E. Peterson, Biogenesis and metabolism of 5 hydroxy-indole compounds, J. Biol. Chem. 219:335 (1956).
18. R. M. Hardesty, and R. S. Stacey, 5 hydroxy-tryptamine in normal platelets, J. Physiol. 130:711 (1955).
19. J. I. Gallin, J. S. Bayak, E. Patten, and S. M. Wolff, Granulocyte function in the Chediak-Higashi syndrome of mice, Blood 403:201 (1974).
20. R. Clark, and H. Kimball, Defective granulocyte chemotaxis in the Chediak-Higashi syndrome, J. Clin. Invest. 50:2645 (1971).
21. J. M. Oliver, Impaired microtubule function correctable by cyclic GMP and cholinergic agonists in the Chediak-Higashi syndrome, Am. J. Path. 85:395 (1976).
22. R. B. Zurier, G. Weissman, S. Hoffstein, S. Kammerman, and H. H. Tai, Mechanisms of lysosomal enzyme release from human leukocytes. II. Effects of cAMP & cGMP autonomic agonists and agents which affect microtubule function, J. Clin. Invest. 53:297 (1974).
23. L. A. Boxer, D. F. Albertini, R. L. Baehner, and J. M. Oliver, Impaired microtubule assembly and polymorphonuclear leukocyte function in the Chediak-Higashi syndrome correctable by ascorbic acid, Br. J. Hemat. 43:207, 1979.
24. J. A. Sandler, R. E. Clyman, V. C. Mangamillo, and M. Vaughan, The effect of serotonin (5 hydroxy-tryptamine) and derivatives on guanosine 3'5' monophosphate in human monocytes, J. Clin. Invest. 55:431 (1975).
25. J. A. Sandler, J. I. Gallin, and M. Vaughan, Effects of serotonin carbonylcholine and ascorbic acid on leukocyte GMP and chemotaxis, J. Cell. Biol. 67:480 (1975).

DISCUSSION
Edited by R. B. Johnston, Jr.

DeChatelet: The observations of Boxer and Oliver regarding cyclic nucleotides in Chediak-Higashi patients have not been reproduced by Gallin at the NIH. Thus there is a minority report.

Kaplan: Boxer found that cyclic AMP was increased in cells and could be reduced with ascorbic acid, with improved host defense. Oliver's work primarily elucidated the role of microtubules in cell motility. She found Chediak-Higashi microtubule formation defective and related this to lack of sufficient cyclic GMP. Mediators of cyclic GMP formation corrected function. Gallin did not reproduce Boxer's finding of elevated cyclic AMP and did not show consistent correction of the functional defect with ascorbate. It is likely that the relative quantity of cyclic AMP and cyclic GMP may be of greater importance to cell function than the absolute quantities, and it may be that not all Chediak-Higashi patients have increased cyclic AMP. These latter patients may not respond to ascorbic acid.

DeChatelet: Would you please compare the disease in the beige mouse with that in the human? It appears as though the bactericidal defect in the animal is quite marginal, yet the disease in humans is deadly. Might this not suggest that the most significant pathologic defect is in chemotaxis rather than intracellular killing?

Kaplan: I agree that the chemotactic defect probably is more relevant to the compromised host defense than is the microbicidal defect. Our beige mice do survive well, but they are housed in laminar flow racks and are, therefore, protected from infection. However, patients are not similarly protected and therefore get infected more often. Also, human patients often die with the accelerated phase or lymphoid dyscrasia. Chediak-Higashi mice also have abnormal lymphocyte function.

Roos: Do you have any idea how serotonin might improve the leukocyte function of the beige mouse? Have you measured levels of cyclic nucleotides in the leukocytes, chemotaxis, or the release of myeloperoxidase (MPO) from the granules?

Kaplan: Serotonin probably improves function through its capacity to raise cyclic GMP, and this promotes the polymerization of tubulin, thereby enhancing lysosomal movement within cells. It is likely that it will improve chemotaxis as well as bacterial killing. However, we have not yet evaluated chemotaxis or release of MPO from granules. Preliminary studies have shown abnormal cyclic GMP levels in leukocytes from beige mice.

Johnston: Dr. Kaplan's proposal, that there is an abnormality in the supplying of serotonin to neutrophils by platelets, appears to tie together the various functional and biochemical abnormalities of beige mice neutrophils particularly well. Unfortunately, the systems involved are not easy to dissect, and there are gaps in our knowledge, especially with the human.

PHYSICOCHEMICAL SURFACE CHANGES ON PHAGOCYTIC CELLS DURING DIFFERENTIATION IN RELATION TO CHEMOTAXIS AND PHAGOCYTOSIS*

Olle Stendahl, Claes Dahlgren and Jan Hed

Department of Medical Microbiology
Linköping University Medical School
S-581 85 Linköping, Sweden

INTRODUCTION

Established cell lines with appropriate functional markers for cell differentiation offer simple methods for study of the regulation of cell differentiation. A human promyelocytic leukemia cell line, HL-60, was recently established from the peripheral blood of a patient with acute promyelocytic leukemia[1]. The majority of the HL-60 cells are promyelocytic in morphology and histochemistry, but 5-10% show morphological characteristics of more mature cells. Induction of morphologic and functional differentiation can be induced by dimethylsulfoxide (DMSO) or other polar solvents[1]. In addition phorbol diester tumor promoters[2] and conditioned media[3] can induce differentiation. Several investigations have recently demonstrated that differentiation is accompanied by functional changes such as increased migration and chemotactic responsiveness[4], increased oxidative metabolism, phagocytosis and bacterial killing[4,5]. Furthermore, the cell surface glycoprotein pattern is altered with the expression of a 130.000 dalton glycoprotein[6].

In the present investigation we have studied alterations of the physicochemical surface properties such as surface charge and liability to hydrophobic interaction during differentiation of HL-60 cells induced by DMSO. Functional differentiation was monitored by studying chemotactic and oxidative responsiveness and IgG- and C3-dependent phagocytosis in relation to these surface property changes.

*This study was supported by grants from The Swedish Medical Research Council (16X-05968) and King Gustav V Foundation.

MATERIALS AND METHODS

The human promyelocytic HL-60 cells were grown in suspension culture in RPMI-1640 (Flow Lab. Ltd., Irvine, Scotland) supplemented with 10% heat-inactivated fetal calf serum (Flow Lab.), penicillin (100 units/ml) and streptomycin (100 μg/ml). The cells were induced to differentiate by addition of DMSO to a final concentration of 1.3%. Human peripheral blood granulocytes were separated from EDTA blood after dextran sedimentation and Hypaque-Ficoll purification[7]. The HL-60 cells were harvested by centrifugation and quantitated in a Coulter Counter ZF with a 100 channel pulse-height analyser (Coulter Electronics Ltd.).

Physicochemical surface changes were assayed in aqueous two-polymer phase systems containing dextran, polyethyleneglycol, bis-trimethylamino ($(CH_3)_3N^+$)-PEG (TMA-PEG) or palmitoyl -PEG (P-PEG) . Migration and chemotactic responsiveness was assayed by a modification of the agarose method described by Nelson et al.[9], using zymosan-activated human serum as chemoattractant. Phagocytosis of IgG- and C3b-sensitized FITC-labelled yeast particles was assayed with "The Fluorescence Quenching Method" described by Hed[10], where the number of cells ingesting one or more yeast particles was calculated. Chemiluminescence was assayed with a modified Luminometer 1250 (LKB-Wallac, Stockholm, Sweden) in the presence of 0.1 mg/ml luminol (Sigma Chemical Co.).

RESULTS

When the HL-60 cells were induced to differentiate in the presence of DMSO, the cell volume decreased (Fig. 1) and after 3 days less than 30% of the cells had the same volume as the undifferentiated cells. Control granulocytes showed a volume distribution further to the left (not shown) indicating a smaller average volume than the differentiated HL-60 cells. During granulocyte differentiation several surface properties are modulated[11], such as adhesiveness, deformability and self aggregation. To assay physico-chemical surface changes related to these properties, we employed different two-phase systems. By introducing positively charged TMA-PEG or hydrophobic palmitoyl-PEG, negative surface charge and hydrophobic interaction can be monitored[8]. Table 1 shows that TMA-PEG influence the partition of undifferentiated, differentiated and control PMNL to the same extent. No detectable difference in surface charge is thus evident between the cell populations. In contrast, both the DMSO-treated HL-60 cells and the mature granulocytes showed increased hydrophobic interaction with the P-PEG (40% and 35% respectively) compared to immature HL-60 cells (11%). Using this two-phase system it is thus possible to separate more mature cells from myeloblasts and promyelocytes during differentiation and in these subpopulations link different functional characteristics to different stages of differentiation.

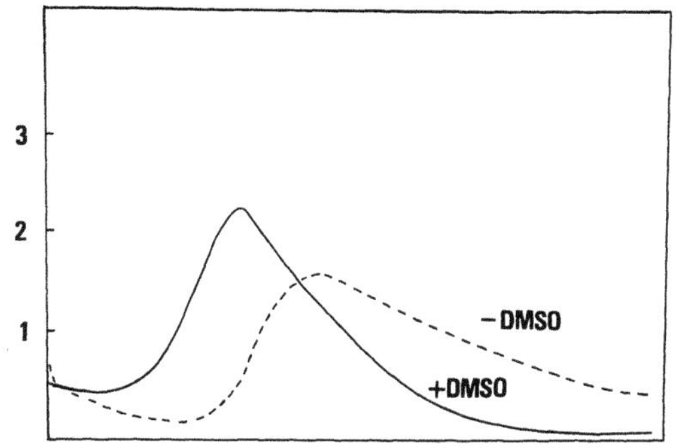

Fig. 1. The volume distribution of HL-60 cells cultured for 4 days in RPMI-1640 in the absence (---) or presence of 1.3% DMSO (——). The ordinate expresses the number of cells (x10^3) and the abscissa the relative volume distribution.

Table 1. Partition of HL-60 cells and PMNL in the top phase of different aqueous two phase systems.

Cells	% cells in the top phase of systems[a] containing:		
	TMA-PEG	PEG	Palmitoyl-PEG
HL-60	44	4	11
HL-60 (DMSO)[b]	62	9	40
Blood PMNL	48	14	35

a) Phase composition: 4.4% dextran, 6.2% PEG substituted with 1% TMA-PEG or 0.2% palmitoyl-PEG in 0.03 M Tris, pH 7.2.
b) The cells were cultivated for 4 days in 1.3% DMSO.

Table 2. Migration and chemotactic response in HL-60 and human PMNL.

	Cells		
	PMNL	HL-60	HL-60 (DMSO)[a]
Stimulated locomotion[b]	66 ± 5	0	60 ± 5
Random locomotion	32 ± 6	0	28 ± 5
Attraction index[c]	2.13		2.21

a) The HL-60 cells were cultivated for 6 days in 1.3% DMSO.
b) Towards zymosan-activated human-serum.
c) Calculated as the ratio between stimulated and random locomotion.

When 3-6 days cultures of HL-60 cells in the presence of DMSO were tested for migration and chemotactic response, the DMSO-treated cells migrated and responded to the chemoattractant as well as the control PMNL (table 2). The attraction index was 2.21 for the DMSO-treated cells and 2.13 for the PMNL. However, fewer cells migrated in the HL-60 population than in the PMNL, indicating that only the most mature cells were responsive. In contrast, the undifferentiated HL-60 cells showed no migratory activity.

The ingestion of IgG- and C3-opsonized yeast particles was assayed during the course of differentiation. Within two days, the C3-dependent ingestion increased from less than 10 to 25%, whereas the IgG-dependent ingested increased more slowly. Figure 2 shows that C3-phagocytosis reaches its maximum within 4-6 days, whereas IgG-phagocytosis increases up to 6-8 days in the presence of DMSO. The different capacity to ingest C3- and IgG-opsonized particles may reflect formation of complement and IgG receptors at different rates. However, since the IgG-particles are more effective than the C3 ones in stimulating superoxide production[12] and elicit chemoluminescence (table 3) the discrepancy may rather reflect different triggering mechanisms necessary for pseudopod formation and ingestion.

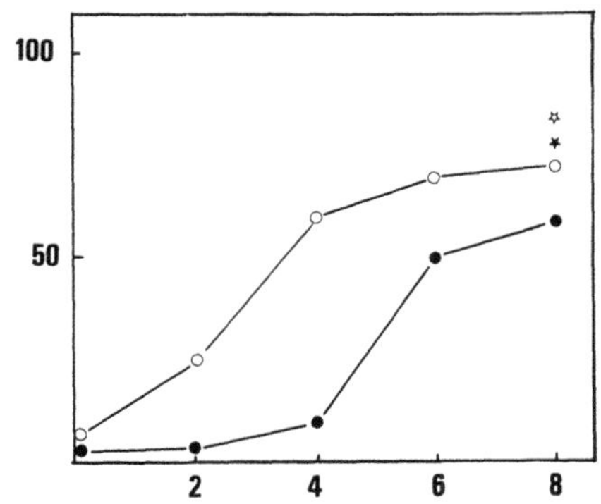

Fig. 2. Phagocytosis of IgG-(●) and C3-(○) sensitized FITC-labelled yeast particles by HL-60 cells (●,○) and PMNL (✫,✶). The ordinate depects the % of cell phagocytosing one or more yeast particles, and the abscissa depicts number of days in culture in the presence of 1.3% DMSO.

Table 3. Chemiluminescence in HL-60 cells and PMNL.

Cells	Chemiluminescence response[a] after stimulation with:		
	IgG-yeast	C3-yeast	PMA
HL-60	2.5	0.3	2.5
HL-60 + DMSO[b]	8.0	0.8	10.5
PMNL	12.5	13.0	20.0

a) The reaction mixture contained in 1.0 ml $5x10^6$ cells, $5x10^7$ yeast particles or 10^{-6} M phorbol myristate acetate (PMA), 0.1 mg luminole in Krebs-Ringer phosphate buffer, pH 7.2 with 10 mM glucose. The chemiluminescence response was monitored continuously for 30 min and expressed as the peak value in arbitrary units.

b) The cells were cultivated for 5 days in 1.3% DMSO.

CONCLUSION

During cell differentiation induced by DMSO, the HL-60 cells exhibit certain surface alterations such as increased hydrophobic interaction. Similar changes have been observed in activated alveolar macrophages[13], and granulocytes exposed to chemoattractant *in vivo* and *in vitro*[14]. Whether these surface changes are directly linked to the enhanced chemotactic, oxidative and phagocytic responsiveness is unclear, although increased liability to hydrophobic interaction promotes phagocytic recognition in several systems[15]. From the present data it is evident that oxidative and phagocytic responsiveness are aquired at different stages of differentiation. Detecting physicochemical differences in the surface properties of the HL-60 cells during differentiation may furthermore provide a useful tool for studying subpopulations of cells during differentiation and activation.

REFERENCES

1. S. J. Collins, R. C. Gallo, and R. E. Gallagher, Continuous growth and differentiation of human myeloid leukemia cells in suspension culture, Nature 270:347 (1977).
2. E. Huberman and M. F. Callaham, Induction of terminal differentiation in human promyelocytic leukemia cells by tumor-promoting agents, Proc. Nat. Acad. Sci. USA, 76:1293 (1979).
3. I. Olsson and T. Olofsson, Induction of differentiation in a human promyelocytic leukemia cell line by chemical and physiological inducersproduction of granule protein, Exp. Cell Res. in press.
4. S. J. Collins, F. W. Ruscetti, R. E. Gallagher, and R. C. Gallo, Normal functional characteristics of cultured human promyelocytic leukemia cells (HL-60) after induction of differentiation by dimethylsulfoxide, J. Exp. Med. 149:969 (1979).
5. P. E. Newburger, M. E. Chovaniec, J. S. Greenberger, and H. J. Cohen, Functional changes in human leukemic cell line HL-60. A model for myeloid differentiation. J. Cell Biol. 82:315 (1979).
6. C. G. Gahmberg, K. Nilsson, and L. C. Andersson, Specific changes in the surface glycoprotein pattern of human promyelocytic leukemia cell line HL-60 during morphologic and functional differentiation. Proc. Nat. Acad. Sci. USA, 76:4087 (1979).
7. A. Bøyum, Separation of leukocyte from blood and bone marrow, Scand. J. Clin. Lab. Invest. 21:suppl. 97 (1968).
8. C. Dahlgren, E. Kihlström, K.-E. Magnusson, O. Stendahl, and C. Tagesson, Interaction of liposomes with polymorphonuclear leukocytes. II. Studies on the consequences of interaction, Exp. Cell Res. 108:175 (1977).

9. R. D. Nelson, R. T. McCormack, and V. D. Fiegel, Chemotaxis under agarose. A new and simple method for measuring chemotaxis and spontaneous migration of human polymorphonuclear leukocytes and monocytes, J. Immunol. 115:1650 (1975).
10. J. Hed, The extinction of fluorescence by crystal violet and its use to differentiate between attached and ingested microorganisms in phagocytosis, FEMS letters, 1:357 (1977).
11. M. L. Lichman and R. I. Weed, Alteration of the cell periphery during granulocyte maturation: Relationship to cell function, Blood, 39:301 (1972).
12. O. Stendahl and C. Dahlgren, unpubl. obs.
13. O. Stendahl, unpubl. obs.
14. C. Dahlgren, B. Normann, and O. Stendahl, unpubl. obs.
15. O. Stendahl, C. Dahlgren, M. Edebo, and L. Öhman, Recognition mechanisms in mammalian phagocytosis, in: "Endocytosis and Exocytosis in host defense", L. Edebo, L. Enerbäck and O. Stendahl, eds. Karger, Basel, in press.

DISCUSSION
Edited by R. B. Johnston, Jr.

Breton-Gorius: Before exposure of the HL60 cells to DMSO they do not have lactoferrin. After DMSO treatment, which induces maturation, is lactoferrin present?

Stendhal: There is no synthesis of lactoferrin in the DMSO-induced cells.

Basford: Can the effect of DMSO on the cell line be due to an increased permeability of the plasma membrane?

Stendhal: We have no evidence of increased permeability caused by DMSO treatment. DMSO treatment for short times gives no changes in permeability or cell volume.

Gallily: Did the cell line following maturation with DMSO lose its tumorigenic properties?

Stendhal: The cell line is tumorigenic in nude mice (Gallagher et al. Blood 54:713 (1979)). Whether the DMSO-induced cells are tumorigenic is uncertain. I would suspect that the effect would be reduced since the life span of the mature cells is limited.

Gallily: Did you assay some of the immunological functions of the cells after maturation, such as presentation of antigen to lymphocytes in antibody production?

Stendhal: No, we have not performed such studies, and I am not aware of any other reports on this matter.

Harkness: Have you checked glucose consumption by the cells?

Stendhal: No.

MECHANISM OF HEXOSE TRANSPORT IN HUMAN POLYMORPHONUCLEAR LEUKOCYTES

Charles E. McCall, David A. Bass, Michael Thomas,
Joseph T. O'Flaherty, and Lawrence R. DeChatelet

Departments of Medicine and Biochemistry
Bowman Gray School of Medicine
Winston-Salem, NC 27103

Cellular functions of human polymorphonuclear leukocytes (PMNL) including phagocytosis, motility, and bactericidal activity require energy derived from glucose. Whereas insulin does not stimulate hexose transport in human PMNL, we have recently reported that chemotactic factors such as complement-derived C5a, the synthetic oligopeptide N-formyl-methionyl-leucyl-phenylalanine (fMLP), or the calcium ionophore A23187 stimulates transport of deoxyglucose (DOG) into human PMNL[1,2]. FMLP with an EC_{50} of 1.2 nM, C5a with an EC_{50} of 1 nM, and A23187 with an EC_{50} of 10 nM all cause at least a 5-fold stimulation of DOG uptake.

The chemotactic factors C5a and fMLP and the calcium ionophore A23187 also cause PMNL *in vitro* to aggregate, degranulate, and generate reactive oxygen radicals[3]. Each of these responses may be associated with important aspects of PMNL function *in vivo*. Aggregation correlates with increased surface membrane adhesiveness and the capacity of PMNL to attach to endothelial or foreign surfaces, whereas degranulation and oxygen radical generation correlate with the ability of PMNL to degrade extrinsic substances and to kill certain microbes.

The metabolism of arachidonic acid (AA) has been implicated in aggregation, degranulation, oxygen radical generation, and hexose uptake in PMNL. For example, PMNL metabolize the stores of AA in response to fMLP and A23187[4,5]. Arachidonic acid itself causes increased hexose uptake in human PMNL[2], it causes rabbit PMNL to degranulate and rabbit and human PMNL to aggregate[6,7]. Moreover, PMNL manufacture derivatives of arachidonic acid[8]. If the lipoxygenase mediated arachidonic acid metabolism is blocked, aggregation, degranulation, and hexose transport of PMNL are inhibited. Such blockage also prevents the generation of reactive oxygen radicals of guinea pig PMNL to fMLP[9]. Thus, AA metabolism may play an important role in diverse PMNL responses.

This report reveiws the stimulation of hexose transport by soluble mediators and further establishes a linkage between the stimulation of hexose transport and arachidonic acid metabolism in human PMNL.

Deoxyglucose uptake was measured by incubating 2 x 10^5 purified PMNL in 0.3 ml glucose-free phosphate buffered saline (PBS) containing 0.9 mM calcium for 15 min at 37°C. Addition of 0.5 μCi ^{3}H-DOG initiated the reactions. The cells were then incubated for 60 min at 37°C and the reaction was stopped by addition of 1 ml iced PBS. The cells were rapidly centrifuged and the pellets washed and the radioactivity determined. Expression of deoxyglucose uptake was as c.p.m./2 x 10^5 PMNL[1].

Deoxyglucose was employed as a measure of glucose transport since it enters cells at the same sites as glucose by a specific, saturable mechanism which conforms to the characteristics of carrier-facilitated transport. Uptake of deoxyglucose in PMNL was found to be saturable with increasing concentrations of either stimulant or substrate, inhibited by cytochalasin B, and stereospecific (inhibited by D-glucose but not L-glucose)[1,2]. Thus, the uptake of deoxyglucose in PMNL simulates carrier-facilitated transport of hexose stimulated by insulin in other tissues[10]. During the 60 min of incubation DOG uptake was essentially linear. Most of DOG is irreversibly phosphorylated within the cell and this is dependent upon hexokinase activity of resting or stimulated PMNL. Also, inhibitors used in these studies did not effect hexokinase activity[1,2].

C5a (partially purified as described previously)[11] or fMLP stimulated deoxyglucose uptake in a concentration dependent manner (Figure 1). Stimulation occurred at concentrations similar to or slightly below those which have been reported to stimulate chemotaxis and degranulation of PMNL. The calcium ionophore A23187 caused a similar concentration-related enhancement of DOG uptake (Figure 1). These stimuli have also been reported to cause release of arachidonic acid and/or arachidonic acid metabolites from PMNL[4,5] presumably by activation of a calcium-dependent phospholipase A_2[12]. Accordingly, we examined the postulate that arachidonic acid might be associated with stimulation of deoxyglucose uptake in PMNL.

Deoxyglucose uptake was stimulated by arachidonic acid in a concentration-related manner (Figure 2). The concentrations were similar to those known to cause aggregation of human PMNL. The stimulation was structurally specific, as 8,11,14-eicosatrienoic acid, 11,14,17-eicosatrienoic acid, 5,8,11,14-eicosatetraynoic acid, and arachidonyl alcohol had no stimulatory effect[2]. Similar structural specificity

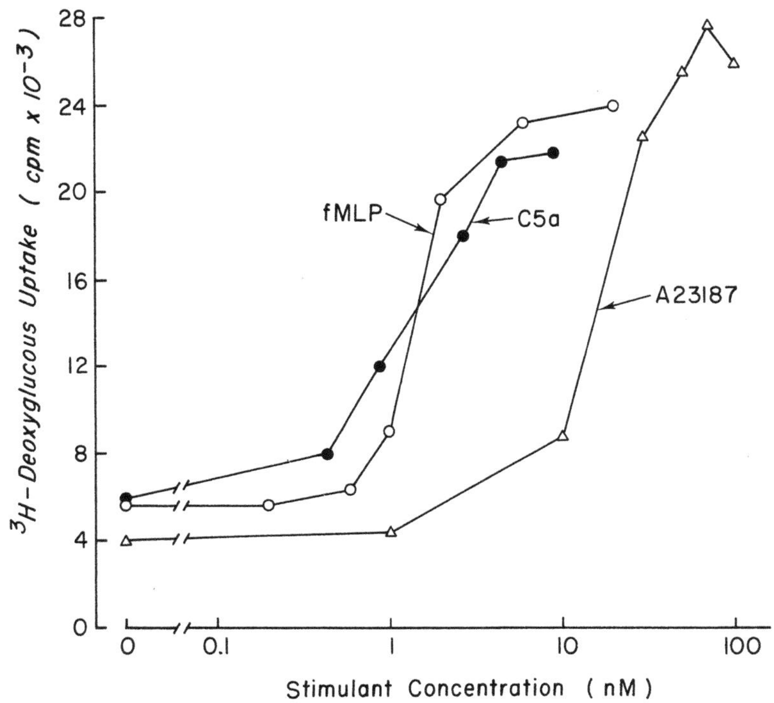

Fig. 1. Stimulation of deoxyglucose uptake by PMNL, by C5a, and the calcium ionophore A23187. Data are from representative experiments. Each point is the mean of closely-agreeing quadruplicate determinations. Similar data have been obtained with at least three separate PMNL preparations. (From Bass, D. A., et al, Proc. Nalt. Acad. Sci. In press).

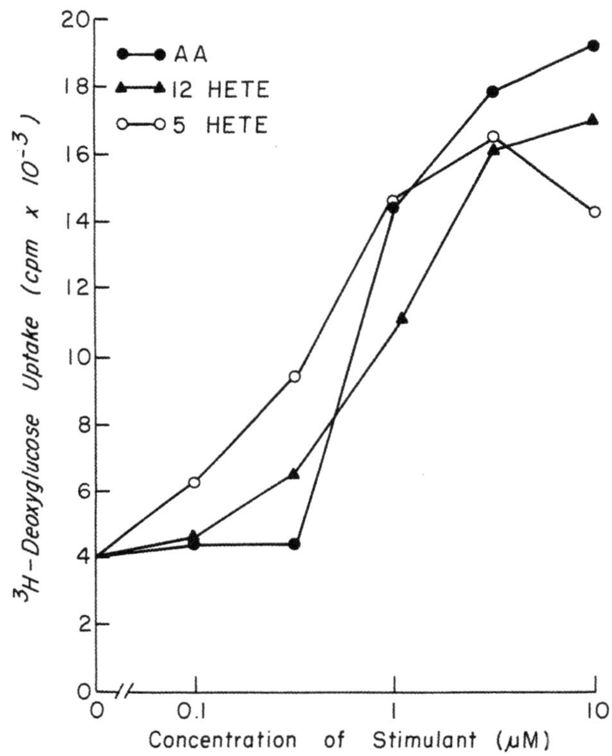

Fig. 2. Stimulation of deoxyglucose uptake by arachidonic acid (AA), 5-HETE, and 12-HETE. Data are means of closely-agreeing quadruplicate determinations from representative experiments. Similar data have been obtained with at least three separate PMNL preparations.

has been reported for the aggregation of human PMNL and degranulation of rabbit PMNL[6,7].

Inhibitors of arachidonic acid metabolism were employed to provide further evidence for the involvement of arachidonic acid metabolism in the stimulation of deoxyglucose transport. 5,8,11,14-eicosatetraynoic acid (ETYA) inhibits both lipoxygenase and cyclooxygenase mediated arachidonic acid metabolism, and it has been found to inhibit the formation and release of mono-hydroxy-eicosatetraenoic acids (HETEs) by PMNL at an EC_{50} of approximately 8 µM[13]. Nordihydroguaiaretic acid (NDGA), an inhibitor of lipoxygenase activities, blocks the formation and release of HETEs by PMNL at an EC_{50} of approximately 5 µM[11]. 30 µM ETYA and 10 µM NDGA markedly inhibited the stimulation of DOG uptake by PMNL at 2.7 nM C5a, 2 nM fMLP, 50 nM A23187 or 3 µM arachidonic acid (Table 1). Furthermore, 10 µM indomethacin which blocks cyclooxygenase activity, did not inhibit the stimulation of DOG uptake. Any firm conclusion regarding the specific pathway of arachidonic acid metabolism involved in the stimulation of deoxyglucose transport is tempered in these studies by the possibility of nonspecific effects of the inhibitors[14]. However, the data are compatible with the hypothesis that stimulation of DOG transport in PMNL is associated with metabolism of arachidonic acid by the lipoxygenase pathway. Further data have recently been obtained to substantiate this hypothesis. Two lipoxygenase derivatives of arachidonic acid, 5-HETE and 12-HETE, were generated through auto-oxidation of arachidonic acid and purified on silicic columns and reverse phase high pressure liquid chromatography. 5-HETE and 12-HETE both of which can also be generated biologically from neutrophil lipoxygenase mediated metabolism of arachidonic acid[8], stimulated uptake of DOG (Figure 2).

Table 1. Effect of Inhibitors of Arachidonic Acid Metabolism

	^{3}H-deoxyglucose Uptake (c.p.m./2x10^5 PMNL)*			
Stimulant	No Inhibitor	30 µM ETYA	10 µM NDGA	10 µM Indo
None	5,879	5,606	5,972	7,456
C5a, 2.7 nM	21,470	4,987	5,721	19,952
fMLP, 2 nM	20,437	8,710	6,719	21,093
A23187, 50 nM	23,052	6,152	5,247	20,027
Arachidonic Acid, 3 µM	23,292	5,297	6,530	22,891

*Data are means of closely-agreeing quadruplicate determinations. Similar data have been obtained on at least three PMNL preparations with each stimulus and inhibitor. (From Bass, D.A., et al, Proc. Natl. Acad. Sci. In press).

The data reported here suggests that carrier-facilitated hexose transport and metabolism of arachidonic acid through the lipoxygenase pathway may be linked, at least in PMNL. This concept is further substantiated in that two of the numerous products of lipoxygenase metabolism of arachidonic acid so far studied, 5-HETE and 12-HETE, also stimulate hexose transport in concentrations similar to that of arachidonic acid. Furthermore, chemotactic factors such as C5a and fMLP and the calcium ionophore A23187 enhance hexose transport in PMNL by pathways which may be linked with arachidonic acid metabolism.

The linkage of carrier-facilitated hexose transport and metabolism of arachidonic acid is probably not limited to PMNL. For example, we have found similar effects to those described herein in human eosinophils (unpublished observations). Furthermore, arachidonic acid metabolism could be associated with stimulation of hexose transport in insulin sensitive tissues. For instance, hexose uptake by adipocytes has been reported to be increased by phospholipase A_2, which is an enzyme causing release of arachidonic acid metabolism in hexose transport in diverse tissues is worthy of further studies.

ACKNOWLEDGEMENTS

We would like to thank Ms. Susan Cousart, Ms. Pamela Szejda, and Ms. Pam Shirley for their excellent technical assistance, and Ms. Anne Winburn for typing the manuscript. This work was supported in part by NIH grants AI 14929, AI 09169, HL 16769, and a grant from the National Foundation March of Dimes.

REFERENCES

1. C. E. McCall, D. A. Bass, S. Cousart, and L. R. DeChatelet, Enhancement of hexose uptake in human polymorphonuclear leukocytes by activated complement C5a, Proc. Natl. Acad. Sci. 76:5896 (1979).
2. D. A. Bass, J. T. O'Flaherty, P. Szejda, L. R. DeChatelet, and C. E. McCall, Role of arachidonic acid in stimulation of hexose transport of human polymorphonuclear leukocytes, Proc. Natl. Acad. Sci. 77:5206 (1980).
3. J. T. O'Flaherty and P. A. Ward, Chemotactic factors and the neutrophil, Semin. Hematol. 16:163 (1979).
4. P. Borgeat and B. Samulesson, Arachidonic acid metabolism in polymorphonuclear leukocytes: effects of ionophore A23187. Proc. Natl. Acad. Sci. USA 76:2148 (1979).
5. F. Hirata, B. B. Corcoran, K. Venkatasubramanian, E. Schiffmann, and J. Axelrod, Chemoattractants stimulate degradation of methylated phospholipids and release of arachidonic acid in rabbit leukocytes, Proc. Natl. Acad. Sci. 76:2650 (1979).

6. J. T. O'Flaherty, H. J. Showell, E. L. Becker, and P. A. Ward, Neutrophil aggregation and degranulation. Effect of arachidonic acid, Am. J. Pathol. 95:433 (1970).
7. P. H. Naccache, H. J. Showell, E. L. Becker, and R. I. Sha'afi, Arachidonic acid induced degranulation of rabbit peritoneal neutrophils, Biochem. Biophys. Res. Commun. 87:292 (1979).
8. E. J. Goetzl and F. F. Sun, Generation of unique mono-hydroxy-eicosatetraenoic acids from arachidonic acid by human neutrophils, J. Exp. Med. 150:406 (1970).
9. G. M. Bockoch and P. W. Reed, Inhibition of the neutrophil oxidative response to a chemotactic peptide by inhibitors of arachidonic acid oxygenation, Biochem. Biophys. Res. Commun. 90:481 (1979).
10. T. P. Ciaraldi, O. G. Kolterman, J. A. Siegel, and J. M. Olefsky, Insulin-stimulated glucose transport in human adipocytes, Am. J. Physiol. 236:E621 (1979).
11. D. A. Bass, T. A. Gonwa, P. Szejda, S. Cousart, L. R. DeChatelet, and C. E, McCall, Eosinopenia of acute infection: Production of eosinopenia by chemotactic factors of acute inflammation. J. Clin. Invest. 65:1265 (1980).
12. R. Franson and M. Waite, Relation between calcium requirement. substrate charge, and rabbit polymorphonuclear leukocyte phosphilipase A_2 activity. Biochemistry. 17:4029 (1978).
13. E. J. Goetzl, A. R. Brash, A. I. Tauber, J. A. Oates, and W. C. Hubbard, Modulation of human neutrophil function by monohydroxyeicosatetraenoic acids, Immunology. 39:491 (1980).
14. R. J. Flower, Drugs which inhibit prostaglandin biosynthesis. Pharmacol. Rev. 26:33 (1974).
15. M. Blecher, Effects of insulin and phosphalipase A on glucose transport across the plasma membrane of free adipase cells, Biochim. Biophys. Acta. 137:557 (1967).
16. G. Dietze, M. Wicklayr, I. Bottger, and L. Mayer, Insulin action on glucose uptake into skeletal muscle: Inhibition of endogenous biosynthesis of prostaglandins. FEBS Lett. 92:294 (1978).

DISCUSSION

Edited by Charles E. McCall, M.D.

"Arachidonic Acid and Hexose Transport in Human Polymorphonuclear Neutrophils (PMNs)"

Presented by Dr. Charles E. McCall

Question 1 - Dr. Esmann

These are very important experiments giving credit to the thought of a carrier-mediated glucose transport. What was the position of the radiolabel and what was the method of measuring glucose uptake? Have you tried to distinguish between transport and phosphorylation by the use of 3-O-methyl glucose?

Answer - Dr. McCall

The radiolabel is in the number one carbon position but we have found little loss of the label. The method we employ to measure hexose transport is the uptake of deoxyglucose which necessarily is a measurement of both carrier-facilitated transport and phosphorylation. At present we cannot completely distinguish between the two but we have no data to indicate that hexokinase is the rate limiting step. We have been unable to employ 3-O-methyl glucose since concentrations as high as 10 mM do not compete for the glucose carrier in the human PMN.

Question 2 - Dr. Borregaard

What is your indication that the cells are alive? They are incubated in glucose-free medium and it appears possible that glycolysis is blocked by 2-deoxyglucose.

Answer - Dr. McCall

Only tracer amounts of deoxyglucose are employed and these are insufficient to inhibit glycolysis or reduce ATP during the time of incubation. The cells are greater than 95% viable as determined by Trypan blue exclusion.

Question 3 - Dr. Schell-Frederick

Is the effect of arachidonic acid on deoxyglucose transport dependent on the presence of calcium in the medium?

Answer - Dr. McCall

No, the absence of extracellular calcium and/or magnesium with the addition of EDTA does not alter hexose transport. Similar findings have been reported in the hexose transport into fat cells when the fat cells are stimulated by insulin. This course does not exclude a possible role of intracellular calcium in hexose transport.

Question 4 - Dr. Romeo

With some of the stimuli you might cause alteration in the plasma membrane after a 60 minute incubation time. Can you exclude the possibility that the increase in cell associated counts were not due to trapping of the isotope in pinocytic vacuoles?

Answer - Dr. McCall

We have observed no significant uptake of radiolabeled sucrose or radiolabeled lysine under the same conditions. We feel this negates the concept of significant trapping of isotopes during the procedure. Secondly, the reaction is stereospecific for d-glucose.

Question 5 - Dr. Basford

Dr. Lynn and others have shown that insulin causes a release of O_2^- and H_2O_2 from adipocytes. Does the addition of arachidonate cause the release of either these reduced O_2 products from either adipocytes or neutrophils?

Answer - Dr. McCall

We have found that arachidonate acts as a "sink" for O_2^- and therefore we cannot measure the effect of arachidonate on superoxide production.

TUFTSIN AND SUBSTANCE P AS MODULATORS OF PHAGOCYTE FUNCTIONS

Zvi Bar-Shavit and Rachel Goldman

Department of Membrane Research
The Weizmann Institute of Science
Rehovot 76100, Israel

A. TUFTSIN

The basic tetrapeptide tuftsin, Thr-Lys-Pro-Arg, was described by Najjar and his collaborators as the entity responsible for the phagocytosis-enhancing activity of leukophilic γ-globulins. Since its discovery and synthesis, tuftsin was shown to affect many of the known functions of phagocytic cells. Tuftsin enhances phagocytosis, bactericidal activity, motility, cytotoxicity towards tumor target cells and immunogenic functions of macrophages. The tetrapeptide occupies the location 289-292 (Fc-region) in the amino acid sequence of all IgG subclasses and is released from the protein by a sequential two-step mechanism, involving a splenic carboxypeptidase and a surface leukokininase found on polymorphonuclear leukocytes (PMNL). The tetrapeptide exerts its activity only after its complete release from its parent molecule. (For reviews, see 1 and 2.)

Since tuftsin, a natural relatively-simple and biodegradeable peptide, enhances macrophage function, it is regarded as a potential candidate for clinical use. The manifold expression of its interaction with phagocytes motivated us to study the affinity, specificity and mechanism of tuftsin action.

Tuftsin-Binding Sites

In an attempt to understand the mechanism by which tuftsin augments the biological activities of phagocytes, and to assess the specificity of action both in terms of structural requirements and cell types involved, it is necessary to determine the binding c characteristics of tuftsin. Direct probing of the binding site became feasible after the synthesis of radiolabeled tuftsin. Stabinsky *et al.*[3] have synthesized [^{3}H]tuftsin by incorporating [^{3}H]Arg into the sequence during synthesis. Using this preparation, they were able to quantify the binding sites for tuftsin on human PMNL ($5x10^4$/cell) and monocytes (10^5/cell), and to determine the respective dissociation constants ($K_D=1.3x10^{-7}M$) and ($K_D=1.25x10^{-7}M$).[3] Further studies have shown that [^{3}H]tuftsin binds to thioglycollate-stimulated macrophages in a rapid and saturable fashion, with an equilibrium dissociation constant of $5.3x10^{-8}M$ (Fig. 1A). The number of binding sites per cell amounts to $7.2x10^4$. Binding competition studies with various analogs revealed that a rather high degree of specificity is involved in the binding of the tetrapeptide (Fig. 1B). [^{3}H]Tuftsin was shown to bind to all the cells of the granulocyte-macrophage cell lineage tested : human PMNL;[3] human peripheral blood monocytes;[3] mouse normal and stimulated macrophages;[4] mouse peripheral blood monocytes;[5] bone marrow cells differentiated *in vitro* into mono-

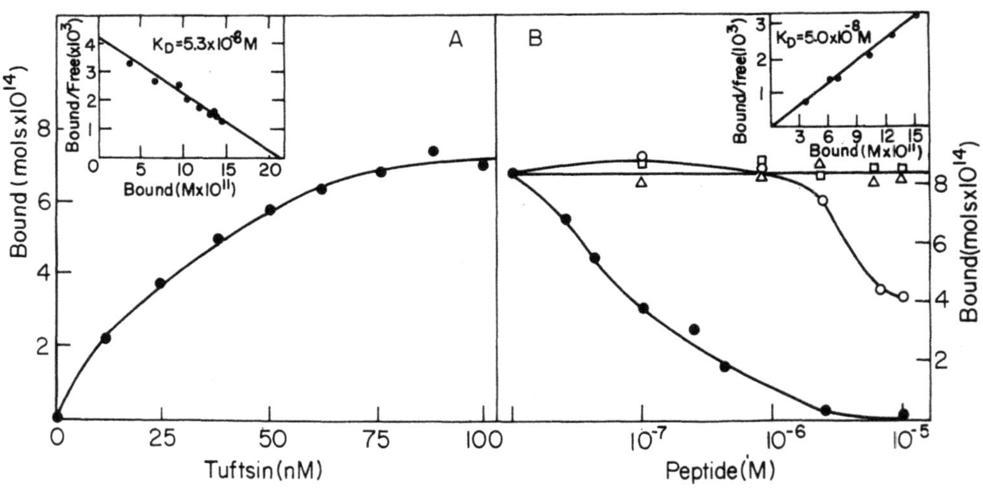

Fig. 1. A - Specific binding of [^{3}H]tuftsin to thioglycollate-stimulated macrophages. (Insert - Scatchard analysis of the binding data.) B - Effect of unlabeled tuftsin (●—●) and its analogs (O—O, [D-Arg^4]tuftsin; Δ—Δ, [N-Acetyl-Thr^1]tuftsin; □—□, Thr-Lys-Pro) on the binding of [^{3}H]-Tuftsin to thioglycollate-stimulated macrophages. (Insert - Scatchard analysis of binding inhibition by unlabeled tuftsin.) (Adopted from Bar-Shavit *et al.*[4])

nuclear phagocytes (BMDMP);[5] as well as to a macrophage-like cell line P388D1.[5] [^{3}H]Tuftsin failed to bind to erythrocytes, lymphocytes and fibroblasts.[3,4] All the cells that were positive in the binding assay bound comparable amounts of tuftsin per cell, except for the BMDMP which bound about three times more. Binding was always accompanied by an augmentation of the phagocytic response.[4,5] The dose-dependence of tuftsin-binding and its phagocytic stimulatory activity reveal that both processes reach saturation at similar tuftsin concentrations.

Tuftsin-Mediated Modulation of Cyclic Nucleotide Levels

Intracellular cyclic nucleotide levels have been shown to influence the phagocytic response of mammalian leukocytes. cAMP and pharmacological agents that elevate intracellular cAMP levels, inhibit phagocytosis,[6] whereas cGMP and cholinergic agonists stimulate it.[7] We have, therefore, explored the possibility that tuftsin-binding may affect the intracellular levels of cAMP and cGMP. As shown in Fig. 2, tuftsin exerts a dose-dependent increase

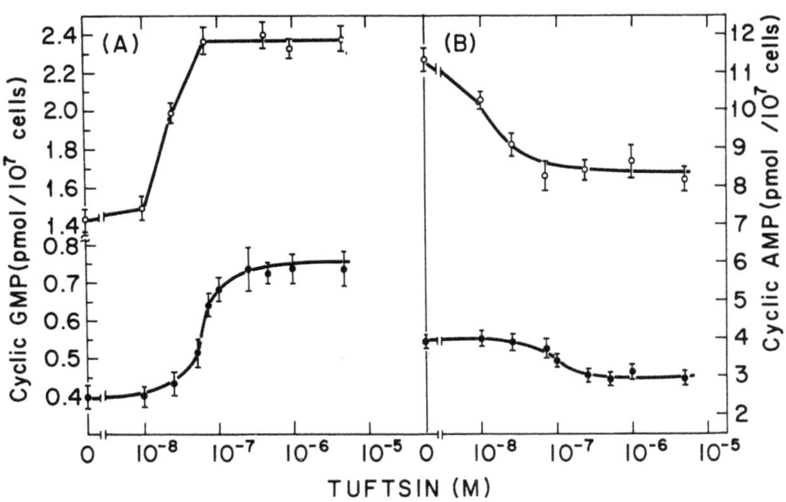

Fig. 2. The dependence of intracellular levels of cGMP (A) and cAMP (B) in normal human PMNL (●) and thioglycollate-stimulated mouse peritoneal macrophages (O) on tuftsin concentration. Tuftsin at the indicated concentrations was incubated with cells for 20 min at 37°C and cyclic nucleotides were then assayed. (From Stabinsky *et al.*[8])

Table 1. The Effect of Exogenous Compounds That Affect the Intracellular Levels of Cyclic Nucleotides on the Phagocytosis of Heat-Killed Yeast by Macrophages.

Compound	Concentration (μM)	Yeast/100 Macrophages	P	%Δ
-	-	321±14	-	-
8-Br cAMP	1.0	237± 9	<0.01	-26
8-Br cGMP	0.1	418±16	<0.02	+30
8-Br cGMP	1.0	467±24	<0.01	+45
PMA	0.015	456±26	<0.02	+42
Carbamyl Choline	10	421±16	<0.01	+31
Tuftsin	0.1	528±48	<0.01	+62

Macrophage monolayers were incubated for 5 min with 0.4 ml of either PBS (control) or the specified compound at 37°C. Yeast in PBS (100 μl) were then added (final concentration $2x10^7$/ml) for another 30 min of incubation at 37°C. (From Stabinsky *et al.*[8])

in intracellular cGMP levels (80-90%) and a decrease in cAMP levels (20-25%) in both human PMNL and thioglycollate-stimulated mouse peritoneal macrophages. The half maximal response of human PMNL ($8\text{-}9x10^{-8}$M) and thioglycollate-stimulated macrophages ($3\text{-}4x10^{-8}$M) reflects the difference in the K_D found for both cell types and is in accord with the concentrations required for elicitation of half the maximal phagocytic stimulation response. Thus, for maximal response to tuftsin, there is a need for a high degree of saturation of the tuftsin-binding sites. The changes in the levels of cyclic nucleotides are detectable already after 4 min of interaction with tuftsin, and persist in the presence of the peptide for at least 60 min.[8]

In Table 1 it is shown that intracellular changes in cyclic nucleotide levels, brought about by exogenous known modulators of the latter, can in fact affect the phagocytic response of mouse peritoneal macrophages. The augmentation of phagocytosis by agents that increase cGMP levels is comparable to that brought about by tuftsin.

Tuftsin-Mediated Redistribution of Intracellular Ca^{+2}

In order to assess the effects of tuftsin on gross changes in Ca^{+2} distribution in phagocytic cells, we examined the influx and outflux of $^{45}Ca^{+2}$ in the presence and absence of tuftsin.[8] No changes in the influx of $^{45}Ca^{+2}$ could be detected in both human polymorphonuclear leukocytes and thioglycollate-stimulated mouse macrophages upon interaction with tuftsin. The initial outflux of $^{45}Ca^{+2}$ (first 5 min) from preloaded phagocytes was, however, significantly faster in the presence of tuftsin, as compared to controls (Fig. 3). After 20 min of incubation, there was no difference between the amount of cell-associated $^{45}Ca^{+2}$ in control and tuftsin-treated cells, suggesting that tuftsin enhances release of $^{45}Ca^{+2}$ only from readily exchangeable intracellular pools. Structurally related analogs of tuftsin, that neither bind specifically nor augment phagocytosis, have no effect on $^{45}Ca^{+2}$ efflux indicating the dependence of the effect on tuftsin-binding. The failure to detect enhanced $^{45}Ca^{+2}$ influx and the observation that $^{45}Ca^{+2}$ efflux is in fact augmented do not exclude the possibility that increase in $^{45}Ca^{+2}$ exchange rate reflects transient elevation in free $^{45}Ca^{+2}$ in the cells, due to release from cell membrane or other storage domains.

The apparent complex regulatory network, by which the different components cAMP, cGMP and $^{45}Ca^{+2}$ affect the level of each other and

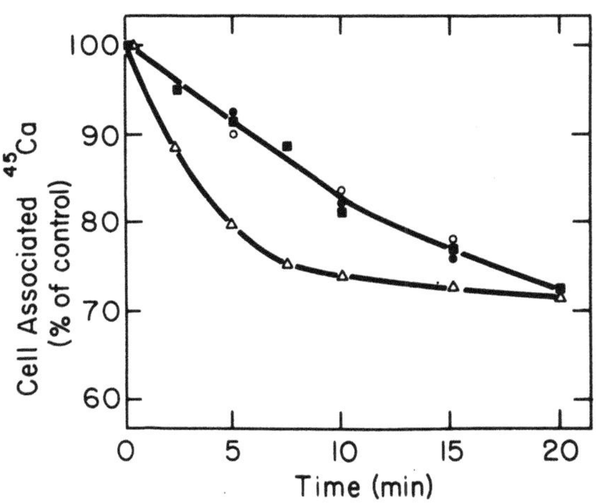

Fig. 3. Time course of the effect of tuftsin and its analogs on outflux of $^{45}Ca^{+2}$ from thioglycollate-stimulated mouse peritoneal macrophages. Macrophages (in monolayers) preloaded with $^{45}Ca^{+2}$ were incubated in the absence (■-control) or presence of the tested peptide (2.5×10^{-7}M) for the indicate time intervals. (Δ - tuftsin; O - [N-acetyl-Thr[1]]tuftsin; ● - [Des-Thr[1]]tuftsin).(Stabinsky *et al.*[8])

cellular function, makes it difficult to ascribe a separate function to either of the components in the enhancing effect that tuftsin exerts on phagocyte function. Of relevance to our findings is the report of Barthelemy *et al.*[9] showing that in PMNL there is an energy dependent resting calcium efflux and that this efflux increases upon addition of phagocytic particles. Hoffstein[10] has provided evidence suggesting that Ca^{+2} is associated with the plasma membrane of resting PMNL in concentrations sufficiently high to precipitate pyroantimonate and that these membrane-associated cations are released from sites of receptor-ligand interaction (segmental loss of Ca^{+2} in the case of particulate stimuli and circumferential loss in the case of soluble stimuli).

Cell activation by various external stimuli was shown to be associated with changes in the concentrations of cAMP, cGMP and Ca^{+2} (for review see 11). Depending on the system being considered, Ca^{+2} was demonstrated to increase or decrease cAMP and cGMP levels by activating or inhibiting adenylate cyclase and guanylate cyclase, respectively, and of phosphodiesterase. cAMP can in turn affect the cytoplasmic concentration of Ca^{+2} either by altering the inward Ca^{+2} flux or by releasing the cation from intracellular reservoirs. cGMP could be linked to cAMP through interactions with Ca^{+2}, as there is evidence suggesting that guanylate cyclase is regulated by the intracellular distribution or concentration of Ca^{+2}.[11] Furthermore, the ionophore, A23187, which promotes the transport of divalent cations across biological membranes, and increases intracellular Ca^{+2}, can promote increases in cGMP levels in various tissues.[11]

B. SUBSTANCE P

The amino acid sequence of the undecapeptide Substance P (SP), Arg-Pro-Lys-Pro-Gln-Gln-Phe-Phe-Gly-Leu-Met-(NH_2), was established by Chang *et al.* in 1971.[12] The sequence of the N-terminal tetrapeptide has certain features in common with that of tuftsin, i.e. two positively-charged amino acids separated by a proline residue. This structural similarity motivated us[13] to explore the possibility that SP and its N-terminal tetrapeptide may exhibit tuftsin-like activity.

Augmentation of Phagocytosis Mediated by Substance P

The effect of SP and related peptides on the phagocytic activity of mouse peritoneal macrophages and human PMNL is shown in Fig. 4. The phagocytosis of heat-killed yeast by macrophages was stimulated in a dose-dependent manner by SP and its N-terminal tetrapeptide, the maximal stimulation being reached at a peptide concentration of 40 nM. The C-terminal hepta- and octapeptide fragments of SP, as well as Arg-Pro and Lys-Pro, did not show statistically significant enhancing effects. The same pattern was obtained, using sheep erythrocytes coated with IgG (EIgG). A higher dose of SP and the N-terminal tetrapeptide was needed for expression of

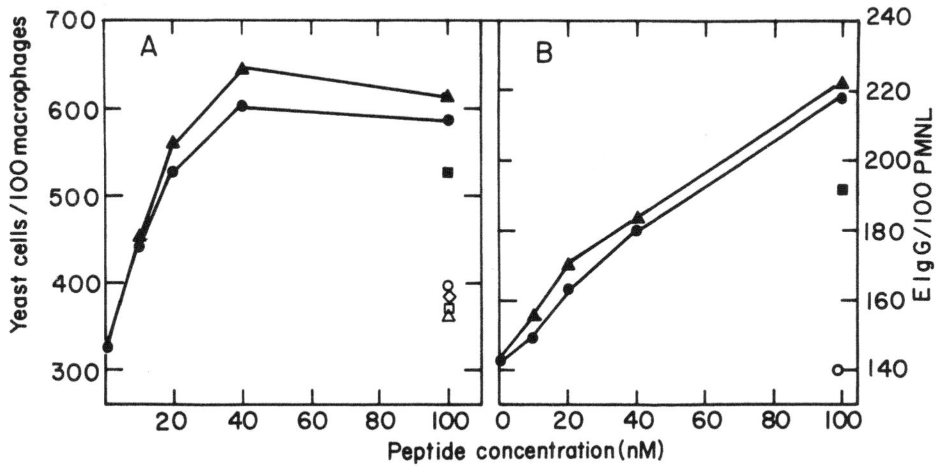

Fig. 4. The effect of SP and related peptides on the phagocytosis of A - Yeast cells by normal macrophages and B - EIgG by human PMNL. ●, SP; ▲, N-terminal tetrapeptide; ■, tuftsin; O, C-terminal heptapeptide; □, Arg-Pro; ◇, Lys-Pro. (From Bar-Shavit *et al.*[13])

half the maximal enhancing effect on PMNL phagocytosis than needed for mouse macrophages (compare 4A and 4B). This is in line with the observed differential affinity and response to tuftsin exhibited by these two cell types.

Substance P Binding Sites

Competitive binding experiments revealed that in both cell types, mouse macrophages and human PMNL, SP and its N-terminal tetrapeptide are competitive inhibitors of [^{3}H]tuftsin-binding, while the C-terminal heptapeptide was non-inhibitory. SP and its N-terminal tetrapeptide have a comparable binding constant to macrophages, as that observed with tuftsin.[13] Using ^{125}I-labeled SP, we have recently been able to demonstrate directly specific SP-binding to macrophages and to assess the K_D (Bar-Shavit *et al.*, unpublished).

Duality of Function of Substance P

SP is widely distributed in the vertebrate nervous system.[14] In the periphery, SP is recognized for its hypotensive, vasodilatory and smooth muscle-contracting properties, whereas in the central nervous system, SP is suggested to play a role in sensory neurotransmission (for review see 15).

Conservation of the sequence of the C-terminal six or seven amino acids is required for biological activity and the C-terminal heptapeptide can replace SP in a variety of biological assay systems.[16,17] Recent work has shown that the N-terminal tetrapeptide of SP is important for the stability of the peptide and protects the molecule against degradation by aminopeptidases.[18] The SP molecule can be cleaved by a high-affinity post-proline endopeptidase, which splits it into an N-terminal tetrapeptide and a C-terminal heptapeptide.[19]

The fact, that phagocytic cells possess receptors which recognize SP via its N-terminal tetrapeptide and respond to SP-binding in augmentation of biological activities [phagocytosis (Fig. 4) and enhanced motility (not shown)], is highly suggestive of a genuine dual function of the SP molecule. The possible action on the level of phagocytic cells is of special interest in view of the notion that SP might play a role not only in neurotransmission but also in inflammatory processes of neural origin.

CONCLUDING REMARKS

Expression of tuftsin-binding sites has been shown to be restricted to phagocytic cells of the granulocyte-macrophage cell lineage. In these cells, tuftsin-binding appears to initiate a comparable sequence of events shown to be activated by various activation-secretion stimuli. The fact, that tuftsin binding modulates intracellular Ca^{+2} distribution and cyclic nucleotide levels, parameters that have an important task in general in regulation of cellular activity, affords probably the best explanation to the ability of the peptide to affect the highly diverse functions that macrophages have been described to perform. Since SP appears to bind to the same binding sites as tuftsin and has been shown to affect at least one function in a pattern comparable to that observed with tuftsin, it is plausible that we will find that the same mechanism of action is involved.

We have recently found that neurotensin is endowed with a phagocytosis enhancing activity on macrophages and that the activity is probably related to the presence of the tripeptide Lys-Pro-Arg in its sequence (Bar-Shavit, Blumberg, Terry and Goldman, unpublished results). Cells of the granulocyte-macrophage lineage and their precursors have been shown to possess receptors to a variety of hormones, neurotransmitters and neurohormones. Ligand-receptor interaction results in a variety of functional changes, such as changes in the proliferative pattern of bone-marrow precursor cells[20] or those involved in enhancement of phagocytosis, secretion of lysosomal enzymes and enhanced motility.[6,7] To the various receptors described, i.e. H_2-histamine,[21] β-adrenergic,[22] nicotinic cholinergic,[23] muscarinic cholinergic[24] and insulin,[25] we now add a receptor(s) for biologically active peptides. Although it may appear to share a mechanism of transduction similar to that of the muscarinic cholinergic receptor, the exact mechanism and the possible interaction between the receptors remain to be explored.

REFERENCES

1. V. A. Najjar, Biological and biochemical characteristics of the tetrapeptide tuftsin, Thr-Lys-Pro-Arg, in:"Macrophages and Lymphocytes : Nature, Functions and Interaction", Part A, R. Escobar and F. Friedman, eds., Plenum Press, New-York and London, pp.131-147 (1980).
2. M. Fridkin and P. Gottlieb, Tuftsin, Thr-Lys-Pro-Arg : Anatomy of an immunologically active peptide, Mol.Cell.Biochem., in press (1980).
3. Y. Stabinsky, P. Gottlieb, V. Zakuth, Z. Spirer and M. Fridkin, Specific binding sites for the phagocytosis stimulating peptide tuftsin on human polymorphonuclear leukocytes and monocytes, Biochem.Biophys.Res.Commun. 83:599-606 (1978).
4. Z. Bar-Shavit, Y. Stabinsky, M. Fridkin and R. Goldman, Tuftsin-macrophage interaction : Specific binding and augmentation of phagocytosis, J.Cell.Physiol. 100:55-62 (1979).
5. Z. Bar-Shavit, I. Bursuker and R. Goldman, Functional tuftsin binding sites on macrophage-like tumor line P388D1 and on bone marrow cells differentiated in vitro into mononuclear phagocytes, Mol.Cell.Biochem. 30:151-155 (1980).
6. J.D. Cox and M.D. Karnovsky, The depression of phagocytosis by exogenous cyclic nucleotides, prostaglandins, and theophylline, J.Cell.Biol. 59:480-490 (1973).
7. L. J. Ignarro and S.Y. Cech, Bidirectional regulation of lysosomal enzyme secretion and phagocytosis in human neutrophils by guanosine 3',5'-monophosphate and adenosine 3',5'-monophosphate, Proc.Soc.Exp.Biol.Med. 151:448-452 (1976).
8. Y. Stabinsky, Z. Bar-Shavit, M. Fridkin and R. Goldman, On the mechanism of action of the phagocytosis-stimulating peptide tuftsin, Mol.Cell.Biochem. 30:71-77 (1980).
9. A. Barthèlemy, R. Paridaens and E. Schell-Frederick, Phagocytosis-induced ^{45}calcium efflux in polymorphonuclear leukocytes, FEBS Letters 82:283-287 (1977).
10. S. T. Hoffstein, Ultrastructural demonstration of calcium loss from local regions of the plasma membrane of surface stimulated human granulocytes, J.Immunol. 123:1395-1402 (1979).
11. M. J. Berridge and P. Rapp, Cyclic nucleotides, calcium and cellular control mechanisms, in: "Cyclic 3',5'-Nucleotides : Mechanism of Action", H. Cramer and J. Schultz, eds., J. Wiley & Sons, Chapter 5, pp.65-76 (1977).
12. M. M. Chang, S.E. Leeman and H.D. Niall, Amino acid sequence of substance P, Nature New Biol. 232:86-87 (1971).
13. Z. Bar-Shavit, R. Goldman, Y. Stabinsky, P. Gottlieb, M. Fridkin, V.I. Teichberg and S. Blumberg, Enhancement of phagocytosis : A newly-found activity of substance P residing in its N-terminal tetrapeptide sequence, Biochem.Biophys. Res.Commun. 94:1445-1451 (1980).

14. T. Hökfelt, O. Johansson, J.-O. Kellerth, Å. Ljungdahl, G. Nilsson, A. Nygards and B. Pernow, Immunohistochemical distribution of substance P, in: "Substance P", U.S. von Euler and B. Pernow, eds., Raven Press, New-York, pp.117-145 (1977).
15. R. A. Nicoll, C. Schenker and S. E. Leeman, Substance P as a transmitter candidate, in: "Ann.Rev.Neurosci.", W.M. Cowan, Z.W. Hall and E.R. Kandel, eds., Vol.3, pp.227-268 (1980).
16. S. Rosell, U. Björkroth, D. Chang, I. Yamaguchi, Y.-P. Wan, G. Rackur, G. Fisher and K. Folkers, Effects of substance P and analogs on isolated guinea pig ileum, in: "Substance P", U.S. von Euler and B. Pernow, eds., Raven Press, New-York, pp.83-88 (1977).
17. R. W. Bury and M. L. Mashford, Biological activity of C-terminal partial sequences of substance P, J.Med.Chem. 19:854-856 (1976).
18. S. Blumberg and V. I. Teichberg, Biological activity and enzymic degradation of substance P analogs : Implications for studies of the substance P receptor, Biochem.Biophys.Res.Commun. 90: 347-354 (1979).
19. S. Blumberg, V. I. Teichberg, J. L. Charli, L. B. Hersh and J. F. McKelvy, Cleavage of substance P to an N-terminal tetrapeptide and a C-terminal tetrapeptide by a post-proline cleaving enzyme from bovine brain, Fed.Proc. 38:350 (1979).
20. J. W. Byron, Pharmacodynamic basis for the interaction of cimetidine with the bone marrow stem cells (CFUs), Exp.Hemat. 8:256-263 (1980).
21. J. W. Byron, Mechanism for histamine H_2-receptor-induced cell cycle changes in the bone marrow stem cell, Agents and Actions 7:209-213 (1977).
22. J. W. Byron, Evidence for a β-adrenergic receptor initiating DNA synthesis in haemopoietic stem cells, Exp.Cell Res. 71: 228-232 (1972).
23. J. W. Byron, Drug receptors and the haemopoietic stem cell, Nature (N.B.) 241:152-154 (1973).
24. J. I. Kurland, J. W. Hadden and M. A. S. Moore, Role of cyclic nucleotides in the proliferation of committed granulocyte-macrophage progenitor cells, Canc.Res. 37:4534-4538 (1977).
25. R. H. Schwartz, A. R. Bianco, C. R. Kahn and B. S. Handwerger, Demonstration that monocytes rather than lymphocytes are the insulin-binding cells in preparations of human peripheral blood mononuclear leukocytes : Implications for studies of insulin-resistant states in man, Proc.Natl.Acad.Sci.USA 72: 474-478 (1975).

THE EFFECT OF HYALURONIC ACID ON NEUTROPHIL FUNCTION IN VITRO AND IN VIVO

Per Venge, Lena Håkansson and Roger Hällgren

Departments of Clinical Chemistry and Internal Medicine
University Hospital, S-750 14 Uppsala, Sweden

INTRODUCTION

Defective neutrophil function is an established cause of increased infection propensity. In addition defects may also predispose to other diseases such as SLE (to be published) and reumatoid arthritis[1] due to an impaired elimination of immune-complexes from the circulation. Even the control of tumor-cell growth[2] or granuloma formation[3] has been suggested to involve the activities of the neutrophil granulocyte.

The underlying causes of a defective neutrophil function are multiple and the therapeutic approach in situations with an abnormal neutrophil function must consider this fact in order to be successful and beneficial to the patient. Basically the defects may affect every aspect of the neutrophil function and includes both intrinsic cellular defects and humoral influences on the cell. If possible a distinction between primary and secondary defects should also be considered. In the case of a humoral factor impairing PMN function one therapeutic approach could be the elimination of the inhibitory principle by means of plasmapheresis.[4] The normalization of an intrinsic abnormality, however, provides a much more delicate problem and today few therapeutic principles are available to accomplish this.

By coincidence we discovered for some years ago that hyaluronic acid (HA), a major constituent of the connective tissue, was able to enhance PMN phagocytosis in vitro at very low concentrations.[5,6] This discovery initiated our investigation on the effects of HA on PMN function both in vitro and in vivo and has today led us to the suggestion that HA, in some patients with defective intrinsic abnormalities, may be used to normalize and regulate the PMN function.[7]

The Effects of HA on Neutrophil Function in Vitro

In Fig. 1 the different parts of the neutrophil function are depicted on which HA has proved to be operative. PMN adherence was studied by means of the nylon-wool method. Heparinized whole blood was preincubated for 30 minutes at room temperature with various concentrations. Adherence of the PMN in the whole blood was increased with optimal HA-concentrations of 10-50 μg/l. At higher concentrations (100-500 μg/l) adherence returned to values as in the absence of HA.

PMN-migration was measured in a modified Boyden-chamber by means of the leading-front method. In these experiments heparinized whole blood was preincubated with and without HA as above for adherence but after preincubation the PMN:s were isolated and washed free of plasma. Chemotactic movement towards a gradient of either casein[5] or activated serum (unpublished) was invariably stimulated with optimal HA-concentrations varying from 5-100 μg/l. Also random migration was stimulated[5] at these concentrations although not so

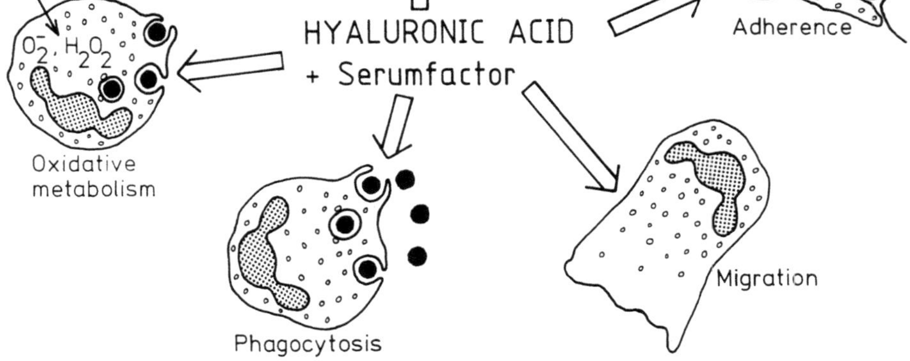

Fig. 1. Neutrophil functions affected by hyaluronic acid.

consistently as for chemotaxis. At higher concentrations i.e. at 100-500 μg/l the stimulatory effect of HA was reversed to levels in the absence of HA.

Phagocytosis was measured kinetically by means of a previously described method[8] which enables the distinction between the Fc-receptor dependent and serum-opsonin (C3b, C4b) dependent particle uptake. Whole heparinized blood was preincubated with HA as for migration experiments. The PMN:s were washed free from plasma and further enriched to a purity of 85-95 % PMN:s by means of sedimentation at +4 °C at a high cell-density, The rate of phagocytosis of both IgG-coated and serum opsonin-coated (unpublished) particles was enhanced with optimal concentrations varying from 10-500 μg/l.

Phagocytosis of IgG-coated particles was also evaluated in whole blood after the preincubation with HA. Morphological destinction was made between PMN:s having more than 10 particles associated per cell and those PMN:s having 1-10 particles/cell. In addition the percentage of PMN:s having no particles associated was calculated. In whole blood without the addition of HA on average 15 per cent of the PMN:s had no particles associated to the cell. This population of PMN:s may be identical to the Fc-receptor negative population which has been described recently.[9] After preincubation with HA, 1-50 μg/l, the proportion of these non-responsive PMN:s decreased to on average 7 per cent which could indicate that these tentative Fc-receptor negative PMN:s had become receptor-positive. Also the percentage of PMN:s with more than 10 particles/cell increased significantly at the expense of those having 1-10 particles/cell which means that phagocytosis was stimulated in this system by HA either due to "receptor-enhancement" or due to an enhanced internalization by other mechanisms.

Enhancement of the oxidative metabolism was also achieved after incubation with HA at concentrations similar to those above.[5] It should be emphasized that the oxidative metabolism was measured indirectly by the production of luminol-enhanced chemiluminescence during phagocytosis of either opsonized zymosan or latex-particles. Since there in general appears to be a very close relationship between the production of chemiluminescence and the phagocytic activity of the PMN:s (unpublished) the observed enhanced production of chemiluminescence in these experiments could be a phenomenon secondary to the HA-enhanced phagocytic activity. The experiments do, however, support the suggestion that the activation of the PMN-phagocytosis brought about by HA is an active process and not just a passive phenomenon at the membran level facilitating the adherence of the particles.

From the above review it appears that HA has the capacity to affect virtually every aspect of the PMN function. To accomplish all these effects it seems reasonable to suggest that HA may affect some very basic mechanisms within the cell. One clue to our understanding of the HA-effect may be the finding of an increased ATP-concentration within resting PMN:s exposed to HA in whole blood. This means that the PMN:s may be furnished with a higher energy-reserve in the form of ATP either by the accomplishment of an increased synthesis of ATP or due to a block in the ATP-catabolization. Since ATP is the energy-source for most if not all PMN-activities a blocked utilization of ATP would be expected to lead to a decrease in activity which is the opposite of our actual findings and therefore a less likely explanation. An increased synthesis of ATP induced by HA, however, has gained some support by our recent findings that HA in whole blood triggers resting PMN:s to an increased glucose-metabolization (unpublished). This effect on glucose-metabolization was induced at similar HA-concentrations as above and started almost immediately (1-2 minutes) after the addition of HA to whole blood. For comparison it should be noted that, under identical experimental conditions, an increased functional activity of the PMN:s was obtained first 10-30 minutes after the HA-addition. Our preliminary interpretation of these data is therefore that HA by some mechanism triggers glucose-metabolization which eventually leads to an increased cellular ATP-concentration and that this higher ATP-level gives rise to an enhanced functional capacity of the PMN irrespective of what function is tested for.

As indicated in figure 1 there seems to be an absolute requirement for a co-factor to obtain stimulation of PMN-function by HA. Most of our present data suggest that this co-factor is present in fresh serum and that it is destroyed by heat-treatment (56° 30 min). The molecular structure of HA which is responsible for the described effects on PMN:s is not entirely clear. In all the above experiments HA of a molecular size of 2×10^{6} has been used. The optimal molar concentration of this preparation has been approximately 5-50 pmol/l. Limited degradation of HA has not destroyed the activity and in preliminary experiments even a tetrasaccharide of HA has been found to be active at a similar molar concentration i.e. at a concentration of about 10 ng/l. From these figures it can also be calculated that the number of HA-molecules required per PMN to achieve optimal effect is about 500.

The Effects of Hyaluronic Acid on Neutrophil Function in Vivo

To investigate the possibility that HA might have effects on neutrophil function in vivo, HA was injected subcutaneously to healthy volunteers.[5] Blood was drawn every or every second day for at least one week and PMN function was tested with respect to phagocytosis. Also the production of chemiluminescence during phagocytosis and the intracellular content of ATP was tested at some occasions. PMN function was enhanced after HA administration and the effect maintained for almost one week after the injection of 10 mg HA.

More recently the effect on migration has been measured and also found to be enhanced after HA administration (unpublished). No side effects were observed either locally at the site of injection or systemically. These results clearly indicated that HA by some mechanism affected PMN function in vivo in normals. The next step was to test the possibility that HA could also enhance PMN function in patients with an abnormal PMN function.[5,7] Until now 15 patients with undue susceptibility to bacterial infections and with defective neutrophil functions have received subcutaneous injections of HA, 5-20 mg once a week. In 13 of these patients an enhanced function with respect to phagocytosis has been recorded after either 5, 10 or 20 mg HA. The optimal dose varied between individuals. In one extremely obese patient and in one child no effect was seen with either of these doses. Higher or lower doses, respectively, were not given to these two patients. Three patients have received HA once a week for more than 6 months and for two of these patients the clinical situation has dramatically improved. One of these patients suffers from a severe chronic bronchitis with several febrile periods every months. Since the start of HA-therapy the number of acute exacerbations has been reduced considerably and the patient who previously had been hospitalized on average two periods every month has not been hospitalized since the commencement over one year ago of the injections. The other patient has since seven years a chronic, deep leg-ulceration which has given rise to repeated severe infectious complications such as sepsis and meningitis and also to a more or less chronic fever. The last 1 1/2 years before HA-theraphy she had been confined to bed and hospitalized. Since the start of HA-injections two years ago no major infectious complications have been noted. Therapy has been discontinued at three occasions and all three times various infectious complications including her febrilness supervened 2-3 weeks after the last HA-injection. These complications vanished again at all three occasions 1-2 weeks after the therapy had been reinstituted. The other patients have received HA for two weeks up to three months. In several of these patients clinical improvement has been noted during HA-therapy. The course of the disease of patients with undue susceptibility to bacterial infections is, however, often highly unpredictable. With the possible exception,therefore, of the two patients described in some detail above , it should be clearly stated that a definitive and positive therapeutic effect of HA has not been proven by this preliminary clinical trail. Firm conclusions as to the possible beneficial effect of HA in therapy of such patients must await regular controlled clinical trials.

CONCLUSION

HA affects neutrophil function both in vitro and in vivo at picomolar concentrations. The present data suggest that the activity of HA is directed against some very basic metabolic mechnisms in the cell involving the increased synthesis of ATP and increased glucose metabolization. A preliminary clinical trail suggests that subcutaneous injections of HA may be used in the therapy of patients with undue suceptibility to bacterial infections but firm conclusions as to this point has to await regular controlled clinical trials.

ACKNOWLEDGEMENTS

This work was supported by grants from Fortia, Pharmacia, the Swedish Medical Research Council, Kung Gustaf V:s 80-års fond.

REFERENCES

1. R. Hällgren, L. Håkansson, and P. Venge, Kinetic studies of phagocytosis. I. The serum independent particle uptake by PMN from patients with rheumatoid arthritis and systemic lupus erythematosus. Arthritis and Rheumatism 21:107 (1978).
2. R. A. Clark, and S.J. Klebanoff, Neutrophil-mediated tumor cell cytotoxicity: Role of the peroxidase system. J. Exp. Med. 141:1442 (1975).
3. A. Segal, The function of the neutrophil leukocyte and its clinical relevance, in: "Blood Leukocytes. Function and use in therapy," C. F. Högman, K. Lindahl-Kiessling and H. Wigzell, ed., Almqvist & Wiksell International, Stockholm (1977).
4. P. Venge, G. Artursson, and C. Högman, Serum inhibitors of PMN migration in severly burned patients. Plasmapheresis a therapeutical possibility?, in: "Blood Leukocytes. Function and use in therapy," C. F. Högman, K. Lindahl-Kiessling and H. Wigzell, ed., Almqvist & Wiksell, Stockholm (1977).
5. L. Håkansson, R. Hällgren, and P. Venge, Regulation of granulocyte function by hyaluronic acid. In vitro and in vivo effects on phagocytosis, locomotion and metabolism. J. Clin. Invest. 66:298 (1980).
6. L. Håkansson, R. Hällgren, and P. Venge, Effect of hyaluronic acid on phagocytosis of opsonized latex particles. Scand. J. Immunol. 11:649 (1980).
7. L. Håkansson, R. Hällgren, P. Venge, G. Artursson, and S. Vedung, Hyaluronic acid stimulates neutrophil function in vitro and in vivo. Scand. J. Infect. Dis. suppl. In press.
8. R. Hällgren, L. Jansson, and P. Venge, Kinetic studies of phagocytosis of IgG-coated latex particles with a thrombocyte counter. J. Lab. Clin. Med. 90:786 (1977)
9. M. S. Klempner, and J.I. Gallin, Separation and functional characterization of human neutrophil subpopulations. Blood 51:659 (1978)

DISCUSSION
Edited by Charles E. McCall, M.D.

"Influence of Hyaluronic Acid on Phagocyte Function _In Vitro_ and _In Vivo_"

Presented by Dr. P. Venge

Question 1 - Dr. Kavai
During the treatment of patients does the hyaluronic acid have antigen activity? What is the turnover time of the hyaluronic acid?

Answer - Dr. Venge
Hyaluronic acid is not antigenic. There are at present no data available on the turnover rate of hyaluronic acid after subcutaneous injection into man.

DISCUSSION
Edited by R. B. Johnston, Jr.

Basford: You have shown a very nice increase in ATP of neutrophils after exposure to hyaluronic acid. It would be nice to know the effect of hyaluronic acid on cyclic AMP and cyclic GMP.

Venge: We have not studied this.

Esmann: We have found it very difficult to increase the ATP concentration in metabolizing leucocytes, and your finding of a large increase by hyaluronic acid addition is very interesting. Normally, phosphofructokinase activity is considered rate-limiting for glycolysis and ATP formation. Do you care to speculate about the mechanism--increased production or decreased degradation?

Venge: We have no explanation for this ATP increase. As you point out, it is very difficult to increase the level by other means. An increased ATP level in the resting cell is, however, a very common finding during acute infection, at least in the burn patients we have been studying. Therefore in vivo increase may be a counterpart to what we observed in vitro after HA stimulation.

EXPRESSION OF Fc AND C3b RECEPTORS AND

INTRACELLULAR DISTRIBUTION OF BACTERIA IN RAT MACROPHAGES

Bernard Vray, Andras Hrabak and Alain Coquette

Laboratory of Microbiology and Immunology, Free University of Brussels (Belgium) and Semmelweiss University (Hungaria)

INTRODUCTION

Peritoneal rat macrophages obtained after thioglycollate stimulation are known to express C3b receptors mediating adsorbtion and ingestion of bacteria (1,2). It was also demonstrated that after one or two days in culture, the macrophages express functionnal Fc receptors able to mediate the internalization of antibodies coated bacteria.

In this report, we summarize recent observations on the C3b and Fc mediated phagocytosis by peritoneal rat macrophages after stimulation with thioglycollate. To differenciate the two types of opsonization, badges of bacteria (Micrococcus lysodeikticus) were labelled with rhodamine or fluorescein and coated either with specific antibodies or with complement. Internal distribution of pahgocytosed bacteria was observed with a microscope equipped for fluorescence.

MATERIAL AND METHODS

Male Wistar rats weighing 150 to 200 g were injected with 5 ml of thioglycollate broth (Difco) in sterile and distillated water. Four days later, the peritoneal cavity was washed with Hanks balanced salt solution without calcium and magnesium. The cellular suspension was purified by centrifugation on Percoll (Pharmacia, density 1.090).

Fluoresceine isothiocyanate (FITC) 0.5 mg or rhodamine isothiocyanate (RITC) 0.5 mg was added to an amount of bacteria in suspension (5 mg/ml) in $NaHCO_3$ buffer (pH 9, 0.1 M). Direct fluorescent labelling was obtained by incubation overnight in the dark at 4°C. After extensive washings in phosphate buffered saline, the fluoresceinated (M1-FITC) and the rhodaminated bacteria (M1-RITC) were distributed in vials and stored at -20°C until use.

C3b coated bacteria were obtained by incubating M1-RITC or M1-FITC with fresh normal rat serum (final dilution 1/5) at 37°C during 15 minutes. IgG coated bacteria were obtained by incubating M1-RITC or M1-FITC with a subagglutinating dilution (1/20) of heat inactivated serum containing anti-M. lysodeikticus specific antibodies. In some experiments, bacteria were simultaneously coated with C3b and IgG.

Kinetic assays were performed according to the method previously described (3). For microscopic observations, the macrophages were allowed to adhere to cover slip in Leighton tubes. Phagocytosis was stopped by addition of paraformaldehyde (0.5% final) at different time intervals. The non ingested bacteria were removed by lysozyme treatment (0.01% solution, 15 minutes at 37°C.

RESULTS

Adherence of the macrophages

85-95 % of the macrophages adhered to the cover slips checked by DNA-determination (4). This percentage was found to depend slightly on the number of the cells added in the Leighton tubes. In typical cases when $1\text{-}2.10^6$ cells were allowed to adhere, the amount of adhering macropahages was at least 90%.

Synergic effect

A suspension of macrophages was divided in three parts. The first one was added to IgG coated M1-FITC. The second one to C3b coated M1-FITC and the third one to IgG and C3b coated M1-FITC. The kinetic assay (Fig. 1) shows that only a weak phagocytosis was mediated through Fc receptors after one hour of incubation. The C3b coated bacteria were readily well ingested and the IgG-C3b coated bacteria are the most intensively phagocytosed.

Internal distribution of bacteria

Macrophages adhering on cover slips received a mixture of IgG coated M1-RITC and C3b coated M1-FITC. In this case, four populations of phagocytes were distinguished :

1. A non phagocytosing population (died macrophages or too low interval of time);

2. Macrophages ingesting only Fc coated M1-RITC (red bacteria);

3. Macrophages ingesting only C3b coated Ml-FITC (green bacteria);

4. Macrophages ingesting both green and red bacteria. Furthemore, in this case, we observed a separate intra-cytoplasmic distribution of green and red bacteria.

Control experiments with Fc coated Ml-FITC and C3b Ml-RITC showed the same pattern of distribution.

The addition of a mixture of C3b coated Ml-RITC and C3b coated Ml-FITC resulted in the absence of any separation of the bacteria inside the cytoplasm. The same results were obtained with IgG coated Ml-RITC and IgG coated Ml-FITC.

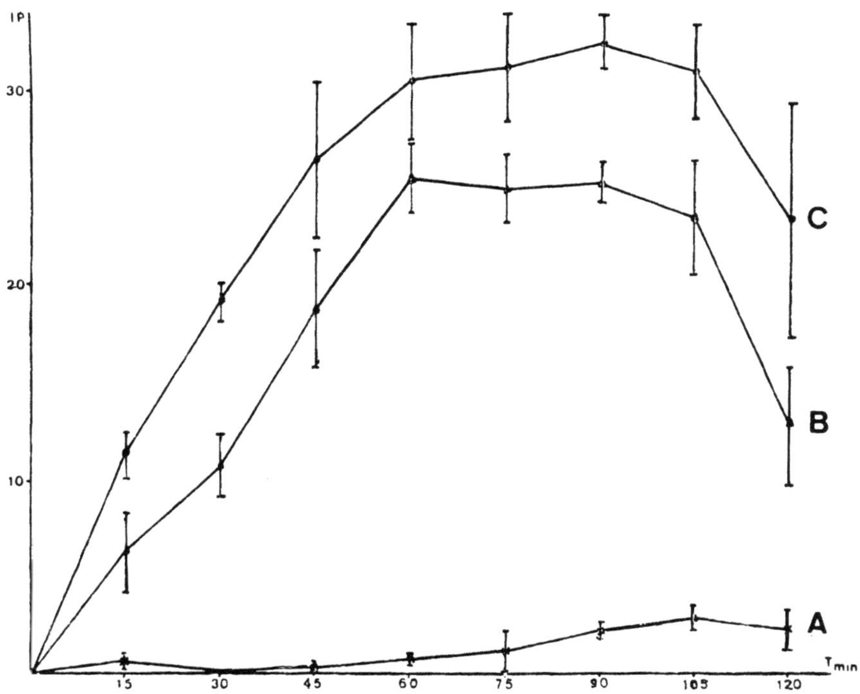

Fig. 1. Kinetic assays with fluorescent M lysodeikticus.

IP : phagocytic index
A : IgG coated bacteria
B : C3b coated bacteria
C : IgG and C3b coated bacteria.

DISCUSSION

The experiments reported here confirm that C3b coated bacteria are quickly ingested by thioglycollate stimulated peritoneal macrophages (1,2,5,6). The phagocytosis of IgG opsonized bacteria is depending on time where the phagocyte are in suspension.

With adhering macrophages, the phagocytic process occurs with C3b and IgG coated bacteria very quickly.

A surprising result is the fact that two populations of macrophages ingest IgG or C3b coated bacteria. A third one ingests the two types of opsonized bacteria and an internal distribution in two distinct areas is visible. These observations suggest that separate phagosomes are induced by area of membrane covered with one of the two receptors as schematically represented by Fig. 2 and 3. This internal distribution in independent of the label (RITC or FITC). Although the separation of the two (C3b and IgG opsonized) different populations of ingested bacteria, inside the cells was not due to the fluorescent labelling, some effect of that could not be ruled out on the phagocytic process because two populations ingesting only RITC or FITC-labelled bacteria were found in these cases too. The same opsonization with green and red bacteria results in an intracytoplasmic mixture suggesting that bacteria are bound to their receptors independently of the label.

These experiments suggest that some macrophages are able to phagocyte IgG bacteria in one area of the membrane through the Fc receptor mediation and to ingest the C3b coated bacteria in another area not necessary far from the first one. Probably each type of opsonized bacteria trigger a movement of the receptor in two specialized areas inducing phagocytosis and separate internal distribution.

Studies are in progress to elucidate the role of the cap-formation in this distribution and we intend to investigate the effect of various factors on the two types (C3b and IgG-mediated) of the phagocytosis.

REFERENCES

1. F.M. Griffin, C. Bianco and S.C. Silverstein, Characterisation of the macrophage receptor for complement and demonstration of its functionnal independance from the receptor for the Fc portion of immunoglobulin, J. Exp. Med., 141:1269 (1975)
2. F.M. Griffin, J.A. Griffin and S.C. Silverstein, Studies on the mechanism of phagocytosis II The interaction of macrophages with anti-immunoglobulin IgG-coated bone marrow-derived lymphocytes, J. Exp. Med., 144:788 (1976)

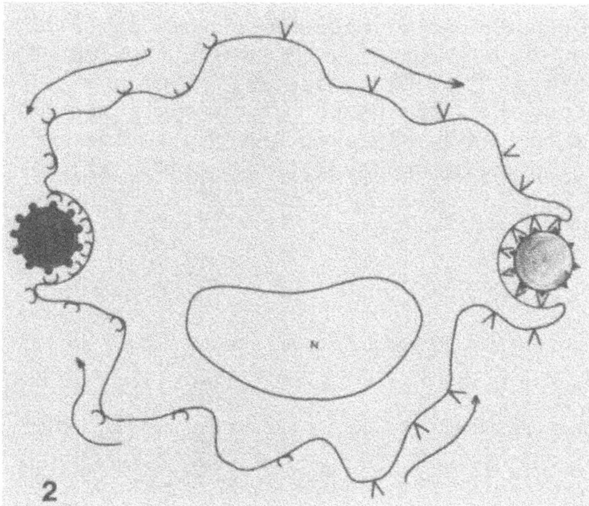

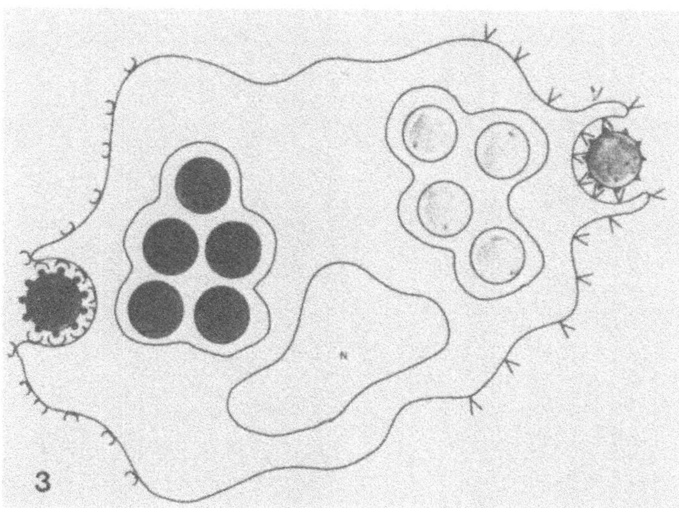

Fig . 2 and 3 : For legends, see text.
Phagocytosis of bacteria, coated with IgG or C3b and labelled with FITC or RITC.

3. B. Vray, J. Hoebeke, M. Saint-Guillain, R. Leloup and A.D. Strosberg, A new quantitative fluorimetric assay for phagocytosis of bacteria, Scand. J. Immunol., 11: 147 (1980)
4. K. Burton, A study of the conditions and mechanism of the diphenylamine reaction for the colorimetric estimation of deoxyribonucleic acid, Biochem. J. 62:315 (1956)
5. W.H. Lay and V. Nussenzweig, Receptors for complement on leukocytes, J. Exp. Med., 128:991 (1968)
6. M.Y. Reynolds, J.P. Atkinson, H.H. Newball and M.M. Frank, Receptors for immunoglobulin and complement on human and rabbit alveolar macrophages, Clin. Res., 22:427 (1974)

DISCUSSION

Edited by R. B. Johnston, Jr.

Gallily: Does the disappearance of complement receptors during prolonged in vitro culturing affect the engulfment of either complement-coated bacteria or IgG-coated ones?

Vray: In the experiments performed so far, the in vitro culture period has never exceeded 24 hours. So for the moment we have no answer to your question, but longer culturing times are under investigation.

Ka'vai: What is your opinion in regard to whether there is some collaboration between the Fc and C3b receptors on the surface of macrophages?

Vray: The phagocytic index is always higher for bacteria coated with both IgG and complement than for the complement-coated ones, and IgG-coated bacteria are poorly ingested. We think that the macrophages obtained after thioglycollate stimulation do not constitute a homogeneous population, at least for the expression of Fc and C3b receptors. Probably, in one particular subpopulation of macrophages, the ingestion of bacteria can be triggered if both C3b and Fc receptors are stimulated, explaining the enhancement of ingestion of bacteria. This hypothesis is suggested to us by the fact that the increased phagocytosis is obtained not only with IgG-complement-coated bacteria but also with a mixture of IgG-coated bacteria and complement-coated ones.

Romeo: Have you tried to first incubate the cells with bacteria to promote binding, or to increase the temperature, to induce phagocytosis? If your hypothesis of receptor involvement in the plane of the membrane is correct, it is likely that the experiment I suggest should lead to a "mosaic type" of particle internalization.

Vray: We have not yet performed the experiment that you suggest, but we think that it is a critical one. Until now, the assays have been performed at 37°C. If the hypothesis suggested is correct, we should observe a "mosaic type" of particle internalization.

Waldvogel: On one of your first slides, you showed increased phagocytosis of opsonized *M. lysodeikticus* as a function of time, followed by a decrease after further incubation. Were the bacteria also preopsonized for prolonged periods of time? Do you have any explanation for this phenomenon?

Vray: We observed a linear increase of the phagocytic process followed by a plateau due to stuffing of the macrophages with organisms. The decrease observed at the end of the kinetic assay is not yet well understood and is under further investigation.

MONOCYTE ACTIVATION BY IMMUNE COMPLEXES OF PATIENTS WITH SLE

Maria Kávai, Attila Zsindely, Ildiko Sonkoly, Aniko Bányai and Gyula Szegedi

Dept. of Pulmonary Disease, Div. of Medicine, Dept. of Biochemistry, University Medical School of Debrecen
Debrecen, 4004, Hungary

INTRODUCTION

Earlier we have observed[1] that monocytes of patients with SLE had greater FcR activity than the controls. The expression of C3b receptors was equivalent to those of controls. In most cases the patients with SLE have a higher level of immune complexes (IC) in sera than those of the controls[2]. There is a correlation between the higher FcR activity of monocytes and the IC content of the same patients' sera[3]. It is possible that the in vivo-bound IC are ingested by the monocytes, so they may activate the FcR function after their ingestion.

The characteristics of monocytes activated by IC may be the increased number of FcR[4] and the secretion of lysosomal enzymes[5,6]. In this work, we have investigated the in vitro influence of IC precipitated by polyaethylenglykol (PEG) from patients with SLE on the expression of FcR and C3bR as well as on the exocytosis of hydrolases of monocytes from healthy volunteers.

METHODS AND EXPERIMENTS

The monocytes were separated from the blood of healthy volunteers on Ficoll-Hypaque. The cells were suspended in medium 199 containing 20 mM HEPES pH 7.3. The immune complexes were precipitated from 5 ml sera of 16 patients with SLE by 3 % PEG 6000[7]. From pooled sera of healthy volunteers a precipitate as control was also obtained by 3 % PEG treatment. The composition of the PEG precipitates were analysed with enzyme linked immunosorbent assay[8] (Table 1). In the most cases the precipitates are free of C3 content or it is a very low extent.

Table 1. Composition of 2 ml PEG Precipitate Obtained from 5 ml Sera

Patients	Protein μg/ml	IgG μg/ml	IgA μg/ml	IgM μg/ml	C3 μg/ml
B.L.	4040	56.0	26.4	65.2	-
M.K.	3120	148.0	20.2	18.2	-
Sz.J.	5120	158.6	13.0	24.0	-
P.E.	232	23,4	12.8	88.2	-
K.M.	1040	35.2	18.0	23.2	-
T.S.	3520	136.4	48.4	28.8	-
Sz.J.	6000	360.0	27.5	55.5	5.0
B.L.	7040	120.0	45.0	177.0	0.25
V.M.	6000	337.7	125.0	66.0	1,0
G.J.	5360	605.0	-	172.0	5.0
R.A.	5000	250.0	35.0	206.5	-
A.Gy.	504	13.0	-	8.0	-
M.M.	540	48.0	4.0	12.0	0.3
M.I.	790	21.2	-	7.0	-
V.J.	428	17.0	-	9.0	0.1
H.S.	732	18.1	5.1	10.2	-
Norm.Sera (Pool)	315	6.2	-	28.3	-

$1\text{-}5 \times 10^5$ adherent monocytes in a monolayer were incubated with the precipitated IC for 30 min at room temperature. Then one part of the monolayers was incubated with sSRBC (EA) for 30 min at 37 °C, and the other part of them with yeast coated with complement for 60 min at 37 °C[1]. Increasing the IC content we measured the quantities of the IC which caused a 50 % inhibition on the rosette formation.

The Fig. 1 shows that very low quantity of IC inhibit the reaction of the monocytes with the sSRBC. Every IC prepared from different patients inhibited the rosette formation in the same way but in a different concentration. At the same time higher concentration of IC caused 50 % inhibition of C3b receptor mediated reaction (Fig. 2). The IC which do not contain C3 don't inhibit the uptake of sensitized yeast by monocytes (Table 2).

In the further investigations 2×10^6/ml monocytes of healthy volunteers were incubated with 100 μg IC precipitated from sera of patient with SLE for 22 hours at 37 °C. In these experiments the monocytes were separated on Ficoll-Hypaque and on microexudate-coated surface[9]. The cells (85-90 % monocytes) were resuspended in medium 199 containing 20 mM HEPES pH 7.3, 4 mM glutamin, penicillin and streptomycin.

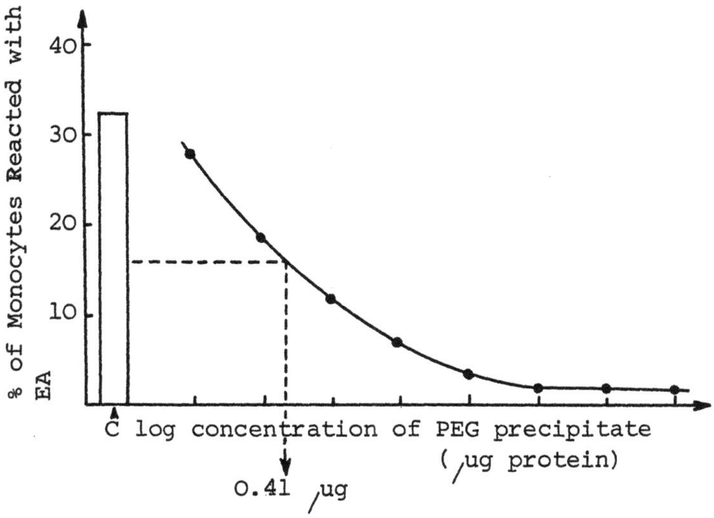

Fig. 1. Inhibiting effect of PEG precipitate on the opsonization of sSRBC (EA). (C) without inhibition.

After this period the surface receptors of the monocytes and in parallel the β-glucoronidase[10] and the lactate dehydrogenase (LDH)[11] enzyme activities were measured in the cell supernatants and in the pelletts.

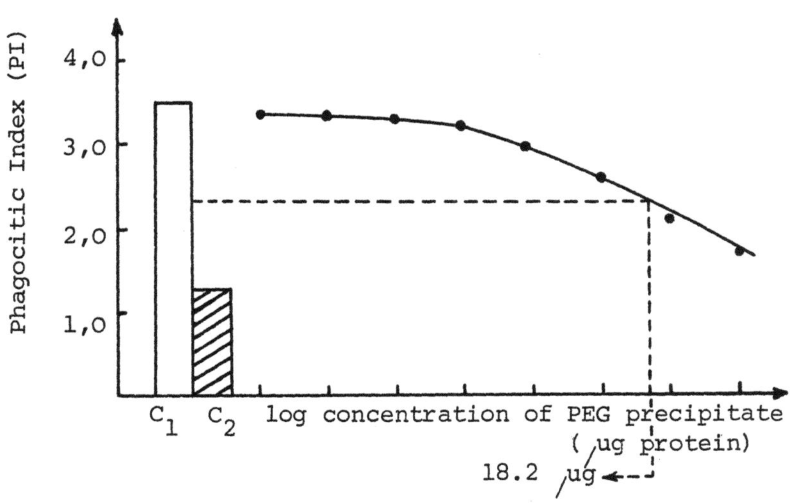

Fig. 2. Inhibiting effect of PEG precipitate on the opsonization of yeast coated with complement. (C_1) without inhibition; (C_2) without sensitization of yeast.

Table 2. Inhibition of C3 Dependent Opsonization by PEG Precipitates (n=16)

μg C3/ml Precipitate	Precipitates from the Total	μg Protein/ml Precipitate Giving 50 % Inhibition
0.00-0.25	11	no inhibition
0.25-0.5	2	40-50
1	1	18.2
5	2	6-10

The β-glucuronidase activity was assayed by a modification of the method described by Brittinger et al.[10], with 0.2 ml cell supernatant and cell suspension, respectively, and 0.1 ml saponine (2 % w/v) using 0.1 ml phenolphtalein-glucuronic acid Na-salt, 0.01 M, as substrate. After 5 hrs of incubation at 37 °C and dilution, the optical density was read at 550 nm. The LDH activity was determined by means of Warburg's optical test with 1 μmol Na-piruvate and 1 μmol NADH Na-salt in phosphate buffer, 0.1 M, pH 7.5 at 25 °C[11]. Using 0.2 ml samples, and 0.1 ml saponine (2 % w/v), the decrease of optical density was measured at 340 nm.

Table 3 summarizes enzyme activities of the cells cultivated without IC for different periods. Though the spontaneous liberation of extracellular enzyme activities were much lower after 6 hrs cultivation, in further experiments they were measured after 22 hrs, as

Table 3. Enzyme Activities of $2x10^6$ Monocytes Cultivated for Different Periods

Time Hours	β-Glucuronidase			Lactate Dehydrogenase (LDH)		
	Total	Extracell.	%	Total	Extracell.	%
6[a]	19.92 ± 5.55	1.57 ± 0.27	7.88	20.40 ± 5.51	1.90 ± 0.46	9.31
22[b]	19.76 ± 4.08	2.94 ± 0.85	14.91	21.41 ± 4.47	4.09 ± 1.06	19.06

[a]n = 3, [b]n = 7, β-Glucuronidase: $\frac{\mu mole/hour}{10^9 \text{ cells}}$, LDH: $\frac{\mu mole/min}{10^9 \text{ cells}}$, ±SD

Table 4. Activating Effect of IC on Monocytes Cultivated for 22 Hours[a]

IC /ug	M-EA %	Uptake[b] of s.Yeast	Extracell. β-Glucuronidase	Extracell. LDH
Ø	33.1	4.90	4.82	5.80
25	39.6	5.50	5.35	5.15
50	49.8	5.31	5.65	4.83
100	55.3	5.60	8.04	6.12

[a]Enzyme activity units shown in Table 3., [b](PI)

an increase of the FcR-dependent reaction caused by IC could be observed after a longer incubation period.

It is shown in Table 4., changing the IC content, the reaction of the monocytes with the sSRBC increased (M-EA %) gradually after 22 hrs, but the uptake of sensitized yeast (s.yeast) did not altered. At the same time the extracellular β-glucuronidase activity gradually increased in parallel with the rosette formation, and the LDH activity practically remained at the original low level.

In Fig. 3 it is shown the results of uptake of sensitized yeast by monocytes incubated with IC precipitated from 16 patients with SLE. The C3b receptor dependent opsonization of the monocytes did not altered after 22 hrs incubation compared to the controls (n=8).

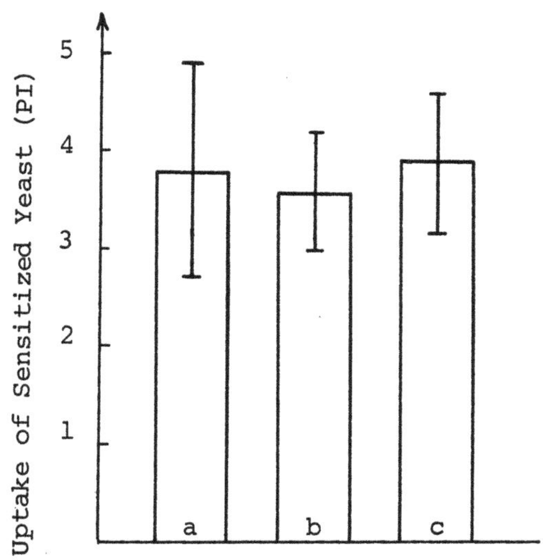

Fig. 3. Effect of IC on the C3bR dependent opsonization. (a) 100 /ug IC precipitated from sera of 16 patients; (b) 100 /ug protein precipitated from sera of healthy volunteers; (c) without IC

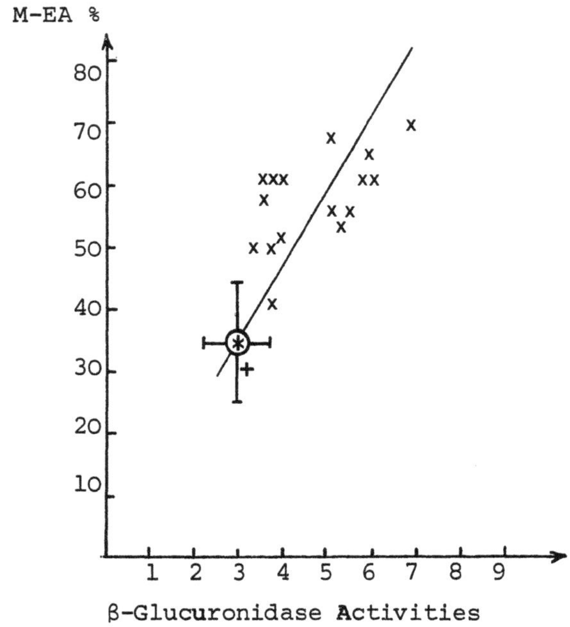

Fig. 4. Effect of IC on the FcR dependent reaction of the monocytes and on the secretion of β-glucuronidase from the cells. (*) without IC (n=8); (+) 100 μg protein precipitated from sera of healthy volunteers; (x) 100 μg IC precipitated from sera of 16 patients.

The FcR dependent reaction of the monocytes increased in parallel with the extracellular β-glucuronidase activity (Fig. 4). As controls were applied monocytes incubated with proteins precipitated from sera of healthy persons, and monocytes without IC.

The results of our experiments can be summarized as follows.

SUMMARY

1./ IC precipitated by PEG from patients with SLE inhibit in vitro the FcR dependent reaction of normal monocytes with sSRBC, while the C3bR dependent reaction of the cells with sensitized yeast is reduced only by some of them. The monocytes were preincubated with the IC for 30 min at room temperature.

2./ When the monocytes were incubated with the IC for 22 hours at 37 ^{o}C the reaction of FcR with sSRBC increased, while the C3bR dependent reaction did not altered.

3./ Simultaneously with the increasing FcR dependent reaction, the secretion of lysosomal β-glucuronidase of monocytes cultivated with IC is greater than those of the controls.

The data of these in vitro experiments suggest that the IC in sera of SLE patients may activate also in vivo the monocytes in the circulation.

REFERENCES

1. M. Kávai, K. Lukács, I. Sonkolý, I. Jr. Jókay, A. Berndt, and Gy. Szegedi, Loss of monocyte membrane receptors in patients with SLE, Clin. exp. Immunol. 40:66 (1980).
2. M. Kávai, K. Dankó, E. Kalmár, I. Francia, and Gy. Szegedi, Radioassay of soluble immune complexes using their uptake by macrophage Fc receptors, Immunology, 32:617 (1977).
3. M. Kávai, K. Lukács, I. Sonkoly, K. Pálóczi, and Gy. Szegedi, Circulating immune complexes and monocyte Fc function in autoimmune diseases, Ann. Rheum. Dis. 38:79 (1979).
4. W. P. Arend, and M. Mannik, The macrophag receptor for IgG: number and affinity of binding sites, J. Immunol. 110:1455 (1973).
5. C. J. Cardella, P. Davies, and A. C. Allison, Immune complexes induce selective release of lysosomal hydrolases from macrophages, Nature, 247:46 (1974).
6. J. Pestel, M. Joseph, F. Santoro, and A. Capron, The β-glucuronidase release from macrophages activated by immune complexes of varying antigen/antibody ratio, Ann. Immunol. /Inst. Pasteur/ 130C:507 (1979).
7. D. Chia, E. V. Barnezz, J. Yamagata, D. Knutson, C. Restivo, and D. Furst, Quantitation and characterization of soluble immune complexes precipitated from sera by polyethylene-glycol (PEG), Clin. exp. Immunol. 37:399 (1979).
8. E. Engvall, and P. Perlmann, Enzyme-linked immunosorbent assay (ELISA). Quantitative assay of immunoglobulin G, Immunochem. 8:871 (1971).
9. S. K. Ackerman, and S. D. Douglas, Purification of human monocytes on microexudate-coated surface, J. Immunol. 120:1372 (1978).
10. G. Brittinger, R. Hirschhorn, S. D. Douglas, and G. Weissmann, Studies on lysosomes. XI. Characterization of a hydrolase-rich fraction from lymphocytes, J. Cell. Biol. 37:394 (1968).
11. N. Dioguardi, A. Agostoni, G. Fiorelli, and B. Lomanto, Characterization of lactic dehydrogenase of normal human granulocytes, J. Lab. Clin. Med. 61:713 (1963).

DISCUSSION

Edited by: R.A.Harkness

VENGE: Your data are consistent with our data on PMN. We have extended this to show normal or reduced recognition of C3b coated particles which might be due to autooxidative mechanisms.

GALLILY: Do your monocytes show activation after in vitro culture?

KAVAI: The FcR dependent reaction of the monocytes cultured for 22 h with no IC present was not assessed. The β-glucuronidase activity of the cell supernatant was not increased.

NUCLEOTIDE CONCENTRATIONS IN LEUCOCYTES AND THEIR USE IN CONTROLLING THE QUALITY OF CELL PREPARATIONS

R.A. Harkness*, S.B. Coade*, M. Mansell† and R.J. Simmonds*

*Divisions of Perinatal Medicine and † Inherited Metabolic Disease
Clinical Research Centre, Harrow, HA1 3 UJ, Middx, UK

The association between a reduction in cellular ATP and reduced cell survival has been established for the erythrocyte by many groups (see Seegmiller 1980) and more recently for the kidney (Buhl, 1976a, b). The ATP concentrations in the erythrocyte have guided a search for better condictions of erythrocyte storage for blood transfusion. The factors altering intracellular ATP concentration are therefore of practical interest. A large number of factors reduce ATP concentration but it is difficult to raise it. One condition associated with raised ATP concentration in erythrocytes is renal failure (Mansell and Grimes, 1979).

In order to gain further insight into the mechanisms controlling intracellular ATP the concentrations of nucleotides and their metabolites in erythrocytes and polymorphonuclear neutrophil leucocytes, (PMN) have been studied in renal failure in combination with the activities of some of the enzymes involved in purine metabolism. Changes in activity of such enzymes, can be associated with changes in nucleotide concentrations and in overall cellular function (Seegmiller, 1980).

The adenine nucleotides and related compounds were estimated in trichloracetic acid extracts of samples of about 10^6 cells by high pressure liquid chromatography. Such methods have high specificity and sensitivity (Harkness et al., 1979). The cellular homogeneity of such preparations was also checked. In the more easily studied erythrocyte the ATP, ADP and AMP concentrations obtained by high pressure liquid chromatography agree well with the results from enzymic methods.

In erythrocytes a clear pattern of significant increases in nucleotide, nucleoside and purine base concentrations and in related enzyme activities in cells have been found in renal failure; these will be the subject of another report. In PMN as in erythrocytes the ATP is largely maintained by anaerobic glycolysis (Seitz, 1965) and similar changes might therefore be detectable in PMN in renal failure to those found in erythrocytes. However, the initial mean results were similar in both groups and suggested that there were no changes.

When results from leucocyte preparations with unphysiologically low relative ATP concentrations, an energy charge of < 0.5, (Atkinson, 1968) were excluded, the mean ATP concentration in PMN in renal failure was higher although this was not significant statistically ($0.1 > P > 0.05$) in the limited sample available (Table 1). Cell preparations with an energy charge below 0.5 were excluded because living cells generally have a value of 0.8 - 0.9 (Chapman et al., 1971). These results suggested that artefactual change was possibly obscuring a pattern of pathological change.

Since unphysiological energy charge values were obtained from some PMN preparations it appeared advisable to examine the conditions affecting 'relative' and 'absolute' ATP concentration in leucocyte preparations. It may be necessary to compromise between cellular homogeneity and metabolic state (Pretlow et al., 1975). Conditions have been examined using mainly blood from umbilical cord since this is a readily available source of fresh blood.

Table 1. ATP, Total Adenine Nucleotides and Energy Charge in PMN (98%) in Renal Failure
(excluding energy charge < 0.5 samples)

Mean ± SEM μmol/l cell vol

	ATP	Total Ade Nucleotides	Energy Charge	n
Controls	3402 ± 436	3631 ± 437	0.95 ± 0.008	7
Renal Failure	5044 ± 1089	5571 ± 1142	0.93 ± 0.019	9

Table 2. Adenine Nucleotide Concentrations in Dextran Sedimentation Leucocyte Preparations, 'PMN' (70-80%) µmol/l cell volume

'Normal' people	ATP	ADP	AMP	Total	Energy Charge
1	5452	786	6156	12394	0.47
3	3905	752	3671	8328	0.51
3	2971	781	3055	6807	0.49
4	3359	696	3328	7383	0.50
Mean	3922	754	4053	8728	0.49
SEM	±545	±21	±713	±1262	±0.01

Table 3. Adenine Nucleotide Concentrations in Ficoll/ Methylcellulose Hypaque PMN preparations (c 95%) Analysed after 30 min Incubation at 37°C in Phosphate Buffered Saline Mean ± SEM µmol/l cell volume (n = 7 'normal' people)

	ATP	ADP	AMP	Total	Energy Charge
Mean	3071	150	79	3300	0.95
SEM	±305	±17	±13	±305	±0.01

The simplest dextran sedimentation method of leucocyte 'PMN' preparation (about 70-80% PMN) gave results in which the energy charges were too low (Table 2). In another laboratory similar methods were used by MM and the effect of 30 min incubation in phosphate buffered saline repeatedly shown to result in high energy charge (Table 3) but the total adenine nucleotide concentrations were lower than with the simpler method. Some of the mechanisms responsible for such adjustments have been defined, for example the stimulation of AMP deaminase activity (Chapman and Atkinson, 1973; Schramm and Leung, 1973). It may therefore be difficult to define physiological concentrations of adenine nucleotides in cells like PMN which can be metabolically labile.

Alterations in the method of cell preparation were studied. The loss of adenine nucleotides from human erythrocytes has been detected within 40 min (Dean et al., 1978); the 'leakage' of the purine base, hypoxanthine, from erythrocytes occurs within 15 mins of sampling (O'Connor et al., 1980). Such changes are likely to occur at least as rapidly from the less stable PMN. McCullough (1977) has shown that on storage of granulocytes for 24 h at 4°C intravascular half life and chemotaxis are reduced but hexose monophosphate shunt and bactericidal activity are preserved.

When PMN were stored in blood at room temperature, 20°C, for 24 h, their mean concentration of ATP was decreased and that of AMP increased compared to results from a similar set of samples processed immediately (Table 4). A reduction of ATP concentration may therefore be responsible for the fall in leucocyte viability and chemotaxis noted by McCullough (1977). However, the reductions are not large enough to account for the marked changes previously observed (Table 2).

Processing at 4°C yielded higher adenine nucleotide concentrations than in cells processed at 20°C but (Table 5) the differences were not statistically significant. Another variable was the period of time for which the cells were exposed to hypotonic conditions during the differential haemolysis of erythrocytes; exposure to hypotonic conditions for periods of at least 5 minutes did not markedly affect adenine nucleotide concentrations either at 20°C or at 4°C when compared to concentrations after a 30 second exposure to hypotonic conditions.

We have now shown that a major factor in affecting energy charge appears to be storage at low temperature as a pellet. Storage of leucocytes at -10°C as a pellet produced large reductions in energy charge compared to results from the same blood sample analysed at the same time (Table 6). Subsequent resuspension of such cells would reduce total adenine nucleotide concentrations but increase energy charge (Table 2 and 3).

A variety of indirect evidence suggests that variations in the quality of PMN preparations may be common. Nussbacher et al., (1977) note that granulocyte preparations sometimes give cells whose integrity and function are poor. Altered nitroblue tetrazolium reduction can be due to a wide variety of *in vitro* and *in vivo* factors showing the sensitivity of the granulocyte to metabolic alteration (Klebanoff and Clark, 1978). Our evidence suggests that variable degrees of reduction of ATP concentration are especially liable to occur in PMN preparations.

Table 4. Effects of Storage as Blood at 20°C for 24 h on Adenine Nucleotide Concentrations in PMN and Lymphocytes (Dextran, Ficoll)
Mean ± SD µmol/l cell volume

		ATP	ADP	AMP	Total	Energy Charge
Fresh						
PMN	(4)	8125 ± 2100	1575 ± 332	99.4 ± 36	9799 ± 2450	0.91 ± .005
Lymphocytes	(5)	7536 ± 1382	3243 ±1154	208 ± 95	10988 ± 2615	0.83 ± 0.03
24 h						
PMN	(5)	5798 ± 1674*	1666 ± 632	395 ± 246**	7858 ± 2099	0.84 ± 0.06
Lymphocytes	(4)	6693 ± 2204	3035 ±1610	511 ± 191	10238 ± 3440	0.80 ± 0.06

* $p < 0.01$ ** $p > 0.002$ compared with fresh.

Table 5. Leucocyte Adenine Nucleotide Concentrations Effect of Leucocyte Preparation at 4°C Compared to 20°C
Mean ± SD μmol/l cell volume (n = 4)

	ATP	ADP	AMP	Energy Charge
PMN 4°C	10900 ± 4175	3188 ± 1048	218 ± 131	0.89 ± 0.02
Lymphocytes 4°C	12966 ± 1653	5395 ± 1539	348 ± 129	0.84 ± 0.03
PMN 20°C	7951 ± 2056	1728 ± 521	92 ± 25	0.90 ± 0.02
Lymphocytes 20°C	10132 ± 3595	3707 ± 1357	201 ± 61	0.85 ± 0.02

Table 6. Effect of Storage of Dextran, Ficoll Preparations of PMN and Lymphocytes as Pellets at -10°C for 18 h (n = 5)
Mean ± SEM μmol/l

		ATP	ADP	AMP	Energy Charge
PMN	Fresh	7462 ± 1027	1524 ± 463	322 ± 111	0.89 ± 0.02
	Stored	2888 ± 694	4145 ± 586	3726 ± 704	0.45 ± 0.06
Lymphocytes	Fresh	5219 ± 595	1367 ± 201	404 ± 171	0.84 ± 0.04
	Stored	1619 ± 434	2045 ± 518	1386 ± 299	0.51 ± 0.03

From our exploratory experiments it has been possible to locate an area in which more systematic studies can usefully be made. However, we must admit that at present we can suggest only one major method of manipulation affecting adenine nucleotide concentrations in PMN preparations. Since we probably cannot define all the important factors for high quality preparations it seems justifiable to suggest that we should perhaps measure the quality of PMN preparations by estimating adenine nucleotide concentrations. We might by such quality control eliminate many apparently conflicting sets of results and possibly reduce the "normal" range of the many variables studied in leucocyte preparations.

REFERENCES

Atkinson, D.E., 1968, Energy charge of the adenylate pool as a regulatory parameter. Interaction with feedback modifiers, Biochemistry, 7:4030.

Buhl, M.R., 1976a, Oxypurine excretion during kidney preservation: an indicator of ischaemic damage, Scand, J. Clin. Lab. Clin. Invest., 36:169.

Buhl, M.R., 1976b, The postanoxic regeneration of 5'-nucleotides in rabbit kidney tissue during *in vitro* perfusion, Scand. J. Clib. Lab. Invest., 36:175.

Chapman, A.G., Fall, L., and Atkinson, D.E., 1971, Adenylate energy charge in *Escherichia coli* during growth and starvation, J. Bacteriol., 108:1072.

Chapman, A.G., and Atkinson, D.E., 1973, Stabilization of the adenylate energy charge by the adenylate deaminase reaction, J. Biol. Chem., 248:8039.

Dean, B.M., Perrett, D., and Sensi, M., 1978, Changes in nucleotide concentrations in the erythrocytes of man, rabbit and rat during short term storage, Biochem. Biophys. Res. Commn., 80:147.

Harkness, R.A., Simmonds, R.J., O'Connor, M.C., and Webster, A.D.B., 1979, Purine metabolism studied with high pressure liquid chromatography, Biochem. Soc. Trans., 7:1021.

Klebanoff, S.J., and Clark, R.A., 1978, "The Neutrophil : Function and Clinical Disorders", North Holland, Amsterdam.

Mansell, M., and Grimes, A.J., 1979, Red and white cell abnormalities in renal failure, Brit. J. Haematol., 42:169.

McCullough, J., 1977, Liquid preservation of granulocytes for transfusion, Progr. Clin. Biol. Res., 13:185.

Nussbacher, J., McCullough, J., and Huestis, D., 1977, Granulocyte collection and processing, Progr. Biol. Res., 13:175.

O'Connor, M.C., Harkness, R.A., Simmonds, R.J., and Hytten, F.E., 1980, The measurements of hypoxanthine, xanthine, inosine and uridine in umbilical cord blood and fetal scalp blood samples as a measure of fetal hypoxia, Brit. J. Obstet. Gynaecol., in press.

Pretlow, T.G., Weir, E.E., and Zettergren, J.G., 1975, Problems connected with the separation of different kinds of cells, Internat. Rev. Exp. Path., 14 91.

Schramm, V.L., and Leung, H., 1973, Regulation of adenosine monophosphate levels as a function of adenosine triphosphate and inorganic phosphate - proposed metabolic role for adenosine monophosphate nucleosidase from Azobacter vinelandü, J. Biol. Chem., 248:8313.

Seegmiller, J.E., 1980, Diseases of purine and pyrimidine metabolism, in: "Metabolic Control and Disease", P.K. Bondy and L.E. Rosenberg, eds., Saunders, Philadelphia.

Seitz, I.F., 1965, Biochemistry of normal and leukemic leukocytes, thrombocytes and bone marrow cells, Adv. Cancer Res., 9:304.

DISCUSSION

Edited by: R.A. Harkness

ROOS: Have you studied the energy charge of PMN during phagocytosis?

HARKNESS: No, but we have planned experiments on the subject. Changes in energy charge could switch on anaerobic glycolysis.

GEE: Does uraemic serum affect nucleotide metabolism in PMN?

HARKNESS: Dialysable factors in serum affect nucleotide metabolism in erythrocytes within 2 days of renal failure.

BASFORD: Would your hplc methods be suitable for measurement of cyclic nucleotides?

HARKNESS: We would still consider radioisotopic methods as more suitable for these compounds. Our methods are sensitive to nmol/l or more of purine and pyrimidine based compounds.

RESTING AND STIMULATED CHEMILUMINESCENCE OF POLYMORPHONUCLEAR LEUKOCYTES: A CLINICAL APPROACH

P. De Sole, S. Lippa and G.P. Littarru

Institute of Biological Chemistry. Università Cattolica del Sacro Cuore - Facoltà di Medicina e Chirurgia - Via Pineta Sacchetti, 644 - 00168 ROMA - Italia

It is the aim of this paper to present our findings about the possibility of determining both the resting and the stimulated human Polymorphonuclear (PMN) Leukocyte activity by means of the Chemiluminescence (Cl) tecnique for a practical clinical purpose.

It is at present very well known that the biochemical events underlying the "Respiratory Burst" of PMN can be disclosed and studied by means of Cl, as extensively shown by Allen.[1,2,3,4] As we can see schematically in fig.1, the PMN generates several "activated forms of Oxygen" which assume the very well known role of biochemical weapons involved in the bactericidal process.[5,6] Such

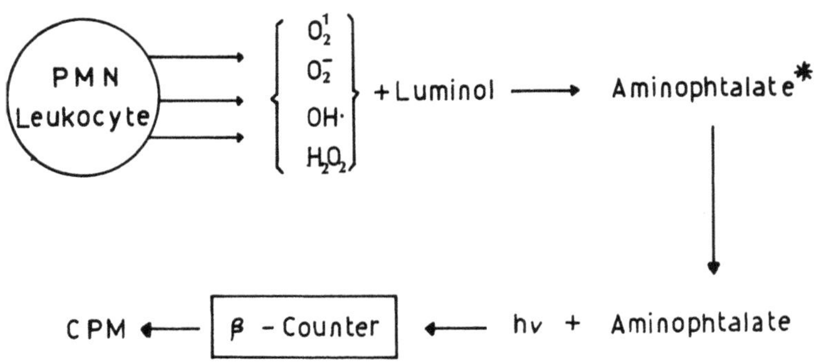

Fig.1 SCHEMATIC DIAGRAM OF THE PMN CHEMILUMINESCENCE.

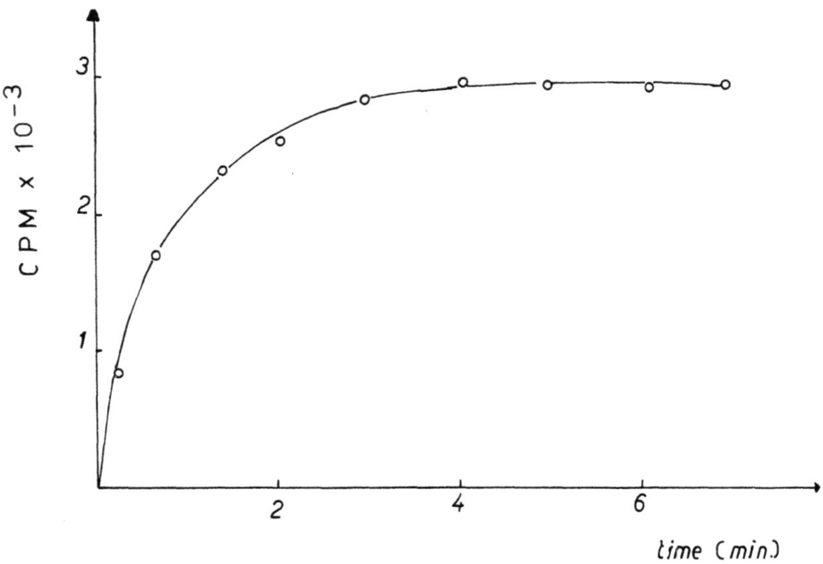

Fig. 2 TIME COURSE OF BASAL PMN CHEMILUMINESCENCE.
PMN leukocytes were separated from whole blood at room temperature by means of ficoll-hypaque and dextran sedimentation tecnique according to the Böyum standard separation.[18] Hypotonic lysis (30 seconds in 0,23%NaCl) was performed and PMN were suspended in a modified Krebs-Ringer phosphate medium pH 7.4 (without $CaCl_2$ and added with 0.33 mM Ca $(NO_3)_2 \cdot 4 H_2O$ and 5.5 mM glucose) to 4.0 x 10^5 cells/ml final concentration. To 5.0 ml of the cellular suspension 5 nmoles of luminol were added and soon after Cl recorded at room temperature with a Packard scintillation counter mod. 2335 with the coincidence circuit inserted, in the "tritium mode".The Cl activity is expressed as counts per minute (CPM).

reactive species of Oxygen, which are likely to be made at a certain level even in the resting stage, are produced at a great extent upon stimulation with phagocytizable particles or with soluble agents both capable of evocating the respiratory burst.[7,8,9,10]

The oxidative process can lead to a "native" or to an "amplified" Cl when one takes advantage of some exogenously added chemical compounds such as luminol (5-amino-2,3 dihydro-1,4-phtalazinedione) and lucigenin, which however do not react at the same extent to different oxygen species.[+]The light emitted in the oxidative process may be detected by different photometers[11] or by Beta-counters. The second ones show a higher sensitivity but, to record the emission of single photons the coincidence circuit must be off: in this condition howe-

[+]see R.C. Allen, same book.

ver basal cellular activities are hardly discernible from the high counter background. For this reason we tried to study the PMN Cl with the coincidence circuit inserted, in the tritium mode. In this condition we obtain the obvious advantage of reducing the blank activity to almost zero CPM values. As a consequence we have the possibility of a sensitive and precise evaluation of the cellular activity in the resting stage, before the phagocytic stimulus. Of course such experimental approach requires the use of a higher number of PMN than when the coincidence circuit is off: the standard amount of PMN that we employed was $4x10^5$/ml. This amount can nonetheless be easily obtained starting from a few ml of whole blood.

Furthermore an inconvenience is likely to occur: in effects with the coincidence circuit inserted, the low cellular activities are underestimated, because photons which are not detected simultaneously by the two phototubes are missed and this fact leads to possible alterations of the kenetic pattern.

In fig. 2 the Cl pattern in the basal condition, i.e. in the cellular suspension right after the luminol addition, appears. A plateau is reached within four to five minutes and this value lasts for the next ten to fifteen minutes. In the following experiments we refer to the Cl value five minutes upon the addition of luminol.

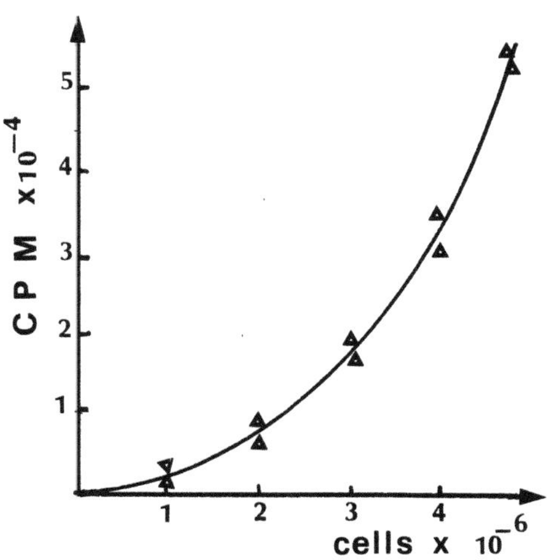

Fig.3 PMN Chemiluminescence as a function of cell number for vial. Experimental details as in figure 2.The Cl values are the CPM at the plateau (4 to 6 minutes after luminol addition).

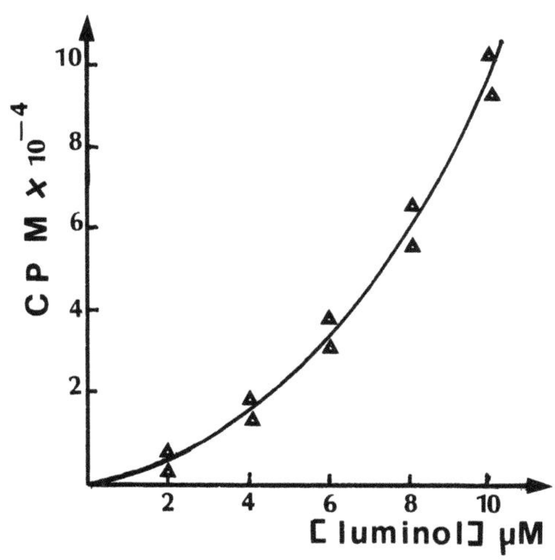

Fig.4 PMN Chemiluminescence as a function of luminol concentration. Experimental details as in figure 2.The Cl values are the CPM at the plateau (4 to 6 minutes after luminol addition).

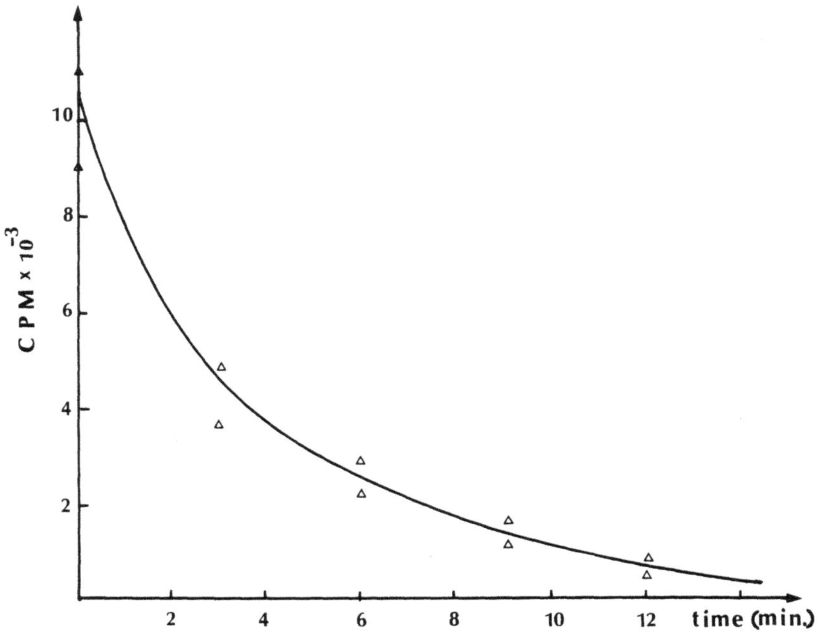

Fig. 5 REQUIREMENT OF CELLULAR INTEGRITY FOR Cl.
Aliquots of a PMN suspension obtained as in figure 2 were subjected to a hypotonic shock (0.20% NaCl) for different times and then isotonicity was restored. Luminol was then added and Cl recorded.CPM refer to the values recorded at the plateau after luminol addition.

Next figures show the variation of Cl as a function of cell number (fig.3) and of luminol concentration (fig.4). The lack of linearity is evident, and, as we mentioned above, this behavior is due to an underestimation of low cellular activities. In this condition the Cl is strictly dependent on cellular integrity, as we can see in fig. 5, where a lysis experiment is reported. A 50% inhibition of Cl activity is obtained by exposing the cells to the hypotonic shock for 2-3 minutes, whereas in the case of metilene blue exclusion test the same degree of inhibition is reached after one hour.[12]

Fig. 6 shows the Cl response before and following the addition of phagocytizable particles, in the presence of luminol. As phagocitizable agent latex particles of 1.1 μ diameter were used, in order to minimize the possible variability which could arise from the use of less standardizable particles. We can clearly see that a striking burst of chemiluminescence is evocated by the phagocytic stimulus: so we define "stimulation index" the ratio between the Cl value obtained following the addition of latex and the basal value obtained before the addition of latex. These are the main features of our Cl analytical model.

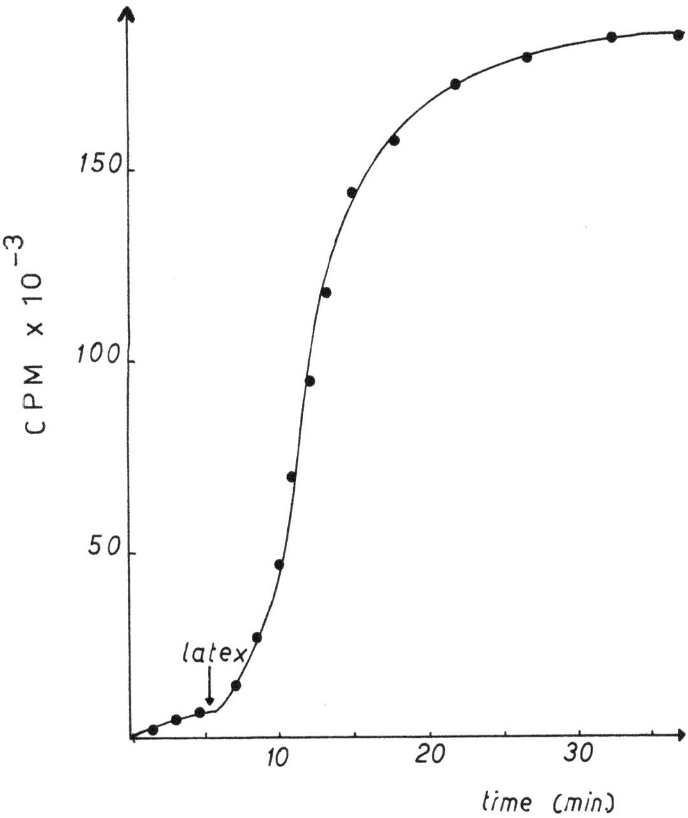

Fig.6 TIME COURSE OF PMN CHEMILUMINESCENCE BEFORE AND AFTER LATEX ADDITION.
Experimental details as in figure 2. Latex particles/leukocytes ratio was 500:1. Each value is the mean of three different determinations.

Let us consider now the data relative to both basal and stimulated Cl in a group of normal subjects. Constant values of Ci activities were obtained with normal donor samples analyzed in different days or weeks, whereas the same samples showed highly different values when analyzed in different months.

Fig.7 schematically indicates the data obtained in two different periods of the year. As one can notice, in the semptember-october period we have a relatively high resting activity (several thousand CPM) together with a stimulated activity of about 200,000 CPM, whereas in the may-june period the resting activity is much lower (a few hundred CPM) but the stimulated activity is much higher than in the previous period, and so the stimulation indexes are ranging

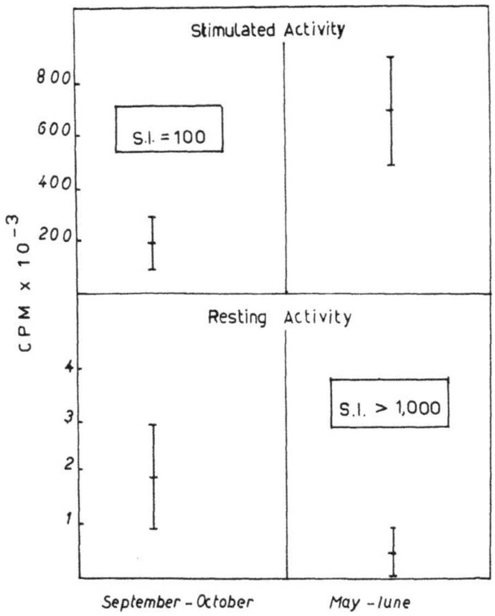

Fig.7 Resting and stimulated PMN Chemiluminescence from normal donors in two different year periods. S.I.=Stimulation index (see text). The CPM values are the mean±2SD for 20 subjects/each period.

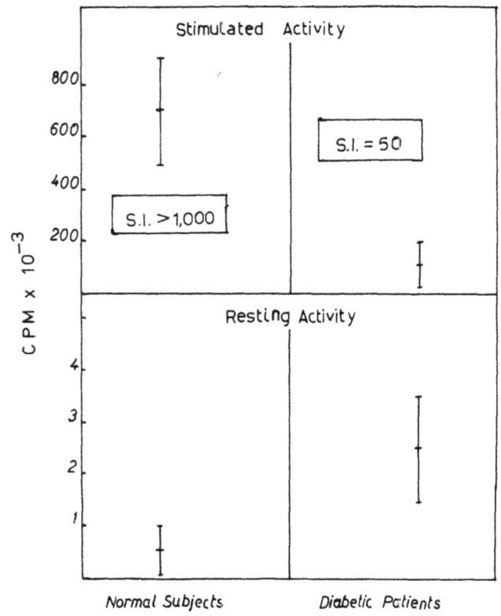

Fig.8 Resting and stimulated PMN Chemiluminescence from normal subjects and from diabetic patients. S.I.=Stimulation index (see text). The CPM values are the mean±2SD for 20 subjects/each group.

from about 100 in the September-October period, to more than 1,000 in the May-June period. With a small number of subjects very frequent determinations were performed within a two year-period and synchronous variations of the same type as the ones we see in Figure 7 were obtained, which seem to be of chronobiological type.Because of this variability, a single assay of Cl in a single case has certain limitations, with the obvious exception of the situation wherein one finds an absolutely lacking activity, which has been shown so far in CGD.[13] The limit imposed to this assay by this variability during the year can be neglected if we compare in the same period of the year two different populations or if we perform single determinations that can be referred to an internal "blank" activity, such as we can obtain for instance, when we look for defects in opsonization by means of cross-reactions[4], or when we check in general the effect on Cl of a certain therapeutical action.

Table I: Effect of General Anesthesia (and Surgery?) on Human PMN Chemiluminescence

ANESTHETIC	PMN STIMULATED CL (CPM X 10^{-3}) PRE ANESTHESIA	DURING ANESTHESIA	% INHIBITION
HALOTANE	240	87	74
"	240	118	51
"	710	390	45
"	990	105	89
"	185	92	51
"	860	280	68
"	310	190	39
ENFLURANE	15	0.6	96
"	850	540	37
"	600	840	NO INHIBITION
"	620	450	28
"	50	0.4	99
"	575	310	46
"	1550	1700	NO INHIBITION
"	615	1100	80% STIMULATION

Experimental details as in figure 2. The CPM values are the mean of at least three different determinations.

As an example of comparison between two populations and because of the very well known susceptibility of diabetic patients to infections [14] we made a comparison between a group of normal and a group of diabetic patients. On the other hand as an example of the second type we just mentioned, that is the effect of drugs, we tested Cl before and after the induction of general anesthesia.

In fig. 8 we can see the Cl behavior of a group of diabetic patients, compared to a control group. Impaired or abnormal granulocyte functions have so far been reported in diabetes [15,16]. In our experimental conditions we found that PMN from diabetic patients show considerably high resting activity compared to normals, together with a lower stimulated activity. The combination of these two findings leads to stimulation indexes much lower than for normals and this might be in agreement with the clinical aspect that we mentioned.

In table 1 the Cl values of stimulated PMN from patients undergoing general anesthesia (and surgery) are reported. Data are referred to two types of general anesthetics: halotane and

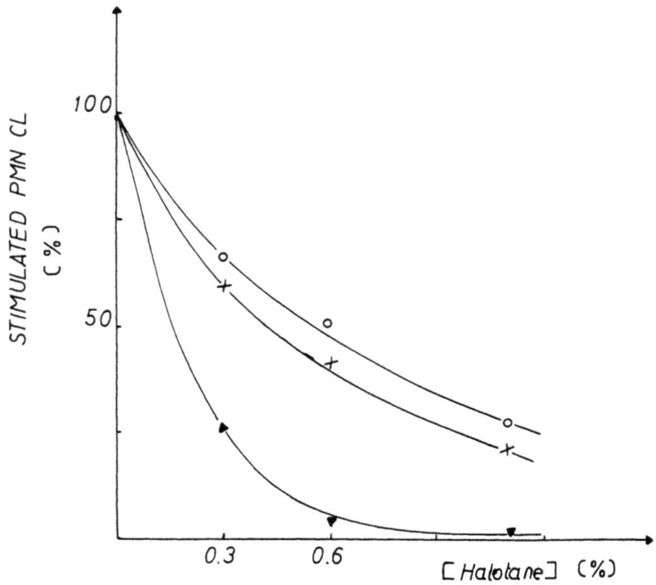

Fig.9 IN VITRO EFFECT OF HALOTANE ON PMN CHEMILUMINESCENCE.
PMN suspensions, obtained as described in fig.2, were subjected for 10 minutes to different percentages of halotane (in air). The different curves are from three different normal subjects. Each value is the mean of three different determinations.

enflurane (ethrane). We generally had an inhibitory effect during anesthesia, from 30 to 90%, which would be in agreement with certain hypothesis that ascribe also to anesthesia the incidence of post-surgery infections[17]. But in the case where enflurane was used we can also find some patients where no inhibition occurred and in one case we had a clear stimulation. The inhibitory effect that we just saw in vivo is confirmed by a clear inhibition in vitro.

Fig.9 shows the inhibitory effect of Halotane at different concentrations on a suspension of isolated PMN. Data are referred to leukocytes from three different normal subjects. The different pattern of inhibition of one of the three subjects and the different responses in vivo might suggest a possible use of this test for assessing variable individual sensitivity to the anesthetic agent.

In conclusion we think that this way of assessing both the resting and the stimulated Cl in various diseases or clinical situations supports the practical value of this sensitive tecnique.

REFERENCES

1. R. C. Allen, Dissertation, Tulane University, New Orleans, LA. University microfilms, 74-291 (1973).
2. R. C. Allen, Reduced, radical and excited state of oxygen in leukocyte microbicidal activity in: "Lysosomes in applied biology and therapeutics" p.197 J.T. Dingle and P.J. Jacques, ed. North Holland, Amsterdam (1979).
3. R. C. Allen, Chemiluminescence from eukariotic and prokariotic cells; reducing potential and oxygen requirements, Photochem. Photobiol. 30: 157 (1979).
4. R. C. Allen, Chemiluminescence: an approach to the study of the humoral-phagocyte axis in the host defense against infection, in "Liquid scintillation counting, recent applications and development"; vol.2, 377, Academic Press, New Yrok (1980).
5. A. J. Sbarra and M. L. Karnovsky, The biochemical basis of phagocytosis I. Metabolic changes during the ingestion of particles by polymorphonuclear leukocytes. J. Biol. Chem. 234: 1355 (1959).
6. F. Rossi and M. Zatti, Changes in the metabolic pattern of polymorphonuclear leukocytes during phagocytosis, British J. Exp. Pathol. 45: 548 (1964).
7. G. Rottini, P. Dri, M. R. Soranzo, P. Patriarca, Correlation between phagocytic activity and metabolic response of polymorphonuclear leukocytes toward different strains of E. Coli, Infect Imm. 11: 417 (1975).
8. D. B. Drath, M. L. Karnovsky, Superoxide production by phagocytic leukocytes, J. Exp. Med. 141: 257 (1975).
9. G. L. Mandell, Influence of type of ingested particle on human leukocyte metabolism, Proc. Soc. Exp. Biol. Med. 137: 1228 (1971).
10. J. E. Repine, J. G. White, C. C. Clawson, B. H. Holmes, The influence of Phorbol Myristate Acetate on oxygen consumption by polymorphonuclear leukocytes, J. Lab. Clin. Med. 83: 911 (1974).
11. G. L. Picciolo, J. W. Deming, D. A. Nibley, E. W. Chappelle, Characteristics of commercial instruments and reagents for luminescent assays, in "Methods in enzymology" vol.57, De Luca ed., p.550, Ac. Press, New York (1978).
12. A. Ferrata and E. Storti: Le malattie del sangue, p.68 and p.536 Vallardi, - Milano (1958).
13. R. L. Stjrnholm, R. C. Allen, R. M. Steele, W. W. Warring and J. A. Harris, Impaired chemiluminescence during phagocytosis of opsonized bacteria, Inf. Imm. 7: 313 (1973).
14. J. D. Bagdade, R. K. Root, R. J. Bulger, Impaired leukocyte function in patients with poorly controlled diabetes, Diabetes 23 9 1974

15. A. G. Mowat, J. Baum, Chemotaxis of polymorphonuclear leukocytes from patients with diabetes mellitus. New Engl. J. Med. 248: 621 (1971).
16. J. S. Tan, J. L. Anderson, C. Watanakunakorn, J. P. Phair, Neutrophil dysfunction in diabetes mellitus. J. Lab. Med. 85: 26 (1975).
17. B. F. Cullen, R. B. Hume, P. B. Chretien, Phagocytosis during General Anesthesia in man, Anesthesia and Analgesia Current Researches 54: 501 (1975).
18. A. Böyum, Isolation of Leukocytes from human blood, Scand. J. Clin. Lab. Invest. 21 (suppl.97): 9 (1968).

DISCUSSION

Edited by: R.A.Harkness

PALMBLAD: You showed considerable variation of light emission by resting PMN. We have found that light emission can be doubled by lysis of erythrocytes, probably due to interactions of erythrocyte membranes with luminol. Have you looked for such effects?

DE SOLE: We did not look specifically for this effect but we always washed the PMN suspension after lysis in our standardised procedure.

VENGE: What method of cell preparation do you use? We found that the Ficoll-Hypaque method gives unpredictable high resting activity.

DE SOLE: The method of cell preparation is Ficoll-Hypaque (Boyum 1968), dextran at a final concentration of 1% and hypotonic by lysis for 30 sec in 0.25% saline. Chemiluminescence is measured at final concentrations of PMN 4 x 10^5 cell/ml, luminol 1 μmol/l, Krebs-Ringer-phosphate buffer with a calcium concentration of 0.05 mmol/l and glucose 1 g/l. The final vol is 5 ml. A scintillation counter is used with coincidence counting at settings for tritium. A source of unpredictable variability is luminol concentrations greater than micromolar.

DEFICIENT PHAGOCYTOSIS SECONDARY TO BREAKDOWN OF OPSONIC FACTORS IN INFECTED EXUDATES

F.A. Waldvogel, P. Vaudaux, P.D. Lew, A. Zwahlen, S. Suter and U. Nydegger
Infectious Disease Division, Departments of Medicine, Microbiology and Pediatrics, University Hospital, 1211 Geneva 4, Switzerland

INTRODUCTION

Abscess formation or development of a purulent exudate within a closed cavity (empyema) remains an unpredictable complication of pyogenic infections. The mechanisms leading to abscess formation are largely unknown. Once established abscess or empyema are characterized by the presence of a large number of viable bacteria, despite the ingress and the persistence of a large population of polymorphonuclear leukocytes (PMNLs). Here again, we do not have adequate data to explain this biological paradox : How can bacteria persist in a surrounding rich in phagocytic cells ? Obviously, one or several of the mechanisms controlling attachment, ingestion or killing of microorganisms must be deficient in such a situation. Biophysical factors such as low pH and/or low pO_2, as measured occasionally in inflammatory exudates, have been reported to play an inhibitory role on phagocytosis (1,2). This does not seem to be constantly the case since unpublished observations from our laboratory on experimental abscesses in the guinea pig have shown biochemical+biophysical parameters compatible with normal PMNL function (3).

Since the killing of pyogenic organisms by PMNL can be conveniently subdivided into different phases amenable to separate analysis, we have tried to explore the problem of defective bacterial killing in abscesses and empyemas by measuring the functional integrity of the cellular phase and of the supernatant (fluid phase) of such infectious exudates. This has been performed by measuring the number and the functional status of PMNLs isolated from purulent exudates, and evaluating the opsonic activity of the supernatant of various types of pleural exudates.

MATERIAL AND METHODS

Pleural exudates and abscess aspirations

Pleural fluids were classified according to well known criteria (4). Empyema and abscesses were characterized by abundant growth of microorganisms, and by PMNL counts with a median value of 1.5 x 10^7/ml. After thoracocentesis, all the samples were immediately mixed with EDTA (10-20 mM final concentration), centrifuged at 1500 g for 15 min and stored at -70°C. Abscess aspirations were processed identically. Serum was obtained from the same patients by allowing blood to clot at room temperature for 60 min, and by centrifugation as described above. Plasma was obtained by collection in tubes containing EDTA at 20 mM final concentration and by centrifugation. All tubes were kept at -70°C until used.

Preparation of PMNL

Leukocytes from peripheral blood and pleural empyemas or abcesses were obtained by dextran sedimentation, repeated washing and resuspension in Krebs-Ringer-Phosphate buffer, pH 7.4.

Phagocytic tests

A phagocytic bactericidal assay was performed using S. aureus Wood 46 var. as the test organism as previously described (5). This strain has been shown to depend on complement for effective opsonisation (5). The Oil Red-O test was performed as described by Stossel (6).

Complement studies

They were all carried out on EDTA treated material. Hemolytic activity (CH50) was measured in a continuous flow system (7). C3, C4 and factor B levels were quantitated by single radial immunodiffusion (8) and C3d levels measured as described previously (5).

Breakdown of human C3 was determined by using human, purified, labeled C3 bound to sepharose, and by measuring liberation of the radioactivity in the supernatant in Mg and Ca free medium (9). Proteolysis of C3 was measured by mixing labeled, sepharose bound C3 with EDTA supernatants of pleural empyema. Timed samples were suspended in SDS-buffer, heated to 100°C for 3 min and run on polyacrylamide gels according to well established techniques.

RESULTS

Activity of PMNLs obtained from purulent exudates (pleural empyemas and abscesses

Various tests extending from phase contrast microscopy to trypan blue exclusion, and to the phagocytic bactericidal test previously described were used in order to evaluate the integrity of the PMNLs obtained from purulent exudates, taking peripheral PMNLs from the same patients as controls. The complement mediated opsonic capacity of the fluid phase (supernatant) of purulent exudates was also tested with the same phagocytic bactericidal test.

As shown in table 1, the 3 clinical specimens obtained from abscess or empyema, all contained more than 10^7/ml identifiable PMNLs, more than 60% of them excluding trypan blue. Thus, the number of viable PMNLs isolated from purulent exudates was at least 10 times higher than the number used in conventional phagocytic assays.

In another case study, PMNLs were obtained from a pleural empyema, washed and incubated with the same patient's serum, or with the supernatant of the empyema in a phagocytic bactericidal assay using the Wood 46 strain as the test organism. The integrity of the C3 receptors and of the phagocytic bactericidal activity of these PMNLs obtained from an empyema could be documented by a 99% killing rate when compared with the peripheral PMNL from a normal subject, which showed a bactericidal activity of 90%. The empyema supernatant, as

Table 1. Purulent exudates : Concentration of viable PMNL, as determined by trypan blue exclusion

Origin, Microbiology	Total leukocyte count	% PMNL	% PMNL excluding trypan blue
Perianal abscess (*S. aureus*)	10×10^7/ml	95	90
Pleural empyema (mixed anaerobic flora)	4×10^7/ml	95	60
Pleural empyema (mixed anaerobic flora)	15×10^7/ml	95	60

expected, promoted less opsonization than autologous serum used at the same concentration. Similar results have been obtained by our group (unpublished results) with PMNL obtained in an experimental abscess model (3). Thus, these experiments performed on a limited number of clinical specimens suggest that decreased bactericidal activity in empyemas is not due primarily to deficient cellular phagocytic function.

<u>Opsonic activity of the supernatants obtained from purulent exudates (empyemas)</u>

Recently, we have reported a marked decrease in complement mediated opsonic activity and in classical pathway mediated hemolytic activity, coinciding with the appearance of C3 breakdown products, in empyema fluids as opposed to pleural effusions of other origins (5). These results were not only obtained with a phagocytic bactericidal test, but were confirmed when the opsonic sources were tested with the Oil Red-O uptake assay mentioned previously. Taken together, these data suggested that continuous consumption of complement dependent immune reactants could occur within infected foci such as empyemas. We therefore undertook to measure the levels and catabolism of complement in culture positive infected pleural effusions as opposed to culture negative, so-called metapneumonic, pleural effusions (10). The results demonstrated a marked decrease in classical and alternative hemolytic activities, as well as factor B, C4 hemolytic, and native C3 levels in infected effusions (10). High levels of C3d and Ba, the breakdown products of C3 and factor B respectively, in the same infected pleural effusions argued in favor of consumption of complement components in this condition.

<u>Evidence in favor of enzymatic, non-complement mediated breakdown of opsonic factors in purulent exudates (empyemas)</u>

We have recently examined the possibility of enzymatic IgG breakdown in empyemas by immunoelectrophoresis using rabbit antisera directed against human IgG, Fc and Fab. IgG breakdown products, demonstrated by multiple precipitation lines, could only be found in the culture positive pleural effusions (11/14 cases). These results were confirmed by adding 125I-labeled IgG to pleural fluids, incubating the mixture at 37°C for 4 hr and determining the radioactivity in the pellets and supernatants after 40% ammonium sulphate precipitation (10). 7 out of 14 empyemas cleaved between 50 and 92% of the added IgG, as opposed to metapneumonic effusions, which were totally devoid of IgG cleaving activity.

Direct enzymatic breakdown of complement differing from physiological activation of classical or alternative pathway has been

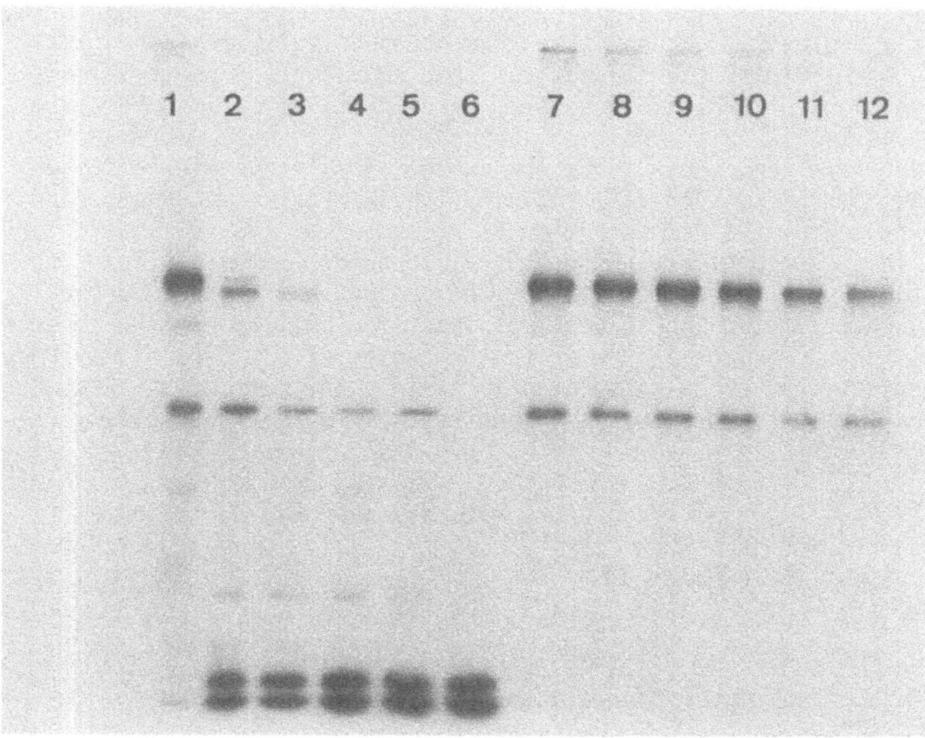

Figure 1. Polyacrylamide gels of labeled C3 incubated with (Tracks 1-6) and without (Tracks 7-12) pleural empyema supernatant in Ca and Mg free medium. Note breakdown of C3 when incubated with empyema supernatant. For details see Material and Methods.

recently investigated in our laboratory using purified human C3 labeled with 125I as a substrate. This compound was incubated in presence of the EDTA supernatant of a culture positive pleural effusion for 6 hr at $37^{o}C$ (9). Timed samples were prepared as described in Material and Metbds, and proteins revealed by autoradiography. As seen in figure 1, incubation led to a progressive, preferential breakdown of the alpha-chain of C3 with simultaneous appearance of small breakdown products (Tracks 1-6), as opposed to the pattern obtained when C3 was incubated alone (Tracks 7-12).

CONCLUSIONS

Pleural empyema, a clinical entity characterized by the simultaneous presence of large number of PMNLs and viable bacteria, is a biological paradox which has not been fully explained yet. Our preliminary studies suggest that receptor and bactericidal functions of PMNL isolated from purulent exudates, can be close to normal in this condition. Supernatants of these empyemas however have been shown to be low in heat labile opsonic activity and complement hemolytic activity. These observations have been extended by the demonstration of breakdown of IgG, C3 and factor B in infected pleural effusions as opposed to pleural fluids obtained under other conditions. The breakdown of Ig and C3 seems to be enzymatic and to occur,at least for C3, even in the absence of Ca and Mg ions : thus, direct cleavage of C3, possibly by PMNL enzymes, has to be postulated to explain these results. Present work in our laboratory is trying to explore this possibility.

REFERENCES

1. R.W. Light, I. MacGregor, W.C. Ball and P.C. Luchsinger. Diagnostic significance of pleural fluid pH and pCO_2. Chest 64:591 (1973).
2. D.E. Potts, D.C. Levin and S.A. Sahn. Pleural fluid pH in parapneumonic effusions. Chest 70:328 (1976).
3. W. Zimmerli, P. Vaudaux and F.A. Waldvogel (manuscript in preparation).
4. R.W. Light, M.I. MacGregor, P.C. Luchsinger and W.C. Ball. Pleural effusions : the diagnostic separation of transudates and exudates. Ann. Intern. Med. 77:507 (1972).
5. P.D. Lew, R. Zubler, P. Vaudaux, J.-J. Farquet, F.A. Waldvogel and P.H. Lambert. Decreased heat-labile opsonic activity and complement levels associated with evidence of C3 breakdown products in infected pleural effusions. J. Clin. Invest. 63: 326 (1979).
6. T.P. Stossel. Evaluation of opsonic and leukocyte function with a spectrophotometric test in patients with infection and with phagocytic disorders. Blood 42:121 (1973).
7. U.E. Nydegger, L.M. Ackermann, P.H. Lambert and P.A. Miescher. A simple automated method for complement estimation in a continuous flow system. J. Immunol. 109:910 (1972).
8. L.H. Perrin, P.H. Lambert, U.E. Nydegger and P.A. Miescher. Quantitation of C3PA (properdin factor B) and other complement components in diseases associated with a low C3 level. Clin. Immunol. Immunopathol. 2:16 (1973).

9. S. Suter, U. Nydegger, L. Roux and F.A. Waldvogel (manuscript submitted for publication).

10. D.P. Lew, J.-P. Despont, L.H. Perrin, M.T. Aguado, P.H. Lambert and F.A. Waldvogl. Demonstration of a local exhaustion of complement components and of an enzymatic degradation of immunoglobulins in pleural empyema : a possible factor favouring the persistence of local bacterial infections. (Clin. exp. Immunol. 42 (1980) (in press).

This work was supported in part by the Fonds National suisse de la recherche scientifique (no 3.836-0.79) and by a special grant from Beecham Pharmaceuticals, Berne, Switzerland.

DISCUSSION

Edited by: R.A.Harkness

GALLILY: Did you try to produce antibody against the enzyme which degrades C3 and thus to prevent the defect in opsonization?

WALDVOGEL: No, we have not performed any experiments trying either to identify or to inhibit specifically this enzymatic reaction.

CONGENITAL AND ACQUIRED LACTOFERRIN DEFICIENCIES IN NEUTROPHILS

Jean-Louis Vildé[+], Janine Breton-Gorius[°], Jacques Hakim*, Diego Buriot[++], Claude Griscelli[++]

+ Hopital Claude Bernard. 75019 Paris, F.
° INSERM U91 Hopital Henri Mondor, 94010 Créteil, F.
*Hopital Bichat. 75018 Paris , F.
++ Hopital des Enfants-Malades. 75015 Paris, F.

INTRODUCTION

Human polymorphonuclear leukocytes (PMN) contain two main types of granules : the primary or azurophil granules and the secondary or specific granules [1]. Each one can be recognized by specific markers of their content : myeloperoxidase (MPO) for the former and lactoferrin (LF) for the latter. The presence of LF, an iron-binding protein [2], in human PMN was demonstrated by Masson [3] in 1969 and its localization in specific granules by Baggiolini [4] in 1970 and Mason [5]. Until today, four cases of absence of specific granules in human PMN have been reported [6-9]. Our purpose is to describe the morphologic, immunocytochemical and functional features of PMN from a child which were shown to lack specific granules and LF, and from some patients who had the same but acquired defects, and to discuss the role of specific granules and LF in the resistance to infection.

MATERIALS AND METHODS

Detection of LF on blood smears or spread PMN was achieved by an immunocytochemical procedure using a rabbit anti LF serum labelled with horse radish peroxidase after inhibition of MPO by H_2O_2 [5,10] ; peroxidase (PO) was then detected by diaminobenzidine and H_2O_2 . Ultrastructural analysis was performed using a classical method with PO substrate in order to mark MPO and azurophil granules [11]. These procedures were applied to PMN from an Egyptian, 6 year-old boy suffering of recurrent infections located on skin, lymph nodes and lungs since the age of six months [12] , and from four adult patients with a preleukemic state and refractory anemia.

RESULTS

Congenital Deficiency

Immunocytochemical detection of LF can easily demonstrate the presence of this protein in normal PMN from human blood. LF was not present in normal monocytes and was not detected in promyelocytes from bone marrow. This confirms that the production of specific granules starts at the promyelocyte stage. LF was absent from the child's PMNs, but was present in his father's PMNs. The patient's PMNs showed also an abnormal lobulation of the nucleus. Ultrastructure of the patient's PMNs displayed normal PO positive granules, but a complete absence of PO negative specific granules in comparison to normal PMNs (Fig.1)The number of azurophil granules appeared greater in patient's PMNs than in control PMNs. After phagocytosis of *Klebsiella pneumoniae* by patient's PMNs, normal degranulation of azurophil granules occured and lysis of bacteria at 10 min appeared similar to control.

Basal O_2 consumption at rest was normal but was reduced to one third of normal value after stimulation by zymosan. Similarly the production of H_2O_2 and superoxide anion after phagocytosis was decreased.

Chemotactic activity, phagocytosis and bactericidal activities were within normal values. The percentage of killed *Serratia marcescens* at 90 min was normal at the two tested ratios bacteria/ PMN : 1/10 and 10/1. (Table 1)

Acquired Deficiency

In four patients with a preleukemic state, partial deficiency of LF in PMN was observed. But a great heterogeneity from one PMN to another can be seen : some PMNs had a normal content of LF, some a decreased LF content and in others the absence of LF was complete. The same heterogeneity was also observed with specific granules, as it has been described for azurophil granules in such patients[13]. Some PMN were agranular and totally devoided of granules either specific or azurophil. They were capable to ingest bacteria, to form a phagosome, but bacteria were not lysed at 90 min.

→

Fig. 1. Peripheral blood PMN after cytochemical demonstration of peroxidase activity. Top : PMN from a control ; azurophil granules contain a dense reaction product due to their MPO content ; numerous peroxidase negative granules (arrows) represent specific granules. Bottom : patient's PMN. Azurophil granules with MPO content appear numerous while no negative granules can be identified.

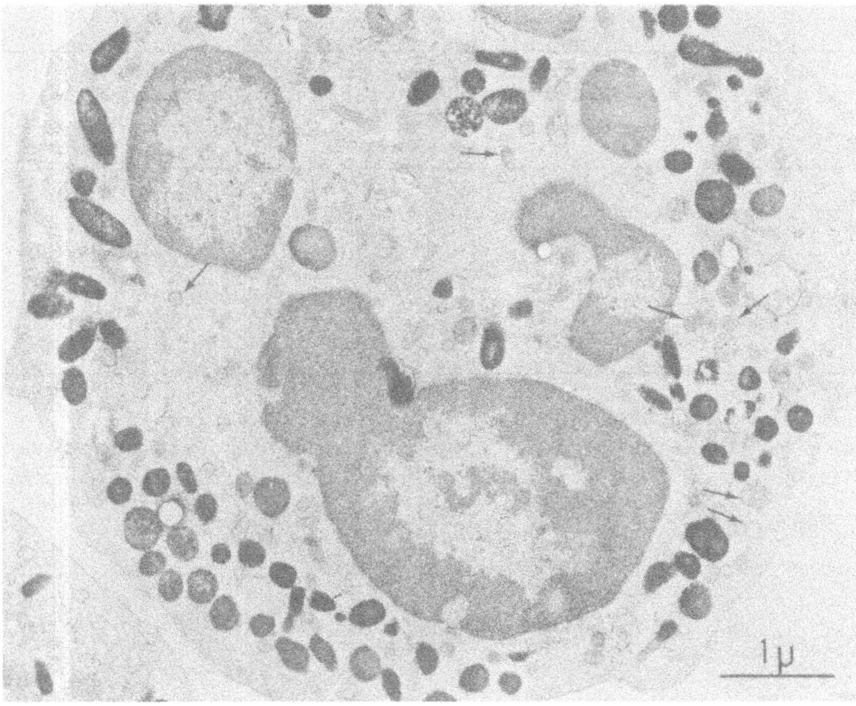
1μ

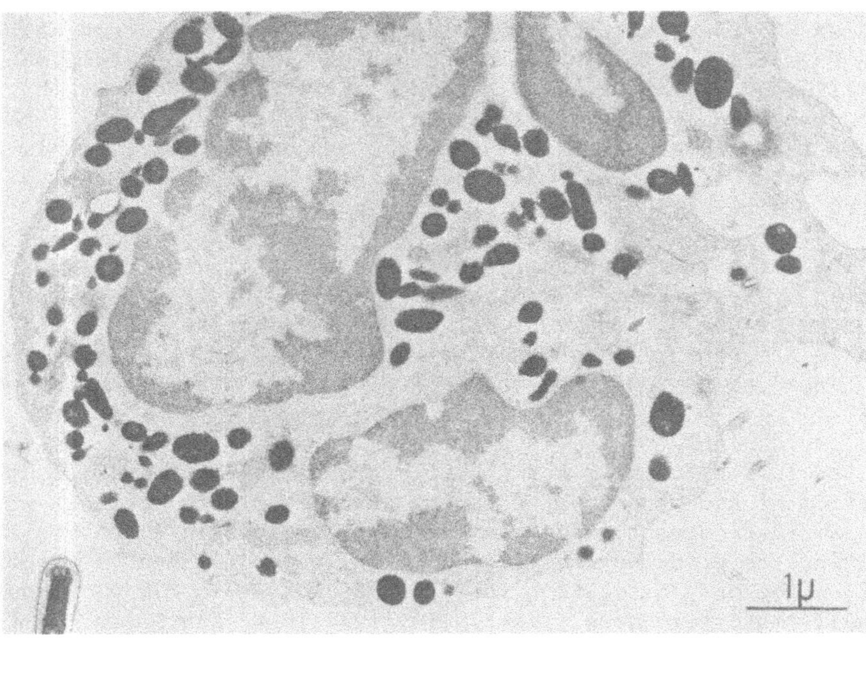
1μ

Table 1 - Bactericidal activity of LF-deficient PMNs

Bacterial strain	Bacteria/PMN	% viable remaining bacteria	
		30min	90min
Serratia marcescens	1/10	4.8	1
" "	10/1	2	0.2
		60 min	120 min
Staphylococcus aureus	1/2	4.5	1.3

PMNs ($2x10^7$/ml) + normal fresh human serum (20%)
Results are expressed as percentage of viable bacteria Inoculum=100%
Values for normal PMNs with S.marcescens ;30min:32.3$\pm$17.5% (n=11)
90min: 2.2$\pm$2.05% (n=11)

DISCUSSION

The PMNs of this child suffering of recurrent bacterial infections were shown to be totally deficient in specific granules and LF. The four reported patients with absence of specific granules had recurrent bacterial infections. One had severe neutropenia [8] . LF was also absent in the patient where the detection was performed[6] . The defect was congenital, except probably in the first described case[6] , as it was in the case reported here : this child's sister died of infection at the age of one year and his parents were first-cousin. This suggests an autosomal recessive pattern of inheritance.

The metabolic respiratory burst was normal in the cases reported by Spitznagel[6] and Strauss[7] , and on the contrary was decreased in our patient. One of the possible explanation could be found in the production of hydroxyl radicals which has been shown to be enhanced by LF in the presence of a particulate fraction of PMN [14] .

The link between LF and specific granules leads easily to the detection of such cases by screening LF in PMNs from patients with unexplained recurrent infection, or with hematologic diseases involving the granulocytic cell. However, we do not know if specific granules are only a location for synthesis of LF in PMN or if some of their content is more directly involved in LF synthesis. In this child, LF was normally present in saliva and this would favour the first hypothesis.

Degranulation of azurophil granules was normal in specific granules deficient PMNs ; this suggests a separate mechanism for degranulation of both types of granules. In normal PMN, the degranulation of the two types of granules is sequential[15] , and the effect of phorbol myristate leads to the degranulation of specific granules only [16] .

The most intriguing question is the role of LF in bactericidal activity of PMN and more generally, in resistance to bacterial infection. Conversely to other studies, PMN of this child were shown to have a normal bactericidal activity. We could believe that this function would have been decreased if a greater ratio bacteria/PMN have been used. But looking at the precise results of bactericidal activity of PMN in other cases of specific granules deficiency, we can see that the differences between patients and control are very slight [7,9]. Whatever the reasons of this discrepancy , the role of LF in antimicrobial activity of PMN, if any, is not known and has not been demonstrated[2] . In vitro, LF is mainly bacteriostatic and could act by its chelating effect on iron. In vitro studies of antimicrobial activity of LF have given various results [17,18] . In PMN , LF is quickly released outside the cell after phagocytosis[19].

This suggests that LF could exert its antimicrobial activity outside PMN, directly within the infectious focus for example. This can be a possible explanation for the recurrent bacterial infections observed in these patients. But, today, the link between the clinical status, recurrent infections and these defects of PMN remains obscure and needs further studies of such patients to be cleared.

SUMMARY

Lactoferrin, a marker of specific granules of neutrophils , can be detected by an immunocytochemical method. Absence of lactoferrin and specific granules has been demonstrated in neutrophils from a child with recurrent bacterial infections. Azurophil granules were present and their degranulation after phagocytosis was normal. Phagocytosis and bactericidal activity were normal but the respiratory burst was decreased. The role of lactoferrin in the bactericidal activity of neutrophils and in defence against bacterial infections is not completely understood, and is discussed. Acquired defects of specific granules have been detected in four patients with refractory anemia, but neutrophils exhibited a great heterogeneity in such cases.

REFERENCES

1. D.F. Bainton, J.L. Ullyot, and M.G. Farqhar, The development of neutrophilic polymorphonuclear leukocytes in human bone marrow. Origin and content of azurophil and specific granules, J. Exp.

Med.134 907 (1971).
2. S.J. Klebanoff and R.A. Clark, "The neutrophils : function and clinical disorders" North Holland Publishing Company, Amsterdam (1978).
3. P.L. Masson, J.F. Heremans, and E.Schonne, Lactoferrin, an iron-binding protein in neutrophilic leukocytes, J. Exp. Med. 130 : 643 (1969).
4. M. Baggiolini, C. de Duve, P.L. Masson , and J.F. Heremans, Association of lactoferrin with specific granules in rabbit heterophil leukocytes, J. Exp. Med. 131 : 559 (1970).
5. D.Y. Mason, C. Farrell, and C.R. Taylor, The detection of intracellular antigens in human leukocytes by immunoperoxidase staining, Br. J. Haematol. 31 : 361 (1975).
6. J.K. Spitznagel, M.R. Cooper, A.E. Mc Call, L.R. Dechatelet, and I.R.H. Welsh, Selective deficiency of granules associated with lysozyme and lactoferrin in human polymorphs (PMN) with reduced microbicidal capacity, J. Clin. Invest. 51 : 93 a (1972).
7. R.G. Strauss, K.E. Boue, J.F. Jones, A.M. Mauer, and V.A. Fulginiti, An anomaly of neutrophil morphology with impaired function, N. Engl. J. Med. 290 : 478 (1974).
8, R.T. Parmley, M. Ogawa , C.P. Darby Jr , and S.S. Spicer, Congenital neutropenia : neutrophil proliferation with abnormal maturation, Blood 46 : 723 (1975)
9. A. Komiyana, H. Morosawa , T. Nakahata, Y. Miyagawa, and T. Akabane , Abnormal neutrophil maturation in a neutrophil defect with morphologic abnormality and impaired function, J. Pediatr. 94 : 19 (1979).
10.J.Breton-Gorius, D.Y. Mason, D.Buriot, J-L. Vildé, and C. Griscelli, Lactoferrin deficiency as a consequence of a lack of specific granules in neutrophils from a patient with recurrent infections, Am. J. Path. 99 : 413 (1980).
11.J. Breton-Gorius, D. Houssay, and B. Dreyfus , Partial myeloperoxidase deficiency in a case of preleukemia : I Studies of fine structure and peroxidase synthesis of promyelocytes, Brit. J. Haematol. 30 : 273 (1975).
12.D. Buriot, J. Breton-Gorius, C. Griscelli, I. Olsson, and J. Hakim, Specific granules and lactoferrin deficiencies of neutrophils in a child with repeated infections (submitted for publication).
13.J. Breton-Gorius, D. Houssay, J-L. Vildé, and B. Dreyfus, Partial myeloperoxidase deficiency in a case of preleukemia. II Defects of degranulation and abnormal bactericidal activity of blood neutrophils, Brit. J. Haematol. 30 : 279 (1975).
14.D.R. Ambruso, and R.B. Johnston Jr, Lactoferrin enhances hydroxyl radical (.OH) production by a particulate fraction (PF) of neutrophils, Clin. Res. 28 : 490 A (1980).
15.D.F.Bainton, Sequential degranulation of the two types of polymorphonuclear leukocyte granules during phagocytosis of microorganisms, J. Cell. Biol. 58 : 249 (1973)
16.J.G.White, and R.D. Estensen, Selective labilization of specific

granules in polymorphonuclear leukocytes by phorbol myristate acetate, Am. J. Path. 75 : 45 (1974).
17. R.R. Arnold, M.F. Cole, and J.R. Mc Ghee, A bactericidal effect for human lactoferrin, Science 197 : 263 (1977).
18. J.J. Bullen, and J.A. Armstrong, The role of lactoferrin in the bactericidal function of polymorphonuclear leukocytes, Immunology 36 : 781 (1979).
19. M.S. Lefell, and J.K. Spitznagel , Fate of human lactoferrin and myeloperoxidase in phagocytosing human neutrophils : effects of immunoglobulin G subclasses and immune complexes coated on latex beads, Infect. Immun. 12 : 813 (1975).

DISCUSSION

Edited by: R.A.Harkness

VALERIUS: Were the parents related and were you able to show a decrease in the lactoferrin content of PMN from the parents?

VILDE': The parents were first cousins. The lactoferrin in the father's PMN by an immunocytochemical method was normal; it was not possible to study the mother.

JOHNSTON: The normal killing by cells from your patient is different to the results from the other three patients. Was there anything special about the organisms you or the other investigators used: how consistently were you able to determine the infecting organism in your patient's infections?

VILDE': The intracellular killing of the other 3 patients were near to normal values: 90% killing with Staph aur, 84% for Komiyana's patients, 86% for Streuss' patient. We used Staph aur and Strep marescens from human infections. The infecting organisms in our patient were mainly Staph aur.

DENSEN: Since these cells show an increase in the number of azurophil granules could the defect be in cell maturation before the development of specific granules and maturation of the cell membrane thus accounting for the depressed respiratory burst?

VILDE': The different pattern of degranulation, protein and enzymic content of the azurophil and specific granules suggest that this is unlikely. There could be a defect in the maturation of the cell membrane accounting for the depressed respiratory burst.

ROSSI: Have other authors studied the respiratory burst? Did you try a soluble stimulant like PMA? Have you any information on the oxidase activity?

VILDE': In Spitznagel's case NBT reduction was normal. In Strauss' patient NBT reduction, O_2 consumption, O_2^- and H_2O_2 production were normal. We did not try PMA stimulation nor did we study oxidase activity since the child stayed only a short time.

WEENING: Since Boxer and his colleagues suggested that lactoferrin plays a role in the adhesiveness of PMN it would be interesting to know whether the aggregation and adhesion was normal. In normal cells the iron saturated lactoferrin might react with O_2^- and/or H_2O_2 so that some increase in the respiratory burst might have been predicted.

VILDE': Aggregation and adhesion was not studied quantitatively but spreading was normal. Lactoferrin can increase the production of hydroxyl radicals from the particulate fraction of PMN (Clin. Res. 28, 499A, 1980). In this case it is possible that there is a dissociated partial defect of oxidase.

DISCUSSION

Edited by: R.A.Harkness

BABIOR: Were chromosomal preparations examined from these patients?

VILDE': They were normal.

BABIOR: What is the biochemical explanation for a situation in which a single mutation leads to the loss of a complex organelle with all its constituents? Do you have a hypothesis?

VILDE': It seems difficult to believe that a single mutation (and its consequence the absence of lactoferrin) leads to the loss of secondary granules. Absence of lactoferrin is observed in PMN, but the level of lactoferrin in saliva was normal in this case. Thus we can believe that the absence of lactoferrin in PMN is a consequence of the absence of its natural bag, i.e. secondary granules. The gene responsible for lactoferrin synthesis at least in saliva is normal. The absence of secondary granules is probably the first and main step in these cases. The reason for this is unknown: genetic absence of a signal at the myelocyte stage or of the envelope of secondary granules? Another possibility is that lactoferrin from PMN, plasma, and superficial mucosa are quite different and are synthesized by different genes.

BABIOR: In view of the currently held idea that lysosomes are made by the internalization and packaging of specific secreted proteins containing a sugar marker which signals the cell to take them back up, maybe the mutation leads to a loss of the signal from the group of proteins destined to form the contents of the specific granule. This could perhaps be tested by looking for lactoferrin or TCIII secretion from marrow cultures from the patients.

VILDE': Specific granules are quite distinct from lysosomes; the nature of their envelope and the mechanism of their formation are unknown. It is quite possible that the synthesis of lactoferrin by PMN, as the synthesis of other proteins of specific granules, is normal per se, and that the defect involves a metabolic signal from the protein of the future specific granules. In this hypothesis, lactoferrin would be synthetized, but not packed and would escape from granulocytes. The assay of lactoferrin which is in progress in this case of congenital lactoferrin deficiency, will probably bring the answer. It has been shown that PMN derived from normal colony-forming cells and obtained by culture of normal bone marrow do not have specific granules (Parmley,R.T., Ogawa,M., Spicer, S.S., Wright,J. Ultrastructure and cytochemistry of human granulocytic cells in culture. J. Histochem. Cytochem. 23, 316 (1975)). This defect is probably related to the artificial conditions of maturation. Thus, it will be impossible to interprete the expected negative result for the presence of specific granules and lactoferrin in PMN obtained from the bone marrow culture of such patients.

FUNCTIONAL AND METABOLIC ABNORMALITIES OF DIABETIC MONOCYTES

Harry R. Hill, Nancy A. Hogan, Marvin L. Rallison, Jose I. Santos, Richard P. Charette, Mitsuo Kitahara

Departments of Pediatrics, Pathology and Medicine
University of Utah College of Medicine
Salt Lake City, Utah 84132, U.S.A.

INTRODUCTION

There is considerable controversy about whether diabetic individuals suffer more infections than do normal individuals[1]. There is no question, however, about the increased severity of infection in the diabetic patient or the fact that infection greatly complicates the diabetic state. Although numerous studies have examined the functional activity of polymorphonuclear leukocytes (PMNs) in diabetic patients, almost none have investigated monocyte activity. The present studies were undertaken to assess the functional and metabolic activity of monocytes from patients with diabetes mellitus.

RESULTS

Metabolic Studies

All patients studied had uncomplicated diabetes mellitus. Blood sugar levels ranged from 215 mg % to 712 mg % while BUNS and creatinines were in the normal range. None of the patients were suffering from acute or chronic infections or other inflammatory disorders. Monocyte metabolic activity was assessed by determining chemiluminescence generation upon exposure to opsonized zymosan or $^{14}CO_2$ production from ^{14}C-1-glucose by previously described methods[2]. As illustrated in Figure 1, mononuclear cell preparations from diabetic individuals generated significantly more chemiluminescence than did controls[2]. Among the 27 diabetic individuals studied, 22 had monocytes which significantly exceeded the chemuluminescence produced by the simultaneously tested control monocytes. The peak in CL generation in the diabetics (32.3 ± 8.1 x 10^3 cmp) was significantly greater than that observed in the controls (25.8 ± 3.8 x 10^3 cpm, p 0.005)[2].

Additional studies indicated that hexose monophosphate shunt activity as determined by $^{14}CO_2$ generation from ^{14}C-1-glucose was also increased in the diabetic monocytes (Figure 2). Upon

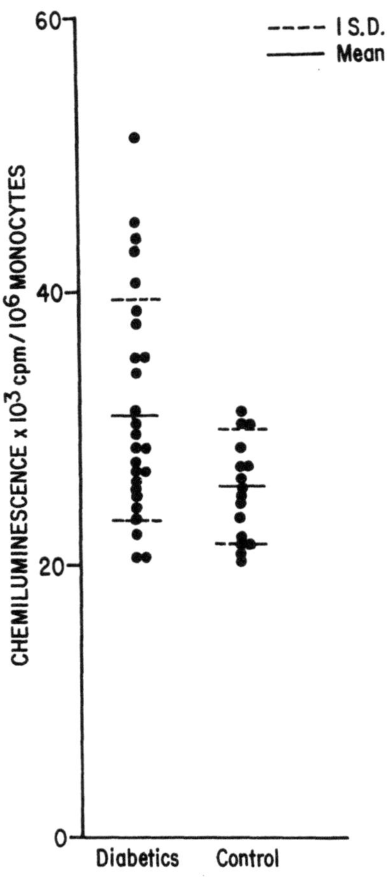

Figure 1. Chemiluminescence generation by diabetic and control monocytes exposed to opsonized zymosan. From Diabetes 29:251, 1980.

stimulation with opsonized zymosan the monocytes of the diabetic patients generated a larger increase in $^{14}CO_2$ than did controls (667% versus 425%, p=0.018)[2]. Similarly superoxide dismutase (SOD) inhibitable reduction of cytochrome c was also increased in the diabetics (9.77 ± 3.06 n mol/10^8 mono/20 min) when compared to the controls (6.36 ± 1.10 n mol, p=0.004)[2]. Thus

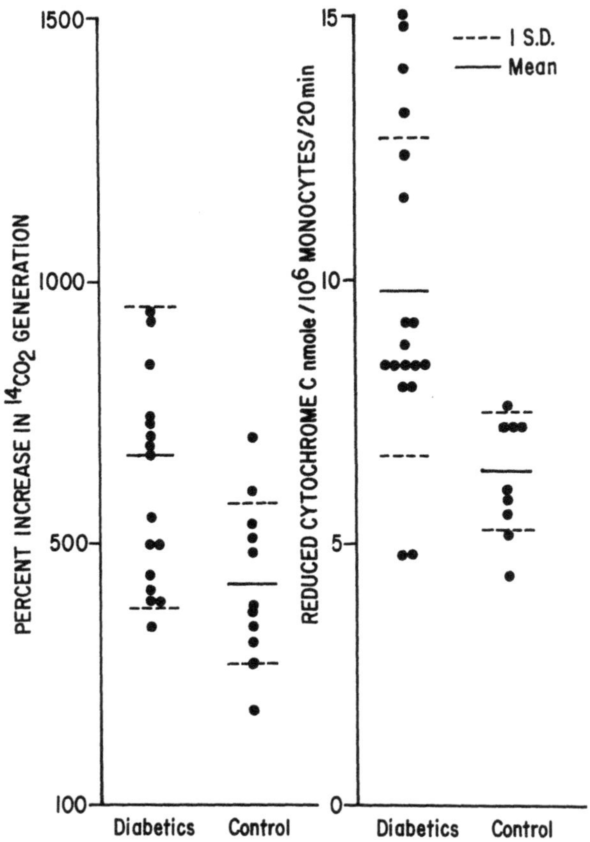

Figure 2. Hexose monophosphate shunt activity and superoxide generation by diabetic and control monocytes. From Diabetes 29:251, 1980.

chemiluminescence generation, hexose monophosphate shunt activity and superoxide production all seem to be enhanced in the monocytes of patients with diabetes mellitus.

In order to determine the mechanism of enhanced metabolic activity in the diabetic monocyte, we next determined the effect of glucose on monocyte chemiluminescence production. Addition of glucose to whole blood samples from diabetics and controls so that the final concentration ranged from 140 to 500 mg % significantly increased CL generation in each instance. Similarly glucose, but

Table 1. Effect of glucose and 2 deoxyglucose on monocyte superoxide generation.

Phagocytic Mixture	SOD Inhibitable cytochrome reduction (n mol/10^6 cells/20 min)
MNS + Opsonized Zymosan (OZ)	11.45 ±.1.12
MNS + OZ + Glucose (100 mg/dl)	28.03 ± 3.40
MNS + OZ + 2 Deoxyglucose (100 mg/dl)	7.33 ± 0.34

not 2-deoxy-glucose or fructose, increased superoxide generation from diabetic and normal monocytes (Table 1). Thus, the concentration of glucose appears to markedly affect the respiratory burst of mononuclear phagocytes and result in increased generation of toxic oxygen products.

Chemotactic Studies

Defective chemotactic activity of diabetic PMNs has been observed by several investigators[3,4,5]. The mechanisms involved in abnormal motility of these cells has not been determined. Furthermore, no studies of monocyte chemotactic responsiveness have been reported in diabetics. We have examined the chemotactic responsiveness of monocytes from 32 juvenile (Type I) and 13 adult-onset (Type II) diabetics in an under agarose system[6]. All patients were free from ketoacidosis at the time of testing and only 3 suffered from diabetic complications. One had diabetic retinopathy, 2 neuropathy and 1 nephropathy. The under agarose system was similar to that described by Nelson and coworkers[7]. Briefly 35 mm x 10 mm culture dishes were flooded with 3 ml volumes of 1.5% agarose (Type II, Sigma Checmical Co., St. Louis, Mo.) in McCoys media with 20% fetal calf serum. Three mm wells were then cut in the agar 5 mm apart. Ten ul of a Ficoll-Paque (Pharmacia) separated mononuclear cell suspension containing 2 x 10^7 monocytes/ml were added to the center well. Media alone or zymosan activated serum (10 mg zymosan/ml serum rotated at 37°C for 1 hour) was placed in the peripheral wells. The plates were then incubated for 18 hours at 37°C in a humidified atmosphere with 5% CO_2. The plates are then inverted and examined microscopically employing a micrometer and 5 x 5 mm ocular grid. The distance migrated by the leading front of three cells moving toward the chemoattractant or media in two 5 x 5 mm grids, placed immediately adjacent to the center cell well, was determined.

The data from 45 diabetics and 25 controls is summarized in Table 2[7]. No differences between diabetic and normal monocytes were noted in the total distance migrated toward the chemoattractant or media alone (data not shown). In contrast, many fewer diabetic monocytes moved out of the central cell well into the area covered by the two adjacent 5 mm x 5 mm grids toward the chemoattractant or to media alone. Thus, the overall chemotactic responses as well as random motility of the diabetic monocytes were markedly depressed when compared to the controls.

Table 2. Chemotaxis and random migration of diabetic and control monocytes.

	Distance Migrated To Activated Serum	Number of MNS Moving to AS	Number of MNS Moving to Media
Controls (25)	61.6 ± 4.2*	446.1 ± 38.3	225.6 ± 24.3
Diabetics (45)	60.7 ± 3.4	292.7 ± 28.2	121.1 ± 13.6
p**	0.77	0.0056	0.0004

* Mean ± SEM
** Unpaired t test

Diabetic monocytes appear, therefore, to have enhanced respiratory burst activity, increased hexose monophosphate shunt activity and increased generation of superoxide upon stimulation. At the same time they have depressed chemotactic activity. For these reasons, we next sought to determine if antioxidants might have an effect on chemotactic responsiveness. As indicated in Table 3, incubation of diabetic monocytes with 2, 3-dihydroxybenzoic acid markedly increased their chemotactic responsiveness while not affecting that of controls. Additional studies were carried out following oral vitamin E therapy. Vitamin E was administrered in a dose of approximately 25 IU per kilogram per day over a 2 to 3 week period. Chemotaxis and random motility of the patients' and control monocytes were studied before and after such therapy. In all, 8 patients and 6 controls

Table 3. Effect of 2,3-dihyroxybenzoic acid on diabetic and control monocyte chemotaxis.

	Chemotaxing MNs				
Subject	Media	10^{-6}M DHB	10^{-5}M DHB	10^{-4}M DHB	10^{-3}M DHB
Diabetic S.C.	156±36	224±41	260±14	219±38	174±54
Control E.C.	332±25	291±26	314±32	263±23	281±34

were included in this phase of the study. Monocyte chemotactic responsiveness and random motility increased to approximately 200% of pretreatment values in the patients[6]. The controls also increased but only to a level 132% of pretreatment values.

DISCUSSION

These studies indicate that diabetic monocytes have enhanced respiratory burst activity and generate increased levels of superoxide and other toxic oxygen products[1]. Furthermore, addition of increasing concentrations of glucose to whole blood or normal leukocyte preparations results in enhanced activity similar to that seen with the diabetic cell[2,8]. Diabetic monocytes were also found to have profound defects in both random motility and chemotaxis[6].

A number of investigators have observed abnormalities in the cellular components of the acute inflammatory response in diabetics. DaCosta and Beardsley[9] first reported that diabetic plasma lowered the phagocytic activity of normal polymorphonuclear leukocytes (PMNs). Bagdade et al.,[10] later observed a decrease in PMN phagocytic activity that could be correlated with the level of fasting blood sugar. Mowat and Baum[4], Miller and Baker[5] and Hill and coworkers[3] found that the PMNs of juvenile and adult onset diabetics had defective chemotactic responses. The mechanism of abnormal phagocyte function in these studies was not precisely defined, however.

Baehner and coworkers[11] have shown that H_2O_2 or a superoxide -H_2O_2 generating system of xanthine-xanthine oxidase decreases directed movement and ingestion by PMNs. Moreover, administration of vitamin E to volunteers prior to leukocyte testing abolished the depressive effect of H_2O_2 on phagocytic uptake. Boxer et al.,[12] showed that 2,3-dihydroxybenzoic acid, an effective iron chelator and scavenger, acted <u>in vitro</u> to increase both chemotaxis and phagocytic uptake by PMNs. Later studies by Boxer and coworkers[13] showed that oral vitamin E therapy promoted normal PMN phagocytosis and killing and improved disordered microtubule assembly in a patient with glutathione synthetase deficiency. These studies suggest that autooxidative damage may profoundly affect phagocytic activity and motility and that scavengers of toxic oxygen products may reverse these abnormalities <u>in vitro</u> and <u>in vivo</u>.

In the present studies, monocytes from patients with diabetes mellitus were found to have enhanced respiratory burst activity and increased generation of superoxide anion and chemiluminescence. Markedly defective chemotactic responses were also observed which were correctible <u>in vitro</u> with 2,3-dihydroxybenzoic acid and <u>in vivo</u> with oral vitamin E therapy.

These results suggest that auto-oxidative membrane damage may have a role in the pathogenesis of abnormal diabetic phagocyte function.

In addition to offering at least one explanation for defective PMN and monocyte function in diabetes, the effect of hyperglycemia on toxic oxygen radical generation may have even more profound biologic significance. Several reports have reviewed the potential ill effects of free radical oxidation and have suggested that auto-oxidation is at least one contributing factor to the process of aging[14,15]. It is conceivable, therefore, that chronic hyperglycemia may affect auto-oxidative mechanisms throughout the body and contribue to the early development of degenerative processes which are often observed in the diabetic patient.

REFERENCES

1. Hill, H. R., 1979, Western J. Med. 130:547.
2. Kitahara, M., Eyre, H.J., Lynch, R.E., Rallison, M.L. and Hill, H.R., 1980, Diabetes 29:251.
3. Hill, H.R., Saule, H.S., Dettloff, J.L. and Qui ,P.G., 1973, Clin. Immunol. and Immunopath. 2:395, 1974.
4. Mowat, A.G., and Baum, J. 1971, New Eng. J. Med. 284:621.
5. Miller, M.E., and Baker, L., 1972, J. Pediatr. 81:979.
6. Hill, H.R., Hogan, N.A., Rallison, M.L. and Santos, J.I. submitted.
7. Nelson, R.D., Quie, P.G., Simmons, R.L., 1975, J. Immunol. 115:1650.
8. Kitahara, M., Lynch, R.E., Eyre, H.J. and Hill, H.R., submitted.
9. DeCosta, J.C. and Beardsley, E.J.G., 1908, Amer. J. Med Sci. 126:361.
10. Bagdade, J.D., Nielson, K.L. and Bulger, R.J., 1972, Amer. J. Med. Sci. 263:452.
11. Baehner, R.L., Boxer, L.A., Allen, J.M., and Davis, J., 1977, Blood 50:327.
12. Boxer, L.A., Allen, J.M. and Baehner, R.L., 1978, J. Lab. Clin. Med. 92:730.
13. Boxer, L.A., Oliver, J.M., Spielberg, S.P., Allen, J.M. and Schulman, J.D., 1979, New Engl. J. Med. 301:901.
14. McCord, J.M. and Fridovich, I., 1978, Am. Dnt. Med. 89:122.
15. Harman, D. and Piette, L.H., 1966, J. Gerontol, 21:560.

DISCUSSION

Edited by: R.A.Harkness

KELLER: It seems possible that the diverging results obtained with the two different procedures for assessing random and directional locomotion are clinically relevant; sufficient numbers of cells may need to accumulate rapidly at the site of an infection. Did you find other examples for such differences in clinical studies?

HILL: No one knows the biological significance of the various assays of 'chemotaxis'. The leading front of diabetic cells moved as well as controls but there was a reduction in the total number of cells moving. Correlations with in vivo assays are needed to determine the biological relevance of the in vitro assays.

ESMANN: In diabetic leucocytes G-6-P is increased; did you measure this metabolite in diabetic monocytes?

The results of the experiments with both cytochalasin and glucose could be interpreted on the basis of known changes in PMN in which irrespective of additional glucose, glycogen is always degraded on phagocytosis. G-6-P concentration is thus maintained.

At this meeting it has been repeatedly said that diabetic patients are prone to infection. However, a recent review of the literature showed that when patients with late diabetic complications are left out no increased incidence remains with the exception of tuberculosis, for example in Sweden in 1945.

HILL: Although metabolic variables in the diabetic cell are altered we have not yet measured G-6-P or ATP.

The concentration of cytochalasin B used should almost totally block phagocytosis but particle contact with the cell could trigger glycogen release.

I agree that the data on infections in diabetics is conflicting. Most of our patients were suffering from recurrent or serious infections; their chemotactic defect was not so profound as that seen in patients with serious infections like hyperimmuno-globulinaemia E. Infection in diabetics does tend to be severe.

ENZYMATIC DEFICIENCY IN MONOCYTES FROM PATIENTS WITH CHRONIC GRANULOMATOUS DISEASE

Marie-Anne Gougerot-Pocidalo[+], Diego Buriot[++], Claude Griscelli[++] and Jacques Hakim[+]

[+] - Laboratoire d'Immunologie et d'Hématologie - CHU Xavier Bichat - Université Paris VII, and
[++] - Unité d'Immunologie et d'Hématologie Pédiatriques - Département Pédiatrie et INSERM (U 132), Hôpital des Enfants Malades - PARIS - F R A N C E

Chronic granulomatous disease (CGD) is by convention defined as a disorder involving a high susceptibility to bacterial infection related to the inability of the patient's neutrophils to increase oxygen consumption, despite normal phagocytosis and normal degranulation[1,2]. CGD appears to be heterogeneous ; most cases show an inherihance pattern typical of X-linkage, but the remainder are of non X-linked variety[2]. CGD heterogeneity has been further suggested by the lack of membrane components in some CGD patients as kx Antigen in neutrophils or red blood cells[3], and b-like cytochrome in neutrophils[4,5]. In two different patients with CGD it has been possible to induce the oxidative burst, with stimuli other than those used in standard analysis procedure[6,7].

Monocytes from CGD patients have been less studied than neutrophils but have been shown to be also defective in their ability to kill bacteria[8,9].

Our purpose here is 1) to study monocytes' functions in parallel with those of neutrophils. In this preliminary work we studied cytochemical NBT reduction.
2) to investigate if a deficiency of any of the monocytes' enzymes could allow to recognize a new form of CGD.

We show in this work that α-naphtyl butyrate esterase activity, a well known cytoplasmic monocyte marker[10], may be absent in some cases of CGD and that it should be added to the other molecular markers already employed to distinguish the different forms of CGD.

PATIENTS AND METHODS

The diagnosis of CGD in six unrelated children : 5 boys and one girl with high susceptibility to infection was based on the results of functional studies of their neutrophils. Using previously described techniques[11], neutrophils from the patients were unable, after stimulation with zymosan, to reduce nitroblue tetrazolium dye (quantitative NBT test) and cytochrome c, to increase oxygen consumption, generate H_2O_2, or to iodinate. The percentage of patients' neutrophils, which reduced NBT upon stimulation with either endotoxin or opsonized zymosan was below 5 %[12]. In all patients, the rate at which neutrophils ingested heat-killed non virulent ^{14}C-labelled Klebsiella (100 micro-organisms per neutrophil) measured as previously described[13], was above control values[14].

Mononuclear cells, isolated by Ficoll-Hypaque gradient centrifugation[15] were washed three times with phenol-free Hank's balanced salt solution, pH 7.40 (HBSS), and then counted in a hemocytometer with Turk's solution. The monocytes of this cell population were enumerated after being identified by their α-naphtyl butyrate esterase activity or by peroxydase staining[16,17] in fluid medium using Technicon Hemalog D reagents. The monocyte concentration was adjusted to 10^6 per ml in HBSS containing 20 % human AB serum. A suspension of monocytes (1 ml) was deposited in 35 x 10 mm tissue culture dishes (3001 Falcon, Oxnard,California), and incubated for 45 min at 37°C in a 5 % CO_2 atmosphere, to allow adherence of the monocytes. Fifty μl of opsonized zymosan A from <u>Saccharomyces cerevisiae</u> (2.5 mg/ml of HBSS) and 200 μl of filtered NBT (2 mg/ml of HBSS) where then added to the suspending fluid. The tissue culture dishes were further incubated at 37°C with 5 % CO_2 for 30 min. Adherent cells were washed three times and treated by either α-naphtyl butyrate esterase[18,19] or peroxidase[20] reactions. Cells were further incubated for 10 min at 37°C, washed with distilled water and air dried. At least 200 adherent cells per dish were examined under a microscope by two cytologists who determined the percentages of each of the four following cell types : cells containing zymosan, those with a deeply-stained red cytoplasm after the α-naphtyl butyrate esterase reaction, cells containing dark formazan precipitates, and cells containing fine scattered precipitates formed by the peroxidase reaction.

RESULTS AND DISCUSSION

Most control cells adherent to the culture dishes were monocytes, as shown (Tab.1) by their characteristic ability to ingest zymosan particles and to reduce NBT, or by the positive

Table 1 : Cytochemical features of adherent mononuclear cells from CGD patients and controls.

	Cells containing zymosan	Cells containing formazan precipitates	α-naphtyl esterase positive cells	peroxidase positive-cells
- CONTROLS (N = 10)				
Range	76-90	66-84	79-86	60-73
M ± 1 SD	82 ± 5	76 ± 6	83 ± 3	67 ± 5
- CGD PATIENTS				
1 : F	79	2	79	69
2 : M	96	0	96	64
3 : M	70	1	71	70
4 : M	78	0	75	70
5 : M	90	0	0	76
6 : M	86	0	0	72

Results for mononuclear cells containing zymosan, cells containing formazan precipitates, α-naphtyl esterase-positive-cells, and peroxidase-positive-cells are expressed as the percentage of the mononuclear cells which adhered to the culture dishes.
F = female ; M = male.

α-naphtyl butyrate esterase and peroxidase reactions[10,20]. For the latter reaction, the monocyte staining pattern is easily distinguishable from that of the neutrophils[10,20].

As expected in CGD patients[18,19,21], the percentage of cells ingesting zymosan was normal whereas NBT was not reduced. The peroxidase staining pattern of CGD monocytes was similar to that of control monocytes. Surprinsingly, the monocytes of two boys out of the five children with CGD were α-naphtyl butyrate esterase-negative (Tab.1). This was obvious both when monocytes were examined in cell suspensions during the isolation procedure, and after their adherence. The lack of α-naphtyl butyrate esterase activity in the monocytes from these two patients was confirmed twice at intervals of 10 days and one month. In contrast, the neutrophils and monocytes from these two boys, as well as those from the three other children with CGD, were normally stained by the granular naphtol AS-D-acetate esterase reaction[22].

In conclusion, monocyte oxidative burst, as expected, was deficient in all patients with CGD. The new information reported in this study is that monocytes from some patients with CGD may be deficient in α-naphtyl butyrate esterase activity. The reason for this deficiency remains unexplained. It cannot be related to chronic infection or to drugs, because monocyte α-naphtyl butyrate esterase activity was found to be normal in control children with infection and because our patients were neither being treated nor infected when they were studied. Several recent reports suggest that esterases are involved in O_2^- production by human neutrophils and monocytes[23-25]. In all our patients, neutrophil and monocyte granular naphtol AS-D acetate esterase activities were normal, and monocyte cytoplasmic α-naphtyl butyrate esterase staining was deficient in two patients only. It is therefore difficult as the present stage, to assign to this enzyme a role in the oxidative burst of the phagocytes.

Our findings suggest that the monocyte α-naphtyl butyrate esterase reaction may help to define new subgroups of CGD patients. Further studies are needed, however, to define CGD patients by all their known molecular markers viz Kx antigen in neutrophils and red blood cells[3], b-like cytochrome in neutrophils[4,5], and monocyte α-naphtyl butyrate esterase.

Acknowledgements : The authors wish to thank Miss C. Delpont for her skilled secretarial help. This study was supported by INSERM grant N°ATP 61.78.93.

REFERENCES

1. B. Holmes, A.R. Page, and R.A. Good, Studies of the metabolic activity of leukocytes from patients with a genetic abnormality of phagocytic function, J. Clin. Invest. 46:1422 (1967).
2. R.B. Jr. Johnston, and I.L. Newman, Chronic granulomatous disease, Pediatr. Clin. North Am. 24:365 (1977).
3. W.L. Marsh, R. Oyen, and M.E. Nichols, Kx antigen, the McLeod phenotype, and chronic granulomatous disease : further studies, Vox Sang. 31:356 (1976).

4. A.W. Segal, O.T.G. Jones, D. Webster and A.C. Allison, Absence of a newly described cytochrome b from neutrophils of patients with chronic granulomatous disease, Lancet 2:446 (1978).
5. N. Borregaard, K.S. Johansen, and V. Esman, Quantitation of superoxide production in human polymorphonuclear leukocytes from normals and 3 types of chronic granulomatous disease, Biochem. Biophys. Res. Commun. 90:214 (1979).
6. D.S. Weening, D. Roos, C.M.R. Weemaes, J.W.T. Homan-Müller, and M.L.J. VanSchaik, Defective initiation of the metabolic stimulation in phagocytozing granulocytes : a new congenital defect, J. Lab. Clin. Med. 88:757 (1976).
7. L. Harvath, and B.R. Andersen, Defective initiation of oxidative metabolism in polymorphonuclear leukocytes, N. Eng. J. Med. 300:1130 (1979).
8. N.C. Davis, H. Huber, S.D. Douglas, and H.H. Fudenberg, A defect in circulating mononuclear phagocytes in chronic granulomatous disease of childhood, J. Immunol.101:1093 (1968)
9. G.E. Rodey, B.H. Park, D.B. Windhorst, and R.A. Good, Defective bactericidal activity of monocytes in fatal granulomatous disease, Blood 33:813 (1969).
10. F. Schmalzl, and H. Braunsteiner, The cytochemistry of monocytes and macrophages, Ser. Haemat. III:93 (1970).
11. E. Cramer, C. Auclair, J. Hakim, E. Feliu, J. Boucherot, H. Troube, J.F. Bernard, E. Bergogne, and P. Boivin, Metabolic activity of phagocytozing granulocytes in chronic granulocytic leukemia : ultrastructural observation of a degranulation defect, Blood 50:93 (1977).
12. D.B.Windhorst, A.R. Page, B. Holmes, P.G. Quie, and R.A. Good, The pattern of genetic transmission of the leukocyte defect in chronic granulomatous disease, J. Clin. Invest. 47:1026 (1967).
13. M. Torres, D. DeProst, J. Hakim, and M.A. Gougerot, Metabolic activity of human polymorphonuclear leucocytes. Relation to ingestion rate, Europ. J. Clin. Invest. 9:209 (1979)
14. D.W. Biggar, Phagocytosis in patients and carriers of chronic granulomatous disease, Lancet 1:991 (1975)
15. A. Boyum, Isolation of mononuclear cells and granulocytes from human blood, Scand. J. Clin. Lab. Invest. 21(Suppl.97):77 (1968).
16. L. Ornstein, H. Awsley, and A. Saunders, Improving mammal differential white cell counts with cytochemistry,Blood cells 2:557 (1976).
17. S.B. Tucker, R.V. Pierre, and R.E. Jordon, Rapid indentification of monocytes in a mixed mononuclear preparation, J. Immunol. Meth.14:267 (1977).
18. H.W. VonHeyden, and D. VonHeyden, Characteristics of macrophages in vitro derived from peripheral blood cells, Blut 29:37 (1974)
19. J.K. Lace, J.S. Tan, and C. Watanakunakorn, An appraisal of the nitroblue tetrazolium test, Amer. J. Med. 58:685 (1975).

20. P.T. Fan, D.T.Y. Yu, C.M. Pearson, and R. Bluestone, Human monocyte-lymphocyte interaction : a new technique, J. Immunol. 119:156 (1977).
21. A.L. Sagone, G.W. King, and E.N. Metz, A comparison of the metabolic response to phagocytosis in human granulocytes and monocytes, J. Clin. Invest. 57:1352 (1976).
22. G. Flandrin, and M.T. Daniel, Practical value of cytochemical studies for the classification of acute leukemias, in : "Nomenclature,Methodology and Results of Clinical Trials in Acute Leukemias", G. Mathe, P. Pouillart, and L. Schwarzenberg, eds., Heinemann, London (1973).
23. S. Kitawaga, F. Takatu, and S. Sakamoto, Evidence that proteases are involved in superoxide anion production by human polymorphonuclear leukocytes and monocytes, J. Clin. Invest. 65:74 (1980).
24. L. Simchowitz, J. Mehta, and I. Spilberg, Chemotactic factor-induced superoxide radical generation by human neutrophils. Requirement for proteinase (esterase) activity. J. Lab. Clin. Med. 94:403 (1979).
25. B.D. Goldstein, G. Witz, M. Amaruso, and W. Troll, Protease inhibitors antagonize the activation of polymorphonuclear leukocyte oxygen consumption, Biochem. Biophys. Res. Commun. 88:854 (1979).

DISCUSSION

Edited by: R.A.Harkness

KELLER: Did lymphocytes also lack the -naphtylbutyrate esterase in those patients who had no enzyme in their monocytes?

HAKIM: This was not investigated.

JOHNSTON: Was there any difference clinically in the two patients with esterase deficiency? Mandell (University of Virginia) has some patients about 40-50 years of age who are infrequently infected. Their monocytes function normally but their PMN are abnormal as in CGD. Could this esterase deficiency be a marker for this milder form of CGD?

HAKIM: There was no obvious clinical difference in our patients but they were young.

BABICR: Did you study formazan reduction and α-naphthyl butyrate esterase in the mothers?

HAKIM: The mother of one affected child showed defective NBT reduction but α-naphthyl butyrate esterase staining of the monocytes was normal.

USE AND RESULTS OF NEUTROPHIL FUNCTION TESTING IN PEDIATRIC IMMUNOLOGY

Antonio G.Siccardi[1], Alberto G.Ugazio[2], Fulvio Sacchi[2], and Suresh D.Jayakar[3]

[1]Istituto di Genetica and [2]Clinica Pediatrica, Università di Pavia, [3]Laboratorio di Genetica Biochimica ed Evoluzionistica, CNR, Pavia

A working group (Table 1) has been set up in Pavia to evaluate the relevance of neutrophil function in patients suffering from recurrent infections.

We have been studying neutrophil function tests for the last four years in addition to the more standardized technologies of lymphocyte function tests and measurement of immunoglobulin and complement in routine laboratory work on patients suffering from immunodeficiency, or in general from proneness to severe, prolonged or untreatable bacterial and fungal infections.

The aim of this group was to set up a complete and reliable battery of routine analyses which uses a reasonably small volume of blood and allows a clear definition of the limits of normality in

Table 1. Collaborators

Istituto di Genetica:	A.Clivio, C.Dos Santos, A.Fortunato, F. Grassi and R.Spalla
Clinica Pediatrica:	A.Beretta, M.Duse, F.A.Ferrari, G.Maggiore, M.Marconi, G.Marsaglia, A.Martini, L.Nespoli and A.Pagani
Clinica Dermatologica:	A.Giannetti and R.Stefanoni

order to make the identification of the defects easier and unequivocal. This paper presents 1) an estimate of the reliability of the methods in use, the range of variation observed on about 200 normal controls and a definition of the limits of normality; 2) a preliminary estimate of the incidence of neutrophil defects among patients suffering from recurrent infections; 3) a summary of the most interesting cases we have studied in the last years.

1. Reliability and range of normality of the neutrophil tests in use.

All the tests in use (table 2) were adapted from existing methods with minor modifications designed to work on a small sample (5-6 ml of heparinized blood) and to obtain as complete a picture as possible of neutrophil function. Nine measurements are so obtained which define neutrophil function in its most important aspects: locomotion, phagocytosis and microbicidal activity. Only 8 million

Table 2. Neutrophil function tests in use

Test	Description	Neutrophils x 10^6/test (replicas)	Ref.
PHF, PHI, NRF	Phagocytosis frequency, phagocytosis index, NBT reduction frequency. The three measures are performed on plastic-adhered cells, exposed to serum-treated yeasts in the presence of NBT.	0.1 (x2)	6,7
PDMA	Phagocytosis-dependent metabolic activation. Leucocytes in suspension are exposed to latex beads. HMS stimulation (P/R) is evaluated by measuring the conversion of ^{14}C-1-glucose into $^{14}CO_2$.	0.25 (x4)	2,10
SaK, SfK	Killing of opsonized S.aureus and S.faecalis. Leukocytes in suspension are exposed to ca.0.1 bacteria/neutrophil to maximize phagocytosis. A colony count method.	0.6 (x2) 0.6	1,7
CaK	Killing of opsonized Candida albicans. A microscopic method based on dye exclusion by live yeasts.	1.0 (x2)	3
RFM, CFM	Random- and Chemotactic-Filter Migrations. Leading-front evaluation of the migration in Millipore filters. Serum+LPS as chemotactic stimulus.	0.5 (x2) 0.5 (x2)	12,13

neutrophils are required for the whole set of tests.

Leukocytes are purified by dextran-erythroprecipitation and low-speed spin washings.

Most of these tests allow to recognize both rate and capacity defects as it is detailed in ref[11].

To standardize the methods, we have analyzed a large number of healthy controls. To assess the reliability of each measurement, experiments were carried out to dissect the total observed variance into its main components, namely experimental error, variation in repeated measurements on the same individual, and variation among individuals (tab.3).

This analysis has provided us with a number of indications:
1) all nine tests are reliable since the experimental error is, in all cases, a minor component of the variation;
2) for five of the tests (PHI, NRF, Cak, SaK and RFM), since the major component of variation is inter-individual difference, the normalization of the data with reference to a "control for the day" would not be useful and would actually introduce an error;
3) for three tests (PHF, SfK and CFM) the between-repetition component is rather high but this could be an artifact of the extremely low overall variance of the three tests. In fact the deviations from the means that are likely to be due to the between-repetitions variation are only 5,6 and 8% of the means, respectively;
4) the case of PDMA is peculiar since this test shows a high overall variance (a coefficient of variation of 30%), a very low experimental error component (2%) and a high between-repetitions variance

Table 3. Components of variance

Test	Mean	Total	VARIANCE: Between individual	(%)	Between repetitions	(%)	Experimental error	(%)
PHF	90.88	31.8	2.6	(8.2)	23.1	(72.6)	6.1	(19.2)
PHI	3.28	0.846	0.457	(54.0)	0.188	(22.2)	0.201	(23.8)
NRF	91.85	26.9	14.8	(55.0)	5.9	(21.9)	6.2	(23.0)
PDMA	5.20	2.46	1.26	(51.2)	1.4	(42.3)	0.16	(6.5)
CaK	66.28	114.3	65.1	(57.0)	27.9	(24.4)	21.3	(18.6)
SaK	84.60	80.3	55.1	(68.6)	13.9	(17.3	11.3	(14.1)
SfK	90.52	48.6	14.1	(29.9)	28.9	(59.5)	5.6	(11.5)
RFM	34.49	143.5	83.6	(58.3)	56.0	(39.0)	3.9	(2.7)
CFM	98.67	124.1	45.6	(36.7)	61.7	(49.7)	16.8	(13.5)

(13%). This is probably the only case where the data should be considered with reference to "controls for the day".

To define pathological thresholds for each measurement we have studied the form of the distributions and compared them to normal defects; only one phagocytosis defect was found; the 12 microbicidal activity defects were very heterogeneous, comprising selective defects in the killing of Candida alone, Aspergillus alone, both Staphylococcus and Streptococcus, Staphylococcus and Candida but not Streptococcus (the CGD cases).

The most frequent defects are the locomotion defects and the majority of them (20 cases out of 24) were isolated defects (all other neutrophil functions studied being normal).

From these studies, it is clear that the single most informative test to perform on "recurrent infection" patients is CFM, followed by a panel of microbicidal activity tests (on Staphylococcus, Streptococcus, Candida, Aspergillus).

2. Case reports

In the last years, our group has reported on a number of cases of neutrophil functional defects which differed from the "classical", well documented, defects (such as CGD, myeloperoxidase deficiency, lazy leukocyte syndrome, etc.).

a. A familial defect in bactericidal activity[10]. A 4-year-old boy with recurrent infections and his clinically healthy father showed a severe, isolated defect in bactericidal activity of peripheral neutrophil leukocytes (the mother and the only sister were normal). Lymph nodes, spleen and liver of the child presented a massive infiltration by macrophages. Such infiltration and the segmentary albinism of the hair resemble traits of the Chediak-Higashi syndrome, but some of the most relevant traits of this syndrome are absent, since all other neutrophil functions were normal in our patient.

b. Disorders of candidacidal activity and other neutrophil functions in three cases of chronic candidiasis[7].

Three unrelated cases of recurrent and untreatable candidiasis have been followed for about one year. A persistent in vitro defect of neutrophil candidacidal activity characterized the three cases, while other neutrophil functions were shown to be only transiently impaired during the period of observation. None of the three cases could be identified with previously described granulocytopathies. In the first case, most of the neutrophil functional defects were transient and apparently secondary to the fungal infection. In fact, it is possible that the transient leukocyte defects were a consequence of an iron deficiency.

In the second case, we observed an unusual association of defective motility and reduced phagocytic capacity of objects beyond certain dimension. The defect in candidacidal activity should then be due only to the limited phagocytic capacity.

In the third case, the defect in candidacidal activity persisted even after transfer factor therapy, which however caused a normalization of the previously defective delayed type hypersensitive reaction to candidin.

c. A defect in neutrophil motility in two siblings with recurrent infections[8].

The same association of infections (otitis media, bronchitis, chronic diarrhoea) has caused seven fatalities in the paternal side of the family, suggesting genetic implications.

Association between otitis media, chronic diarrhoea, defective chemotaxis and hyper-IgE have been described; however, in the reported cases the defects were transient (or intermittent) and disappeared at clinical remission; no indication was found of the possible involvement of genetic factors.

Conversely, our patients suffer from a very similar association of symptoms, but the motility defects have persisted at clinical remission and the very remarkable family history suggests a familial prevalence of the condition. The mode of genetic transmission, however, is not obvious.

d. Functional neutrophils defects in Viscera Larva Migrans syndrome[4].

Severe defects of neutrophil function were found in a patient suffering from larva viscera migrans syndrome. Some of the defects (metabolic activation, bactericidal activity) were transient and normalized at remission, whereas a chemotactic defect persisted throughout the observation period.

The patient had persistently elevated IgE, even at remission; thus we could not distinguish between an intrinsic cellular motility defect (which could have played a role in the establishement of the parasitic disease) and a hyper-IgE syndrome.

e. Defective Aspergillus Killing by neutrophils in a case of systemic aspergillosis[5].

A persistent defect of Aspergillus killing was observed in the neutrophils of a 6-year-old patient with a systemic A.fumigatus infection which was highly refractory to anti-mycotic therapy. Aspergillus phagocytosis in vitro was normal, but nearly 80% of the ingested organisms (versus 30% in the controls) survived intracellularly during the 2-hr assay period. The patient's neutrophils showed a subnormal frequency of nitroblue tetrazolium reduction and a subnormal hexose-monophosphate-shunt activation in response to that of chronic granulomatous disease neutrophils tested for comparison.

The immune status of the patient and the following properties of his neutrophils were found to be normal: random and distribution by a goodness of fit test. Only two measurements (CFM and RFM) and the log-transform of a third one (PDMA) distribute normally and can thus be used as such. The other six measurements had distributions with a truncated-normal form.

A computer method was devised in order to estimate the means and standard deviations of the complete normal distributions of which the observed truncated distribution would be a portion. This method consisted of trying out ten equidistant "trial" mean values in a suitable range on each side of the raw mean (that between this raw mean and the extreme values), estimating the standard deviation from the data on the non-truncated side of the mean for each trial value, and testing for goodness of fit of a normal distribution for the observed data. The best estimate of the mean (and standard deviation) is ideally that which gives the best fit.

Table 4 shows, for each measurement, the raw mean, the type of distribution fitted, and the "trial" mean chosen, e.g. +7 indicates that the mean chosen is: raw mean +7/10 (right extreme value - raw mean). It also shows the estimate of the standard deviation and the P value for the goodness of fit.

Having estimated the appropriate means and standard deviations for all the variables, we can define pathological thresholds (PT) as the mean minus 1.96 times the standard deviation. The pathological thresholds for the nine measurements are also shown in Table 4.

Table 4. Raw means, type of distribution, estimated means and SD and Pathological Thresholds (PT) of the nine variables.

Test	No.	Raw Mean	Distr.	Trial No.	Mean	S.D.	P	PT
PHF	162	90.88	TR	+7	95.75	8.46	0.198	79.17
PHI	145	3.28	TL	-1	3.28	1.03	0.340	1.26
NRF	163	91.85	TR	+4	94.00	7.10	0.069	80.08
PDMA	176	5.20	L	0	5.01	1.07	0.488	2.92
CaK	69	66.28	TL	-2	64.55	12.85	0.558	41.09
SaK	139	84.60	TR	+3	87.48	11.08	0.164	65.76
SfK	88	90.52	TR	+3	92.22	8.90	0.230	74.78
RFM	86	34.49	N	0	39.49	11.98	0.347	11.01
CFM	87	98.67	N	0	98.67	11.14	0.180	76.84

The clinical value of this panel of tests is far from being established; for example, it is not yet clear whether all these tests are really necessary, or whether some of them are uninformative and represent partial duplications of other tests. This information can only come from the analysis of data obtained on populations of patients; in fact, we have been using this panel of tests to study neutrophil functions in various kinds of patients (affected by "recurrent infections", or by respiratory viroses, hepatopathies, leukemia).

3. Incidence of neutrophil defects among "recurrent infections" patients.

The operational definition of "recurrent infections" to which we have made reference is: five infectious episodes in one year, of polytopic nature and with a tendency to recurrence at the suspension of the appropriate antimicrobic therapy[9].

Patients already diagnosed as affected by diseases known to predispose to recurrent infection (such as cystic fibrosis, diabetes, chromosomal aberrations, leukemia, etc.) were excluded from this study. We have selected for this study 94 children (61 males and 33 females; age range 10 mo to 12 y; mean age 7 and 5/12) who fitted the above definitions and for whom no diagnosis was available at the time of their selection.

For 27 patients (group A) immunological diseases were later diagnosed, including Bruton' disease, (4 cases; 3 of which brothers), severe combined immunodeficiency (2 cases), common variable immunodeficiency (1 case), T cell deficiency (2 cases), selective IgA deficiency (14 cases), hypogammaglobulinemia (1 case), lupus erythematosus systemicus (1 case) and two cases of chronic myeloid leukemia.

In other 9 cases (group B) other pathologies were present, chronic hepatopathies, hyperinsulinism, Viscera Larva Migrans Syndrome, one Turner's Syndrome, and one adrenogenital syndrome.

In the remaining 58 cases no well defined diagnosis was possible, with the exception of 2 cases of chronic granulomatous disease (correlated with the diagnosis of CGD-carriers for the mothers).

In table 5 neutrophil functional defects have been grouped as phagocytosis (Pha), microbicidal activity (M.A.) and locomotion (CFM) chemotactic motility, killing of S.aureus and C.albicans, and the contents of several granula enzymes.

These findings suggest the existence of neutrophils factors or functions, which are selectively required for killing particular microorganisms (e.g. Candida or Aspergillus) and in general, point to the evidence that the study of neutrophil function is still a

Table 5. Neutrophil functional defects in patients with recurrent infections (see text).

Associated Pathology	No.	Defective Neutrophil function		
		Pha	M.A.	CFM
None	58	1	8	17
Group A	27	0	2	4
Group B	9	0	3	3
	94	1	13	24

very young and open discipline where most of the problems (never mind of the solutions) still have to be defined. As in other fields of immunology, and as it has already happened in this field with CGD, the investigations of the "experiments of nature", the rare defective mutants, will probably turn out to be a fruitful approach to set the questions correctly and thus look for the right, gnoseologic, diagnostic and, hopefully, therapeutic, answers.

REFERENCE LIST

1. Keusch,G.T. and Douglas,S.D. (1975) J. Infect. Dis. 131:584-588.
2. Keusch,G.T., Douglas,S.D., Mildvan,D. and Hirschman,S.Z. (1972) Infect. Immun. 5: 414-415.
3. Lehrer,R.I. and Cline,M.J. (1969) J. Bacteriol. 98: 996-1004.
4. Maggiore,G., Sacchi,F., Marconi,M., Bianchi,E., Deicas,A. and Siccardi,A.G. (1980) Infection 8: 171-173
5. Pagani,A., Spalla,R., Ferrari,F.A., Duse,M., Lenzi,M., Bretz,U., Baggiolini,M. and Siccardi,A.G. (1981) Clin. exp. Immunol. 43: (in press)
6. Preisig,E. and Hitzig,W.H. (1971) Eur. J. Clin. Invest. 1: 409-412
7. Sacchi,F., Cisternino,M.A., Clivio,A., Ferrari,F.A., Fortunato,A., Nespoli,L., Beretta,A. and Siccardi,A.G. (1979) Helv. Paediat. Acta 34: 607-616
8. Sacchi,F., Ferrari,F.A., Fortunato,A., Maggiore,G., Marconi,M., Pagani,A. and Siccardi,A.G. (1979) Infection 7: 45-47
9. Siccardi,A.G. (1980) Biologia e patologia dei Fagociti Umani. pp. 1-253, Edizioni Mediche Italiane, Pavia.
10. Siccardi,A.G., Bianchi,E., Callegari,A., Clivio,A., Fortunato,A., Magrini,U. and Sacchi,F. (1978) Helv. Ped. Acta 33: 401-412

11. Siccardi,A.G., Ferrari,F.A., Pagani,A., Marconi,M., Sacchi,F., Fortunato,A., Clivio,A., Marseglia,G., Stefanoni,R., Dos Santos, C., Ugazio,A.G. and Jayakar,S.- Submitted to: Clin. exp. Immunol.
12. Wilkinson,P.C. (1977) In Thompson,R.A. (ed.): Techniques in Clinical Immunology. pp.201-218. Blackwell Scientific Publications, Oxford.
13. Zigmond,S.H. and Hirsch,J.G. (1973) J. Exp. Med. 137: 387-410.

DISCUSSION

Edited by: R.A.Harkness

HILL: Did the parents of the patients with diarrhoea and defective chemotaxis show increased infections or allergies because this is shown by some of our patients?

SICCARDI: Both parents were healthy and their tests were all normal.

WEENING: Did your patient with the Chediak-Higashi like syndrome have normal release of MPO?

SICCARDI: β-Glucuronidase secretion was normal.

KAPLAN: Did your patient with the Chediak-Higashi like syndrome have giant granules? Did you examine chemotaxis under agarose or using the filter technique?

SICCARDI: Giant granules were absent on light and electron microscopy. Chemotaxis was studied with the filter technique.

THE EFFECT OF INFLUENZA VIRUS ON OXYGEN-DEPENDENT METABOLISM OF HUMAN NEUTROPHILS

Jan Verhoef, Elaine L. Mills, Yvette Debets-Ossenkopp, and Henri A. Verbrugh

Department of Microbiology
State University of Utrecht
The Netherlands

INTRODUCTION

Many viral infections predispose to bacterial superinfections[1]. The increased susceptibility to bacterial infections has been explained by the effect of viruses on the phagocytic host defense[2-9]. Controversy exits about the mechanism of action of influenza virus on the phagocytic cell function. Some studies showed a defect in uptake of bacteria by virus-treated PMN with intact O_2 dependent microbicidal activity[9], while other studies showed the opposite[7,10]. Because of these discrepancies we studied the effect of influenza virus on the respiratory burst of human neutrophils.

MATERIALS AND METHODS

Preparation of Leucocyte Suspensions

PMN were isolated by dextran sedimentation and Ficoll-Hypaque separation as previously described[11]. The final leucocyte pellets were adjusted to a concentration of $5x10^6$ PMN and $5x10^6$MN/ml Hanks Balanced salt solution. The preparation as evaluated by Wright stained smears never contained greater than 1% mononuclear cells.

Preparation of Virus

Influenza virus, type A, Texas 77 (H_2N_2) obtained from Prof. N. Masurel (Rotterdam, The Netherlands) was harvested from allantoic fluid of 11-day-old chicken embryos after incubation at 37°C for 2 days. For all experiments excepts that of oxygen consumption, the virus used was purified and concentrated by continuous flow ultracentrifugation on sucrose density gradients according to the method of Schild et al.[12]. The final titre of the stock virus suspended in phosphate buffered saline was 10^8 egg infectivity dose $(EID)_{50}$ per ml. Purified virus was stored in liquid nitrogen and virus in allantoic fluid was stored at -70°C; both were thawed at 37°C just prior to use.

Phagocytosis Assay

Phagocytosis was measured according to a method previously described[11,13]. Briefly, S. aureus EV (a clinical isolate) was radioactively labeled by growing overnight in ^{3}H-thymidine containing Mueller Hinton Broth. After 18 hr growth and 3 washings, 0.3 ml of the bacterial suspension (2.5 x 10^8 colony forming units/ml) were incubated with 1.2 ml of 5% normal human serum at 37°C for 30 min. The opsonized bacteria were washed and resuspended into 1.5 ml HBSS. Of the opsonized bacteria 0.2 ml was added to 0.2 ml of the PMN suspension in 4 polypropylene vials (Biovials, Beckman, Chicago, Ill., USA). The final bacteria to PMN ratio was 10:1. These mixtures were incubated in a shaking water bath at 37°C (150 rev/min). After 2, 6 and 12 min, phagocytosis was stopped by adding 2.5 ml ice-cold PBS to one of the vials. Leucocyte-associated bacteria, i.e. bacteria ingested by and bacteria attached to the PMN were then washed free of non-leucocyte associated bacteria by 3 cycles of differential centrifugation (160 g at 4°C). Leucocyte-associated radioactivity was measured in a liquid scintillation counter and expressed as a percentage of the total radioactivity added, i.e. per cent uptake of S. aureus. Total added radioactivity representing both leucocyte-associated and non-leucocyte associated bacteria was determined in the pellet of the fourth vial after centrifugation at 1600 g for 15 min.

Oxygen Consumption

Oxygen consumption was measured with a Clark type electrode (Yellow Springs Instrument co, Yellow Springs, Ohio; Model YSI 5331), as described by Weening et al[14].

PMN Chemiluminescence

Chemiluminescence was measured as previously described[15] with the following modifications: the chemiluminescence mixtures were prepared by adding to polypropylene vials 1 ml of 5x10^6 virus-treated PMN/ml, 0.1 ml of a 15 µM luminol solution and enough HBSS to make a final volume of 1.3 ml. The vials were incubated for 10 min at 22°C in a shaking water bath and the reaction was initiated at zero time by adding opsonized zymosan to the biovials and agitating the phagocytic mixture for 10 sec before vials were placed inside glass scintillation vials in the counter for a 0.1 min chemiluminescence measurement and then returned to the water bath. Measurements were obtained every 60 sec over a 10-30 min period. To study the stimulation of chemiluminescence by virus-treated PMN, 1 ml of 5×10^6 PMN/ml, 0.1 ml of a 15 μM luminol solution and 0.2 ml HBSS were incubated at 22°C for 10 min. Chemiluminescence was then initiated by adding 10^8, 5×10^7 and 10^7 EID_{50} of the virus suspension.

RESULTS

The Effect of Influenza Virus on the Phagocytic Function of PMN

To test the effect of influenza virus on the capacity of PMN to phagocytize staphylococci, PMN were incubated with various concentrations of virus suspension and after incubation the phagocytic function was determined. When PMN were washed free of virus after 1 hr incubation and were further incubated in RPMI for 18 hr a significant decrease in the capacity to take up staphylococci was observed (fig. 1). An inhibitory effect on the phagocytic function of the PMN was observed when about 10^6 EID_{50} units of the virus suspension were used.

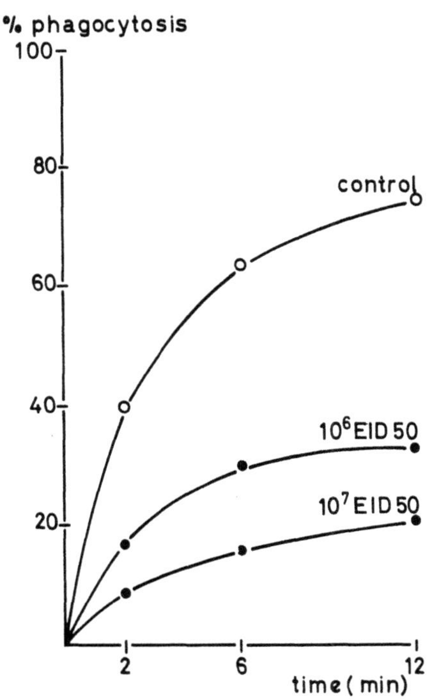

Fig. 1. The effect of influenza virus on the phagocytic function of PMN. PMN were incubated with influenza virus. After incubation, uptake of radiolabeled bacteria by PMN was measured.

Influence of Influenza Virus on the Stimulation of Chemiluminescence in PMN

After incubation of PMN with influenza virus ($EID_{50}:10^7$ units), generation of chemiluminescence by zymosan was studied. In each of 4 experiments, influenza virus had a markedly suppressive effect on chemiluminescence. The representative pattern of chemiluminescence observed with influenza virus treated PMN is presented in fig. 2. When PMN were incubated with influenza virus for 1 hr and then washed and further incubated for 18 hr, only 5% of chemiluminescence was produced after stimulation of PMN with opsonized zymosan as compared to normal control PMN.

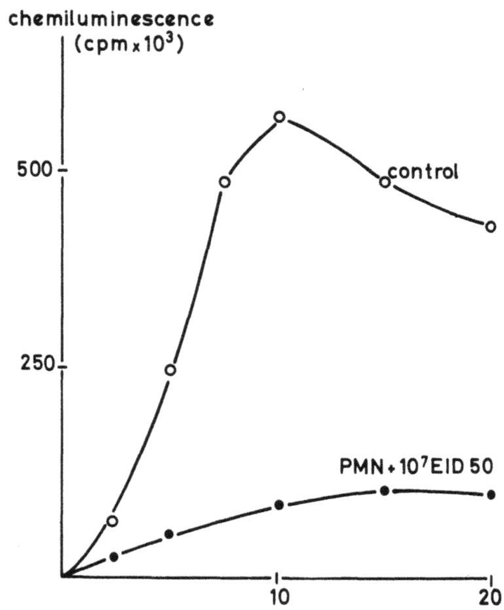

Fig. 2. Effect of influenza virus on the generation of chemiluminescence. Generation of chemiluminescence was measured during incubation of virus treated PMN with opsonized zymosan.

Initiation of the Respiratory Burst by Influenza Virus

It was concluded that influenza virus diminished the phagocytic uptake by PMN and the activity of the respiratory burst upon stimulation by opsonized zymosan. It could be possible that the PMN were damaged by toxic oxygen products generated from the PMN during incubation with influenza virus. Therefore, we studied the possible initiation of the respiratory burst by influenza virus. Virus particles (10^7 EID_{50}) were added to the PMN. As can be seen in fig. 3 (left panel) oxygen was taken up by the PMN at an average rate of 0.76 nmol/10^7 cells per min. After 10 min a maximum of 1.54 nmol/ 10^7 cells per min oxygen consumption was reached. When PMN were incubated with zymosan instead of virus particles about three times as much oxygen was consumed. Phagocytizing cells incubated with appropriate stimuli emit chemiluminescence which can be measured in a scintillation counter. PMN incubated with live virus particles also emit chemiluminescence (fig. 3, right panel). Light emission in the presence of luminol was dose-dependent and peaked at 5 min. Heat-inactivated virus particles did not generate light in this system.

DISCUSSION

The observation reported here, showed that influenza virus had an inhibitory effect on the phagocytic capacity of the PMN and on the generation of chemiluminescence by these cells. Our findings are partly in agreement with findings of others. Already in 1956, Fisher & Ginsberg[3,4] showed a defect in phagocytosis of heat-killed

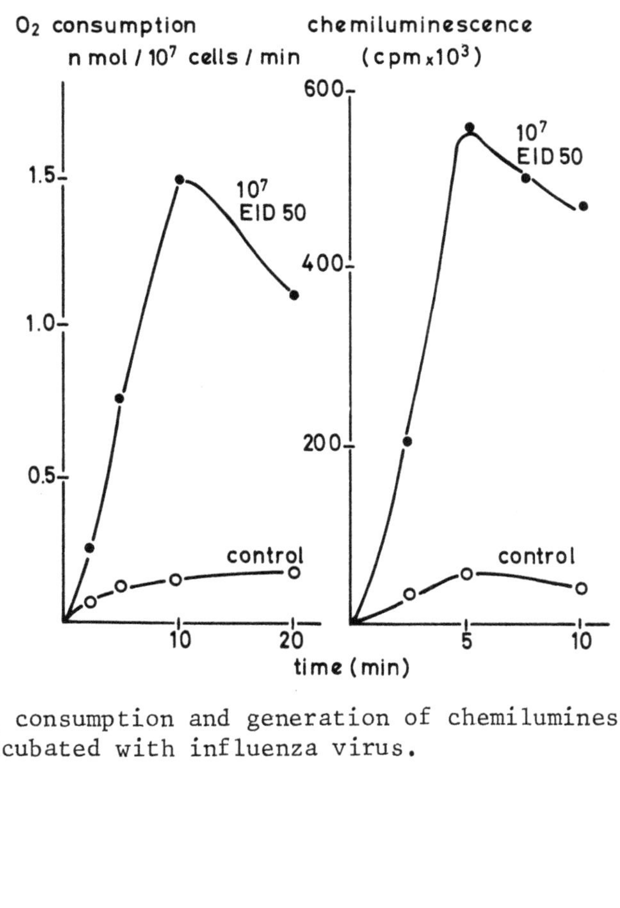

Fig. 3. O_2 consumption and generation of chemiluminescence by PMN incubated with influenza virus.

yeast cells by influenza-treated guinea-pig PMN. It was postulated that the effect of the virus was due to a block in the energy yielding mechanisms of virus-treated cells. On the other hand, Larson & Blades[8] and Larson et al.[9] found that the uptake of yeast particles by human PMN was decreased when these cells were treated with influenza virus, but no decrease in CO_2 production by the PMN was found in virus treatment. Metabolic activity of the PMN remained normal. Recently, using Newcastle disease, herpes simplex, vaccinia, encephalomyocarditis, vesicular stomatitis and rheo virus, Faden and coworkers[10] showed the opposite. They found a normal phagocytic function of PMN but decreased generation of chemiluminescence. Since PMN respond to appropriate particulate and soluble stimuli with a burst in oxygen consumption, production of superoxide and H_2O_2 and generation of chemiluminescence[16], it was concluded that viruses affected the oxygen-dependent antimicrobial system of PMN.

Since polymorphonuclear leucocytes are an important part of the body defences against bacterial and fungal infection, the demonstration of defective function during episodes of viral infection would help to explain the occasional occurrence of secondary bacterial infection.

The mechanism by which the influenza virus depresses PMN function is not yet defined. It could conceivably be due to a direct interaction between the virus and the leucocyte membrane, causing disruption of the membrane with subsequent decreased phagocytic uptake of particles. It is possible that the enzyme neuraminidase plays an important role in this respect. This enzyme is a surface antigen of influenza virus, and is believed to destroy cellular surface components necessary for migration and ingestion[5,17,18].

It is also possible that the direct initiation of the respiratory burst by the influenza virus particles generates toxic oxygen radicals which damage the PMN. During viral infection the capacity of phagocytes to produce antibacterial oxygen products may then be depleted and through auto-oxidation toxic metabolites may accumulate causing decreased phagocytic function and accelerated cell death. Since PMN are an important part of the body defences against bacterial and fungal infection, the demonstration of defective function during episodes of viral infection would help to explain the occasional occurrence of secondary bacterial infection.

REFERENCES

1. C. M. Martin, C. M. Kunin, L. S. Gottlieb and M. Findland, Asian influenza A in Boston 1957-1958. II Severe staphylococcal pneumonia complicating influenza, Arch. Intern. Med. 103:532 (1959).
2. D. J. Merchant and H. R. Morgan, Inhibition of phagocytic action of leucocytes by mumps and influenza viruses, Proc. Soc. Exp. Biol. Med. 74:651 (1950).
3. T. N. Fisher and H. S. Ginsberg, The reaction of influenza viruses with guinea-pig polymorphonuclear leucocytes. II The

reduction of white blood cell glycolysis by influenza viruses and receptor destroying enzyme (RDE), Virology 2:637 (1956a).
4. T. N. Fisher and H. S. Ginsberg, The reaction of influenza viruses with guinea-pig polymorphonuclear leucocytes. III Studies on the mechanism by which influenza viruses inhibit phagocytosis, Virology 2:656 (1956b).
5. W. D. Sawyer, Interaction of influenza virus with leucocytes and its effect on phagocytosis, J. Inf. Dis. 125:541 (1969).
6. E. S. Kleinerman, R. Snijderman and C. A. Daniels, Depression of human monocyte chemotaxis by herpes simplex and influenza viruses, I. Immunol. 113:1562 (1974).
7. A. W. Craft, M. M. Reid and W. T. Low, Effect of virus infections on polymorph function in children, Birt. Med. J. 1:1570 (1976).
8. H. E. Larson and R. Blades, Impairment of human polymorphonuclear leucocyte function by influenza virus, Lancet 1:283 (1976).
9. H. E. Larson, R. P. Parry, C. Gilchrist, A. Luquette and D. A. J. Tyrrell, Influenza viruses and staphylococci in vitro some interactions with polymorphonuclear leucocytes and epithelial cells, Br. J. Exp. Path. 58:281 (1977).
10. H. Faden, P. Sutyla and P. L. Ogra, Effect of viruses on luminol dependent chemiluminescence of human neutrophils, Infect. Immunity 24:673 (1979).
11. H. A. Verbrugh, R. Peters, P. K. Peterson and J. Verhoef, Phagocytosis and killing of staphylococci by human polymorphonuclear and mononuclear leucocytes, J. Clin. Path. 31:539 (1978).
12. G. C. Schild, W. D. Winter and C. M. Brand, Serological diagnosis of human influenza infections by immunoprecipitin techniques, Bull. Wld Hlth Org. 45:465 (1971).
13. J. Verhoef, P. K. Peterson and P. G. Quie, Kinetics of staphylococcal opsonization, attachment, ingestion and killing by human polymorphonuclear leucocytes: a quantitative assay using ^{3}H-thymidine labeled bacteria, J. Immunol. Meth. 14:303 (1977).
14. R. S. Weening, D. Roos and J. A. Loos, Oxygen consumption of phagocytizing cells in human leucocyte and granulocyte preparations a comparative study, J. Lab. Clin. Med. 83:570 (1974).
15. E. L. Mills, K. S. Rholl and P.G. Quie, X-linked inheritance in females with chronic granulomatous disease, J. Clin. Invest. 66:332 (1980).
16. B. M. Babior, Oxygen-dependent microbial killing by phagocytes, N. Engl. J. Med. 298:659 and 721 (1978).
17. J. Woodruff and B. Gesner, The effect of neuraminidase on the fate of tranfused lymphocytes, J. Exp. Med. 129:551 (1969).
18. J. F. Woodruff and J. J. Woodruff, Virus induced alterations of lymphoid tissue modification of the circulating pool of small lymphocytes. Cell. Immunol. 1:333 (1970).

DISCUSSION

Edited by: J. Verhoef

ROOS: Do the viruses have to be opsonized before reacting with the PMN?

VERHOEF: The viruses do not have to be opsonized, but the PMN response is higher when they are.

WALDVOGEL: Did you try to determine whether the increased chemiluminescence was accompanied by increased virucidal effect, by measuring the recovery of CFU at the end of incubation?

Did you try to infect the PMNs with attenuated influenza virus?

VERHOEF: The answer to both questions is: "No, we did not".

ROSSI: Did you do EM studies to see whether or not viruses were endocytized or attached to the membrane of leucocytes?

VERHOEF: Yes, we did EM studies in collaboration with Dr. P.Brederode (Leiden). These studies showed that most of the influenza viruses were endocytized and only a few were attached to the membrane.

CHRONIC GRANULOMATOUS DISEASE, Kx NEGATIVE NEUTROPHILS AND LINKAGE WITH Xg

Peter Densen, Susan Wilkinson, Gerald L. Mandell, Gail Sullivan, Ragnhild Øyen, and W.L. Marsh

Evans Memorial Department of Clinical Research Department of Medicine, Boston University Medical Center, Boston, MA; Division of Infectious Diseases Departments of Internal Medicine and Pathology University of Virginia, Charlottesville, VA; and The Lindsley F. Kimball Research Institute of The New York Blood Center, New York, NY

Individuals afflicted with chronic granulomatous disease (CGD) suffer from recurrent infections due to the inability of their neutrophils to kill ingested microorganisms. Defective microbicidal function is a consequence of the inability of these cells to generate metabolites of oxygen which are toxic to ingested bacteria. This defect in oxidative activity is believed to be due to the absence or dysfunction of a membrane bound oxidase. In 1975, Marsh and co-workers (1) reported that neutrophils from patients with X-linked CGD, but not other forms of CGD, lacked a surface antigen called Kx. Kx is an X-linked, Kell-related antigen normally present on the surface of both red cells and granulocytic white cells. Its distribution on these two cell types is determined by any of four alleles that have been described at the Xk locus (2). Erythrocytic Kx is converted to the appropriate Kell antigen by an autosomally determined enzyme. Thus, Kx is believed to serve as a precursor substance for Kell antigens on red cells, whereas on granulocytes it remains unaltered.

Individuals whose red cells lack Kx have a mild compensated hemolytic anemia which is associated with an acanthocytic shape of these cells. The association of this abnormal shape with the absence of Kx from these red cells as well as the association of CGD with the absence of Kx from granulocytes has led to the suggestion that a causal relationship might exist between the absence of

Kx and the defective oxidative response observed in neutrophils from individuals with X-linked CGD.

The present study examines both the association between CGD and absent neutrophil Kx surface antigen and the possibility of linkage between the Xk and Xg loci in a large kindred with X-linked CGD (3). The Xg locus codes for Xg^a, the only other red cell surface antigen known to have an X-linked mode of inheritance.

Red cell Xg^a and Kx typing was performed using the standard direct antiglobulin technique and scoring for hemagglutination as described previously (4,5). The presence of Kx on neutrophils was determined indirectly by the ability of test neutrophils to reduce the titer of a standard antiserum to Kx (1).

The results of erythrocyte Xg^a typing and neutrophil Kx typing for the first two generations of the family pedigree are shown in Table 1 in conjunction with the results of nitroblue tetrazolium (NBT) testing. Several points are apparent from this data. First, the fact that the father and the mother both have Xg^a positive red cells while three of their sons have Xg^a negative red cells indicates that the mother must be heterozygous at the Xg locus. Similarly, the mother must be heterozygous at the Xk locus since she has sons with both Kx positive and negative neutrophils. Thirdly, there is a precise correlation between the Xg^a type and the Kx type, that is, no recombination between these loci was observed. The lod score calculated for the linkage between the Xk and Xg loci is +2.1. A lod score of 2 or greater is customarily accepted as the level of significance for linkage of X-borne characters. Lastly, as reported in other families with X-linked CGD, Kx was absent from the neutrophils of the affected males. However, Kx was also absent from the neutrophils of the last born male who did not have a clinical history to suggest CGD and who had a normal NBT test.

Since the absence of neutrophil Kx antigen had previously been reported only in patients with CGD, the function of this sibling's neutrophils was studied. Neither a quantitative defect nor a kinetic defect in oxidative function could be demonstrated in neutrophils from this individual. Thus, the results of neutrophil Kx typing in this family, taken in conjunction with the results of the function studies on neutrophils from the Kx negative male sibling, appear to indicate that while the absence of neutrophil Kx is closely associated with CGD, the presence of this surface antigen is not required for normal neutrophil function. The data are consistent with the interpretation that separate genes on the X chromosome code for CGD and Kx. If this is the case, then the normal brother with Kx negative neutrophils reflects a recombination between the CGD and Xk loci. However, recombination did not separate the Xk and Xg loci, suggesting that the order of these

Table 1. NBT, Xg^a and Kx Typing in a Family With X-Linked Chronic Granulomatous Disease

	NBT*	Erythrocyte Xg^a Type	Neutrophil Kx Type
Father	+	a+	+
Mother	+(carrier)	a+	+
Daughter	+	a+	+
Son	-	a+	-
Son	+	a-	+
Son	+	a-	+
Son	+	a-	+
Son	-	a+	-
Son	-	a+	-
Son	-	a+	-
Son	+	a-	-

*Nitroblue tetrazolium.

three loci on the X chromosome might be CGD-Xk-Xg. However, this apparent linkage does not account for the high degree of association (linkage disequilibrium) that exists between Kx negative granulocytes and CGD.

In summary, these studies indicate that the absence of Kx antigen from neutrophils is strongly associated with CGD; that the absence of the Kx antigen does not preclude normal neutrophil function; and that the Xk and Xg loci are closely linked on the X chromosome.

References

1. W.L. Marsh, R. Øyen, M.E. Nichols, and F.H. Allen, Chronic granulomatous disease and the Kell blood groups, Brit. J. Haematol. 29:247 (1975).
2. W.L. Marsh, R. Øyen, and M.E. Nichols, Kx antigen, the McLeod phenotype and chronic granulomatous disease: Further studies, Vox. Sang. 31:356 (1976).
3. J.A. Dilworth and G.L. Mandell, Adults with chronic granulomatous disease of "childhood," Amer. J. Med. 63:233 (1977).
4. Technical Methods and Procedures, Amer. Assn. Blood Banks, 7th ed., Washington, D.C. (1977).
5. W.L. Marsh, Scoring of hemagglutination reactions, Transfusion 12:352 (1972).

DISCUSSION

Edited by: R.A.Harkness

SICCARDI: The family you have described shows recombination between CGD and the K_X antigen loci. Can you explain the association between CGD and the absence of K_X?

DENSEN: This is an example of linkage disequilibrium; we do not know why the two are associated. One possible explanation for these results is the sensitivity of the test for neutrophil K_X. If the affected patient had insufficient K_X on his cells to be detected he would not represent a recombination between the Xk and the proposed CGD loci. In that case K_X might still be related to the respiratory burst.

ROUND TABLE ON THE CLINICAL APPLICATION OF LEUCOCYTE FUNCTION TESTS

Harry R.Hill and Pierluigi Patriarca

Patriarca (Trieste,Italy): The idea of organizing this round table originated from the following considerations, 1) leucocyte function tests are performed in an increasing number of laboratories; these are usually specialized laboratories, and most of them have a background of basic research in leucocyte physiology, 2) in general, routine laboratories are not equipped to do the tests, 3) different tests are used by different laboratories to explore a given leucocyte function, 4) the techniques used for a given test vary among laboratories, and 5) most physicians are not appropriately trained in the evaluation of the clinical significance of leucocyte function tests.

Therefore, we thought it useful for both clinical pathologists and physicians to try and gather as much information as possible concerning some important topics in the evaluation of leucocyte function such as 1) clinical indications to perform leucocyte function tests, 2) significance of the most important leucocyte function tests, 3) preparation of a protocol for the monitoring of leucocyte functions, and 4) standardization of the procedures to be used for the most common tests.

May be it is unrealistic to expect that all these aspects of the problem can be settled tonight, but certainly the proceedings of this round table will be an essential reference point for the future work on the evaluation of phagocyte functions.

Dr. Harry Hill, who is the head of the Division of Clinical Immunology and Allergy of the Department of Pediatrics of the University of Utah, has kindly accepted to cooperate in the chairmanship of this Round Table. As far as the format of the Round Table is concerned, Dr. Hill and I have decided to ask specific questions which should hopefully stimulate comments and proposals from the audience.

Hill (Salt Lake City, USA): After such a long hard day of meetings, we'll try to make this Round Table brief, but as you can see, we have elected to cover a number of fairly important topics. When this Round Table was initially discussed, I was sitting down to braekfast with Drs. De Chatelet, McCall, Bass and others who were saying, "Well, we don't know whether or not we could derive a whole lot from such a meeting". They then proceeded to have a rather heated 30 minute discussion about which test should be run on which patients, etc., etc. So I think there is a lot of interest and controversy about the clinical application of leucocyte function tests. Furthermore, the defects in human phagocyte function that have been discovered have added tremendously to our basic knowledge of white cell biochemistry and functions, in general. Perhaps, with those few comments we can begin to attack some of the very difficult questions listed on the handout. The first one is to define the clinical indications to perform phagocyte function tests. I am a clinician and a clinical pathologist and for that reason am often faced with the decision on whether to work up a patient extensively or perform simple screening tests of his phagocyte function. I wonder if the participants have comments about how an individual should be selected for further leucocyte function studies. In our hospital, before any leucocyte tests are run, either I see the patients myself or one of the people who works directly with me sees them. One exception to this rule is when an individual in the Infectious Disease Division of the Department of Pediatrics or Medicine refers to patient in. Thus, only people in clinical immunology or infectious diseases can call up and order white cell functions studies. Are there any comments on the policy in your particular hospitals?

Siccardi (Pavia, Italy): I could tell about the experience in our own Pediatric Department. They had been doing lymphocyte studies for a few years, and this Pediatric Department became somewhat of a reference center for immunodeficiencies. When we added granulocyte function tests and complement tests, there were initially a number of indiscriminate requests. Anybody with a strange patient would come and ask for the tests. That, I believe, is a common experience whenever you set up new tests in a clinic which all doctors may order. So we decided to test new patients only on Mondays and to discuss every Friday all the cases which would be done on Mondays. This means that some member of the immunology team would have seen the patients during the week before. Alternately the referring physicians could attend the clinical meeting and present the case to the immunology group. Usually the kind of patients we study are those with recurrent infections. In order to develop criteria to define recurrent infections we decided to consider only people with three to five episodes of severe infections. We also pay special attention to the kind of infection. Fungal infections are regarded as particularly indicative and especially mucocutaneous candidiasis or recurrent infections by gram negative organisms. We try to get the entire history before examining the cases so as to establish firmly the presence of recurrent infections. A clear cut definition of what constitutes recurrent infections could be very useful in the selection of the cases for phagocyte function tests. I would like to know what other people mean when they talk about recurrent infections.

Johnston (Denver, USA): I don't know how well I will be able to articulate this, but it seems to me that the question resolves around whether there are assays which might be ordered by someone not skilled in making the decision as to whether or not this particular individual is having too many or too severe infections: in other words, screening assays that could be widely available, inexpensive and useful. A decision on the availability of such techniques might be made tonight in these discussions. In terms of deciding whether or not there have been too many infections, I think you start by putting somebody through medical school and you give him or her a good residency either in Pediatrics or Internal Medicine, and then you teach them to take a good history and perform a physical examination. If they get good at that, then you admit them into your infectious disease, immunology or hematology program and teach them about host-parasite interactions. The initial point is, however, that they have to know how to take a good history. No two of these individuals with recurrent infections are the same, so that deciding when to go beyond initial screening tests or determining which tests to order is a decision that has to be made by a well trained person. You just can't have these sophisticated assays available to untrained individuals.

Hill (Salt Lake City, USA): Yes, that's absolutely true. When you come right down to it, it's a matter of economics. We have had the people in our Pathology Department go through with a stop watch and supply costs and determine essentially how much it costs to run each of the leucocyte function tests. For instance, for phagocytosis I think we charge somewhere around $70 to $80 at the present time. Most clinical immunologists in our area don't charge that much for an initial physical exam and history. I think the best plan therefore, would be to have someone well trained in the use of such tests see the individual first.

Patriarca (Trieste, Italy): Excuse me Dick, what did you mean when you said "screening"? What kind of screening? Would this be cheaper than doing all the tests or not?

Johnston (Denver, USA): Oh, absolutely. I don't think that all aspects of leucocyte function need necessarily be available for assay to the untrained person (persons not fully trained in host-parasite interactions). There might be, let's say, an NBT test, or the whole blood superoxide, that one might be able to order as a screen. If you have a resident who knows enough about chronic granulomatous disease, he could order such a test. The availability of screening tests would be based on a cost-effectiveness ratio decision, I believe. Harry is better prepared to speak to that issue than I would be. I think there are a few assays that might be routinely available. But in any system, be it in a university, private practice or any other setting, you have to try to educate the people who are ordering the tests. It takes a great deal of time but all of us who deal with these things attempt to go out to the private practicioners and residents and teach them to be selective and to understand as much as possible about what they are trying to do.

Hill (Salt Lake, USA): I think there are very few places where such tests are done that aren't academic institutions. The purpose of an academic institution is to educate and, therefore, I agree one hundred percent. What screening tests can be offered? The only thing that come to my mind as a screening test would be something like the NBT. The test is inexpensive and generally recognized to be of some value in ruling out chronic granulomatous disease.

McCall (Winston-Salem, USA): My feeling is that, at least in the United States, there is no role for the screening test of host defenses other than the lymphocyte series, unless it is performed at a center which is equipped to evaluate the patient clinically and potentially equipped to evaluate the patient completely from the laboratory standpoint. So what we do as a consultant in the field of infectious diseases is to determine arbitrarily (and it is arbitrarily) that anywhere between three and five pyogenic infections in an adult, without other underlying disease is an indication for further evaluation. Initially we attempt to thoroughly evaluate the patient clinically, which is mundane to say but it is possibly the most important point. Second we do an NBT test, immunoglobulins and an absolute lymphocyte count, and then decide whether to go into the other parameters. Those other parameters which we do include phagocytosis, motility and secretion. So far as I am concerned and historically speaking, there has been no evidence that the cost effectiveness of this approach is really justified. So my point is that basically, I do not think that screening for defects in the host defense mechanism should be separated from the central, academic institution in the community.

Hill (Salt Lake, USA): I think you made a very good point. Certainly one should rule out the routine immunoglobulin abnormalities. I think one of the major functions we perform is to discourage people from the inappropriate use of gammaglobulin.

De Chatelet(Winston-Salem, USA): We operate in my laboratory on a purely research basis as I am not a clinician. We don't have an over abundance of such tests being ordered. In fact, my biggest problem is really trying to educate primarily the pediatricians, to send us samples when they think they have a really bizarre syndrome that they don't understand. We don't charge at all for the work-up that we do.

I like to oxidize glucose (I think it is great fun) and so my own bias is to run a shunt assay on the individual on the first go around. If that shows up abnormal, we go through the entire battery of virtually everything available. But I think, eliminating humoral defects and lymphocyte defects, that leucocyte function tests are basically in a research realm. You tend to gain a lot more than you tend to lose by doing them. We have turned up a number of patients in the course of looking at such individuals that have intrigued the heck out of us. I think the benefit primarily is a reseach one. I don't think that you are going to be inundated with surgeons sending you cells from possible cases of chronic granulomatous disease. My experience has been just the opposite. We have a tough time trying to dig up patients, and we are more than willing to screen anyone who clinically appears to have the possibility of such a defect. I really think you have to differentiate between a clinical laboratory that is set up on a fee for service basis and a research laboratory. My feeling is that most of the leucocyte neutrophil functional tests are very research orientated, not clinical at all.

Hill (Salt Lake City, USAP: I would say, Larry, we do two to three chemotaxis assays a week, one or two chemiluminescence tests and a couple of NBTs. That is the extent of the clinical work that we do, drawing from a million and one-half in our area.

Patriarca (Trieste, Italy): Now let's listen to the experience of a Dutch hospital. Dr. Weening is in a special position. He is a pediatrician and is also in close connection with a research group at the Red Cross of Amsterdam.

Weening (Amsterdam, The Netherlands): First of all, it depends upon what kind of lab you have, of course. We are in a position so that we get patients from all over the country. I have to select patients by phone; I can't see them all. I try to do the best I can by phone and even then, I would miss patients if the referring physicians didn't insist. The patients don't follow rules, I mean they can be completely healthy for say fifteen or sixteen years and then develop an infection and be diagnosed as CGD. So it is very difficult to set rules for selection of patients for phagocyte function tests. I was a little surprised in listening to the discussion on screening tests. In our experience, and we do about 250 patients a year, the most common defects are chemotactic defects, second are killing defects and third are opsonic defects. Phagocytic defects are hardly ever found and are usually secondary to lupus or rheumatoid arthritis. So, we feel that apart from the problems of screening of selected patients, we should perform a chemotactic assay. We don't do phagocytic assays as a routine, since I feel that primary phagocytic defects are extremely rare. We also do killing assays and oxygen consumption tests. In selected cases, we also do opsonic assays.

Siccardi (Pavia, Italy): I don't understand why some people don't charge for these tests, because if you ask for immunoglobulins you have to pay for them. I don't see why neutrophil functions should not be paid by insurance. I don't agree with that at all.

Hill (Salt Lake City, USA): I have a technician that is supplied by the clinical labs and works in my research lab doing the clinical work. The minute we find a problem, we shift it over to the technician who does research work and no further charges are made to the patient. The results of the studies are given to the patients and clinicians and are useful in the subsequent management of the patient. I agree that the chemotactic defects are what we most commonly find. These are more difficult tests and more expensive and, therefore I never have one done unless I or one of my colleagues sees the patient.

Weening (Amsterdam, The Netherlands): Can you decide on the basis of the clinical history whether it is a defect in chemotaxis or not?

Hill (Salt Lake City, USA): I would say most of those individuals have problems like cutaneous abscesses. A few have bad ear infections and other have had cutaneous candidiasis. I have also seen several patients with pneumonia.

Weening (Amsterdam, The Netherlands): It is pretty difficult on the basis of the clinical picture to rule out chemotactic defects, in my opinion.

Elsbach (New York, USA): From what McCall has said and what Weening has said, I get the impression that there is already a major difference of opinion concerning the usefulness of the clinical appraisal. From what Weening says telephone conversations, which give you a reasonable impression of the clinical history of the patients, are not adequate to judge whether or not a defect may be present. In addition, a physical examination would also be of little help. I think this point should be settled in terms of how you want to carry on your further discussion of the usefulness of screening, because if Weening is correct, then there should be no particular usefulness in relying on the criteria that McCall would want to establish. Who is right?

Keller (Bern, Switzerland): I just wanted to say something about the chemotaxis part. If you care to make a distinction between chemotaxis and locomotion, then we find that at least 95% of the patients who have a defect have a defect in locomotion and not a defect in chemotaxis. Now maybe we have been looking at a very special selection of patients, and the proportion may be different in other laboratories. I suspect, however, that it is worth making the distinction between defects in chemotaxis and locomotion. It is certainly useful to distinguish between a patient who is blind and one who is paralized and maybe it will also be helpful to distinguish between a cell which has impaired locomotion or impaired recognition of signals. Another point which is very important in this context is that we should see these defects in connection with others, in mutual clinics. May I stress, that locomotion problems are what we find most frequently as a defect, as Dr. Weening said.

Hill (Salt Lake City, USA): You know the criteria for the use of immunoglobulin therapy in hypogammaglobulinemia include 1) severe recurrent infections; 2) a certain level of immunoglobulin, and 3) no antibody response following immunization. It would seem that one of the first indications for performing phagocyte function tests would be the presence of severe infections.

Verhoef (Utrecht, The Netherlands): I think it all boils down to the question of the frequency with which you find defects. If every test you do shows a defect then you are doing too few assays. If you never find a defect, you are doing too many tests. So my question to Dr. Hill, for example, is: when you screen very carefully, what is the frequency of defects which you find? I should like at the same time to put the same question to Dr. De Chatelet. He asks clinicians to order tests, so what is his frequency in finding defects?

Hill (Salt Lake City, USA): I would say, in our clinic when I see the patients and they have abscesses as well as other evidence that suggests a neutrophil problem, perhaps 20% will have defects. That's a rough estimate, of course.

McCall (Winston-Salem, USA): If one is dealing with diagnostic considerations in medicine, then first we have to classify what is going on. I'm all for the laboratory which is able to do so, without an enormous expenditure to the patient. To be able to classify a large number of motility defects, which probably represents what we see the most of, would be useful scientifically. I'm not aware, in fact, that we can place any specific diagnosis upon these patients, or whether at the present time, we can employ any specific therapy for these patients. And so that's why I say, in my personal opinion, phagocyte testing needs to remain a research tool.

Segal (London, United Kingdom): I think one of our problems is that if a clinician sees a patient, he has no hesitation whatsoever in ordering an X-ray. An X-ray tells him something that is generally defined. I think we really are wasting a bit of time. I think it would be very helpful to define what are useful tests to do in a clinical situation that would help the clinician and the research worker. I think it would be useful to define a number of tests to look at the basic functions we recognize in neutrophils.

Bass (Winston-Salem, USA): I think one major problem is the concept of doing only tests which might result in different therapeutic modalities in the patient. In the last 25 to 40 years a major Function of medicine, in general, has been to provide an accurate diagnosis. This diagnosis could then allow an accurate prognosis irrespective of whether the doctor could provide therapeutic intervention. Only after that capability was attained through a precise definition of the natural course of diseases was therapy possible. I don't think it is totally inappropriate that we should provide tests which allow that degree of analysis - meaning only prognosis, not therapy. In the future we may develop therapies based on the results of such tests. It does not seem unreasonable to charge for such a clinical test. The tests that we should be able to provide at this time should be those that would define defects which we think would then allow an accurate prognostic indication. Now I think that there is no question that metabolic defects of a major degree fall into that category. Chemotactic defects are far more vague and, in particular, chemotactic defects which are detected by the untrained individual who may not have ruled out complicating infections or other systemic diseases. Thus, we are thrown back on the necessity for having an infectious disease specialist or immunologist evaluate the patient before the tests are ordered.

Johnston (Denver, USA): I would just like to extend what David said and offer some enthusiasm for trying to make a diagnosis, because I believe that it really does make a difference. We don't have specific therapy, obviously, for most of these diseases, but proper management makes a big difference. How you manage a patient is modified tremendously if you know he or she has this or that defect. A good example that comes to mind is C2 deficiency. If one makes this diagnosis, one can immunize against pyogenic organisms, promptly institute antibiotic therapy with fever, and be aware of the propensity to develop collagen-vascular disease. The same reasoning can be applied to all of these immunodeficiencies.

Hill (Salt Lake, USA): Look, for instance, how seldom an X-ray actually helps us in selecting the therapy for a patient and yet the patient pays enormous sums for the performance and interpretation of these "tests".

Arslan (Padova, Italy): Concerning this problem of therapy, I want to ask if anyone here has any experience with a problem that is common to the hematologist? They tell me that 50% of leukemic patients die of infections. Is it possible to balance the chemotherapy of the leukemia against the potential for developing infections, based on phagocyte function tests?

Hill (Salt Lake, USA): Can you, in fact, screen for the susceptible patient by looking at leukocyte function tests? I think most of those patients end up with marked neutropenia and that is why they develop infection.

Arslan (Padova, Italy): No, but some of these cases get too much chemotherapy. Can you balance the treatment according to the results of leucocyte tests?

Hill (Salt Lake, USA): Does anybody want to comment on that?

Siccardi (Pavia, Italy): Well, I just want to back up a lot of the suggestions. In particular the one about the frequency of defects that one finds can determine the kind of tests you want to perform. In our laboratory we are probably doing too many but, as other people have said, this is an academic institution and our research interest is to find new abnormalities. Just doing NBT tests would have given us only 5 cases in three years. This would have really been very little to justify the big load of postdoctoral fellows and the research funds utilized. I have to agree with Dr. Weening on his classification. Chemotactic defects are by far the most frequent cases found. For us the second most common abnormality is opsonic defects, defects in activation of the alternative pathway. Now, these abnormalities would never be picked up unless one does very specialized tests on the complement system. These are really very frequent: about 5% of the people with or without recurrent infections have the defect. Some of these patients have very severe problems with recurrent infections such as those with Leiner's syndrome. The therapeutic indications that can come from the tests are very important. If some opsonic abnormalities are present, plasma transfusions are often helpful. So, in fact, what I wanted to say is that we have not been able to cut down on the number of tests. I must agree however, that we have never found anything wrong with phagocytosis, per se. We continue to run phagocytosis assays since they are cheap and simple but we never have to go futher with these assays. We do phagocytosis tests, glucose oxidation, microbicidal activity using 2 to 3 agents and two locomotion tests. Now the important thing that we have found out with the routine is the importance of the microbiology reports. If a patient has recurrent infections always with the same pathogen, like aspergillous, we have found that it was very rewarding to examine aspergillous killing. In many cases we have found a selective defect in microbicidal activity.

Hill (Salt Lake, USA): Let me comment on that and then let's go to the second part which is "determining the significance of the most important leucocyte function tests". In adding up the tests that you do Dr. Siccardi, it probably takes two to three technicians a whole day to perform these assays. I wonder if some tests like chemiluminescence, which we use a great deal, might be employed as a screen both for phagocytosis and for the generation of microbicidal products. Could we have some comments on specific tests that you feel would be good as a superficial screen to be ordered by a specialist, and then on secondary tests that one should pursue if abnormalities are found.

Allen (San Antonio, USA): A blood test for chemiluminescence can be used as a screening technique. All you need is one microliter of blood and you can tell, first of all, whether or not a case is CGD. We perform this screening technique for the local University and we do it for our Pediatric Department. If we fail to elicit a chemiluminescence response then we go on and study glucose oxidation, which is probably the gold standard as Dr. De Chatelet indicated previously.

Hill (Salt Lake, USA): Do you use a luminol enhanced system?

Allen (San Antonio, USA): Yes, this technique requires about one microliter of blood. I might add, it requires whole blood so no separation is required. If you prepare your blood you can do it in less than a hour.

Hill (Salt Lake, USA): We have found the chemiluminescence assay to be so much better than the NBT, and it can probably screen for myeloperoxidase deficiency, also.

Allen (San Antonio, USA): Actually, that is a good point because chemiluminescence probably is a good measure of myeloperoxidase in the presence of a low concentration of luminol.

Hill (Salt Lake, USA): It is also simple and adaptable to a clinical laboratory.

Bass (Winston-Salem, USA): Harry, this comment is probably out of order but it relates to our discussion of a few minutes ago. If specific tests are going to be ordered for determining microbicidal activity, there are now a number of reports indicating that we should assess the ability of leucocytes to kill the organism infecting the patient. It appears that there may be specific organism related defects in some of these patients. When we get past the screening test stage we should not only include the standard *S.aureus* and *E. coli* but the patients own organism as well. These, however, are later tests looking for highly defined defects.

Verhoef (Utrecht, The Netherlands): Speaking of screening tests, I think in addition to the chemiluminescence you need a chemotaxis assay and an opsonization test as well.

Hill (Salt Lake, USA): Have you tried using the chemiluminescence assay to measure opsonins with specific bacteria?

Verhoef (Utrecht, The Netherlands): Yes, but we have not had much experience with it.

Hill (Salt Lake, USA): We have had good results with group B - streptococci.
Clark (Boston, USA): I would like to put in a word for the iodination assay as a screening test which we find very useful. The test screens out most defects other than those of motility. One can use phagocytic or nonphagocytic stimuli. If you use phagocytic stimuli you can compare patient's serum with a normal source of opsonins. A normal response indicates the integrity of opsonization, ingestion, the metabolic burst and secretion of myeloperoxidase, and thus, it really should detect most of the described defects other than those in motility. In combination with an assay of chemotaxis, this assay should constitute a good screening system, I would think. That is generally what we would do on a patient.
Weening (Amsterdam, The Netherlands): To some extent I agree with Dr. Clark that iodination tests would be better to screen for some defects than chemiluminescence alone. I think, however, that we also must be concerned with nonoxidative microbicidal activity. Thus, an actually killing assay should be performed. I guess you will miss a couple of patients if you just keep doing chemiluminescence alone or glucose oxidation alone.
Dri (Trieste, Italy): In our laboratory, we have developed a test that meets several of the requireemnts that have been discussed in this session. It is a measure of the responsiveness of phagocytes to stimulation by particulate substances and also of the opsonizing activity of plasma. It can be done on whole blood, requires only 40 minutes and can be performed in most laboratories since it does not require special instrumentation. The test is based on the measurement of superoxide production during the incubation of whole blood with zymosan. The zymosan is not preopsonized but added directly to the whole blood. A mixture of 0.2 ml of blood, zymosan and cytochrome c in a final volume of 1 ml is required. A parallel assay is also run in the presence of SOD. After 15 minutes, the O.D. of the supernatant is measured to determine superoxide generation. Lack of production of superoxide indicates either a defect in opsonization or in metabolic activation. It is possible to differentiate between these two possibilities by testing normal cells and patient cells with zymosan opsonized with the patient plasma or with the normal plasma, respectively. If the defect is at the cellular level, the leucocytes of the patient fail to produce superoxide in the presence of zymosan opsonized with control plasma, whereas the control cells respond normally to zymosan opsonized with patient plasma. If there is an opsonization defect, both normal and patient leucocytes fail to

produce superoxide in the presence of zymosan opsonized with the patient plasma. No superoxide production by blood from CGD patients detected with this test, while their carrier mothers produced 30 to 60 per cent of the superoxide produced by normal blood. This method proved to be much more sensitive in detecting opsonization defects than the other methods normally employed to test opsonization (for example, ingestion of zymosan).

Hill (Salt Lake, USA): I have one question about that. One thing that always disturbs me is when you measure opsonization by a respiratory burst assay with a large volume of serum or plasma in the preparation, itself. We know, for instance, that a number of soluble factors including several of the complement components can trigger superoxide generation by themselves. How do you know that some of the superoxide that is being generated isn't coming from such a soluble stimulant as C5a?

Dri (Trieste, Italy): We have looked at that. Leucocytes exposed to plasma pretreated with zymosan (which probably contained C5a) generated no more than 10% of the superoxide generated by leucocytes treated with opsonized zymosan. From this experiment we conclude that, under our experimental conditions, the effect of soluble stimulants such as C5a on the superoxide generation by leucocytes is only a minor one and cannot mask a possible phagocytic defect.

Allen (San Antonio, USA): One of the comments I would like to make also to Dr. Hill is if you do any type of opsonic function you have to do serum titrations. The other thing is that for myeloperoxidase deficiency, you might actually see an overabundance of superoxide generation so that you should use another test with it. Either iodination or chemiluminescence with luminol.

Keller (Bern, Switzerland): I think we should make a distinction between basic assay which can be used for screening (by specialists) and secondary assays. Most assays employ filters or the under agarose technique and we do it with or without gradients. This tells us something about locomotion and chemotaxis. I think migration in a gradient may include chemokinetic as well as chemotactic responses. So, if we really suspect on the basis of these primary tests, that there is a defect in chemotaxis then we should do an orientation assay. Otherwise we will never be sure if it's really chemotaxis or not. The second point I want to make is that we have been running adhesion assays along with migration assays. What came out was that we never found defects in adhesion unless there was also a defect in locomotion. Therefore, we are now considering that the adhesion assay is a secondary assay system. I wonder if that is the general experience.

Segal (London, United Kingdom): There is just one point about chemotaxis. I think that chemotaxis is a very nice *in vitro* system for looking at and studying cell motility. I think it is very important in patients to look at the overall inflammatory response, because we don't know what the afferent loop of the neutrophil attraction really is like and the actual mechanism of it. Apart from the history of the accumulation of pus, the only technique that I am aware of employs a skin window. The Rebuck skin window is qualitative only and not very good. A chamber skin window technique is semiquantitative and open to a number of manipulations. It is a very simple technique and is simple to apply to the patient. We have found it extremely useful in a variety of conditions including Crohn's disease and sarcoidiosis where we found major abnormalities of neutrophil migration.

Hill (Salt Lake, USA): Tony, do you think that test is actually adaptable to a number of different centers? Can one really create a standard lesion?

Segal (London, United Kingdom): Absolutely, if you put a plastic template over the skin and you abrade it through the template so you take a standard amount of skin off. Obviously, it is a crude test but it's no more crude than separating cells and placing them in plastic chambers. This is an *in vivo* model of attraction of neutrophils to sites of acute inflammation. If you get normal accumulation of neutrophils, then I would not worry much about what the cells do in an *in vitro* assay.

Hill (Salt Lake, USA): Would everybody agree that it is probably not necessary to run adhesion if you have normal chemotaxis? We obviously do not perform adhesion assays. Does anybody routinely run adhesion? No, so that test probably is not indicated unless abnormalities in motility are detected.

De Sole (Roma, Italy): Concerning the chemiluminescence, I think that it would be useful to determine the activity in the resting state and after stimulation. We observed some chronobiologic variation during the year. It is interesting to note that when we have high resting activity, we have low stimulated activity and vice-versa.

Weening (Amsterdam, The Netherlands): A question for Dr. Dri. Obviously you have no problem with superoxide dismutase in all of the red cells. Do you have to block the SOD?

Dri (Trieste, Italy): We have found that lysis of about 5% of the red cells present in the assay does not significantly alter the superoxide producing activity of the blood. Therefore, blocking of the red blood cell SOD is not needed.

Bass (Winston-Salem, USA): A comment or two about measuring chemotaxis. I have great trouble with the measurement of chemotaxis in vitro. My friend McCall, a few years ago, indicated that patients with acute bacterial infections had depressed chemotaxis. This was done in a series of 14 to 15 patients. Shortly thereafter Harry Hill and Paul Quie came out with roughly the same technique and said that patients with acute bacterial infections have increased chemotaxis. So you are looking at a wide spectrum of responses with the in vitro chemotaxis technique. I agree with Tony Segal that the only way you can get something that is meaningful to the patient, if we are talking about that, is with a skin window technique. I like his idea of using a template to get a certain area of skin abraded, but the other real factor that we've seen is the depth of the abrasion. I have a question for Tony, as to how to control that. The method of abrasion is going to give more or less chemotaxis into a given area. I wonder just how you go about creating a standard lesion.

Segal (London, United Kingdom): It is very difficult to regulate the depth of abrasion. We use a portable dental drill and try not to allow bleeding. We generally do 2 to 4 abrasions and take the mean of emigrating cells in each.

Hill (Salt Lake, USA): How good are your replicates?

Segal (London, United Kingdom): Not good at all. May I make one point. That's the big problem in assessing immune function. With all of us healthy individuals sitting in this room, one of us could get pneumonia tomorrow. If we were assessed today our neutrophil function would probably be better than normal, in view of the hospitality that has been shown to us here. The problem is that it is purely a statistical point as to whether one gets infected or not. The healthiest person may get infected and the person with a serious immune deficiency may not, for instance the CGD patient who goes 16 years without a serious infection. This is where the big problem comes, tying up the clinical abnormalities with the test abnormalities we observed.

Hill (Salt Lake, USA): I think we have to be very careful with chemotaxis. A defect should be a very profound defect to be of any real significance.

Venge (Uppsala, Sweden): Since we have some different experiences compared to many others here in this room about leucocyte function testing, I would like to tell what we are doing. We are doing 50 patients a year and are performing phagocytosis. We are differentiating between IgG uptake and complement uptake. This is a very common defect in many patients. It is at least as common as the chemotactic defects. I think phagocytosis should be done in the kinetic way. I think it is very crude to make an end point assay. As I said, about 50% of our patients don't recognize C3b on a particle. Then we perform chemotaxis assays both with isolated cells and with the patient's serum and normal cells. This is also a very common defect. Let me make a point of warning that most of these defects seem to be secondary. These patients have defects when they are having infection but often one or two years later these defects seem to vanish. The infections also then disappear, especially in children. Then, of course, we do the chemiluminescence test as a qualitative screen. As a last test we quantitate intracellular bactericidal proteins, chymotrypsin like cationic proteins, lactoferrin and myeloperoxidase. We see many cases of myeloperoxidase deficiency but have never seen abnormalities in chymotrypsin-like cationic proteins. Lactoferrin deficiencies are common but usually occur in patients with acute infections or those who have had infection during the last one or two weeks.

Johnston (Denver, USA): I would like to endorse what Dr. Segal said, that an inflammatory skin window gives you information that you can't get from in vitro chemotaxis assays. Not as specific, of course, but you can study a number of cellular functions and mediators. The problem is one of practicality, however. A dermatome can be used to create a standard lesion. You can then put on a plastic device and inject mediators into these. There are many ways you can do this if you go to the trouble to set them up properly. The second point has to do with opsonins. I think as a screen immunoglobulin and complement levels should always be determined first.

Allen (San Antonio, USA): To expand on that point just a little bit further. The Norwegians have published a technique where they use a blistering by suction. They are able to control skin temperature by use of a thermocouple attached to the blistering apparatus. They then just apply the smooth suction. They have used this to study the effect of certain antibiotics on chemotaxis. Although the technique does require special apparatus it would be relatively atraumatic, nonpainful, and would give you quantitative information in that you could sample the blister contents for the different opsonic components as well as cellular components.

Venge (Uppsala, Sweden): The idea of Tony Segal is very attractive but it seems that nobody here has ever tested patients routinely in such a way. Thus, if sophisticated laboratories cannot agree on this method how can we be expected to use it as a screening test.

Hill (Salt Lake, USA): I would personally like to come out for the in vitro chemotactic assay as the most practical one. I personally believe that it really means something, if done correctly, in that we generally find abnormalities in individuals who have infections. You will notice as we discuss things tonight - phagocytosis people like the phagocytosis assays, the chemotaxis people like the chemotaxis assays and the biochemists like the metabolic assays. I think all would agree, however, that you can't just ignore any of the major functions of phagocytes.

Hakim (Paris, France): Each laboratory has its own habits. We know that in our laboratory we do a screening test that is ours. We do movement under agarose. If there is something else we give the cells to someone who is an expert in that area. We do NBT tests, oxygen uptake and myeloperoxidase staining. The remaining cells are frozen at -80°C for further enzyme studies if indicated. Our tests work for us in detecting defects in about 10% of patients and we test about 1 patient per day.

Clark (Boston, USA): It is probably fool-hearted to go up against Segal, DeChatelet, Johnston and Allen and whoever else spoke up for in vivo studies of leucocyte migration, but I would like to go along with Harry Hill and Jacques Hakim in sounding a note of caution about the use of the skin window and the skin chamber techniques, for several reasons. For one thing, I think it is apparent to everybody that these are very difficult techniques to quantitate, no matter how careful you are or how much you attempt to standardize your lesion. These are semiquantitative techniques, at best. Secondly, there is a lack of information correlating abnormalities in these tests with known clinical defects in leucocyte function and recurrent infections. Finally it is not reasonable at all to regard these tests as measures of chemotaxis or any other defined function of neutrophils. There is no basis for thinking that we are looking at directed migration or increased speed of migration or anything that we know about in a reasonably scientific sense. We may simply be observing trapping of cells that happen to be wandering purely by chance. So I think we would be very careful in interpreting these studies. We need to do them more and more because we would like to accumulate information but, I think, at the present time we don't know how to interpret the results.

Gallily (Jerusalem, Israel): I would like to support in vitro chemotaxis and I will give you an example which I think will convince some of you that I am right. We examined a patient with Familial Mediterranean Fever which is common in Israel. We found that person to have high migration during an attack of the disease. Colchicine caused the patient to return toward normal values. This drug greatly helps the patients and prevents them from having more attacks. So here you see that giving a drug that decreases chemotaxis in vitro improved the patients condition in vivo.

Harkness (Harrow, Great Britain): I agree with the foregoing speakers in preferring the chemotaxis assay *in vitro* for reasons which have been mentioned already. The *in vivo* assay is difficult to perform and interpret, especially in severely affected patients. Moreover, you can cause a new infection at the site of the abrasion. I would, therefore, prefer the *in vitro* chemotaxis assay.

Keller (Bern, Switzerland): I have no prejudice with respect to *in vivo* and *in vitro* measurement of leucocyte locomotion. We have tried very hard to do it *in vivo*, but it turned out that the in vitro assay was more effective. The problem I think about *in vitro* assays is: what is a defect? I mean you can easily find decreased locomotion, sometimes 20%, sometimes 50%, but in our experience it only becomes clinically relevant at 80 or 90%. Otherwise in clinical terms you can just forget about it. I think that is one important thing that one has to keep in mind for clinical assays. Now it has been said that you can get very different results *in vitro*, and I think that is very true. What I really want to show you briefly is that this method is entirely reproducible. What we have recently done is to compare the one and two filters Boyden systems with a leading front filter assay. Comparing the three techniques one sees that in the one filter system all of the cells remain attached to the upper filter. In contrast when two filters are used essentially the same number of cells are counted but on the lower filter. Thus, these absolutely reproducible results indicate that both systems are working. Furthermore, when you employ a variety of chemotactic agents you will get differences in chemotaxis to each of the test solutions. Still the activity is there and you can pick it up readily with each of the systems. You can also use different media for the cells. So you see, with each system you get reproducibly different results. We shouldn't be too worried about that situation. We should realize two things; 1) there are differences which are due to the fact that sometimes we do not measure the whole response, that is we measure only part of the cells, and 2) there are differences because we really are using different test systems (maybe different techniques or gradients). We also find, for instance, that we can sometimes get different results between the leading front and our two filter counting. I think these are things which we have to face and attempt to interpret in the long run. If the clinician has to use the technique, I think he should use a technique which counts all cells which have responded or the leading front technique. We have to try and analyze to what extent we really can pick up defects with the different techniques. I have no fixed opinion about that, but I think that is just a fact we have to face.

Johnston (Denver, USA): Just a comment - I don't want to go down in history as suggesting that any form of the Rebuck skin window takes the place of quantitative in vitro chemotactic assays. It adds another dimension or other information, however.

Palmblad (Stockholm, Sweden): When assessing chemotaxis or migration I would like to emphasize that it is important to use several different phagotaxins. In many studies we have found that the response, for instance, to the bacterial factor from E.coli might be quite normal in some patients, while the response to serum might be abnormal.

Patriarca (Trieste, Italy): One of the purposes of this Round Table was to define a common protocol to monitor leucocyte function. It seems however, that this goal is difficult to reach, particularly after Dr. Hakim's comments who remarked that each laboratory has his own habits which cannot be changed from one day to the other. Nevertheless, if anyone of you is willing to indicate the operative protocol he uses, this would certainly be of interest to the audience.

Hill (Salt Lake, USA): All right, I can tell you what we routinely do on the initial visit of a patient. I usually get a CBC, an NBT and if there is evidence of severe eczema I will run an IGE level. Subsequently, if there is evidence of abscesses or mucocutaneous candidiasis I will probably run a chemiluminescence and a chemotaxis on the second visit. In addition, we always determine random motility which I think will probably adequately assess chemokinesis. If I see anything abnormal with the chemiluminescence then I'll do a myeloperoxidase stain, phagocytosis and killing, superoxide generation and glucose oxidation. This is in essence, the sort of workup we carry out.

Schell-Frederick (Bruxelles, Belgium): Sometimes one is faced with a patient that one wants to test rapidly and who may be infected. Which tests are going to be best in all conditions?

Hill (Salt Lake, USA): It has been my experience that most patients with infections, if they don't have some other debilitating disease, tend to turn up their phagocyte functions. Their cells tend to phagocytize better, generate more superoxide upon challenge and chemotax better. Now there are certain subsets of very severely ill patients whose PMN function may be depressed. I would like to hear other comments on this point.

Venge (Uppsala, Sweden): One point is very important in function testing, that is the control for the assays. Shall we use normal controls which could vary a lot between individuals or reference material? I mean do we refer our results to means ($\pm$ 2 SD) for the laboratory? What is the best way to do this?

Weening (Amsterdam, The Netherlands): I think it is very important to define the normals for your patient's different age groups. Our findings as well as the published experience of other indicates that chemotaxis is age dependent. Small children and neonates undoubtedly have depressed chemotaxis. This is not our experience in the killing or phagocytosis assays. Thus, except for the chemotaxis assays, you may rely on normal values for adults. We used $\pm$ 2 SD or the range of normal as the control value.

Unidentified participant: Do you include in your assays cells from a normal person?

Weening (Amsterdam, The Netheralnds): Always.

Venge (Uppsala, Sweden): What do these normal controls really tell you?

Weening (Amsterdam, The Netherlands): They tell me a technical error has not been made on that day.

Unidentified participant: Other than that, what does testing the cells of a "normal" individual tell you?

Hill (Salt Lake, USA): I don't think any of us make the diagnosis of a white cell defect on the results of one test. We test them 2, 3 or 4 times before anything conclusive can be said. And we compare the results with all of the controls.

Siccardi (Pavia, Italy): Well we have, of course, approached the problem of analysis of variance. This is obviously very important because I believe that there are two ways to approach this problem. The first one is to include controls every day and express the data as a function of the controls. The second one is to have a large population of controls over a long period of time that you can use. This is obviously the best method statistically. We have kept the data from two controls for each set of 6 cases that we examine. After three years we have studied all these data. We have also repeated the same test on the same people on multiple occasions over an entire year. This is the most complicated type of variance. One then determines the statistics. You can then decide how much of the total variance is due to variance between repetition and how much is due to variance between individuals. You will find that for most tests including glucose oxidation and microbicidal activity, more than 50% of the total variance is due to individual variation. In no cases is experimental error greater than 20%. There is somewhat more variation due to differences between repetitions and less to that observed between individuals for the tests that are very similar in everybody. For example the frequency of adequate phagocytosis is between 92 and 100% in most subjects. So here it turns out that individual variation is minimal and variation between repetition is higher. I believe that if one collects enough data all of the data can be used as the control. You don't need to normalize for controls of the day. The thing you have to do is check that the controls of the day do fall within the normal variance. If they are grossly out, then you discard the whole day.

Weening (Amsterdam, The Netherlands): I've already said what our protocol is, but I will just sum it up. You have to start with the white blood cell count. We also ask always for IgE levels and we do a chemotaxis test. We use zymosan activated serum and casein and compare appropriate age groups as controls. We don't use superoxide generation or chemiluminescence but we do employ oxygen consumption. I think it would be nice to see what people think of the difference between these three assays. I think they all pick up the same abnormalities. Then we employ opsonization and phagocytosis assays only on further indications, either clinical or laboratory. If there are abnormalities we then proceed with the more sophisticated tests such as superoxide or hydrogen peroxide generation.

Patriarca (Trieste, Italy): It is a pity to stop such an active and most profitable discussion. However, it is getting late, and tomorrow we all will be faced with another hard day of meetings. I don't dare to draw any conclusions. I just want to express my feeling that an entire conference would be needed for an in-depth analysis at the many aspects of the clinical application of leucocyte function tests. Good night.

Appendix to the Round Table

A LIST OF LABORATORIES THAT PERFORM LEUCOCYTE FUNCTION TESTS

NAME OF THE LABORATORY	PERSON IN CHARGE	TESTS PERFORMED
Bowman Gray School of Medicine, Winston Salem, USA	De Chatelet, L.R.	CL, GOX, O_2, EA, ING, MA, IOD, O_2^-
Bowman Gray School of Medicine, Winston Salem USA	McCall, C.A.	NBT, CHEM, RL, EA, OPS, COMP, MA, GOX, O_2^-, IOD, H_2O_2
Central Lab. of the Netherlands Red Cross Blood Transfusion Service, Amsterdam, The Netherlands	Roos, D., Weening, R.S.	Chem, ING, MA, O_2, OPS
Centro Trasfusionale e di Immunologia dei Trapianti Milano, Italy	Tedesco, F., Silvani, C.M.	CHEM, FA, O_2^-
Clinica Medica I, Università di Pavia, Italy	Ricevuti, G.	SW, NBT, EA, CHEM, RL
Clinique des Maladies Infectieuses, Hôpital Claude Bernard, Paris, France	Vildè, J.L.	CHEM, MA, O_2, OPS
Clinical Immunology Unit Linköping University, Sweden	Stendhal, O.	CHEM, IOD, O_2^-
CRC-Perinatal Medicine Division, Harrow, UK	Webster, A.D.B.	standard leucocyte function tests
Department Pediatrics, Nagasaki, Japan	not specified	CHEM, CL, MA, OPS
Department Pediatrics, University of Zurich, Switzerland	Seger, R.	ADH, CHEM, CYT, FA, ING, MA, O_2, SECR

NAME OF THE LABORATORY	PERSON IN CHARGE	TESTS PERFORMED
Dept.Clinical Chemistry University Hospital Uppsala, Sweden	Venge,P.	BP, CHEM, CL, EA, SECR
Dept. Clinical Microbiology Rigshospitalet, Copenhagen Denmark	Valerius,N.H.	CHEM, CL, GOX, MA, NBT
Dept. Biochemistry, Kyushu University, Japan	The staff	O_2^-, H_2O_2
Dept.Hematology, University College Hospital, London, UK	Segal,A.W.	ADH, CHEM, DIG, FA, ING, MA, O_2, SECR, SW
Istituto Chimica Biologica Ferrara, Italy	Spisani,S., Traniello,S.	CHEM, ING, NBT, RL
Istituto Chimica Biologica Università Cattolica, Roma Italy	De Sole,P.	CL
Infectious Disease Division Hopital Universitaire Cantonal, Geneve, Switzerland	Waldwogel,F.A.	COMP, ING, IgL, LI, MA, NBT
Istituto di Patologia Generale, Università di Trieste, Italy	The staff	EA, H_2O_2, O_2, O_2^-, MA, NBT, OPS, SECR
Istituto di Patologia Generale, Università di Padova, Sede di Verona, Italy	Bellavite,P.	CHEM, ING, MA, OPS, RL
Istituto Mario Negri Milano, Italy	Mantovani,A.	MMTA, MR
Institut de Recherches Interdisciplinaire, Free University, Brussels, Belgium	Schell-Frederick, E.	CHEM, IOD

NAME OF THE LABORATORY	PERSON IN CHARGE	TESTS PERFORMED
Immunologic Assessment Lab. National Jewish Hospital Denver, USA	Harbeck,R., Johnston,R.B.	CL, ING, NBT, O_2, O_2^-, H_2O_2
Lab. Immunologia Pediatrica Clinica Pediatrica, Pavia Italy	Ugazio,A.G., Siccardi,A.G.	ING, GOX, MA, NBT, RL
Laboratory for Medical Microbiology, Utrecht The Netherlands	Verhoef,J.	CHEM, CL, ING, OPS
Lab. Cellular Immunology University Hospital, Leiden, The Netherlands	van Furth,R., Leijh, P.C.J., van den Meer,J.	CHEM, CYT, COMP, C-R, Fc-R, ING, IgL, MA, OPS
Laboratoire d'Immunologie et d'Hematologie, CHUV Xavier Bichat, Paris, France	Hakim,J.	CHEM, CYT, EA, NBT, RL, O_2
Marselisborg Hospital Aarhus, Denmark	Borregaard,N.	Cyt b, ADCC, EA, GOX, NBT, O_2, O_2^-, SECR
Neutrophyl Lab., Dept. Medicine, Södersjukhuset Stockholm, Sweden	Palmblad,J.	CL, ADH, MA, SECR

ADH = adherence
ADCC = antibody dependent cytotoxicity
BP = bactericidal proteins
CHEM = chemotaxis
CL = chemiluminescence
C-R = C-receptors
COMP = complement assay
Cyt b = cytochrome b assay in neutrophils
DIG = digestion
EA = enzyme assay
FA = fungicidal activity
Fc-R = Fc receptors
GOX = glucose oxidation

IOD = iodination
ING = ingestion assay
IgL = immunoglobulin levels
LF = lymphocyte function
LI = lymphocyte identification
MMTA = monocyte-mediated tumoricidal activity
MR = monocyte-macrophage responsiveness to endotoxin, interferon and lymphokines
MA = microbicidal activity
OPS = opsonizing activity
H_2O_2 = hydrogen peroxide production
O_2 = O_2 consumption
O_2^- = superoxide production
NBT = nitrobluetetrazolium test
RL = random locomotion
SW = skin window
SECR = secretion

CONTRIBUTORS

Afzelius,B. - Wennergren Institute, University of Stockholm,Sweden
Aglietta,M. - Cattedra di Patologia Medica I, 10126 Torino, Italy
Allavena,P. - Istituto di Ricerche Farmacologiche "Mario Negri", via Eritrea,62, 20157 Milano, Italy
Allen,R.C. - U.S.Army Institute of Surgical Research and Clinical Investigation Service, Brooke Army Medical Center, Fort Sam Houston, Texas 78234, U.S.A.
Arslan,P. - C.N.R. Unit for the Study of Physiology of Mitochondria, Institute of General Pathology, University of Padova, Italy
Babnik,J. - Department of Biochemistry, J.Stefan Institute, Ljubljana, Yugoslavia
Baggiolini,M. - Wander Research Institute, Wander Ltd., P.O. Box 2747, CH 3001 Berne, Switzerland
Banfi,E. - Istituto di Microbiologia, Università di Trieste, Italy
Bannert,H. - Institut für Immunologie und Serologie der Universität Heidelberg, 6900 Heidelberg, West Germany
Bànyai,A. - Division of Medicine, University Medical School, Debrecen 4004, Hungary
Bar-Shavit,Z. - Department of Membrane Research, The Weizmann Institute of Science, Rehovot 76100, Israel
Basford,R.E. - Department of Pathology, Biochemistry, Radiation Biology, University of Pittsburgh School of Medicine, Pittsburgh PA 15261, U.S.A.
Bass,D.A. - Departments of Medicine and Biochemistry, Bowman Gray School of Medicine, Winston Salem NC 27103, U.S.A.
Basset,F. - Hôpital Bichat, 75877, Paris Cedex 18, France
Bassøe,H.H. - Medical Department B, Haukeland sykehuis, University of Bergen, Norway
Batista,U. - Department of Biophysics, Medical Faculty, E.Kardelj University, Ljubljana, Yugoslavia

Bellavite,P. - Istituto di Patologia Generale, Università di Padova, Sede di Verona, Strada Le Grazie, Verona, Italy

Berton,G. - Istituto di Patologia Generale, Università di Padova, Sede di Verona, Strada Le Grazie, Verona, Italy

Biondi,A. - Istituto di Ricerche Farmacologiche "Mario Negri", via Eritrea, 62, 20157 Milano, Italy

Boggs,S.S. - Departments of Pathology, Biochemistry, Radiation Biology University of Pittsburgh School of Medicine, Pittsburgh PA 15261, U.S.A.

Bonina,L. - Istituto di Microbiologia, Università di Messina, Italy

Bonventre,P. - Department of Microbiology, University of Cincinnati College of Medicine, Cincinnati, Ohio, U.S.A.

Boraschi,D. - Istituto di Ricerche Farmacologiche "Mario Negri", via Eritrea,62, 20157 Milano, Italy

Bordignon,C. - Istituto di Ricerche Farmacologiche "Mario Negri", via Eritrea,62, 20157 Milano, Italy

Borregaard,N. - Department of Medicine and Infectious Diseases, Marselisborg Hospital, DK-8000 Århus, Denmark

Bot,A.A.M. - University of Amsterdam, P.O.Box 9190, 1006 AD Amsterdam The Netherlands

Brederoo,P. - Laboratory for Electron Microscopy, University of Leiden 2333 AA Leiden, The Netherlands

Breton-Gorius,J. - INSERM U.91, Hôpital Henri Mondor, 94010 Creteil, France

Brzin,J. - Department of Biochemistry, J.Stefan Institute, Ljubljana, Yugoslavia

Buriot,D. - Hôpital des Enfants Malades, 75015, Paris, France

Bursuker,I. - Department of Membrane Research, The Weizmann Institute of Science, Rehovot 76100, Israel

Bussolino,F. - Laboratorio di Immunopatologia, Cattedra di Nefrologia, 10126 Torino, Italy

Cabilly,S. - The Lautenberg Center for General & Tumor Immunology, Hebrew University, Hadassah Medical School, Jerusalem, Israel

Camussi,G. - Laboratorio di Immunopatologia, Cattedra di Nefrologia, 10126 Torino, Italy

Charette,R.P. - University of Utah College of Medicine, Salt Lake City Utah 84132, U.S.A.

Cinco,M. - Istituto di Microbiologia, Università di Trieste, Italy

Clark,R.A. - Evans memorial Department of Clinical Research and Department of Medicine, Boston University Medical Center, Boston MA 02118, U.S.A.

Coade,S.B. - Division of Perinatal Medicine, Clinical Research Center, Harrow HA1 3UJ Middx, United Kingdom

Coquette,A. - Laboratory of Microbiology and Immunology, Free University of Brussels, Belgium

Cottier,H. - Department of Pathology, University of Bern, 3010 Bern Switzerland

Cross,A.R. - Department of Biochemistry, University of Bristol, University Walk, Bristol, United Kingdom

Cummings,N.P. - National Jewish Hospital and Research Center, University of Colorado, Denver, Colo. 80206, U.S.A.

Dahlgren,C. - Department of Medical Microbiology, Linköping University Medical School, S-581 85 Linköping, Sweden

Debatin,K.M. - Fachklinik Königsthul, Heidelberg, West Germany

Debets-Ossenkopp,Y. - Department of Microbiology, State University of Utrecht, The Netherlands

De Chatelet,L.R. - Departments of Biochemistry and Medicine, Bowman Gray School of Medicine, Winston Salem NC 27103, U.S.A.

de Graaf,C.J. - Laboratory of Biochemistry, B.C.P. Iansen Institute, University of Amsterdam, 1018 TV, The Netherlands

Delfino,D. - Istituto di Microbiologia, Università di Messina, Italy

Densen,P. - Evans Memorial Department of Clinical Research, Department of Medicine, Boston University Medical Center, Boston MA, U.S.A.

De Sole,P. - Istituto di Chimica Biologica, Università Cattolica del Sacro Cuore, Via Pineta Sacchetti,644, 00168 Roma, Italy

Dewald,B. - Wander Research Institute, Wander Ltd., P.O.Box 2747, CH 3001 Berne, Switzerland

Dolzani,L. - Istituto di Chimica Biologica, Università di Trieste, Italy

Dri,P. - Istituto di Patologia Generale, Università di Trieste, Italy

Drobnic-Kosorok,M. - Department of Biochemistry, J.Stefan Institute Ljubljana, Yugoslavia

Elsbach,P. - Department of Medicine, New York University School of Medicine, New York NY 10016, U.S.A.

Frei,J. - Laboratoire Central de Chimie Clinique, Centre Hospitalier Universitaire Vaudois (CHUV), CH 1011 Lausanne, Switzerland

Fusaro,A. - Istituto di Patologia Generale, Università di Bari, 70124 Italy

Gallily,R. - The Lautenberg Center for General & Tumor Immunology, Hebrew University, Hadassah Medical School, Jerusalem, Israel

Gallin,J.I. - Bacterial Diseases Section, Laboratory of Clinical Investigation, National Institute of Allergy and Infectious Diseases, National Institutes of Health, Bethesda M.D. 20205, U.S.A.
Garcia,R. - Department of Hematology, University College London, London WC1, United Kingdom
Gee,J.B.L. - Department of Internal Medicine, Yale University School of Medicine, New Haven Ct. 06510, U.S.A.
Geisow,M. - Department of Biophysics, National Institutes for Medical Research, Mill Hill, London NW7, United Kingdom
Gennaro,R. - Istituto di Chimica Biologica, Università di Trieste, Italy
Giordano,D. - Istituto di Patologia Generale, Università di Bari, 70124 Bari, Italy
Giraldi,T. - Istituto di Farmacologia, Università di Trieste, Italy
Glette,J. - Medical Department B; Haukeland sykehuis, University of Bergen, Norway
Goldman,R. - Department of Membrane Research, The Weizmann Institute of Science, Rehovot 76100, Israel
Gougerot-Pocidalo,A.M. - Laboratoire d'Immunologie et d'Hématologie CHU Xavier Bichat, Paris, France
Griscelli,C. - Hôpital des Enfants Malades, 75015, Paris, France
Grover,W.H. - Bowman Gray School of Medicine, Winston Salem, NC 27103 U.S.A.
Guichard,J. - INSERM U.91, Hôpital Henri Mondor, 94010 Creteil, France
Guthrie,L.A. - National Jewish Hospital, University of Colorado School of Medicine, Denver, Colo. 80206, U.S.A.
Haidaris,C.G. - Department of Microbiology, University of Cincinnati College of Medicine, Cincinnati, Ohio, U.S.A.
Håkansson,L. - Department of Clinical Chemistry and Internal Medicine University Hospital, S-750 14 Uppsala, Sweden
Hakim,J. - Laboratoire d'Immunologie et d'Hematologie, Hôpital Bichat 75018 Paris, France
Hällgren,R. - Department of Clinical Chemistry and Internal Medicine University Hospital, S-750 14 Uppsala, Sweden
Hamers,M.N. - Central Laboratory of The Netherlands Red Cross, University of Amsterdam, 1066 CX, The Netherlands
Harkness,R.A. - Division of Perinatal Medicine, Clinical Research Center, Harrow HA1 3UJ Middx., United Kingdom
Harper,A. - Department of Hematology, University College London, London WC1, United Kingdom
Hartwig,J.H. - Massachusetts General Hospital, Harvard Medical School, Department of Medicine, Boston, MA 02114, U.S.A.

Hed,J. - Department of Medical Microbiology, Linköping University Medical School, S-581 85 Linköping, Sweden

Hess,M.W. - Department of Pathology, University of Bern, 3010 Bern Switzerland

Hill,H.A.O. - Inorganic Chemistry Laboratory, South Parks Road, Oxford OX1 3QR, United Kingdom

Hill,H.R. - Departments of Pediatrics, University of Utah College of Medicine, Salt Lake City, Utah 84132, U.S.A.

Hinman,L.M. - Department of Internal Medicine, Yale University School of Medicine, New Haven, Ct. 06510, U.S.A.

Hogan,N.A. - Departments of Pediatrics, University of Utah College of Medicine, Salt Lake City, Utah 84132, U.S.A.

Hrabak,A. - Semmehweiss University, Hungaria

Hulsing-Hesselink,E. - Department of Infectious Diseases, University Hospital Leiden, The Netherlands

Iannello,D. - Istituto di Microbiologia, Università di Messina, Italy

Introna,P. - Istituto di Ricerche Farmacologiche "Mario Negri", via Eritrea, 62, 20157 Milano, Italy

Jayakar,S.D. - Laboratorio di Genetica Biochimica ed Evoluzionistica, CNR, Pavia, Italy

Johnston,R.B. - National Jewish Hospital and Research Center, University of Colorado School of Medicine, Denver, Colo. 80206, U.S.A.

Jones,O.T.G. - Department of Biochemistry, University of Bristol, University Walk, Bristol, United Kingdom

Kakinuma,K. - The Tokio Metropolitan Institute of Medical Science, Honkomagome, Bunkyo-ku, Tokio 113, Japan

Kaneda,M. - The Tokio Metropolitan Institute of Medical Science, Honkomagome, Bunkyo-ku, Tokio 113, Japan

Kaplan,S.S. - Department of Pathology, Biochemistry, Radiation Biology, University of Pittsburgh School of Medicine, Pittsburg, PA 15261, U.S.A.

Kàvai,M. - Department of Pulmonary Disease, University Medical School Debrecen 4004, Hungary

Keller,H.U. - Department of Pathology, University of Bern, 3010 Bern Switzerland

Kitagawa,S. - Department of Hematology, Jichi Medical School, Minamikawachi-machi, Kawachi-gun, Tochigi-ken 329-04, Japan

Kitahara,M. - University of Utah College of Medicine, Salt Lake City Utah 84132, U.S.A.

Kopitar,M. - Department of Biochemistry, J. Stefan Institute, Ljubljana, Yugoslavia

Korbelik,M. - Department of Biophysics, Medical Faculty, E. Kardelj University, Ljubljana, Yugoslavia

Kragballe,K. - Department of Dermatology, Marselisborg Hospital, DK-8000 Århus C., Denmark

Leijh,P.C.J. - Department of Infectious Diseases, University Hospital Leiden, The Netherlands

Leonard,E.J. - Immunopathology Section, National Cancer Institute, National Institutes of Health, Bethesda, Maryland 20205 U.S.A.

Lew,P.D. - Infectious Disease Division, University Hospital, 1211 Geneva 4, Switzerland

Lewis,J.C. - Bowman Gray School of Medicine, Winston Salem, NC 27103 U.S.A.

Lippa,S. - Istituto di Chimica Biologica, Università Cattolica del Sacro Cuore, Via Pinetá Sacchetti,644, 00168 Roma, Italy

Littarru,G.P. - Istituto di Chimica Biologica, Università Cattolica del Sacro Cuore, via Pineta Sacchetti,644, 00168 Roma, Italy

Mandell,G.L. - Division of Infectious Diseases, Departments of Internal Medicine and Pathology, University of Virginia, Charlottesville, VA, U.S.A.

Mansell,M. - Division of Inherited Metabolic Disease, Clinical Research Center, Harrow, HA1 3UJ Middx, United Kingdom

Mantovani,A. - Istituto di Ricerche Farmacologiche "Mario Negri", via Eritrea,62, 20157 Milano, Italy

Markert,M. - Laboratoire Central de Chimie Clinique, Centre Hospitalier Universitaire Vaudois (CHUV), CH-1011, Lausanne, Switzerland

Marsh,W.L. - The Lindsley F.Kimball Research Institute of the New York Blood Center, New York, NY, U.S.A.

Matsumoto,T. - Department of Biochemistry, Kyushu University School of Medicine, Fukuoka 812, Japan

McCall,C.E. - Department of Biochemistry and Medicine, Bowman Gray School of Medicine, Winston Salem, NC 27103, U.S.A.

Meerhof,L.J. - University of Amsterdam, P.O. Box 9190, 1006 AD, Amsterdam, The Netherlands

Meltzer,M.S. - Immunopathology Section, Laboratory of Immunology, National Cancer Institute, National Institutes of Health Bethesda, Maryland 20205, U.S.A.

Menzel,J. - Fachklinik Königstuhl, Heidelberg, West Germany

Mills,E.L. - Department of Microbiology, State University of Utrecht, The Netherlands

Minakami,S. - Department of Biochemistry, Kyushu University School of Medicine, Fukuoka 812, Japan

Miolo,A. - CNR Unit for the Study of Physiology of Mitochondria, Institute of General Pathology, via Loredan,16, 35100 Padova, Italy

Montecucco,C. - CNR Unit for the Study of Physiology of Mitochondria, Institute of General Pathology, Via Loredan,16, 35100 Padova, Italy

Mottola,C. - Istituto di Chimica Biologica, Università di Trieste, Italy

Nabi,Z.F. - Department of Biochemistry, Kyushu University School of Medicine, Fukuoka 812, Japan

Naef,A. - Department of Pathology, University of Bern, 3010 Bern, Switzerland

Nydegger,U. - Infectious Disease Division, Departments of Medicine, Microbiology and Pediatrics, University Hospital, 1211 Geneve 4, Switzerland

O'Flaherty,J.T. - Departments of Medicine and Biochemistry, Bowman Gray School of Medicine, Winston Salem, NC 27103 U.S.A.

Okolow-Zubkowska,M.J. - Inorganic Chemistry Laboratory, South Parks Road, Oxford OX1 3QR, United Kingdom

Øyen,R. - The Lindsley F.Kimball Research Institute of the New York Blood Center, New York, NY, U.S.A.

Pabst,M.J. - National Jewish Hospital and Research Center, University of Colorado School of Medicine, Denver, Colo. 80206, U.S.A.

Palmblad,J. - Depts. of Medicine IV & Clinical Bacteriology, The Karolinska Institute of Söderjukhuset, S-10064 Stockholm, Sweden

Pantaleo,R. - Istituto di Patologia Generale, Università di Bari, 70124 Bari, Italy

Papa,S. - Istituto di Chimica Biologica, Università di Bari, 70124 Bari, Italy

Piacibello,W. - Cattedra di Patologia Medica I, 10126 Torino, Italy

Pozzan,T. - CNR Unit for the Study of Physiology of Mitochondria, Institut of General Pathology, via Loredan 16, 35100 Padova, Italy

Rallison,M.L. - University of Utah College of Medicine, Salt Lake City, Utah 84132, U.S.A.

Romeo,D. - Istituto di Chimica Biologica, Università di Trieste, Italy

Roos,D. - Central Laboratory of The Netherlands Red Cross Blood Transfusion Service, University of Amsterdam, 1066 CX, The Netherlands

Rossi,F. - Istituto di Patologia Generale dell'Università di Padova, Sede di Verona, Strada Le Grazie, Verona, Italy

Ruco,L. - II Istituto di Anatomia Patologica, Unità di Patologia Sperimentale, Università di Roma, 00161 Roma, Italy

Sacchi,F. - Clinica Pediatrica, Università di Pavia, Italy

Santos,I. - University of Utah College of Medicine, Salt Lake City, Utah 84132, U.S.A.

Sasada,M. - National Jewish Hospital, University of Colorado School of Medicine, Denver, Colo. 80206, U.S.A.

Sava,G. - Istituto di Farmacologia, Università di Trieste, Italy

Schneider,C. - Istituto di Chimica Biologica, Università di Trieste, Italy

Segal,A.W. - University College London, Department of Hematology, London WC1, United Kingdom

Segoloni,G. - Laboratorio di Immunopatologia, Cattedra di Nefrologia, 10126 Torino, Italy

Seligmann,B. - Bacterial Diseases Section, Laboratory of Clinical Investigation, National Institute of Allergy and Infectious Diseases, National Institutes of Health, Bethesda, M.D. 20205, U.S.A.

Shirley,P.S. - Departments of Biochemistry and Medicine, Bowman Gray School of Medicine, Winston Salem, NC 27103, U.S.A.

Siccardi,A. - Istituto di Genetica, Università di Pavia, Italy

Simmonds,R.J. - Division of Perinatal Medicine, Clinical Research Center, Harrow HA1 3UJ, Middx, United Kingdom

Sips,H.J. - Laboratory of Biochemistry, B.C.P. Jansen Institut, University of Amsterdam, 108 TV, The Netherlands

Skrk,J. - Oncological Institute, Ljubljana, Yugoslavia

Sluiter,W. - Department of Infectious Diseases, University Hospital, Leiden, The Netherlands

Sonkoly,I. - Division of Medicine, University Medical School, Debrecen 4004, Hungary

Soranzo,M.R. - Istituto di Patologia Generale, Università di Trieste, Italy

Spisani,S. - Istituto di Chimica Biologica, Università di Ferrara, Italy

Stendhal,O. - Department of Medical Microbiology, Linköping University Medical School, S-581 85 Linköping, Sweden

Stevens,C.A. - Department of Internal Medicine, Yale University School of Medicine, New Haven, Ct.06510, U.S.A.

Stossel,T.P. - Hematology-Oncology Unit, Massachusetts General Hospital, Harvard Medical School, Boston, MA 02114, U.S.A.

Sullivan,G. - Division of Infectious Diseases, Department of Internal Medicine, University of Virginia, Charlottesville, VA, U.S.A.

Suter,S. - Infectious Disease Division, Departments of Medicine, Microbiology and Pediatrics, University Hospital, 1211 Geneva 4, Switzerland

Svetina,S. - Department of Biophysics, Medical Faculty, E.Kardelj University, Ljubljana, Yugoslavia

Szegedi,G. - Division of Medicine, University Medical School, Debrecen 4004, Hungary

Takaku,F. - Department of Hematology, Jichi Medical School, Minamikawaki-machi, Kawaki-gun, Tochigi-ken 329-04, Japan

Takeshige,K. - Department of Biochemistry, Kyushu University School of Medicine, Fukuoka 812, Japan

Tatscheck,B. - Department of Biochemistry, Kyushu University School of Medicine, Fukuoka 812, Japan

Tetta,C. - Laboratorio di Immunopatologia, Cattedra di Nefrologia, 10126 Torino, Italy

Thomas,M. - Departments of Medicine and Biochemistry, Bowman Gray School of Medicine, Winston Salem, NC 27103, U.S.A.

Tomisawa,S. - Immunopathology Section, National Cancer Institute, National Institutes of Health, Bethesda, Maryland 20205, U.S.A.

Torres,M. - Laboratoire d'Immunologie et d'Hematologie, CHU Xavier Bichat, Université Paris VII, 75877 Paris Cedex 18, France

Traniello,S. - Istituto di Chimica Biologica, Università di Ferrara, Italy

Turk,V. - Department of Biochemistry, J.Stefan Institute, Ljubljana, Yugoslavia

Udén,H.M. - Depts. of Medicine IV & Clinical Bacteriology, The Karolinska Institute at Södersjukhuset, S-10064 Stockholm, Sweden

Ugazio,A.G. - Clinica Pediatrica, Università di Pavia, Italy

Vainchenker,W. - Hôpital Claude Bernard, 75019 Paris, France

Valerius,N.H. - Department of Clinical Microbiology, Rigshospitalet Copenhagen, Denmark

van den Barselaar,M.Th. - Department of Infectious Diseases, University Hospital, leiden, The Netherlands

van der Meulen,J. - Laboratory for Electron Microscopy, University of Leiden, 2333AA, Leiden, The Netherlands

van Furth,R. - Department of Infectious Diseases, University Hospital, Leiden, The Netherlands van Schaik,M.L.J. - Pediatric Clinic, Binnengasthuis, Academic Hospital, University of Amsterdam, The Netherlands

van Zwet,Th.L. - Department of Infectious Diseases, University Hospital, leiden, The Netherlands

van Zwieten,R. - Central Laboratory of The Netherlands Red Cross Blood Transfusion Service, University of Amsterdam, The Netherlands

Vaudaux,P. - Infectious Disease Division, Departments of Medicine, Microbiology and Pediatrics, University Hospital, 1211 Geneva 4, Switzerland

Venge,P. - Departments of Clinical Chemistry and Internal Medicine, University Hospital, S-750 14 Uppsala, Sweden

Venizelos,N. - Depts. of Medicine IV & Clinical Bacteriology, The Karolinska Institute at Södersjukhuset, S-10064, Stockholm, Sweden

Verbrugh,H.A. - Department of Microbiology, State University of Utrecht, The Netherlands

Verhoef,H.A. - Department of Microbiology, State University of Utrecht, The Netherlands

Vildè,J.L. - Hôpital Claude Bernard, 75019 Paris, france

Vincenzi,E. - Istituto di Chimica Biologica, Università di Ferrara, Italy

Voetman,A.A. - University of Amsterdam, P.O. Box 9190, 1006AD, Amsterdam, The Netherlands

Vray,B. - Laboratory of Microbiology and Immunology, Free University of Brussels, Belgium

Waldvogel,F.A. - Infectious Disease Division, departments of Medicine, Microbiology and Pediatrica, University Hospital, 1211 Geneva 4, Switzerland

Weening,R.S. - Central Laboratory of The Netherlands red Cross Blood Transfusion Service, University of Amsterdam, 1066CX The Netherlands

Weiss,J. - Department of Medicine, New York University School of Medicine, New York, NY 10016, U.S.A.

Wever,R. - Laboratory of Biochemistry, B.C.P. Jansen Institute, University of Amsterdam, 1018TV, The Netherlands

Wilkinson,S. - Division of Infectious Diseases, Departments of Internal Medicine and Pathology, University of Virginia, Charlottesville, VA, U.S.A.

Zabucchi,G. - Istituto di Patologia Generale, Università di Trieste, Italy

Zdziarski,U.E. - Departments of Pathology, Biochemistry, Radiation Biology, University of Pittsburgh School of Medicine, Pittsburgh, PA 15261, U.S.A.

Zsindely,A. - Division of Medicine, University Medical School, Debrecen 4004, Hungary

Zwahlen,A. - Infectious Disease Division, Departments of Medicine, Microbiology and Pediatrics, University Hospital, 1211 Geneva 4, Switzerland

INDEX

www.ingramcontent.com/pod-product-compliance
Ingram Content Group UK Ltd.
Pitfield, Milton Keynes, MK11 3LW, UK
UKHW051450290926
15471UKWH00008B/1129

* 9 7 8 1 4 6 8 4 8 0 8 9 4 *